MECHANICAL VIBRATIONS

Forced vibration test system. A 28,000 pound-force exciter is shown in the right foreground, control console on the left, and power supply unit in the background. (courtesy MB Electronics, A Division of Textron Electronics, Inc.)

MECHANICAL VIBRATIONS

Francis S. Tse

MICHIGAN STATE UNIVERSITY

Ivan E. Morse

UNIVERSITY OF CINCINNATI

Rolland T. Hinkle

MICHIGAN STATE UNIVERSITY

Allyn and Bacon, Inc. Boston

FIRST PRINTING . . . JANUARY 1963
SECOND PRINTING . . . AUGUST 1964

Library of Congress Catalog Card Number: 63-9257

Printed in the United States of America

PREFACE

The subject of vibration deals with the oscillatory motion of physical systems. The object of a vibration study is to determine the effect of vibrations on the performance and safety of the systems under consideration. The study of oscillatory motion is an important step toward this goal.

The purpose of this book is to present the fundamentals of vibration theory and to provide a background for advanced study in the field. No attempt has been made to cover all phases of vibrations, as the subject is very extensive.

This book is written primarily for mechanical engineering students of senior-year-college and beginning-graduate levels. The reader is assumed to have an elementary knowledge of dynamics, strength of materials, and differential equations. To provide adequate background, differential equations and other mathematical techniques used in the book are reviewed in the appendices.

The first three chapters constitute the core of an elementary terminal course. Chapter 1 describes the general concepts of vibration and simple harmonic motion. Chapter 2 treats systems having one degree of freedom through the study of a single second-order linear ordinary differential equation. The significance of each of the terms in this equation is explained. Then it is shown that this equation is applicable to the study of a large number of physical systems. The concept of vibration modes is introduced in Chapter 3. Although the discussion is primarily centered on the two-degree-of-freedom system, it gives the physical concepts and prepares the groundwork for studying multi-degree-of-freedom systems by other mathematical techniques.

The remainder of the book deals with more advanced topics. The Lagrange equations, introduced in Chapter 4, provide a potent tool for solving problems in vibrations. Chapters 5 and 6 illustrate additional techniques and practical applications. Chapter 7 uses matrix algebra to solve multi-degree-of-freedom systems. It may serve as a foundation for studying vibration problems on the digital computer. Mechanical transients and electromechanical systems are discussed in Chapter 8 by the use of the Laplace transform method. It is believed that transient study is sufficiently important to justify a separate chapter in a course in vibrations.

Chapter 9 describes the application of the electronic analogue computer in solving vibration problems, and it is regarded as an extension of Chapter 8 on analogue study. It is found that the operation of the computer can easily be grasped by mechanical engineering students, and it aids in their understanding of vibration phenomena. The use of the analogue computer as a tool is presented in some detail. Since laboratory work is essential for the understanding of the capabilities of the computer, a list of suggested experiments is included in Appendix A.

The material in this book has been used for a number of years by different instructors in Vibrations I and II taught at the Michigan State University. There is a laboratory section associated with each of these courses. Chapters 1 to 3 and part of Chapter 9 are being used for the first course, the remaining chapters for the second course. Chapter 9 is used principally for the laboratory sections. The analogue computer is used to supplement the vibration laboratory instead of the digital because students can fully participate in the experiments, only an elementary knowledge of electrical circuits is required, and small-size analogue computers are relatively inexpensive. Except for Chapters 1 to 3, the chapters are organized as independently as possible, but without repetition. Thus, after the first three chapters, the selection of topics can be very flexible.

To limit the scope of the book to systems described by linear ordinary differential equations, certain topics, such as vibration of a continuous medium, acoustics, and nonlinear vibrations, are purposely omitted. Nonlinear equations, however, are briefly discussed in Chapter 9.

No attempt has been made to compile a complete bibliography of the literature, which is very extensive. The authors, however, wish to acknowledge their indebtedness to the writers who have contributed to this field of study and to the authors of the texts listed as references. The authors are especially thankful to Dr. C. U. Ip for his valuable suggestions.

CONTENTS

3 SYSTEMS WITH MORE THAN ONE DEGREE OF FREEDOM 95

4 THE LAGRANGE EQUATIONS 151

5 APPLICATIONS 173

9 ELECTRONIC ANALOGUE COMPUTER 333

MECHANICAL VIBRATIONS

A 12,500 pound-force shaker performing lateral vibration test on a Talos missile (courtesy Bendix Products Aerospace Division, South Bend, Indiana)

1 INTRODUCTION

PRIMARY OBJECTIVES

The subject of vibration deals with the oscillatory motion of *dynamic systems*. A dynamic system is a combination of matter which possesses mass and whose parts are capable of relative motion. All bodies possessing mass and elasticity are capable of vibration. The mass is inherent in the body, and the elasticity is due to the relative motion of the parts of the body. The system considered may be in the form of a structure, a machine or its components, or a group of machines. The oscillatory motion of the system may be objectionable, trivial, or necessary for performing a task.

The objective of the designer is to control or minimize the vibration when it is objectionable and to utilize and enhance the vibration when it is desirable. Objectionable vibrations in a device may cause the loosening of parts or the malfunctioning or eventual failure of a machine. On the other hand, shakers in foundries and vibrators in testing machines require vibration. The proper functioning of many instruments depends on the proper control of the vibrational characteristics of the devices.

The primary objective of our study is to analyze the oscillatory motion of dynamic systems and the forces associated with the motion. It should be remembered that the ultimate goal in the study of vibration is to determine its effect on the performance and safety of the system under consideration. The analysis of the oscillatory motion is an important step toward this goal.

Our study begins with the description of the elements in a vibratory system, the introduction of some terminology, and a discussion of simple harmonic motion. These concepts will be used throughout the text. Other concepts and terminology will be introduced in the appropriate places as needed.

1-2. ELEMENTS OF A VIBRATORY SYSTEM

The elements that constitute a vibratory system are illustrated in Fig. 1-1. They are idealized and called (1) the *mass*, (2) the *spring*, (3) the *damper*, and (4) the *excitation* elements. The first three elements are the *parameters* descriptive of the physical system. For example, it can be said that the given system consists of a mass, a spring, and a damper arranged as shown in the figure. Energy may be stored in the mass and the spring and dissipated in the damper in the form of heat. These parameters are called the inactive or *passive elements.* To simplify the mathematics involved in the treatment of the subject, the passive elements are assumed to be *invariant with time.* Energy enters the physical system through the application of an excitation to the system. Hence the excitation is called an *active element*, and its magnitude varies according to a prescribed function of time. As shown in Fig. 1-1, an excitation force is applied externally to the system.

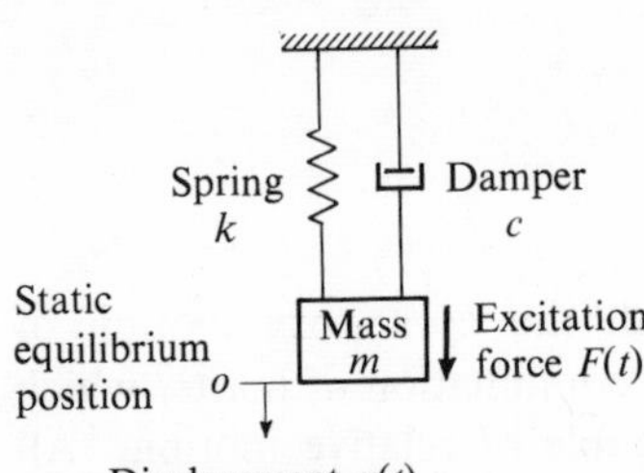

Fig. 1-1. *Elements of a vibratory system*

Furthermore, the parameters are assumed to be "lumped" together and are symbolized by the corresponding elements. Not all physical systems have *lumped parameters*. For example, a coil spring possesses both mass and elasticity. In order to consider it as a spring element, either its mass is assumed to be negligible or an appropriate portion of its mass is lumped together with the other masses of the system. A beam has its mass and elasticity inseparably distributed along its length. Hence the vibrational characteristics of a beam, or more generally of an elastic body, can be studied by this approach only if the elastic body is approximated by a finite number of lumped parameters. This method, however, is a practical approach to the study of some very complicated structures such as that of an aircraft.

In spite of the limitations, the lumped-parameter approach to the study of vibration problems is well justified for the following reasons: (1) Many physical systems are essentially lumped-parameter systems. (2) The concepts can be extended to analyze the vibration of elastic bodies. (3) Many physical systems are too complex to be investigated analytically as elastic bodies, and they are often studied through the use of their equivalent lumped-parameter systems. (4) The assumption of lumped parameters greatly simplifies the analytical effort required to obtain a solution.

The *mass* element is assumed to be a rigid body. It executes the vibrations and can gain or lose kinetic energy in accordance with the velocity change of the body. *Newton's law of motion* may be stated as follows: The product of the mass and its acceleration is equal to the force applied to the mass, and the acceleration takes place in the direction in which the force acts. *Work* is force times displacement in the direction of the force. The work done in changing the *kinetic energy* of a mass is *conserved*. The kinetic energy increases if work is positive and decreases if work is negative.

The *spring* element possesses elasticity and is assumed to be of negligible mass. A spring force exists only if the spring is deformed, such as the extension or the compression of a coil spring. Therefore the spring force exists if there is a *relative displacement* between the two ends of the spring. The work done in deforming a spring is *conserved* and is equal to the *strain energy* stored in the spring. This strain energy is often called the *potential energy*. A *linear spring* is one that obeys *Hooke's law*, that is, the spring force is proportional to the spring deformation. The constant of proportionality, measured in force per unit deformation, is called the spring constant k.

The *damping* element has neither mass nor elasticity. Damping force exists only if there is *relative motion* between the two ends of the damper. The work or energy input to a damper is converted into heat, and therefore the damping element is *nonconservative*. Many types of damping are encountered in engineering, and most of them are nonlinear. For example, the frictional drag of a body moving in a fluid is approximately proportional to the velocity squared, but the exact value of the exponent is dependent on many variables. *Coulomb* or dry friction damping is a function of the normal force between the bodies as well as the materials involved. The Coulomb damping force is generally assumed independent of the relative velocity between the sliding bodies. *Viscous damping*, in which the damping force is proportional to the velocity, is called linear damping. The mathematics for dealing with linear damping is relatively simple. Thus, viscous

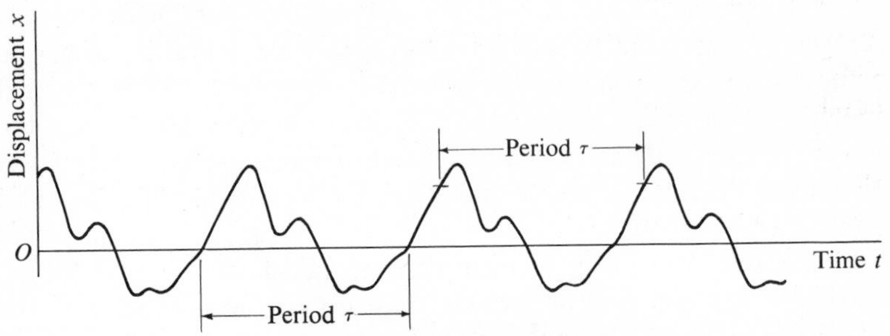

Fig. 1-2. *Periodic motion*

damping or its equivalent is generally assumed in engineering. The viscous-damping coefficient c is measured in force per unit velocity.

Energy enters a system through the application of an *excitation* to the system. Figure 1-1 shows an excitation force applied to the mass. The magnitude of the excitation varies in accordance with a prescribed function of time. Alternatively, if the system is suspended from a support, excitation may be applied to the system through imparting a prescribed motion to the support. In machinery, excitation often arises from the unbalance of the moving components. The vibrations of a dynamic system under the influence of an excitation are called *forced vibrations*. Forced vibrations, however, are often defined as the vibrations that are caused and maintained by a periodic excitation.

If the vibratory motion is *periodic*, the system repeats its motion at equal intervals of time, as shown in Fig. 1-2. The time required for the system to repeat its motion is called a *period* τ, which is the time required to complete one *cycle* of motion. Frequency f is the number of times that the motion repeats itself per unit time. A motion that does not repeat itself at equal time intervals is called an *aperiodic* motion.

If a dynamic system is set into motion by some disturbance and from time equal to zero and thereafter no excitation or disturbance is applied, the vibrations are called *free vibrations*. This represents the behavior of a system as it relaxes from an initial state of constraint to its equilibrium state. The constraints are called the *initial conditions* imposed on the system. For example, the mass of the system shown in Fig. 1-1 may be given an initial displacement and/or an initial velocity to set it into motion. If the system possesses damping, owing to the dissipation of energy in the damper, this vibratory motion will eventually die out. Thus the equilibrium state corresponds to the *static equilibrium* position of the system. Under idealized conditions, if the system does not possess damping, the vibratory motion will not diminish with time. The system will then oscillate about its static equilibrium position.

For simplicity, lumped parameters, namely rigid masses, linear springs, and viscous dampers, will be assumed unless otherwise stated. Systems possessing these characteristics are called *linear systems*, and their equations of motion are ordinary linear differential equations. An important property of linear systems is that they follow the *principle of superposition*. For example, the resultant motion of a linear system due to the simultaneous application of two excitations is a linear combination of the motions due to each of the excitations acting separately. Furthermore, if these parameters are time invariant, the equations become ordinary linear differential equations with constant coefficients. A brief review of this type of equation is given in Appendix B. In the beginning chapters, the excitation is assumed to be periodic. Nonperiodic excitations will be treated by the Laplace transform method in Chap. 8.

So far we have discussed only systems with rectilinear motion. In the case of systems with rotational motions, the system parameters are (1) the mass moment of inertia of a body J, (2) the torsional spring with spring constant k_t, and (3) the torsional damper with damping coefficient c_t. An angular displacement θ is analogous to a rectilinear displacement x, and an excitation torque $T(t)$ is analogous to an excitation force $F(t)$. The two types of systems are compared as shown in Table 1-1. The comparison is shown in greater detail in Tables 2-2 and 2-3, pp. 50 and 51.

It is apparent from this comparison that, with the change of units, the discussion on rectilinear systems can be extended to rotational systems. Furthermore, the differential equations of motion of the two types of systems must be of the same form.

TABLE 1-1

COMPARISON OF RECTILINEAR AND ROTATIONAL SYSTEMS

RECTILINEAR	ROTATIONAL
Spring force $= kx$	Spring torque $= k_t\theta$
Damping force $= c\dfrac{dx}{dt}$	Damping torque $= c_t\dfrac{d\theta}{dt}$
Inertia force $= m\dfrac{d^2x}{dt^2}$	Inertia torque $= J\dfrac{d^2\theta}{dt^2}$

1-3. EXAMPLES OF VIBRATORY MOTIONS

To illustrate different types of vibratory motion, let us choose different combinations of the four elements discussed in the previous section to form various simple dynamic systems.

The spring-mass system of Fig. 1-3(a) serves to illustrate the case of *undamped free vibration.* The mass m is initially at rest at its *static equilibrium* position. It is acted upon by two equal and opposite forces, namely, the spring force, which is equal to the product of the spring constant k and the static deflection δ_{st} of the spring, and the gravitational force mg due to the weight of the mass. Now, assume that the mass is displaced from equilibrium by an amount x_o and then released with zero initial velocity.† As shown in the free-body sketch, at the time the mass is being released, the spring force is equal to $k(x_o + \delta_{st})$ which is greater than the gravitational force on the mass by an amount kx_o. Upon being released, the mass will move toward the equilibrium position. The converse is true should the mass be displaced above the equilibrium position.

Since the mass is initially displaced by an amount x_o from its static equilibrium position, potential energy corresponding to this displacement is stored in the spring. The system is *conservative* because

† Time equal to zero is interpreted to mean the time immediately after the mass is being released, that is, at time $t = 0^+$.

there is no damping element to dissipate the energy of the system. When the mass passes through equilibrium, this potential energy of the spring is zero. Thus the potential energy has been transformed to become the kinetic energy of the mass. As the mass moves above the equilibrium position, the spring is being compressed, thereby gaining potential energy from the kinetic energy of the mass. When all the

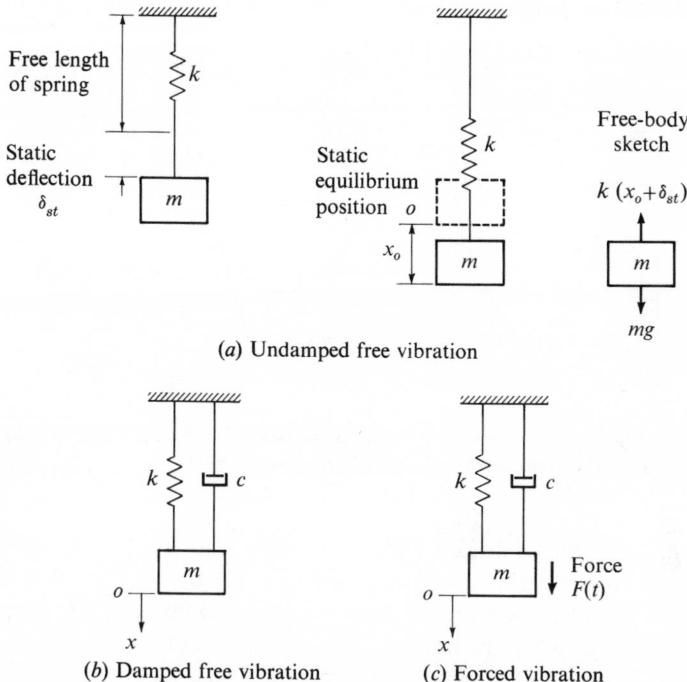

Fig. 1-3. *Simple vibratory systems*

kinetic energy of the mass has been transformed to become the potential energy of the spring, the mass is at its uppermost position. Thus, through the exchange of potential and kinetic energies between the spring and the mass, the system will oscillate periodically at its *natural frequency* about the static equilibrium position. Since the system is conservative, the maximum displacement from equilibrium, or the amplitude of the motion, of the mass will not diminish from cycle to cycle.

It will be shown in Chap. 2 that this periodic motion is sinusoidal or *simple harmonic*. It is implicit in this discussion that the natural frequency of a system is inherent in the system parameters. It is determined by the mass and the spring and is independent of the *initial conditions* imposed on the system or the amplitude of the oscillation.

If damping exists in the system, as shown in Fig. 1-3(*b*), the mass at rest is under the influence of only the spring force and the gravitational force, since the damping force is a function of motion. Now if the mass is displaced by an amount x_o from its static equilibrium

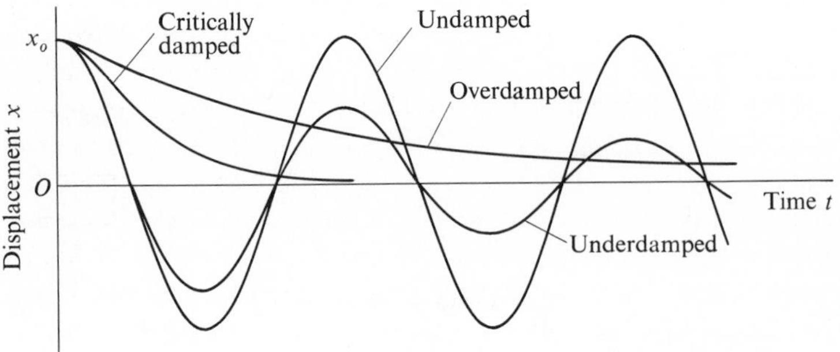

Fig. 1-4. *Free vibration of systems shown in Figs.* 1-3(*a*) *and* (*b*). *Initial displacement* = x_o; *initial velocity* = 0

position and then released with zero initial velocity, the spring force will tend to restore the mass to equilibrium, as before. In addition to the spring force, however, the mass is also acted upon by the damping force which opposes its motion. The resultant motion depends on the amount of damping existing in the system. If the damping is light, the system is said to be underdamped and the motion is oscillatory. The presence of damping will (1) cause the eventual dying out of the oscillation and (2) cause the system to oscillate more slowly than without damping. In other words, the amplitude decreases with each subsequent cycle of oscillation, and the frequency of oscillation with viscous damping is lower than the natural frequency of the system. If the damping is heavy, the system is said to be *overdamped*, and the motion is nonoscillatory or *aperiodic*. The mass, upon being released, will simply tend to return to its static equilibrium position. The system is said to be *critically damped* if the amount of damping is such that the resultant motion is on the border line between the two cases enumerated. The free vibrations of the systems shown in Figs. 1-3(*a*) and (*b*) are illustrated in Fig. 1-4.

All physical systems possess damping to a greater or a lesser degree. When there is very little damping in a system, such as a steel structure or a simple pendulum, the damping may be negligibly small, and the system is considered to be undamped. Damping is often built into a system to give it the desired performance. For example, vibration-measuring instruments are often built with damping corresponding to 70 percent of the critically damped value. In the case of a gun mount, it is desirable to have the return stroke after the recoil critically damped.

If an excitation force is applied to the mass of the system, as shown in Fig. 1-3(*c*), the resultant motion depends on the initial conditions imposed on the system as well as the excitation. For this discussion, let us assume that the excitation force is sinusoidal. Once the system is set into motion, it will tend to vibrate at its natural frequency as well as to follow the frequency of the excitation. If the system possesses damping, the part of the motion not sustained by the sinusoidal excitation will eventually die out. Thus the system will eventually vibrate at the excitation frequency, regardless of the initial conditions or the natural frequency of the system. The sustained vibrations are called the *steady-state vibrations* or *steady-state response* of the system. The motion which is not sustained is called the *transient* motion.

It is obvious that for a given excitation frequency the amplitude of the steady-state response is constant, but this amplitude may be frequency dependent. It is inferred here that only the transient motion is dependent on the initial conditions imposed on the system, and its frequency of oscillation is that of the *damped natural frequency* of the system, that is, the frequency of oscillation of a system with damping under free-vibration conditions. This is true because the transient motion is not sustained by the excitation, and it represents the behavior of the system as it relaxes from a state of constraint to its equilibrium state.

It will be shown in Chap. 2 that the steady-state response is described by the particular integral and the transient motion by the complementary function of the differential equation of the system.

Resonance occurs when the frequency of the excitation is equal to the natural frequency of the system. No energy input is necessary to maintain the vibrations of an undamped system at its natural frequency. Thus any energy input will be used to build up the amplitude of the vibration, and the amplitude at resonance will increase without limit. In a system with damping, the energy input is dissipated in the damper. Thus the amplitude of vibration at resonance is governed by the amount of damping existing in the system.

1-4. SIMPLE HARMONIC MOTION

It was mentioned in the preceding section that the vibration of the system shown in Fig. 1-3(*a*) is simple harmonic. This is the simplest form of periodic motion. It will be shown in Chap. 2 that any complex periodic motion can be considered to be composed of harmonic motions of various amplitudes and frequencies by means of a Fourier series analysis. Furthermore, the analysis of steady-state vibrations can be greatly simplified by the use of vectors to represent the harmonic motions. Hence it is desirable to discuss, in some detail, harmonic motion and the manipulation of vectors.

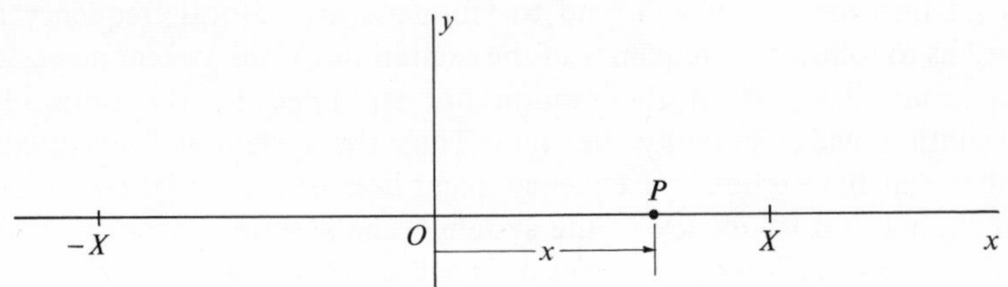

Fig. 1-5. *Simple harmonic motion:* $x(t) = X \cos \omega t$

A simple harmonic motion is a reciprocating motion. It can be represented by the circular functions, sine or cosine. Consider the motion of point P on the horizontal axis of Fig. 1-5. If the distance OP is

$$OP = x(t) = X \cos \omega t \tag{1-1}$$

where $t =$ time, $\omega =$ constant, and $X =$ constant, the motion of P about the origin O is sinusoidal, and the motion is called *simple harmonic.*† Since a circular function repeats itself in 2π radians, a cycle of motion is completed when $\omega\tau = 2\pi$, that is,

† It is obvious that a sine, a cosine, or their combination can be used to represent a simple harmonic motion. For example, let

$$x(t) = X_1 \sin \omega t + X_2 \cos \omega t = X\left(\frac{X_1}{X} \sin \omega t + \frac{X_2}{X} \cos \omega t\right)$$

$$= X(\sin \omega t \cos \alpha + \cos \omega t \sin \alpha) = X \sin (\omega t + \alpha)$$

where $X = \sqrt{X_1^2 + X_2^2}$ and $\alpha = \tan^{-1} \dfrac{X_2}{X_1}$. It is apparent that the motion $x(t)$ is sinusoidal and, therefore, simple harmonic. For simplicity, we shall confine our discussion to a cosine function.

In Eq. (1-1), $x(t)$ indicates that x is a function of time t. Since this is implicit in the equation, we shall omit (t) in all subsequent equations.

$$\text{Period } \tau = \frac{2\pi}{\omega}, \text{ sec-cycle}^{-1} \tag{1-2}$$

$$\text{Frequency } f = \frac{1}{\tau} = \frac{\omega}{2\pi}, \text{ cycle-sec}^{-1} \tag{1-3}$$

ω is called the circular frequency and is measured in rad-sec^{-1}.

If $x(t)$ represents the displacement of a mass in a vibratory system, the velocity and the acceleration of the mass are obtained from the first and second time derivatives of the displacement,† that is,

$$\text{Displacement } x = X \cos \omega t \tag{1-4}$$

$$\text{Velocity } \dot{x} = -\omega X \sin \omega t = \omega X \cos (\omega t + 90°) \tag{1-5}$$

$$\text{Acceleration } \ddot{x} = -\omega^2 X \cos \omega t = \omega^2 X \cos (\omega t + 180°) \tag{1-6}$$

These equations indicate that the velocity and acceleration of a harmonic displacement are also harmonic of the same frequency. Each differentiation changes the amplitude of the motion by a factor of ω and the time *phase angle* of the circular function by 90 deg. The phase angle of the velocity is 90 deg *leading* the displacement, and the acceleration is 180 deg leading the displacement.

Simple harmonic motion is often defined by the relations shown in Eqs. (1-4) and (1-6). When the acceleration of a particle having rectilinear motion is always proportional to the distance of the particle from a fixed point on the path and is directed toward the fixed point, the particle is said to have simple harmonic motion. This relationship can be expressed by the equation

$$\ddot{x} = -\omega^2 x \tag{1-7}$$

where ω^2 is the constant of proportionality. It can be shown that the solution of this differential equation has the form of a sine and a cosine function with circular frequency equal to ω.

The sum of two harmonic motions of the same frequency but with different phase angles is also a harmonic motion of the same frequency. For example, the sum of the harmonic motions $x_1 = X_1 \cos \omega t$ and $x_2 = X_2 \cos (\omega t + \alpha)$ is

$$\begin{aligned} x = x_1 + x_2 &= X_1 \cos \omega t + X_2 \cos (\omega t + \alpha) \\ &= X_1 \cos \omega t + X_2 (\cos \omega t \cos \alpha - \sin \omega t \sin \alpha) \\ &= (X_1 + X_2 \cos \alpha) \cos \omega t - X_2 \sin \alpha \sin \omega t \\ &= X (\cos \beta \cos \omega t - \sin \beta \sin \omega t) \\ &= X \cos (\omega t + \beta) \end{aligned}$$

† The symbols $\dot{x}$ and $\ddot{x}$ represent the first and second time derivatives of the function $x(t)$, respectively. This notation is used throughout this text unless ambiguity may arise.

where $X = \sqrt{(X_1 + X_2 \cos \alpha)^2 + (X_2 \sin \alpha)^2}$ is the amplitude of the resultant harmonic motion, and $\beta = \tan^{-1} \dfrac{X_2 \sin \alpha}{X_1 + X_2 \cos \alpha}$ is its phase angle.

The sum of two harmonic motions of different frequencies is not harmonic. A special case of interest is when the frequencies are slightly different. Let the motions be

$$\begin{aligned} x = x_1 + x_2 &= X \cos \omega t + X \cos (\omega + \varepsilon)t \\ &= X [\cos \omega t + \cos (\omega + \varepsilon)t] \\ &= 2X \cos \frac{\varepsilon}{2} t \cos \left(\omega + \frac{\varepsilon}{2}\right) t \end{aligned}$$

where $\varepsilon \ll \omega$. The resultant motion $x(t)$ may be considered as a cosine wave with circular frequency equal to $\left(\omega + \frac{\varepsilon}{2}\right)$, which is approximately equal to ω, and with varying amplitude equal to $\left(2X \cos \frac{\varepsilon}{2} t\right)$. This resultant motion is illustrated in Fig. 1-6. Every time the amplitude

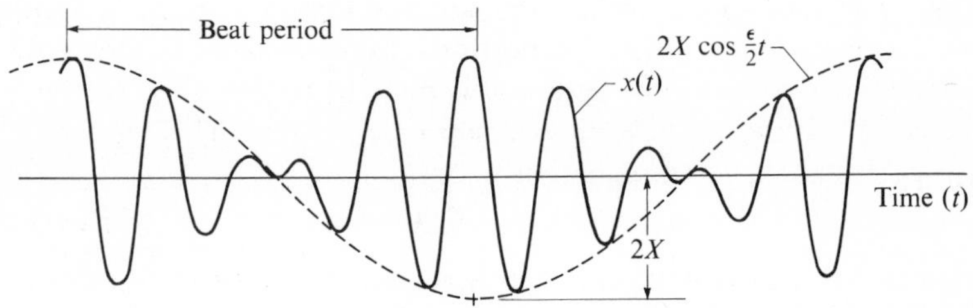

Fig. 1-6. *Graphical representation of beats*

reaches a maximum, there is said to be a *beat*. The beat frequency f_b, as determined by two consecutive maximum amplitudes, is equal to

$$f_b = f_2 - f_1 = \frac{\omega + \varepsilon}{2\pi} - \frac{\omega}{2\pi} = \frac{\varepsilon}{2\pi} \tag{1-8}$$

where f_2 and f_1 are the frequencies of the constituting motions.

The phenomenon of beats is common in engineering. For example, beats can be heard in electric powerhouses when a generator is started and just before it is connected to the line. Since the electrical frequency of the generator and the line may differ slightly, the hum of the generator and the line are of different pitch, and beats can be heard.

1-5. VECTORIAL REPRESENTATION OF HARMONIC MOTIONS

It is convenient to represent a harmonic motion by means of a rotating vector of constant magnitude† X and constant angular velocity ω. In Fig. 1-7(a), using the x-axis as reference, the distance $\dot{O}P = x(t) = X \cos \omega t$ is the projection of the rotating vector $\boldsymbol{X}$ on a diameter along the x-axis. Thus the displacement of P from the center O is sinusoidal and is simple harmonic. Since the projection of the rotating vector on any diameter gives a sinusoidal motion, the projection of $\boldsymbol{X}$ on the y-axis can be represented by $y(t) = X \sin \omega t$. Naming the x-axis as the real axis and the y-axis as the imaginary one, the rotating vector $\boldsymbol{X}$ is represented by the equation

$$\boldsymbol{X} = X \cos \omega t + jX \sin \omega t = Xe^{j\omega t} \qquad \textbf{(1-9)}$$

where X is the length of the vector or its magnitude, and $j = \sqrt{-1}$ is called the imaginary unit.‡

If a harmonic function is given as $x(t) = X \cos \omega t$, it may be expressed as $x(t) = \text{Re}\,[Xe^{j\omega t}]$, where the symbol "Re" denotes the real part of the function $Xe^{j\omega t}$. Similarly, if the harmonic function is

† In dealing with vectors in complex variables, the length of a vector is called the *absolute value* or *modulus*, and the phase angle is called the *argument* or *amplitude*. The length of the vector in this discussion is the amplitude of the harmonic motion. To avoid confusion, we shall use *magnitude* to denote the length of the vector.

‡ A complex number is of the form $z = x + jy$, where x is the real part and y the imaginary part of the complex number z. Both x and y may be time dependent. For a specified time, they assume constant values, and z can be treated as a complex number. Let the vector $\boldsymbol{X}$ in Fig. 1-7(a) be a complex number. The vectorial representation of $\boldsymbol{X}$ is

$$\boldsymbol{X} = x + jy = X(x/X + jy/X) = X(\cos \omega t + j \sin \omega t)$$

where $X = \sqrt{x^2 + y^2}$ is the magnitude of the vector $\boldsymbol{X}$. Defining $\theta = \omega t$ and expanding the sine and cosine functions by Maclaurin's series, we obtain

$$\begin{aligned}\boldsymbol{X} &= X(\cos \theta + j \sin \theta)\\ &= X\left[\left(1 - \frac{\theta^2}{2!} + \frac{\theta^4}{4!} \cdots\right) + j\left(\theta - \frac{\theta^3}{3!} + \frac{\theta^5}{5!} \cdots\right)\right]\\ &= X\left[1 + j\theta - \frac{\theta^2}{2!} - j\frac{\theta^3}{3!} + \frac{\theta^4}{4!} + j\frac{\theta^5}{5!} \cdots\right]\\ &= X\left[1 + \frac{(j\theta)}{1!} + \frac{(j\theta)^2}{2!} + \frac{(j\theta)^3}{3!} + \frac{(j\theta)^4}{4!} + \frac{(j\theta)^5}{5!} + \cdots\right]\\ &= Xe^{j\theta} = Xe^{j\omega t}\end{aligned}$$

The equation $e^{\pm j\theta} = \cos \theta \pm j \sin \theta$ is called Euler's formula.

$y(t) = X \sin \omega t$, it is expressed as $y(t) = \text{Im}\,[Xe^{j\omega t}]$, and the symbol "Im" denotes the imaginary part of the function $Xe^{j\omega t}$. It should be remembered that a harmonic motion is a reciprocating motion, and that all physical quantities, whether they are displacement, velocity, acceleration, or force, are real quantities. The representation of a harmonic function by means of a rotating vector enables the exponential function $e^{j\omega t}$ to be used in equations involving harmonic functions. It will be shown that the use of complex functions and complex numbers greatly simplifies the mathematical manipulations

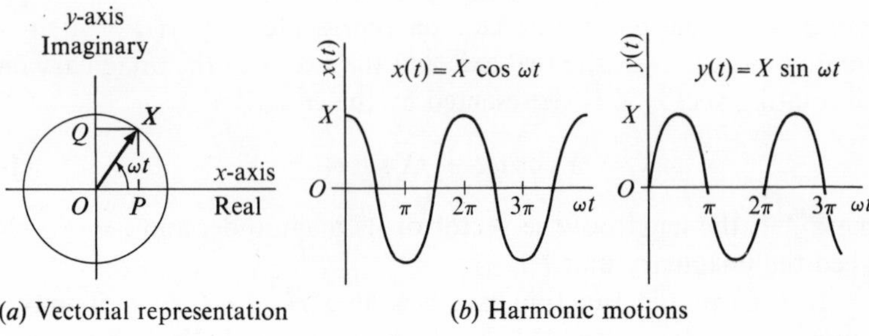

(a) Vectorial representation (b) Harmonic motions

Fig. 1-7. *Representation of harmonic motions by means of rotating vector*

involved in this type of equations. The differentiation of a harmonic function can be carried out in its vectorial form. The differentiation of a vector $\boldsymbol{X}$ is

$$\frac{d}{dt}\boldsymbol{X} = \frac{d}{dt}(Xe^{j\omega t}) = j\omega Xe^{j\omega t} = j\omega \boldsymbol{X} \tag{1-10}$$

$$\frac{d^2}{dt^2}\boldsymbol{X} = \frac{d}{dt}(j\omega Xe^{j\omega t}) = (j\omega)^2 Xe^{j\omega t} = -\omega^2 \boldsymbol{X}$$

Thus each differentiation is equivalent to the multiplication of the vector by $j\omega$. Since X is the magnitude of the vector $\boldsymbol{X}$, ω is real, and $|j| = 1$, each differentiation changes the magnitude by a factor of ω. Since the multiplication of a vector by j is equivalent to advancing it by a phase angle of 90 deg, each differentiation also advances a vector by 90 deg.

If a given harmonic displacement is $x(t) = X \cos \omega t$, the relation between the displacement and its velocity and acceleration is

$$\begin{aligned}
\text{Displacement } x &= \text{Re}\,[Xe^{j\omega t}] = X\cos\omega t &\quad\textbf{(1-11)}\\
\text{Velocity } \dot{x} &= \text{Re}\,[j\omega Xe^{j\omega t}] = -\omega X\sin\omega t\\
&= \omega X\cos(\omega t + 90^\circ)\\
\text{Acceleration } \ddot{x} &= \text{Re}\,[(j\omega)^2 Xe^{j\omega t}] = -\omega^2 X\cos\omega t\\
&= \omega^2 X\cos(\omega t + 180^\circ)
\end{aligned}$$

These relations are identical to those shown in Eqs. (1-4) to (1-6). The representation of displacement, velocity, and acceleration by rotating vectors is illustrated in Fig. 1-8. Since the given displacement $x(t)$ is

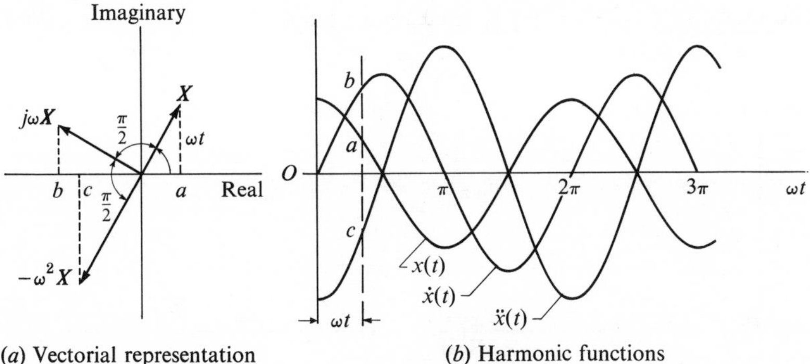

(a) Vectorial representation

(b) Harmonic functions

Fig. 1-8. *Displacement, velocity, and acceleration represented by rotating vectors*

a cosine function, or along the real axis, the velocity and acceleration must be along the real axis. Hence the real parts of the respective vectors give the actual physical quantities.

Harmonic functions can be added graphically by means of vector addition. The vectors $\boldsymbol{X}_1$ and $\boldsymbol{X}_2$ representing the motions $X_1 \cos\omega t$ and $X_2\cos(\omega t + \alpha)$, respectively, are added graphically as shown in Fig. 1-9(*a*). The resultant vector $\boldsymbol{X}$ has magnitude

$$X = \sqrt{(X_1 + X_2\cos\alpha)^2 + (X_2\sin\alpha)^2}$$

and phase angle

$$\beta = \tan^{-1}\frac{X_2\sin\alpha}{X_1 + X_2\cos\alpha}$$

with respect to $\boldsymbol{X}_1$. Since the original motions are given along the real axis, the Re $[\boldsymbol{X}] = X\cos(\omega t + \beta)$ is the sum of the harmonic motions.

The addition operation can easily be extended to include the subtraction operation.

Since both $\boldsymbol{X}_1$ and $\boldsymbol{X}_2$ are rotating with the same angular velocity ω, and usually the relative phase of the vectors is of interest, it is convenient to assign arbitrarily $\omega t = 0$ as a datum of measurement of phase angles. The vectors $\boldsymbol{X}_1$, $\boldsymbol{X}_2$, and their sum $\boldsymbol{X}$ are plotted in this manner, as shown in Fig. 1-9(*b*). It is noted that vector $\boldsymbol{X}_2$ may be expressed as

$$\boldsymbol{X}_2 = X_2 e^{j(\omega t+\alpha)} = (X_2 e^{j\alpha})e^{j\omega t} = (X_2 \cos\alpha + jX_2 \sin\alpha)e^{j\omega t} \quad \textbf{(1-12)}$$

Naming the x- and y-axes of Fig. 1-9(*b*) the real and imaginary axes, the quantity $(X_2 \cos\alpha + jX_2 \sin\alpha)$ is a complex number and is called

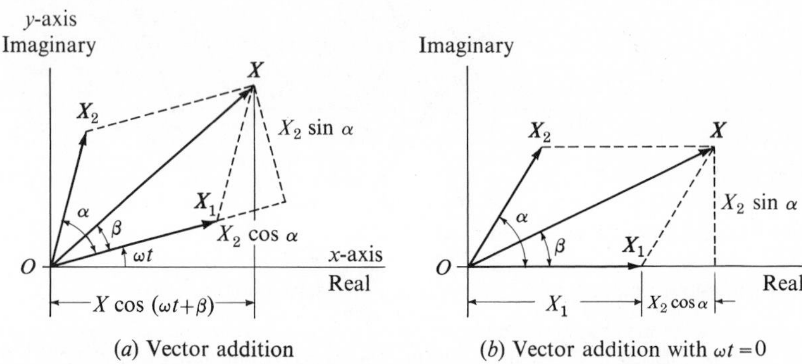

Fig. 1-9. *Addition of harmonic functions: vectorial method*

the *complex amplitude* of $x_2(t)$. Similarly, $Xe^{j\beta}$ is the complex amplitude of $x(t)$. We shall use the symbol $\bar{X}$ to denote the complex amplitude vector $\boldsymbol{X}$.

Harmonic functions can be added algebraically by means of vector addition. Using the same functions $x_1 = X_1 \cos\omega t$ and $x_2 = X_2 \cos(\omega t + \alpha)$, their vector sum is

$$\begin{aligned}\boldsymbol{X} = \boldsymbol{X}_1 + \boldsymbol{X}_2 &= X_1 e^{j\omega t} + X_2 e^{j(\omega t+\alpha)} = (X_1 + X_2 e^{j\alpha})e^{j\omega t} \\ &= (X_1 + X_2\cos\alpha + jX_2 \sin\alpha)e^{j\omega t} = Xe^{j\beta}e^{j\omega t} = Xe^{j(\omega t+\beta)}\end{aligned}$$

where

$$X = \sqrt{(X_1 + X_2\cos\alpha)^2 + (X_2 \sin\alpha)^2}$$

and

$$\beta = \tan^{-1}\frac{X_2 \sin\alpha}{X_1 + X_2\cos\alpha}.$$

Since the given harmonic motions are along the real axis, their sum is

$$x = \text{Re}\,[\boldsymbol{X}] = \text{Re}\,[Xe^{j(\omega t+\beta)}] = X\cos(\omega t + \beta)$$

In representing harmonic motions by means of rotating vectors, it is often necessary to determine the product of complex numbers. This product can be handled easily by expressing the complex numbers in the exponential or vector form. For example, the product of the complex numbers $\boldsymbol{A}$ and $\boldsymbol{B}$ is

$$\boldsymbol{AB} = (a_1 + ja_2)(b_1 + jb_2) = (Ae^{j\alpha})(Be^{j\beta}) = ABe^{j(\alpha+\beta)} \qquad \textbf{(1-13)}$$

where $A = \sqrt{a_1^2 + a_2^2}$, $B = \sqrt{b_1^2 + b_2^2}$ are the magnitudes of the numbers and $\alpha = \tan^{-1}\dfrac{a_2}{a_1}$, $\beta = \tan^{-1}\dfrac{b_2}{b_1}$ are their phase angles. Equation (1-13) indicates that

$$\text{Magnitude of } \boldsymbol{AB} = (\text{magnitude of } \boldsymbol{A})(\text{magnitude of } \boldsymbol{B}) \qquad \textbf{(1-14)}$$

$$\text{Phase of } \boldsymbol{AB} = (\text{phase of } \boldsymbol{A}) + (\text{phase of } \boldsymbol{B}) \qquad \textbf{(1-15)}$$

Obviously, the multiplication operation can be generalized to include the division operation.

Example 1. Manipulation of Vectors

(*a*) $\boldsymbol{A} = 1 + j\sqrt{3} = \sqrt{1+3}\,(\cos 60° + j\sin 60°) = 2e^{j\frac{\pi}{3}} = \underline{2/60°}$

The symbol $2/60°$ is a convenient way of writing $2e^{\frac{\pi}{3}}$. It represents a vector of a magnitude of 2 units and a phase angle of 60° or $\dfrac{\pi}{3}$ rad with the reference x-axis.

(*b*) $\boldsymbol{A} = \boldsymbol{A}_1 + \boldsymbol{A}_2 = (1 + j2) + (4 + j3) = 5 + j5 = \underline{5\sqrt{2}/45°}$

(*c*) $\boldsymbol{A} = \boldsymbol{A}_1\boldsymbol{A}_2 = (1 + j\sqrt{3})(4 + j3) = (2e^{j\frac{\pi}{3}})(5e^{j0.642})$

$$= 10e^{j\left(\frac{\pi}{3}+0.642\right)} = \underline{10/60° + 36.8°} = \underline{10/96.8°}$$

(*d*) $\boldsymbol{A} = \dfrac{\boldsymbol{A}_1}{\boldsymbol{A}_2} = \dfrac{1 + j\sqrt{3}}{4 + j3} = \dfrac{2/60°}{5/36.8°} = \underline{\dfrac{2}{5}/60° - 36.8°} = \underline{0.4/23.2°}$

(*e*) $\boldsymbol{A} = 2j = 0 + 2j = 2\left(\cos\dfrac{\pi}{2} + j\sin\dfrac{\pi}{2}\right) = 2e^{j\frac{\pi}{2}} = \underline{2/90°}$

(*f*) $\boldsymbol{A} = j(1 + j\sqrt{3}) = (e^{j\frac{\pi}{2}})(2e^{j\frac{\pi}{3}}) = \underline{2/60° + 90°} = \underline{2/150°}$

(*g*) $\boldsymbol{A} = \dfrac{1 + j\sqrt{3}}{j} = \dfrac{2/60°}{1/90°} = \underline{2/60° - 90°} = \underline{2/-30°}$

The last two examples indicate that the multiplication of a vector by j advances it counterclockwise by a phase angle of 90 deg, and a division by j retards it by 90 deg.

1-6. UNITS

Newton's law of motion may be expressed as

$$\text{Force} = (\text{mass})(\text{acceleration}) \tag{1-16}$$

Dimensional homogeneity of this equation is obtained when the force is given in pounds lb_f, the mass in slugs, and the acceleration in ft-sec^{-2}. This is called the ft-lb_f-sec system in which the mass has the unit of lb_f-sec^2-ft^{-1}. A body falling under the influence of gravitation has an acceleration of g ft-sec^{-2}, where $g \doteq 32.2$ is the gravitational acceleration. Hence one pound-mass lb_m exerts one pound-force under the gravitational pull of the earth. If the mass is given in pounds lb_m, it must be divided by g to obtain dimensional homogeneity of Eq. (1-16).

The in.-lb_f-sec system is used in the study of vibrations. To use the units of pound-force lb_f, pound-mass lb_m, and in.-sec^{-2}, the gravitational acceleration is $32.2 \times 12 = 386$, and the pound-mass is divided by 386 in order to obtain dimensional homogeneity of Eq. (1-16). We shall assume that the gravitational acceleration is constant unless it is otherwise stated. The units of pound-force and pound-mass are used without their respective subscripts f and m. In the derivation of equations, the mass m is assumed to have the proper units, but the in.-lb_f-sec system is used in the solution of problems.

SUGGESTED READING

Church, A. H., *Mechanical Vibrations* (New York: John Wiley & Sons, Inc., 1957), chap. 1.

Churchill, R. V., *Introduction to Complex Variables and Applications* (New York: McGraw-Hill Book Co., Inc., 1948), chap. 1.

Hansen, H. M., and P. F. Chenea, *Mechanics of Vibration* (New York: John Wiley & Sons, Inc., 1956), chap. 1.

Myklestad, N. O., *Fundamentals of Vibration Analysis* (New York: McGraw-Hill Book Co., Inc., 1956), chap. 1.

Sokolnikoff, I. S., and R. M. Redheffer, *Mathematics of Physics and Modern Engineering* (New York: McGraw-Hill Book Co., Inc., 1958), sec. 1, chap. 7.

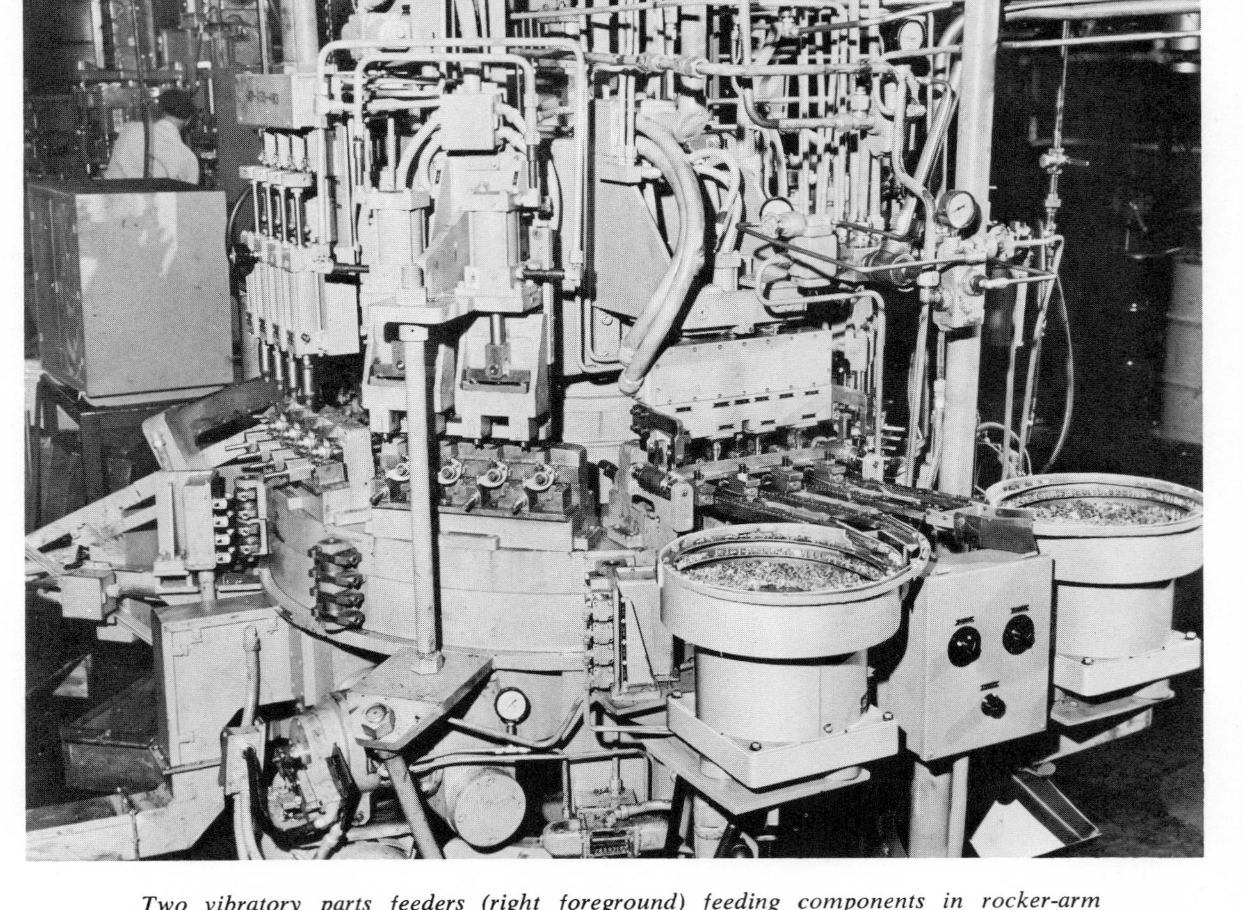

Two vibratory parts feeders (right foreground) feeding components in rocker-arm assembly. Feeders can be designed to feed parts in a specific position, such as heads or tails first, face up or face down, on left or right side, etc. (courtesy Syntron Company, Pennsylvania)

2 SYSTEMS WITH ONE DEGREE OF FREEDOM

2-1. INTRODUCTION

Several simple physical systems and their vibratory motions were described in Chap. 1. A system generally consists of many, or infinitely many, mass particles. If the interrelationship of the masses is such that only one spatial coordinate is required to define the *configuration* of the system, it is said to possess *one degree of freedom*. A configuration is defined as the geometric location of all the masses of a system in space.

Many dynamic systems in engineering can be represented or approximated by systems with one degree of freedom. Through this simplification, the resonance phenomenon in machines, the working principle of vibration-measuring instruments, and the concept of vibration isolation can be explained.

The purpose of this chapter is (1) to examine the one-degree-of-freedom systems through the analysis of a generalized model representative of this class of problems, and (2) to apply the results from the model study to physical systems. The outward appearance of the systems described may be quite different from that of the generalized model. The model serves to unify the class of problems considered, however, and to bring into focus (1) the concept of natural frequency, (2) the role of damping on the oscillatory motion of a system, and (3) the response of the system to excitation. Natural frequency is the most important single characteristic of a vibratory system, for all systems tend to vibrate at their natural frequencies. Damping is necessary for vibration control as well as for the proper operation of many vibration-measuring instruments. The response of a system to excitation is what the engineer tries to control or to enhance for his particular application.

Newton's law of motion is generally employed for this analysis. The equations are developed for systems with rectilinear motion. With a change of units, the theory can be extended to systems with rotational motion. A comparison of the two types of systems is shown in Tables 2-2 and 2-3, pp. 50 and 51. Since vibration is also an energy-interchange phenomenon, some of the concepts are also introduced from energy considerations.

2-2. DEGREES OF FREEDOM

The number of *degrees of freedom* of a physical system is equal to the number of independent spatial coordinates necessary to define the configuration of the system. A rigid body in space has six degrees of freedom; namely, three coordinates to define rectilinear positions and three to define the angular positions. Ordinarily, however, the masses in a system are *constrained* to move only in a certain manner. Thus the constraints limit the degrees of freedom to a much smaller number.

Several one-degree-of-freedom systems are shown in Fig. 2-1. They are briefly discussed as follows:

1. In the spring-mass system shown in Fig. 2-1(a), the mass m is suspended from a coil spring with a spring constant k. If the mass is constrained to move only in the vertical direction and the displacement of the mass is measured from the static equilibrium position o, only one spatial coordinate $x(t)$ is required to define the configuration of the system. The system is said to possess one degree of freedom.

2. Similarly, the torsional pendulum shown in Fig. 2-1(b) consists of a heavy disk J and a shaft with a torsional spring constant k_t. The shaft is assumed to be of negligible mass. If the torsional pendulum is

constrained to oscillate about the longitudinal axis of the shaft, the configuration of the system can be specified by a single coordinate $\theta(t)$.

3. The mass-spring-cantilever system shown in Fig. 2-1(*c*) has one degree of freedom if the cantilever is of negligible mass and the mass m

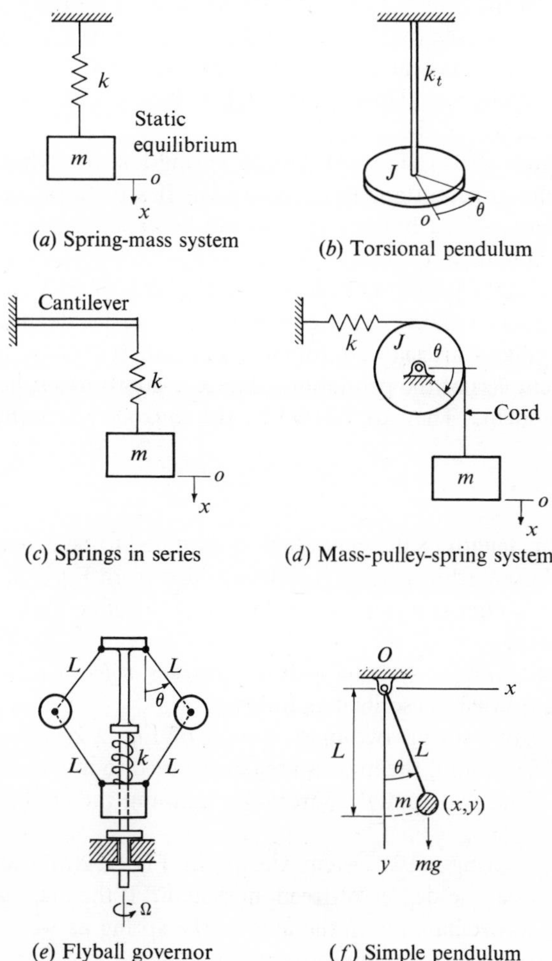

Fig. 2-1. *Systems with one degree of freedom*

is constrained to move vertically. By neglecting the inertial effect of the cantilever and considering only its elasticity, the cantilever is assumed to be a spring which is placed in series with the other spring k of the system. A spring, equivalent to the two springs in series, can be assumed, and the system reduces to the case of the spring-mass system of Fig. 2-1(*a*).

4. The mass-pulley-spring system shown in Fig. 2-1(*d*) has one degree of freedom if it is assumed that there is no slippage between the cord and the pulley J and that the cord is inextensible. Although the system possesses two mass elements m and J, the linear displacement $x(t)$ of the mass and the angular displacement $\theta(t)$ of the pulley are not independent. Thus, either $x(t)$ or $\theta(t)$ can be used to specify the configuration of the system.

5. Figure 2-1(*e*) shows a simple spring-loaded flyball governor rotating with constant angular velocity Ω. If a disturbance is applied to the governor, its vibratory motion can be expressed in terms of the single coordinate $\theta(t)$.

6. A simple pendulum, shown in Fig. 2-1(*f*), is constrained to move in the x–y-plane. This configuration can be defined either by the rectangular Cartesian coordinates $x(t)$ and $y(t)$ or by the angular displacement $\theta(t)$ of the pendulum. The x–y-coordinates, however, are not independent. They are related by the *equation of constraint*

$$x^2 + y^2 = L^2 \tag{2-1}$$

where L, the length of the pendulum, is assumed to be constant. Thus, if $x(t)$ is chosen arbitrarily, $y(t)$ is determined from Eq. 2-1. It is more convenient to choose a single coordinate $\theta(t)$ to define the configuration of this system.

Several dynamic systems with *two degrees of freedom*, as shown in Fig. 2-2, are briefly described as follows:

1. The two-spring–two-mass system of Fig. 2-2(*a*) possesses two degrees of freedom if the masses are constrained to move in the vertical direction. The two spatial coordinates defining the configuration are $x_1(t)$ and $x_2(t)$.

2. The spring-mass system shown in Fig. 2-2(*b*) was described previously as a one-degree-of-freedom system. If the mass m, however, is allowed to oscillate along the axis of the spring as well as to swing from side to side, the system possesses two degrees of freedom.

3. The pendulum in space shown in Fig. 2-2(*c*) can be described by the $\theta(t)$ and $\phi(t)$ coordinates as well as by the $x(t)$, $y(t)$, and $z(t)$ coordinates. The latter are related by the equation of constraint $x^2 + y^2 + z^2 = L^2$. Thus this pendulum has only two degrees of freedom.

It is observed from these illustrations that the number of degrees of freedom can be alternatively defined as the number of coordinates required to define the configuration of a system minus the number of equations of constraint.†

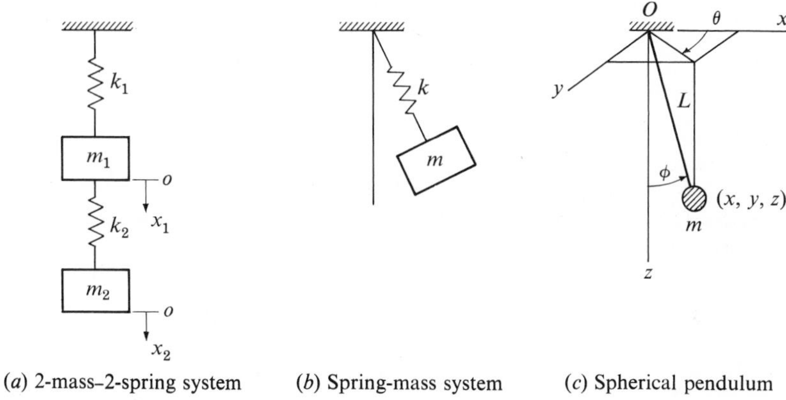

(*a*) 2-mass–2-spring system (*b*) Spring-mass system (*c*) Spherical pendulum

Fig. 2-2. *Systems with two degrees of freedom*

2-3. EQUATION OF MOTION: ENERGY METHOD

The equation of motion of a conservative system can be established from energy considerations. If a conservative system is set into motion, the mechanical energy in the system is partially kinetic and partially potential. The kinetic energy is due to the velocity of the mass, and the potential energy is due to the strain energy of the spring by virtue of its deformation. Since the system is conservative, the total mechanical energy, which is the sum of the kinetic and potential energies, is constant. Furthermore, the time rate of change of the total mechanical energy must be zero. These relations can be expressed as

$$T + U = \text{constant} \tag{2-2}$$

$$\frac{d}{dt}(T + U) = 0 \tag{2-3}$$

where T and U are the kinetic and potential energies, respectively.

† Such a system is called a holonomic system; it is the only type of system considered in this text. For a brief discussion on holonomic and nonholonomic systems, see, for example, H. Goldstein, *Classical Mechanics*, Addison-Wesley Publishing Company, Inc., Reading, Mass., 1957, pp. 11–14.

To derive the equation of motion for the spring-mass system of Fig. 2-3, assume that the displacement $x(t)$ of the mass m is measured from the static equilibrium position of the system and that it is positive in the downward direction. Let us choose the static equilibrium

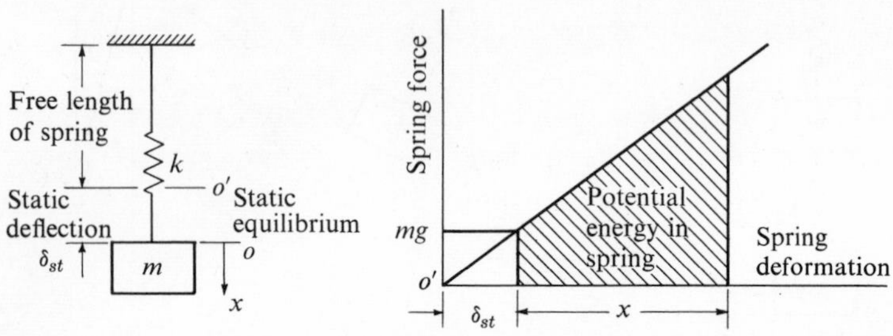

Fig. 2-3. *Potential energy in spring*

position as reference. Since the spring element is assumed to be of negligible mass, the kinetic energy of the system is

$$T = \tfrac{1}{2}m\dot{x}^2 \tag{2-4}$$

The change in potential energy of the system, owing to the displacement $x(t)$, is equal to the strain energy in the spring k minus the potential energy change of the mass due to the difference in elevation. The potential energy is

$$\begin{aligned} U &= \int_o^x (\text{total spring force})\, dx - mgx \\ &= \int_o^x (mg + kx)\, dx - mgx \\ &= \tfrac{1}{2}kx^2 \end{aligned} \tag{2-5}$$

Substituting Eqs. (2-4) and (2-5) in Eq. (2-3), we obtain

$$\frac{d}{dt}(\tfrac{1}{2}m\dot{x}^2 + \tfrac{1}{2}kx^2) = (m\ddot{x} + kx)\dot{x} = 0$$

Since the velocity $\dot{x}(t)$ in this equation cannot be zero for all values of time, clearly

$$m\ddot{x} + kx = 0 \tag{2-6}$$

or

$$\ddot{x} + \omega_n^2 x = 0 \tag{2-7}$$

where $\omega_n^2 = k/m$. The equation of motion of the system can be expressed as shown in Eqs. (2-6) or (2-7).

It can be shown that the solution of this differential equation is of the form

$$x = C_1 \cos \omega_n t + C_2 \sin \omega_n t \tag{2-8}$$

where C_1 and C_2 are arbitrary constants to be evaluated by the *initial conditions*, $x(0)$ and $\dot{x}(0)$, specified for the problem.† It is apparent that ω_n, as defined in Eq. (2-7), is the circular frequency of the harmonic motion $x(t)$. Since the components of the solution are harmonic of the same frequency, their sum is also harmonic and can be written as

$$x = A \sin (\omega_n t + \psi) \tag{2-9}$$

where $A = \sqrt{C_1^2 + C_2^2}$ is the amplitude of the motion, and $\psi = \tan^{-1} C_1/C_2$ is the phase angle.

Equation (2-9) indicates that once this system is set into motion it will vibrate with simple harmonic motion, and the amplitude A of the motion will not diminish with time. The system oscillates because it possesses two energy storage elements, namely the spring and the mass. As was described in Sec. 1-3 (Chap. 1), there is a constant interchange of energy between these elements. The frequency of oscillation is

$$f_n = \frac{\omega_n}{2\pi} = \frac{1}{2\pi}\sqrt{\frac{k}{m}} \tag{2-10}$$

which is the natural frequency of the system.‡ Thus the natural frequency is inherent in a system. It is a function of the system parameters k and m, and it is independent of the amplitude of oscillation or the manner by which the system is set into motion. It should be noted that only the amplitude A and the phase angle ψ are dependent on the initial conditions. Here, the phase angle merely specifies the initial value of $x(t)$.

Example 1. Determine the equation of motion of the simple pendulum shown in Fig. 2-1(f).

† The arbitrary constants C_1 and C_2 can be evaluated by conditions specified other than $t = 0$. It is customary and convenient, however, to use initial conditions.

‡ It is convenient to call ω_n the natural frequency instead of the natural circular frequency. In the subsequent sections of this text, natural frequency will refer to f_n or ω_n unless ambiguity arises. Similarly, frequency may refer to f or ω.

Solution: Assume that the size of the bob is small as compared with the length L of the pendulum and that the rod connecting the bob to the hinge point O is of negligible mass. The mass moment of inertia of the bob of mass m about O is

$$J_o = (J_{cg} + mL^2) \doteq mL^2$$

where J_{cg} is the mass moment of inertia of m about its center of gravity. If the bob is sufficiently small in size, then $J_{cg} \ll mL^2$.

For an angular displacement $\theta(t)$ from the static equilibrium position of the pendulum, the kinetic energy of the system is $T = \frac{1}{2}J_o\dot{\theta}^2 = \frac{1}{2}mL^2\dot{\theta}^2$. The corresponding potential energy is $U = mgL(1 - \cos\theta)$, where $L(1 - \cos\theta)$ is the change in elevation of the pendulum bob. Substituting these energy quantities in Eq. (2-3) gives

$$mL^2\ddot{\theta} + mgL\sin\theta = 0 \qquad \textbf{(2-11)}$$

or

$$\ddot{\theta} + \frac{g}{L}\sin\theta = 0 \qquad \textbf{(2-12)}$$

The equation of motion of the simple pendulum is as shown in Eqs. (2-11) or (2-12). If it is further assumed that the amplitude of oscillation is small, then $\sin\theta \doteq \theta$, and Eq. (2-12) becomes

$$\ddot{\theta} + \frac{g}{L}\theta = 0 \qquad \textbf{(2-13)}$$

This equation is of the same form as Eq. (2-7), and the solution follows. The frequency of oscillation of a simple pendulum is $\omega_n = \sqrt{g/L}$.

It should be noted that, if small oscillations are not assumed, Eq. (2-11) is a nonlinear differential equation, and elliptic integrals will have to be used to solve the equation. The dependent variable $\theta(t)$ and the independent variable t are related by†

$$t = \int_{\theta_o}^{\theta} \frac{d\theta}{\sqrt{\dot{\theta}_o^2 + \dfrac{2mgL}{J_o}(\cos\theta - \cos\theta_o)}} \qquad \textbf{(2-14)}$$

where θ_o and $\dot{\theta}_o$ are the initial conditions at $t = 0$. It is conceivable that, if the pendulum is given a sufficiently large initial velocity to set

† T. von Karman and M. A. Biot, *Mathematical Methods in Engineering*, McGraw-Hill Book Co., Inc., New York, 1940, pp. 115–119.

it into motion, the pendulum will continue to rotate about the hinge point. Thus $\theta(t)$ will increase with time, and the motion is not periodic. The assumption of small oscillation greatly simplifies the effort necessary to obtain the solution. Small oscillations will be assumed throughout this text unless otherwise stated.

Example 2. Figure 2-4 shows a cylinder of mass m and radius R_1 rolling without slippage on a curved surface of radius R. By the energy method, derive the equation of motion of this system.

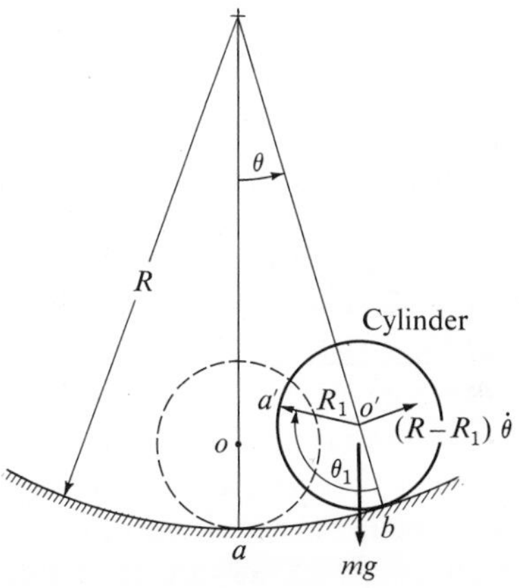

Fig. 2-4. *Oscillation of cylinder on curved surface*

Solution: The kinetic energy of the cylinder is due to the translational and rotational motions of the cylinder. The translational velocity of the center of the cylinder is $(R - R_1)\dot{\theta}$, and the angular velocity of the cylinder is $(\dot{\theta}_1 - \dot{\theta})$. Since the cylinder rolls without slippage, the arc $\widehat{ab} =$ arc $\widehat{a'b}$, and $R\theta = R_1\theta_1$. Therefore, the angular velocity may be written as $\left(\dfrac{R}{R_1} - 1\right)\dot{\theta}$. The total kinetic energy of the cylinder is

$$T = \tfrac{1}{2}m[(R - R_1)\dot{\theta}]^2 + \frac{1}{2}m\frac{R_1^2}{2}\left[\left(\frac{R}{R_1} - 1\right)\dot{\theta}\right]^2$$

where $m\dfrac{R_1^2}{2}$ is the moment of inertia of the cylinder about its longitudinal axis. The potential energy with respect to the static equilibrium position of the cylinder is due to the change in elevation of the center of gravity of the mass; that is,

$$U = mg(R - R_1)(1 - \cos\theta)$$

Substituting the T and U relations into Eq. (2-3), we obtain

$$[\tfrac{3}{2}m(R - R_1)^2\ddot{\theta} + mg(R - R_1)\sin\theta]\dot{\theta} = 0$$

or

$$\ddot{\theta} + \frac{2g}{3(R - R_1)}\theta = 0$$

where $\theta \doteq \sin\theta$ for small oscillations. Comparing this equation of motion with Eq. (2-7), it is noted that ω_n of this system is equal to $\sqrt{2g/3(R - R_1)}$.

Alternatively, using a simplified version of Rayleigh's method, the natural frequency can be deduced by assuming that (1) the motion is simple harmonic and (2) the maximum kinetic energy of the system is equal to its maximum potential energy. During the cyclic motion of the system, there is a constant interchange of energy between the kinetic energy of the mass and the potential energy of the spring. As the mass passes through the static equilibrium position, Eq. (2-5) indicates that the potential energy of the system is zero. Therefore, the kinetic energy is equal to the total mechanical energy of the system. When the mass is at a position of maximum displacement, it is on the verge of changing direction; therefore its velocity, and correspondingly the kinetic energy, is zero. Thus the potential energy is equal to the total energy of the system. If the motion is harmonic, as indicated by Eq. (2-9), the maximum displacement is A and the maximum velocity is $\omega_n A$. Equating the maximum kinetic and potential energies, we have

$$T_{\max} = U_{\max} = \text{total energy of the system} \tag{2-15}$$

$$\tfrac{1}{2}m(\omega_n A)^2 = \tfrac{1}{2}k(A)^2$$

giving

$$\omega_n = \sqrt{k/m}, \quad \text{or} \quad f_n = \frac{\omega_n}{2\pi}$$

which is identical to Eq. (2-10).

Example 3. Equivalent Mass of Spring: Rayleigh's Method

If the mass of the spring in the system shown in Fig. 2-5 is not negligible, determine the natural frequency of the system by Rayleigh's method.

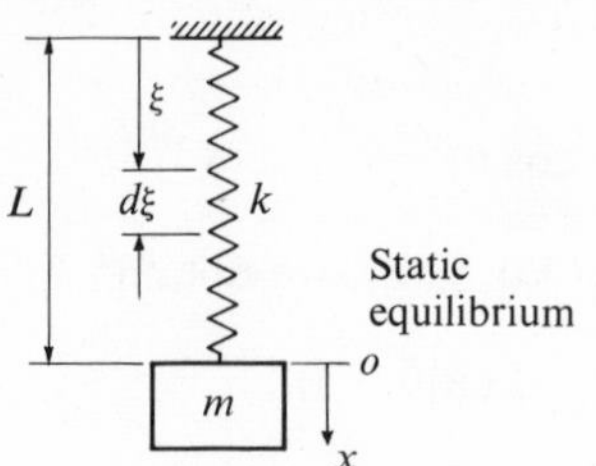

Fig. 2-5. *Equivalent mass of spring: Rayleigh method*

Solution: Let L be the length of the spring k when the system is in its static equilibrium position. Now, if the free end of the spring has a displacement $x(t)$ and it is assumed that an intermediate point of the spring at a distance ξ from the fixed end has a displacement equal to $\frac{\xi}{L}\, x(t)$, then $x(t)$ defines the configuration, and the system has but one degree of freedom.

The kinetic energy of the system is the sum of that of the mass m and the spring. The kinetic energy of an element of the spring of length $d\xi$ is $\frac{1}{2}(\rho\, d\xi)\left(\frac{\xi}{L}\,\dot{x}\right)^2$, where ρ is its density in mass per unit length. Let $x = A \sin \omega_n t$; then the maximum kinetic energy of the system is

$$T_{\max} = \frac{1}{2}\, m\dot{x}_{\max}^2 + \int_o^L \frac{1}{2}\rho \left(\frac{\xi}{L}\,\dot{x}_{\max}\right)^2 d\xi$$

$$= \frac{1}{2}\left(m + \frac{\rho L}{3}\right)\dot{x}_{\max}^2$$

$$= \frac{1}{2}\left(m + \frac{\rho L}{3}\right)(\omega_n A)^2$$

From Eq. (2-5) the maximum potential energy of the system is

$$U_{\max} = \tfrac{1}{2}kx_{\max}^2 = \tfrac{1}{2}kA^2$$

The natural frequency is obtained by equating the maximum kinetic and potential energies; that is,

$$\frac{1}{2}\left(m + \frac{\rho L}{3}\right)(\omega_n A)^2 = \tfrac{1}{2}kA^2$$

giving

$$f_n = \frac{\omega_n}{2\pi} = \frac{1}{2\pi}\sqrt{\frac{k}{m + \frac{\rho L}{3}}}$$

This equation shows that, for the assumptions made, the inertial effect of the spring can be accounted for by adding an equivalent spring mass to the rigid mass. The equivalent spring mass is equal to one third of the entire mass of the spring. The natural frequency can then be calculated as if the system were a massless spring and rigid-mass system.

This approximate method shows that the natural frequency is independent of the mass ratio, $\rho L/m$, that is, the mass of the spring to that of the rigid mass. For a heavy spring with a light mass, a larger fraction of the spring mass would have to be used for the frequency calculation. This error, however, is less than 1 percent, as compared with the exact value, when the spring mass is equal to the rigid mass.† Rayleigh's method will be discussed in greater detail in Chap. 6.

† S. Timoshenko and D. Young, *Vibration Problems in Engineering*, 3rd ed., D. Van Nostrand Co., Inc., New York, 1954, pp. 306–314.

2-4. EQUATION OF MOTION: NEWTON'S LAW OF MOTION

This section employs Newton's law of motion to establish the differential equation of motion of one-degree-of-freedom systems. A generalized model, as shown in Fig. 2-6, is used to represent this class of physical problems.† All four elements discussed in Sec. 1-2 (Chap. 1) are incorporated in this model. The displacement $x(t)$ of the mass m, as measured from the static equilibrium position, is considered positive in the downward direction, and so are the velocity $\dot{x}(t)$ and the acceleration $\ddot{x}(t)$. Referring to the free-body sketch in Fig. 2-6, the forces acting on the body of mass m are (1) the gravitational force mg which is constant, (2) the spring force $k(x + \delta_{st})$ which always opposes the displacement, (3) the damping force $c\dot{x}$ which always opposes the velocity, and (4) the excitation force which, for the present discussion, is assumed to equal $F \sin \omega t$.

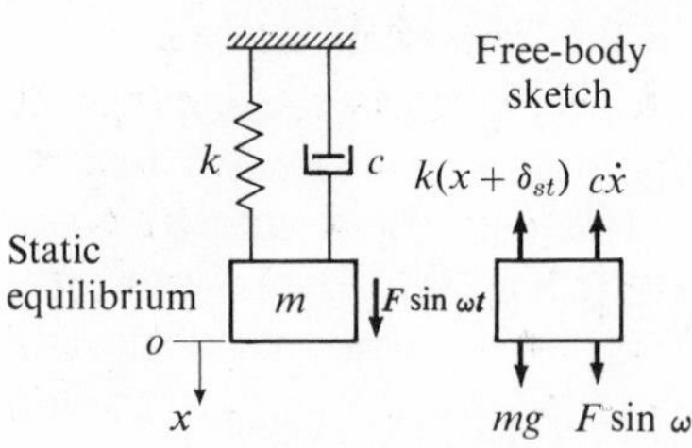

Fig. 2-6. *Generalized model: one-degree-of-freedom system*

Newton's law of motion (second law) may be stated as follows: The rate of change of momentum is proportional to the impressed force and takes place in the direction of the straight line in which the force acts. If the mass is constant, the rate of change of momentum is equal to the mass times its acceleration. Considering the motion in the x-direction, the equation of motion of this system is

$$m\ddot{x} = \sum(\text{forces in the } x\text{-direction}) \tag{2-16}$$

or

$$m \frac{d^2}{dt^2}(x + \delta_{st}) = -k(x + \delta_{st}) - c \frac{d}{dt}(x + \delta_{st}) + mg + F \sin \omega t \tag{2-17}$$

Since the gravitational force mg is equal to the static spring force $k\delta_{st}$, we have

$$m\ddot{x} + c\dot{x} + kx = F \sin \omega t \tag{2-18}$$

† The present discussion is limited to systems that can be represented by a second-order linear differential equation. Some variations of the one-degree-of-freedom systems are represented by third-order differential equations. These systems will be treated in Case 7, Sec. 2-12.

It should be noted that this equation can be derived whether the general position of the mass is considered above or below the static equilibrium position of the system or whether the mass is moving upward or downward. Thus this equation is true for all positions of the mass. The verification of this statement is left as an exercise for the reader.

Alternatively, Eq. (2-16) can be written as

$$\sum(\text{forces in the } x\text{-direction}) - m\ddot{x} = 0 \qquad \textbf{(2-16a)}$$

The equation in this form represents *d'Alembert's principle.* The quantity $-m\ddot{x}$, a product of the mass and negative acceleration, is called the *inertia force.* In other words, introducing the appropriate inertia force, the state of motion of a mass at any instant may be considered as a state of equilibrium: the problem is reduced to an equivalent problem of statics. If the body has rectilinear motion, the inertia force acts at its mass center. When the body has rectilinear and rotational motions, the inertia force must be applied to each particle of the body.

2-5. GENERAL SOLUTION

The equation of motion, Eq. (2-18), is a second-order linear differential equation with constant coefficients. The general solution $x(t)$ is the sum of the *complementary function* $x_c(t)$ and the *particular integral* $x_p(t)$ (see Appendix B); that is,

$$x = x_c + x_p \qquad \textbf{(2-19)}$$

Let us consider the two parts of the solution separately before discussing the general solution.

The *complementary function* satisfies the corresponding homogeneous equation

$$m\ddot{x} + c\dot{x} + kx = 0 \qquad \textbf{(2-20)}$$

The solution of Eq. (2-20) is of the form

$$x_c = Ce^{st} \qquad \textbf{(2-21)}$$

where C and s are constants. Substituting Eq. (2-21) into Eq. (2-20) gives

$$(ms^2 + cs + k)Ce^{st} = 0 \qquad \textbf{(2-22)}$$

Since the quantity Ce^{st} cannot be zero for all values of t, Eq. (2-20) is satisfied if

$$ms^2 + cs + k = 0 \tag{2-23}$$

This is called the *auxiliary equation* or the *characteristic equation* of the system. Equation (2-23) is satisfied if the values of s are

$$s_{1,2} = \frac{1}{2m}(-c \pm \sqrt{c^2 - 4mk}) \tag{2-24}$$

that is, when the values of s are the roots of the characteristic equation. Since Eq. (2-24) gives two roots, the complementary function is

$$x_c = C_1e^{s_1t} + C_2e^{s_2t} \tag{2-25}$$

where C_1 and C_2 are arbitrary constants depending on the initial conditions.

Let us write Eq. (2-24) in a more convenient form by defining

$$\frac{k}{m} = \omega_n^2, \text{ and } \frac{c}{m} = 2\zeta\omega_n \tag{2-26}$$

It is recognized from the discussion in Sec. 2-3 that ω_n is the natural circular frequency of the system. ζ is called the *damping factor*. From Eq. (2-26), the relation between ζ and the system parameters is

$$\zeta = \frac{c}{2\sqrt{km}} \tag{2-27}$$

Since the system parameters are assumed to be positive, ζ is a positive number. It can assume values $\gtreqless 1$. The physical interpretation of ζ will be given in the paragraphs to follow. Substituting Eq. (2-26) in Eq. (2-24), the roots of the characteristic equation become

$$s_{1,2} = -\zeta\omega_n \pm \sqrt{\zeta^2 - 1}\,\omega_n \tag{2-28}$$

If ζ is greater than 1, Eq. (2-28) shows that the roots are real, distinct, and negative, since $\sqrt{\zeta^2 - 1} < \zeta$. Thus Eq. (2-25) indicates that no oscillatory motion can be expected from the complementary function of the equation of motion regardless of the initial conditions imposed on the system, and the motion is *aperiodic*. Since both of the roots are negative, the motion diminishes exponentially with increasing time.

If ζ is equal to 1, Eq. (2-28) shows that both of the roots are equal to $-\omega_n$. The complementary function is of the form

$$x_c = (C_3 + C_4t)e^{-\omega_nt} \tag{2-29}$$

Thus the motion is again aperiodic. Since the $\lim_{t\to\infty} e^{-\omega_n t} = 0$ and $\lim_{t\to\infty} te^{-\omega_n t} = 0$, the motion will eventually diminish to zero.

If ζ is less than 1, the roots are complex conjugates,

$$s_{1,2} = -\zeta\omega_n \pm j\sqrt{1-\zeta^2}\,\omega_n \tag{2-30}$$

where $j = \sqrt{-1}$. Defining

$$\omega_d = \sqrt{1-\zeta^2}\,\omega_n \tag{2-31}$$

and using Euler's formula $e^{\pm j\theta} = \cos\theta \pm j\sin\theta$, the complementary function, Eq. (2-25), becomes

$$x_c = e^{-\zeta\omega_n t}\,(C_1 e^{j\omega_d t} + C_2 e^{-j\omega_d t})$$

$$x_c = e^{-\zeta\omega_n t}\,[(C_1 + C_2)\cos\omega_d t + j(C_1 - C_2)\sin\omega_d t] \tag{2-32}$$

where C_1 and C_2 are arbitrary constants. Since the displacement $x_c(t)$ is a real physical quantity, C_1 and C_2 in Eq. (2-32) must be complex conjugates; that is, the coefficients of the cosine and sine functions in this equation must be real. Rewriting Eq. (2-32), we have

$$x_c = e^{-\zeta\omega_n t}\,(A_1\cos\omega_d t + A_2\sin\omega_d t) \tag{2-33}$$

or

$$x_c = Ae^{-\zeta\omega_n t}\sin(\omega_d t + \psi) \tag{2-34}$$

where A_1 and A_2 are arbitrary constants to be specified by the initial conditions. The two harmonic functions in Eq. (2-33) can be combined to give Eq. (2-34), where $A = \sqrt{A_1^2 + A_2^2}$ and $\psi = \tan^{-1}\dfrac{A_1}{A_2}$. The motion described by Eq. (2-34) consists of a harmonic motion of frequency ω_d and amplitude $Ae^{-\zeta\omega_n t}$ which decreases exponentially with time.

In the three cases enumerated, the type of motion prescribed by the complementary function $x_c(t)$ depends on whether the damping factor ζ is greater than, equal to, or less than 1. The system is said to be *overdamped* if ζ is greater than 1, *critically damped* if ζ is equal to 1, and *underdamped* if ζ is less than 1. Vibratory motion exists only if the system is underdamped. The frequency of oscillation ω_d is lower than the natural frequency ω_n of the system. Because of the exponential decay in all the cases enumerated, the motion described by x_c will eventually die out. Thus the complementary function $x_c(t)$ gives the *transient motion* of the system. As a limiting case, if the system does not possess damping, the amplitude of $x_c(t)$ will not diminish with time, and the frequency of oscillation is equal to the natural frequency of the system.

It is observed in Eq. (2-34) that the rate of the exponential decay and the frequency ω_d of the transient motion are inherent in the system parameters and are independent of the arbitrary constants of the equation. Since these arbitrary constants are specified by the initial conditions, the rate of decay and the frequency of oscillation are independent of the manner by which the system is set into motion.

The foregoing discussions indicate that the mode of vibration is specified by the value of ζ; whether $\zeta \gtrless 1$. If $\zeta = 1$, Eq. (2-27) becomes

$$c_c = 2\sqrt{km} \tag{2-35}$$

where c_c is called the *critical damping coefficient*. It is the amount of damping necessary for a system to be critically damped. Combining Eq. (2-27) and Eq. (2-35) gives

$$\zeta = \frac{c}{c_c} \tag{2-36}$$

Hence the damping factor ζ is a measure of the existing damping as compared with that necessary for a system to be critically damped.

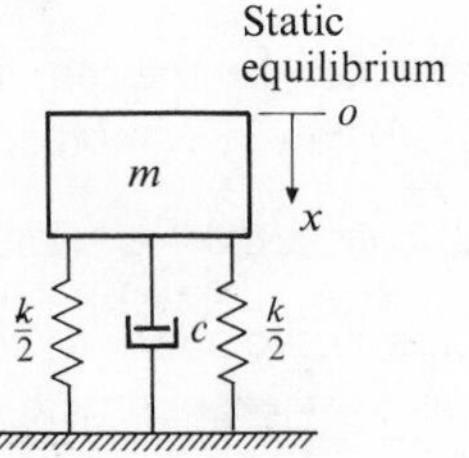

Fig. 2-7. *Damped free vibration*

Example 4. Damped Free Vibration

A machine weighing 40 lb is mounted on springs and dashpots, as shown schematically in Fig. 2-7. The total stiffness of the spring is 50 lb-in.$^{-1}$, and the total damping is 0.75 lb-sec-in.$^{-1}$. If the system is initially at rest and a velocity of 4 in.-sec^{-1} is imparted to the mass, determine (*a*) the displacement and velocity of the mass as a function of time, and (*b*) the displacement at time equal to 1 sec.

Solution: The displacement $x(t)$ is obtained by the direct application of Eq. (2-33). The parameters of the equation are

$$\omega_n = \sqrt{k/m} = \sqrt{kg/w} = \sqrt{(50)(386)/(40)} = 22 \text{ rad-sec}^{-1}$$

$$\zeta = c/2m\omega_n = (0.75)/(2)(40/386)(22) = 0.1645$$

$$\omega_d = \sqrt{1-\zeta^2}\,\omega_n = 22\sqrt{1-(0.1645)^2} = 21.7 \text{ rad-sec}^{-1}$$

(*a*) Substituting these values in Eq. (2-33), we obtain

$$x = e^{-3.62t}(A_1 \cos 21.7t + A_2 \sin 21.7t)$$
$$\dot{x} = -3.62e^{-3.62t}(A_1 \cos 21.7t + A_2 \sin 21.7t) + 21.7e^{-3.62t}(-A_1 \sin 21.7t + A_2 \cos 21.7t)$$

Applying the initial conditions gives

$$x(0) = 0, \quad \therefore\ A_1 = 0$$
$$\dot{x}(0) = 4 \text{ in.-sec}^{-1} \quad A_2 = 4/21.7 = 0.1845$$
$$\therefore\ x = 0.1845e^{-3.62t} \sin 21.7t$$
$$\dot{x} = e^{-3.62t}(-0.667 \sin 21.7t + 4 \cos 21.7t)$$
$$\dot{x} = 4.05e^{-3.62t} \cos (21.7t + 9.5^\circ)$$

(*b*) The displacement at $t = 1$ sec is

$$x(1) = 0.1845e^{-3.62} \sin 21.7 = 0.00131 \text{ in.}$$

The *particular integral* of the equation of motion, Eq. (2-18), is of the form

$$x_p = B_1 \sin \omega t + B_2 \cos \omega t \qquad \textbf{(2-37)}$$

Substituting Eq. (2-37) in Eq. (2-18) and collecting the coefficients of the sine and the cosine terms, we obtain

$$[(k - m\omega^2)B_1 - c\omega B_2] \sin \omega t + [c\omega B_1 + (k - m\omega^2)B_2] \cos \omega t = F \sin \omega t$$

giving

$$(k - m\omega^2)B_1 - c\omega B_2 = F \qquad \textbf{(2-38)}$$
$$c\omega B_1 + (k - m\omega^2)B_2 = 0$$

Solving for B_1 and B_2 and substituting their values in Eq. (2-37), $x_p(t)$ becomes

$$x_p = \frac{F}{(k - m\omega^2)^2 + (c\omega)^2}[(k - m\omega^2) \sin \omega t - c\omega \cos \omega t]$$

$$x_p = \frac{F}{\sqrt{(k - m\omega^2)^2 + (c\omega)^2}} \sin (\omega t - \phi) = X \sin (\omega t - \phi) \qquad \textbf{(2-39)}$$

where

$$\phi = \tan^{-1} \frac{c\omega}{k - m\omega^2} \qquad \textbf{(2-40)}$$

X is the amplitude of the steady-state response and $-\phi$ is the phase lag of $x_p(t)$ with respect to $F \sin \omega t$.

For convenience, the last two equations are often reduced to nondimensional form. Dividing the numerator and denominator by k, defining $X_o = F/k$, and substituting the relations $\omega_n^2 = k/m$ and $2\zeta\omega/\omega_n = \dfrac{c\omega}{k}$, these equations become

$$\frac{X}{X_o} = \frac{1}{\sqrt{\left[1 - \left(\dfrac{\omega}{\omega_n}\right)^2\right]^2 + \left(2\zeta\dfrac{\omega}{\omega_n}\right)^2}} = \frac{1}{\sqrt{(1-r^2)^2 + (2\zeta r)^2}} = \kappa \quad \text{(2-41)}$$

$$\phi = \tan^{-1}\frac{2\zeta\omega/\omega_n}{1 - \left(\dfrac{\omega}{\omega_n}\right)^2} = \tan^{-1}\frac{2\zeta r}{1 - r^2} \quad \text{(2-42)}$$

where $r = \omega/\omega_n$ is the *frequency ratio* of the excitation frequency to the natural frequency of the system. κ is called the *magnification factor*. These equations indicate that κ and ϕ are functions of the frequency ratio r and the damping factor ζ. The equations are plotted as shown in Figs. 2-8 and 2-9 with ζ as a parameter.

The following are observed from the particular integral $x_p(t)$ of the equation of motion:

1. The motion described by Eq. (2-39) is harmonic and is of the same frequency as the excitation. For a given harmonic excitation of constant amplitude and frequency, the amplitude of the response is constant. Thus the motion described by the particular integral is called the *steady-state response* or the *steady-state vibrations*.

2. Since the particular integral does not contain arbitrary constants, the steady-state response of a system is independent of the initial conditions imposed on the system.

3. The amplitude of the steady-state response is a function of the amplitude and the frequency of the excitation. The quantity X_o, defined as F/k, is the response of the system to a static force of magnitude F. Thus the ratio $\dfrac{X}{X_o} = \kappa$ may be regarded as the amplitude ratio of the steady-state response to the static response of the system. This ratio is called the *magnification factor*. As shown in Fig. 2-8, the magnification factor can be considerably greater than or less than unity.

4. At resonance, when $r = \dfrac{\omega}{\omega_n} = 1$, it is noted in Eq. (2-41) that the magnification factor is limited only by the damping factor ζ. In

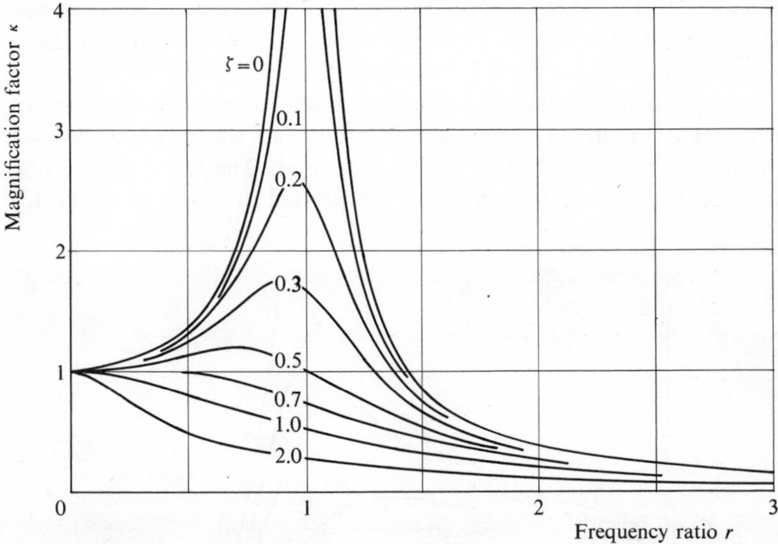

Fig. 2-8. *Magnification factor versus frequency ratio for various amounts of damping; system shown in Fig. 2-6*

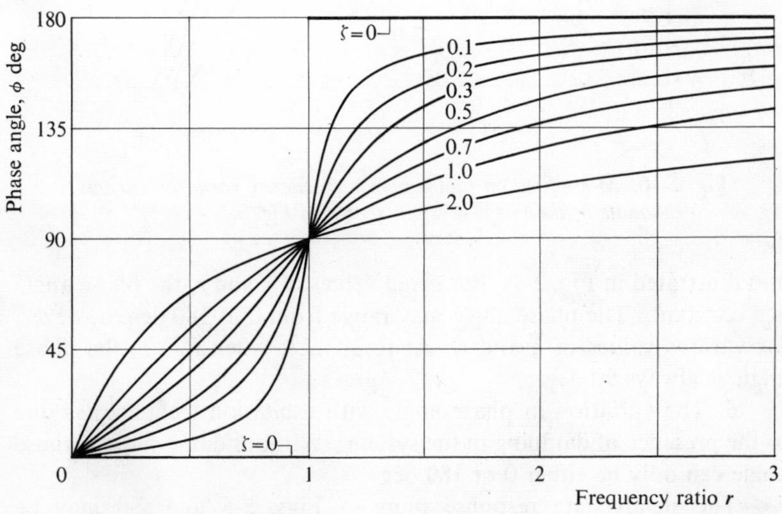

Fig. 2-9. *Phase angle versus frequency ratio for various amounts of damping; system shown in Fig. 2-6*

other words, the amplitude of vibration is limited only by the presence of damping in the system. Without damping, the amplitude is theoretically infinite.

5. Since the excitation is given as $F \sin \omega t$ and the steady-state response as $X \sin (\omega t - \phi)$, the excitation and the response do not attain their maximum values at the same time. The *phase angle* ϕ is a measure of this time difference and is expressed as shown in Eq. (2-42)

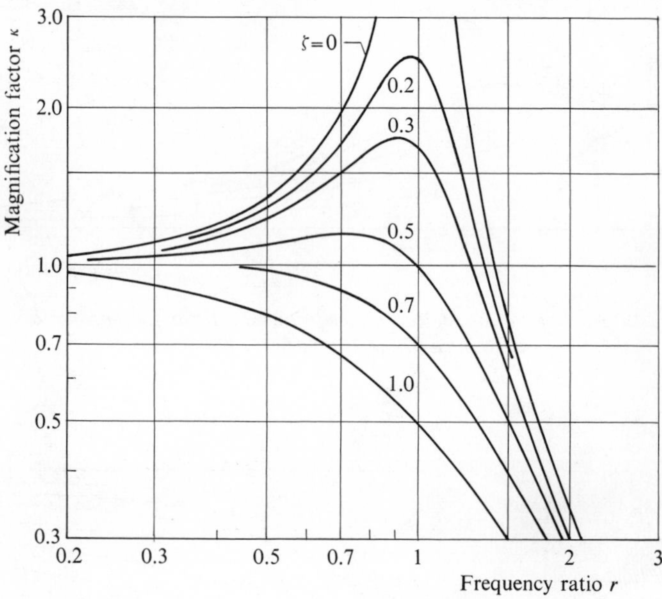

Fig. 2-10. *Magnification factor versus frequency ratio for various amounts of damping; system shown in Fig.* 2-6

and illustrated in Fig. 2-9. For given values of ζ and r, the phase angle is a constant. The phase angle may range from 0 to 180 deg, however, for various values of ζ and r. At resonance, when $r = 1$, the phase angle is always 90 deg.

6. The variation in phase angle with excitation frequency is due to the presence of damping in the system. Without damping, the phase angle can only be either 0 or 180 deg.

The steady-state response plots of Figs. 2-8 and 2-9 may be presented in the form of $\log \kappa$ versus $\log r$ and ϕ versus $\log r$ as shown in Figs. 2-10 and 2-11. These are known as the Bode plot. Alternatively, the steady-state response data may be combined into a single plot as shown in Fig. 2-12, known as the Nyquist plot: the magnification factor κ is represented by the length of a vector, $\left|\frac{X}{X_o} e^{i\phi}\right|$, and the phase angle ϕ by the incline of the vector with the real axis. Both of these plots are used in control theory, a subject closely allied with vibrations.

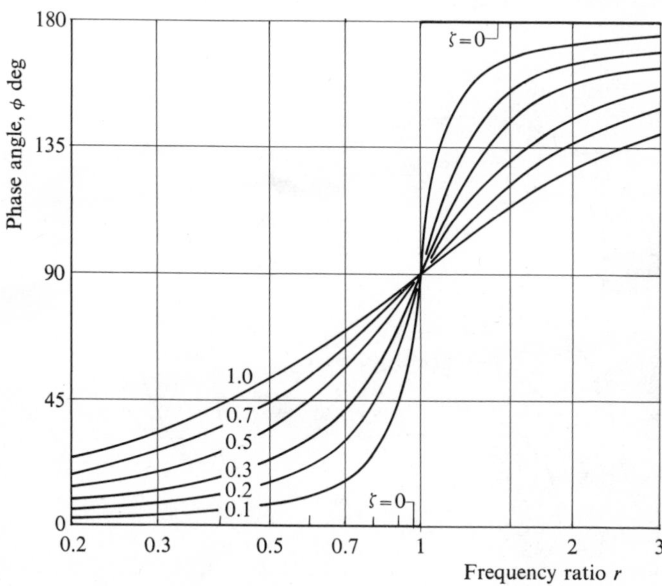

Fig. 2-11. *Phase angle versus frequency ratio for various amounts of damping; system shown in Fig.* 2-6

The *general solution* of the equation of motion, Eq. (2-18), represents the motion of a system in response to a harmonic excitation and a given set of initial conditions. It is a linear combination of the complementary function and particular integral, that is,

$$x = x_c + x_p$$

The complementary function can be expressed as Eqs. (2-25), (2-29), or (2-34), depending on whether the damping factor ζ is greater than, equal to, or less than 1. Assuming that the system is underdamped, which is often encountered in vibration problems, the general solution is

$$x = Ae^{-\zeta\omega_n t} \sin (\omega_d t + \psi) + \frac{F}{k} \kappa \sin (\omega t - \phi) \tag{2-43}$$

where only A and ψ are arbitrary constants.

The physical interpretation of this equation was explained intuitively in Chap. 1. As a system is disturbed from its static equilibrium position through the application of a set of initial conditions and a harmonic excitation, it tends to vibrate at its own natural frequency as well as to follow the excitation frequency. The motion consists of the transient motion and the steady-state vibrations. Since the latter is the motion sustained by the excitation, it must have the same frequency

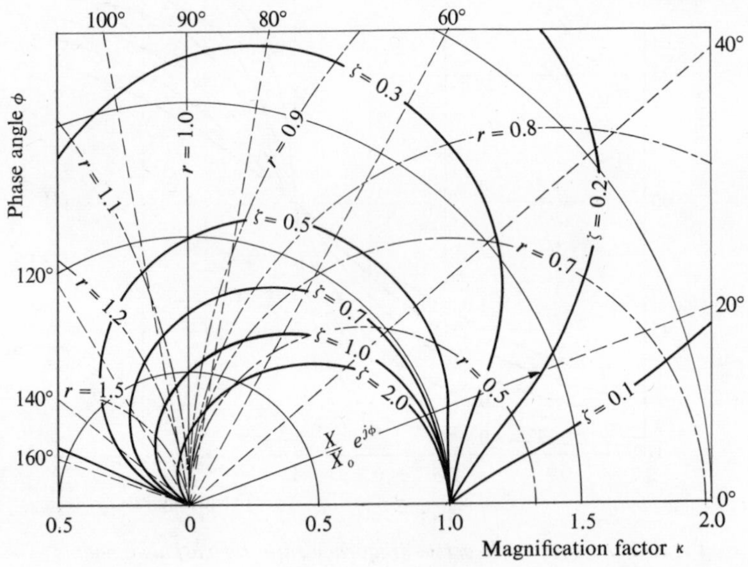

Fig. 2-12. *Nyquist plot of magnification factor and phase angle for various frequency ratio and damping factors; system shown in Fig.* 2-6

as the excitation. On the other hand, the transient motion is not sustained by the excitation. It represents the behavior of the system as it relaxes from an initial state of constraint to its static equilibrium state. Hence, the frequency ω_d of the transient motion corresponds to that of the free vibration of the system, and the transient motion is governed by the initial conditions imposed on the system.

Example 5. Find the steady-state response and the transient motion of the system in Example 4, if, in addition to the initial conditions specified, an excitation force of 6 sin 15t lb is applied to the mass.

Solution: The displacement of the mass m is obtained by the application of Eq. (2-43). The system parameters are identical to those calculated in Example 4. Hence the steady-state response of the system is

$$\begin{aligned} x_p &= \frac{F}{k}\kappa \sin(\omega t - \phi) \\ &= \frac{6/50}{\sqrt{[1-(\frac{15}{22})^2]^2 + [2(0.1645)(\frac{15}{22})]^2}} \sin(15t - \phi) \\ &= 0.207 \sin(15t - \phi) \end{aligned}$$

where

$$\phi = \tan^{-1}\frac{2\zeta r}{1 - r^2} = \tan^{-1}\frac{2(0.1645)(15/22)}{1-(15/22)^2} = 22.7^\circ$$

The general solution is

$$x = Ae^{-3.63t}\sin(21.7t + \psi) + 0.207\sin(15t - 22.7^\circ)$$

The initial conditions give

$$x(0) = 0 = A\sin\psi + 0.207\sin(-22.7^\circ)$$
$$\dot{x}(0) = 4 = A(-3.63\sin\psi + 21.7\cos\psi) + (0.207)(15)\cos(-22.7^\circ)$$

Solving for A and ψ, we obtain $\psi = \tan^{-1} 1.23 = 50.9^\circ$ and $A = 0.103$.

$$x = 0.103e^{-3.63t}\sin(21.7t + 50.9^\circ) + 0.207\sin(15t - 22.7^\circ).$$

This equation is plotted as shown in Fig. 2-13.

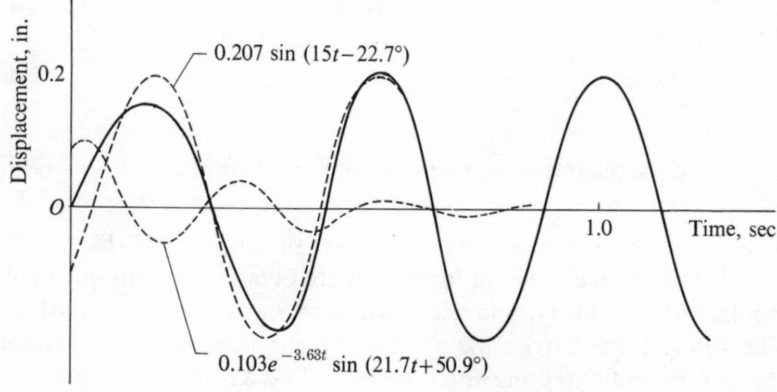

Fig. 2-13. *Displacement versus time curve; Example* 5

2-6. STEADY-STATE ANALYSIS: MECHANICAL IMPEDANCE

It was shown in the previous section that if the differential equation of motion of a single-degree-of-freedom system is linear, its steady-state response to a harmonic excitation is harmonic and of the same frequency as the excitation. This is the basis of the impedance method. It can be shown that if the differential equations of motion of a multiple-degree-of-freedom system are linear, the steady-state response of the system to a harmonic excitation is also harmonic at the excitation frequency. Therefore, the impedance method is readily applicable to more complex systems. This method greatly simplifies the work involved for solving the steady-state response as compared with the classical method discussed in the previous section.

If both the excitation and the steady-state response are harmonic and of the same frequency, they can be represented by rotating vectors with the same angular velocity. (See Sec. 1-5, Chap. 1.) Referring to the system shown in Fig. 2-6, let the excitation force $F \sin \omega t$ be represented by

$$\boldsymbol{F} = Fe^{j\omega t} \tag{2-44}$$

If the response lags the excitation by a phase angle ϕ, the displacement vector is written as $\boldsymbol{X} = Xe^{j(\omega t - \phi)}$. The velocity and acceleration vectors are the first and second time derivatives of the displacement vector, that is,

$$\text{Displacement } \boldsymbol{X} = Xe^{j(\omega t - \phi)} \tag{2-45}$$

$$\text{Velocity } \frac{d}{dt}(\boldsymbol{X}) = j\omega \boldsymbol{X} = \omega X e^{j\left(\omega t + \frac{\pi}{2} - \phi\right)} \tag{2-46}$$

$$\text{Acceleration } \frac{d^2}{dt^2}(\boldsymbol{X}) = (j\omega)^2 \boldsymbol{X} = \omega^2 X e^{j(\omega t + \pi - \phi)} \tag{2-47}$$

The relative positions of these vectors are shown in Fig. 2-14.

The harmonic forces in the system are obtained by multiplying the displacement, velocity, and acceleration by the appropriate constants. The spring force $kx(t)$ always resists the displacement $x(t)$, therefore the corresponding spring force vector is $-k\boldsymbol{X}$. The damping force $c\dot{x}(t)$ always resists the motion, and the inertia force opposes the acceleration: the corresponding damping and inertia force vectors are $-jc\omega\boldsymbol{X}$ and $m\omega^2\boldsymbol{X}$, respectively. These force vectors are shown in Fig.

2-15(a). The corresponding polygon of the force vectors is shown in Fig. 2-15(b).

In the system considered, the actual forces are along the vertical direction. For dynamic equilibrium, the algebraic sum of these forces is zero for all values of time. Therefore, in Fig. 2-15(a), the sum of the projections of the force vectors on the vertical axis is zero. Since the phase angle ϕ is constant at steady state and the vectors have the same angular velocity, the polygon of the force vectors, Fig. 2-15(b), is a closed polygon.

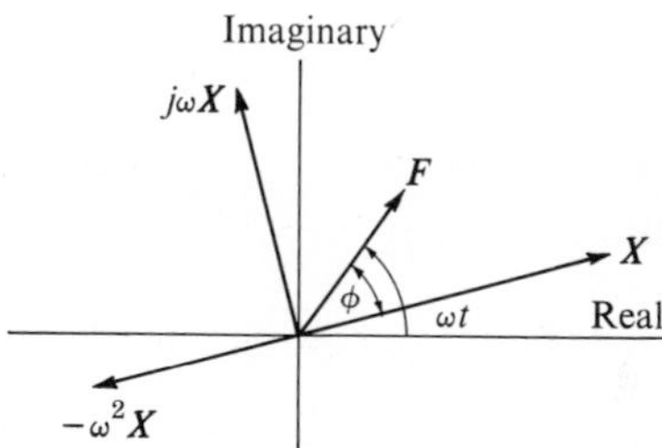

Fig. 2-14. *Force, displacement, velocity, and acceleration represented by rotating vectors*

Figure 2-16 shows the relation of these vectors for the cases of frequency ratio ω/ω_n less than, equal to, and greater than 1. Since the interest is in the relative phase positions of the vectors, these vector polygons are rotated clockwise by an amount $(\omega t - \phi)$. This is equivalent to choosing $(\omega t - \phi)$ as a datum of measurement. The polygons are drawn for an excitation force of constant amplitude but with different frequencies. The phase angles indicated may be compared with those shown in Fig. 2-9.

The impedance method can be deduced directly from the vectorial representation of harmonic forces. The equation of motion of the one-degree-of-freedom system shown in Fig. 2-6 is

$$m\ddot{x} + c\dot{x} + kx = F \sin \omega t \tag{2-48}$$

Substituting Eqs. (2-44) to (2-47) in Eq. (2-48) gives

$$(-m\omega^2 + jc\omega + k)Xe^{j(\omega t - \phi)} = Fe^{j\omega t} \tag{2-49}$$

Factoring out $e^{j\omega t}$ and rearranging, the equation becomes

$$Xe^{-j\phi} = \frac{F}{(k - m\omega^2) + jc\omega} \tag{2-50}$$

Since X is the magnitude of the displacement vector, it is evident from the last equation that

$$X = \left| \frac{F}{(k - m\omega^2) + jc\omega} \right| = \frac{F}{\sqrt{(k - m\omega^2)^2 + (c\omega)^2}} = \frac{F}{k}\kappa \tag{2-51}$$

and

$$\phi = \tan^{-1} \frac{c\omega}{k - m\omega^2} = \tan^{-1} \frac{2\zeta r}{1 - r^2} \tag{2-52}$$

These relations are identical to Eqs. (2-41) and (2-42) derived by the classical method in the previous section. Comparing Eqs. (2-48) and (2-49), it is noted that the steady-state response can be obtained by the substitution of $j\omega$ for the time derivative and a vector for the harmonic motion.

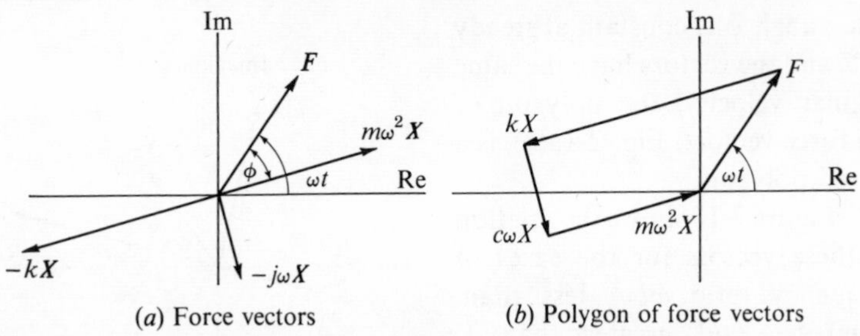

Fig. 2-15. *Spring, damping, inertial, and excitation forces represented by rotating vectors*

Since Eq. (2.49) is a force equation, the quantity $-m\omega^2$ times the displacement is the inertia force, $jc\omega$ times the displacement is the damping force, and k times the displacement is the spring force. The quantities $-m\omega^2$, $jc\omega$, and k are the *impedances* of the mass, damper, and spring elements, respectively. The quantity $(k - m\omega^2 + jc\omega)$ is

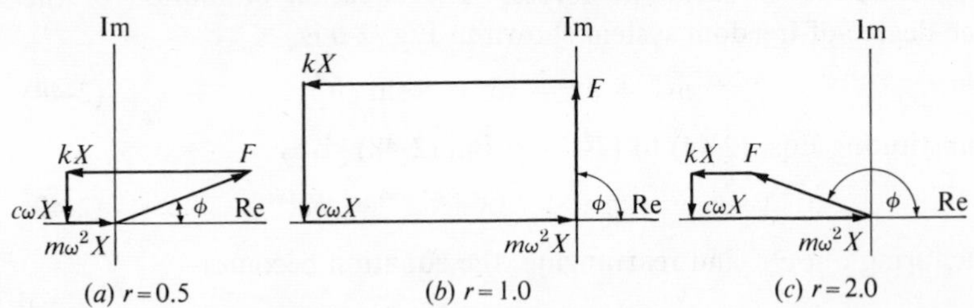

Fig. 2-16. *Polygon of force vectors for various frequency ratios: F = constant, $\zeta = 0{\cdot}25$*

the impedance of the system of Fig. 2-6. Thus *mechanical impedance* is defined as force per unit displacement. The impedances of the elements are tabulated in Table 2-1. The impedance concept is used in electrical engineering for the study of alternating current. Electrical impedance may be defined as voltage drop across an electrical element per unit current flowing through the element.

TABLE 2-1

IMPEDANCE OF SYSTEM ELEMENTS

ELEMENT	SYMBOL	IMPEDANCE
Mass	m	$-m\omega^2$
Damper	c	$jc\omega$
Spring	k	k

Example 6. Impedance Method

Determine the steady-state response of the system described in Example 5.

Solution: The impedance of the system is

$$(k - m\omega^2) + jc\omega = [50 - (\tfrac{40}{386})(15)^2] + j(0.75)(15)$$
$$= 26.7 + j11.3 = 29\underline{/22.7^\circ}$$

From Eq. (2-51) we have

$$Xe^{-j\phi} = \frac{6}{29\underline{/22.7^\circ}} = 0.207\underline{/-22.7^\circ}$$

which is the complex amplitude of the displaeement vector. The displacement vector is

$$\boldsymbol{X} = Xe^{j(\omega t - \phi)} = 0.207e^{j(15t - 22.7^\circ)}$$

Since the given excitation force is $F \sin \omega t$, which is equal to $\text{Im}(Fe^{j\omega t})$, the displacement $x(t)$ is $\text{Im}(Xe^{j(\omega t - \phi)})$, which is $X \sin(\omega t - \phi)$.

$$x = 0.207 \sin (15t - 22.7^\circ)$$

2-7. COMPARISON OF RECTILINEAR AND ROTATIONAL SYSTEMS

The discussion in the previous sections centered on systems with rectilinear motion. The theory and interpretations given are equally applicable to systems with rotational motion. The analogy between the two types of motion and the units normally employed in vibration studies are tabulated in Table 2-2. The responses of the two types of systems to harmonic excitations are compared in Table 2-3.

Extending this analogy concept, it may be said that systems are analogous if they are described by the same type of differential equation: the theory developed for one system is applicable to its analogous systems. Thus our study of the generalized model of one-degree-of-freedom systems is applicable to a large number of physical problems, the appearance of which may bear little resemblance to one another.

TABLE 2-2

ANALOGY BETWEEN RECTILINEAR AND ROTATIONAL SYSTEMS

ITEM	RECTILINEAR SYSTEM		ROTATIONAL SYSTEM	
	Symbol	*Unit*	*Symbol*	*Unit*
Time	t	sec	t	sec
Displacement	x	in.	θ	rad
Velocity	$\dot{x}$	in.-sec^{-1}	$\dot{\theta}$	rad-sec^{-1}
Acceleration	$\ddot{x}$	in.-sec^{-2}	$\ddot{\theta}$	rad-sec^{-2}
Inertia	m	lb-sec^2-in.	J	in.-lb-sec^2
Effort (force or torque)	$F = m\ddot{x}$	lb	$T = J\ddot{\theta}$	in.-lb
Momentum	$m\dot{x}$	lb-sec	$J\dot{\theta}$	in.-lb-sec
Impulse	Ft	lb-sec	Tt	in.-lb-sec
Kinetic energy	$T = \frac{1}{2}m\dot{x}^2$	in.-lb	$T = \frac{1}{2}J\dot{\theta}^2$	in.-lb
Potential energy	$U = \frac{1}{2}kx^2$	in.-lb	$U = \frac{1}{2}k_t\theta^2$	in.-lb
Work	$\int F\,dx$	in.-lb	$\int T\,d\theta$	in.-lb
Spring constant	k	lb-in.$^{-1}$	k_t	in.-lb-rad^{-1}
Damping coefficient	c	lb-sec-in.$^{-1}$	c_t	in.-lb-sec-rad^{-1}
Damping factor	$\zeta = \frac{1}{2}\frac{c}{\sqrt{km}}$		$\zeta = \frac{1}{2}\frac{c_t}{\sqrt{k_t J}}$	
Angular natural frequency	$\omega_n = \sqrt{k/m}$	rad-sec^{-1}	$\omega_n = \sqrt{k_t/J}$	rad-sec^{-1}
Natural frequency	$f_n = \omega_n/2\pi$	cycles-sec^{-1}	$f_n = \omega_n/2\pi$	cycles-sec^{-1}

TABLE 2-3

RESPONSE OF RECTILINEAR AND ROTATIONAL SYSTEMS

ITEM	RECTILINEAR SYSTEM	ROTATIONAL SYSTEM
System	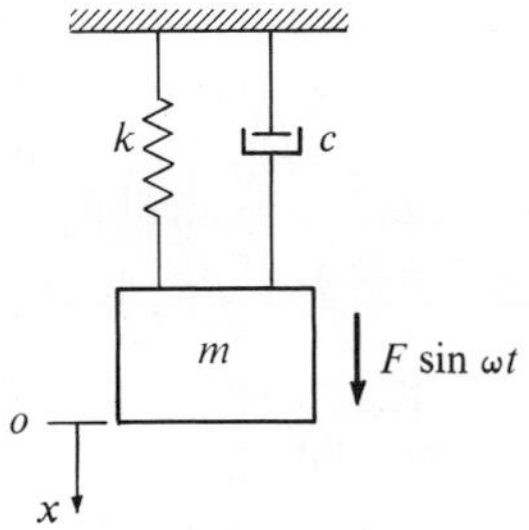 k, c, m, $F \sin \omega t$, o, x	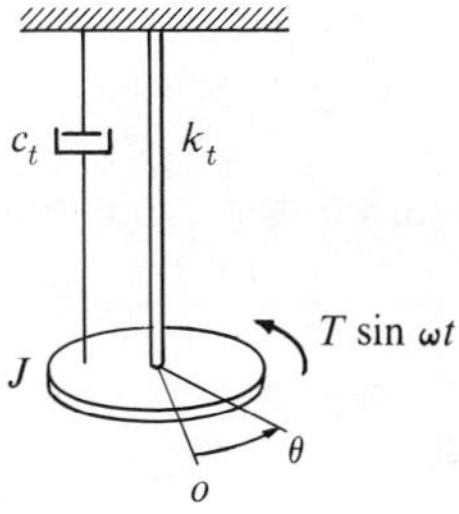 c_t, k_t, $T \sin \omega t$, J, θ, o
Equation of motion	$m\ddot{x} + c\dot{x} + kx = F \sin \omega t$	$J\ddot{\theta} + c_t\dot{\theta} + k_t\theta = T \sin \omega t$
Response	$x = x_c + x_p$	$\theta = \theta_c + \theta_p$
Initial conditions	$x(0) = x_o,\ \dot{x}(0) = \dot{x}_o$	$\theta(0) = \theta_o,\ \dot{\theta}(0) = \dot{\theta}_o$
Transient response	$x_c = Ae^{-\zeta\omega_n t} \sin (\omega_d t + \psi)$	$\theta_c = Ae^{-\zeta\omega_n t} \sin (\omega_d t + \psi)$
	$\omega_d = \sqrt{1 - \zeta^2}\,\omega_n$	$\omega_d = \sqrt{1 - \zeta^2}\,\omega_n$
Steady-state response	$x_p = X \sin (\omega t - \phi)$	$\theta_p = \Theta \sin (\omega t - \phi)$
	$X = \dfrac{F}{k}\kappa$	$\Theta = \dfrac{T}{k_t}\kappa$
	$\phi = \tan^{-1} \dfrac{2\zeta r}{1 - r^2}$	$\phi = \tan^{-1} \dfrac{2\zeta r}{1 - r^2}$
	$r = \dfrac{\omega}{\omega_n}$	$r = \dfrac{\omega}{\omega_n}$

2-8. APPLICATIONS

The general theory for the study of one-degree-of-freedom systems was developed in the previous sections through the analysis of a generalized model representative of this class of systems. The remainder of this chapter is devoted to the application of this theory to physical systems.

The generalized model, as shown in Fig. 2-6, consists of four elements, namely, the mass, the spring, the damper, and the excitation. The systems considered are grouped according to the elements involved. If a system does not possess one of these elements, such as a damper, it is simply omitted from the equation of motion: new techniques are not needed for solving the equation of motion, nor are new concepts required for interpreting the results. We shall begin with systems consisting of the simplest combination of these elements and then proceed to more complex ones.

Because the appearance of a given system may differ appreciably from that of the generalized model, the emphasis in the sections to follow is on the setting up of the differential equation of motion. Once this equation is formulated and converted to the form as shown in Eq. (2.18), the solution of the equation and the interpretation of the results become evident.

2-9. UNDAMPED FREE VIBRATION

The simplest vibratory system is one that consists of a mass and a spring element. All bodies possessing mass and elasticity are capable of vibration. This simple system is important because (1) it brings into focus the concept of natural frequency, and (2) if a system possesses very little damping, its vibratory characteristics can be approximated by that of an undamped system.

A spring-mass system, as shown in Fig. 2-1(*a*), has neither damping nor excitation. The equation of motion, Eq. (2-18), and the resultant motion, Eq. (2-33), become

$$m\ddot{x} + kx = 0 \tag{2-53}$$

$$x = A_1 \cos \omega_n t + A_2 \sin \omega_n t \tag{2-54}$$

Substituting the initial conditions $x(0) = x_o$ and $\dot{x}(0) = \dot{x}_o$ in Eq. (2-54), the resultant motion is

$$x = x_o \cos \omega_n t + \frac{\dot{x}_o}{\omega_n} \sin \omega_n t \tag{2-55}$$

Example 7. Equivalent Spring

A mass m is attached to the end of a cantilever beam of negligible mass, as shown in Fig. 2-17(a). It is observed that the static deflection of the beam is δ_{st}. Determine the frequency of vibration if the system is disturbed from its static equilibrium position.

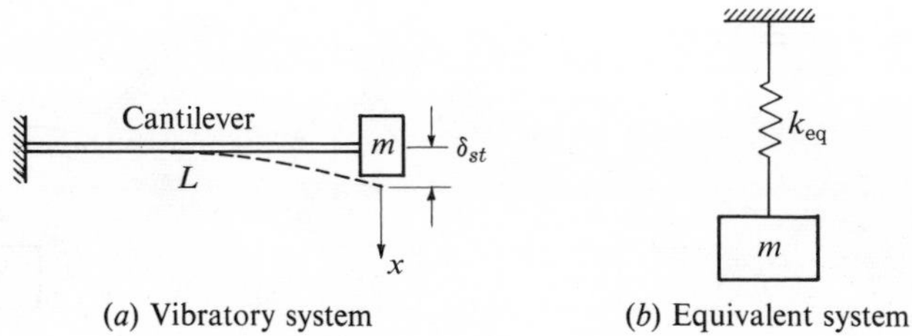

(a) Vibratory system (b) Equivalent system

Fig. 2-17. *Equivalent spring of cantilever*

Solution: If the cantilever is of negligible mass, the system can be represented by a spring-mass system of Fig. 2-17(b). The static deflection δ_{st} at the end of a cantilever beam of length L owing to a concentrated force mg is

$$\delta_{st} = \frac{mgL^3}{3EI}$$

where EI is the flexural stiffness of the beam. The equivalent spring constant k_{eq}, defined as force per unit deflection, is equal to

$$k_{eq} = \frac{mg}{\delta_{st}} = \frac{3EI}{L^3}$$

From the equivalent system shown in Fig. 2-17(b), the natural frequency of the system is

$$f_n = \frac{1}{2\pi}\sqrt{\frac{k_{eq}}{m}} = \frac{1}{2\pi}\sqrt{\frac{3EI}{mL^3}} = \frac{1}{2\pi}\sqrt{\frac{g}{\delta_{st}}}$$

Example 8. Springs in Series

Assuming that the cantilever of the system shown in Fig. 2-18(a) is of negligible mass, determine the equation of motion of the system.

Solution: It was shown in Example 7 that the cantilever can be replaced by an equivalent spring k_1 of spring constant $3EI/L^3$. Thus the system of Fig. 2-18(*b*) is equivalent to the given system. The two springs k_1 and k_2 are said to be *in series*. If the system is at its static equilibrium position, a unit force applied in the x-direction at the mass m will cause the springs k_1 and k_2 to elongate by $1/k_1$ and $1/k_2$, respectively. The

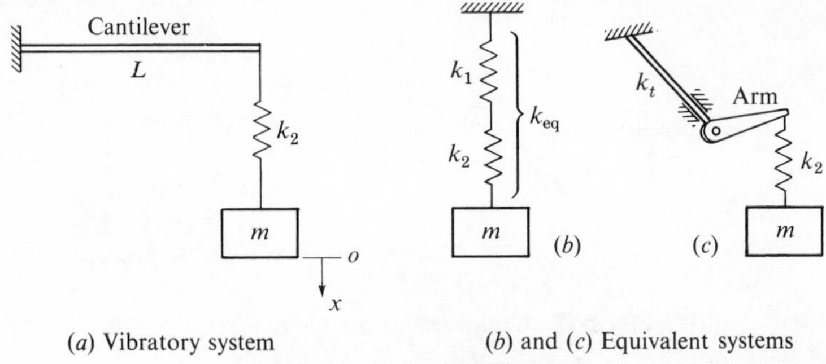

(*a*) Vibratory system (*b*) and (*c*) Equivalent systems

Fig. 2-18. *Springs in series*

elongation of the equivalent spring k_{eq} of the system is equal to the sum of these elongations, that is,

$$\frac{1}{k_{eq}} = \frac{1}{k_1} + \frac{1}{k_2} \tag{2-56}$$

or

$$k_{eq} = \frac{3EIk_2}{3EI + k_2L^3}$$

From the equivalent system shown in Fig. 2-18(*b*), the equation of motion of the system is

$$m\ddot{x} + k_{eq}x = 0$$

The system shown in Fig. 2-18(*c*) consists of a torsional shaft with an extended arm and a spring k_2 in series. If the mass of the shaft and its arm are negligible, this system reduces to that of Fig. 2-18(*b*).

Example 9. Springs in Parallel

A disk J is connected to two shafts, as shown in Fig. 2-19(*a*). Determine the natural frequency of the system for torsional vibration.

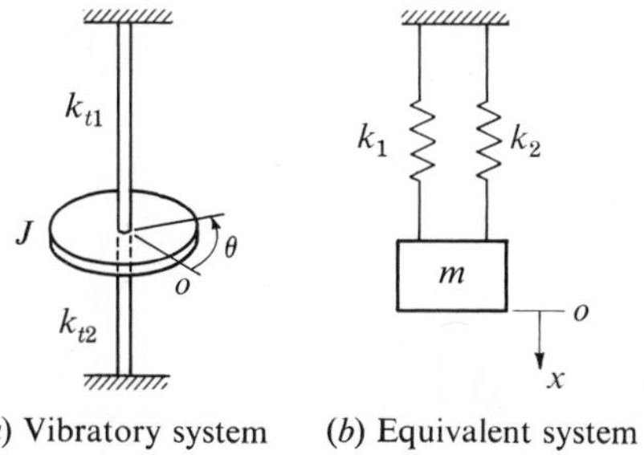

(*a*) Vibratory system (*b*) Equivalent system

Fig. 2-19. *Springs in parallel*

Solution: If the disk is rotated by an angle θ, both shafts tend to restore the disk to its equilibrium position. The two shafts are said to be *in parallel*, and the system is equivalent to that shown in Fig. 2-19(*b*). The restoring torque of a circular shaft is

$$T = \frac{\pi d^4 G}{32L}\theta = k_t\theta$$

where G is the shear modulus and d and L are the diameter and the length of the shaft, respectively. The total restoring torque is the sum of the restoring torques of the individual shafts, that is,

$$T = k_{\text{eq}}\theta = (k_{t1} + k_{t2})\theta$$

or

$$k_{\text{eq}} = k_{t1} + k_{t2} \tag{2-57}$$

$$k_{\text{eq}} = \frac{\pi}{32}\left(\frac{d_1^4 G_1}{L_1} + \frac{d_2^4 G_2}{L_2}\right)$$

Hence the natural frequency of the system is

$$f_n = \frac{1}{2\pi}\sqrt{\frac{k_{\text{eq}}}{J}} = \frac{1}{2\pi}\sqrt{\frac{\pi}{32}\left(\frac{d_1^4 G_1}{L_1} + \frac{d_2^4 G_2}{L_2}\right)\frac{1}{J}}$$

Example 10. Effect of Orientation

Determine the equations of motion of the systems shown in Fig. 2-20.

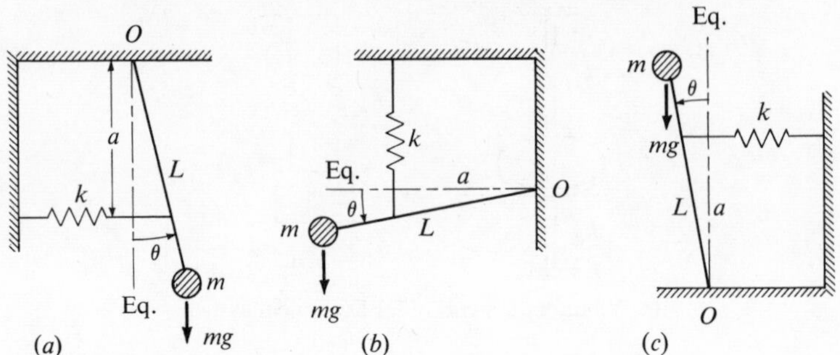

Fig. 2-20. *Effect of orientation of system on motion*

Solution: Because of the differences in the orientation of the systems, the restoring torque owing to the gravitational force on the mass is different for the three systems. Assuming small oscillations, and taking moments about point O, the equations of motion are

(*a*)
$$J_o\ddot{\theta} = \sum(\text{torque})$$
$$mL^2\ddot{\theta} = -mgL\sin\theta - (ka\sin\theta)(a\cos\theta)$$
$$mL^2\ddot{\theta} + (mgL + ka^2)\theta = 0$$

(*b*)
$$mL^2\ddot{\theta} = -(ka\sin\theta)(a\cos\theta) - mgL(1 - \cos\theta)$$
$$mL^2\ddot{\theta} + ka^2\theta = 0$$

(*c*)
$$mL^2\ddot{\theta} = mgL\sin\theta - (ka\sin\theta)(a\cos\theta)$$
$$mL^2\ddot{\theta} + (ka^2 - mgL)\theta = 0$$

Example 11. The equation of motion of the system shown in Fig. 2-4 was derived by the energy method in Example 2. Derive the equation of motion by Newton's law of motion.

Solution: The relation between θ and θ_1 is $R\theta = R_1\theta_1$. Since θ_1 is the rotation of the cylinder relative to the curved surface, the absolute rotation of the cylinder is $(\theta_1 - \theta)$. Taking moments about the instantaneous center of rotation b, the equation of motion is

$$J_b(\ddot{\theta}_1 - \ddot{\theta}) = -mgR_1\sin\theta$$

where $J_b = (J_o + mR_1^2)$ and $J_o = \frac{1}{2}mR_1^2$. Substituting $\theta = \sin\theta$, $\ddot{\theta}_1 = \dfrac{R}{R_1}\ddot{\theta}$, we obtain

$$\left(\frac{3}{2}mR_1^2\right)\left(\frac{R}{R_1} - 1\right)\ddot{\theta} + mgR_1\theta = 0$$

or

$$\ddot{\theta} + \frac{2g}{3(R - R_1)}\theta = 0$$

2-10. DAMPED FREE VIBRATION

To a greater or lesser degree, all physical systems possess damping. In this case, the generalized model includes the mass, spring, and damper elements as shown in Fig. 2-7. Without excitation, the equation of motion, Eq. (2-18), becomes

$$m\ddot{x} + c\dot{x} + kx = 0 \tag{2-58}$$

The displacement $x(t)$ of the mass can be described by Eqs. (2-25), (2-29), or (2-33), depending on whether the system is overdamped, critically damped, or underdamped. Should the system be underdamped, the displacement is

$$x = e^{-\zeta\omega_n t}(A_1 \cos \omega_d t + A_2 \sin \omega_d t) \tag{2-59}$$

Substituting the initial conditions, $x(0) = x_o$ and $\dot{x}(0) = \dot{x}_o$, in this equation, the displacement becomes

$$x = e^{-\zeta\omega_n t}\left(x_o \cos \omega_d t + \frac{\dot{x}_o + \zeta\omega_n x_o}{\omega_d} \sin \omega_d t\right) \tag{2-60}$$

or

$$x = Ae^{-\zeta\omega_n t} \sin (\omega_d t + \psi) \tag{2-61}$$

where

$$A = [(\dot{x}_o + \zeta\omega_n x_o)^2 + (x_o\omega_d)^2]^{1/2}/\omega_d$$

and

$$\psi = \tan^{-1} \frac{x_o\omega_d}{\dot{x}_o + \zeta\omega_n x_o}$$

Example 12. A component in a machine is represented schematically as shown in Fig. 2-21. Derive the equation of motion of this system.

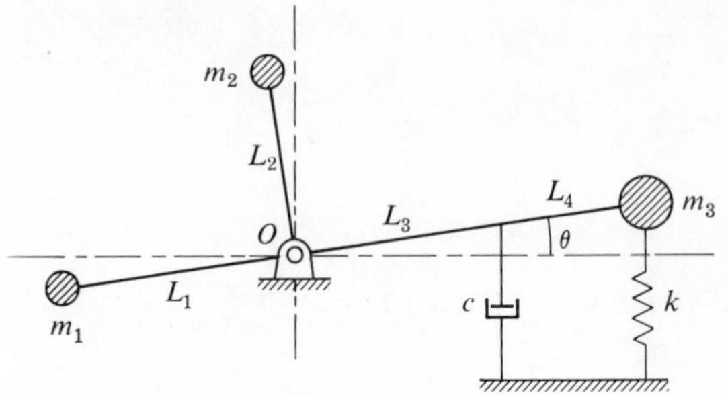

Fig. 2-21. *One-degree-of-freedom system with damping*

Solution: Assuming small oscillations and taking moments about the point O, the equation of motion is

$$J_o\ddot{\theta} = \sum(\text{torque})$$

$$[m_1L_1^2 + m_2L_2^2 + m_3(L_3 + L_4)^2]\ddot{\theta} = m_2gL_2\theta - cL_3^2\dot{\theta} - k(L_3 + L_4)^2\theta$$

or

$$[m_1L_1^2 + m_2L_2^2 + m_3(L_3 + L_4)^2]\ddot{\theta} + cL_3^2\dot{\theta} + [k(L_3 + L_4)^2 - m_2gL_2]\theta = 0$$

Example 13. Logarithmic Decrement

A mass-spring system with viscous damping, as shown in Fig. 2-7, is displaced by an amount x_o from its static equilibrium position and released with zero initial velocity. Determine the ratio of any two consecutive amplitudes.

Solution: From Eq. (2-61), the maximum amplitude in a cycle occurs when the product of $Ae^{-\zeta\omega_n t}$ and $\sin(\omega_d t + \psi)$ is a maximum. Rewriting Eq. (2-61) with $\omega_d t$ as the independent variable and equating $\dfrac{dx}{d\omega_d t}$ to zero, we have

$$x = Ae^{-\frac{\zeta}{\sqrt{1-\zeta^2}}\omega_d t}\sin(\omega_d t + \psi)$$

and

$$\frac{dx}{d\omega_d t} = Ae^{-\frac{\zeta}{\sqrt{1-\zeta^2}}\omega_d t}\left[-\frac{\zeta}{\sqrt{1-\zeta^2}}\sin(\omega_d t + \psi) + \cos(\omega_d t + \psi)\right] = 0$$

Hence the maximum amplitude is obtained when

$$\tan(\omega_d t + \psi) = \frac{\sqrt{1-\zeta^2}}{\zeta}$$

Let the motion $x(t)$ be represented as shown in Fig. 2-22, and let $\omega_d t_1$ and $\omega_d t_2$ be the locations of the maxima x_1 and x_2. The last equation indicates that $\tan(\omega_d t_1 + \psi) = \tan(\omega_d t_2 + \psi)$, that is,

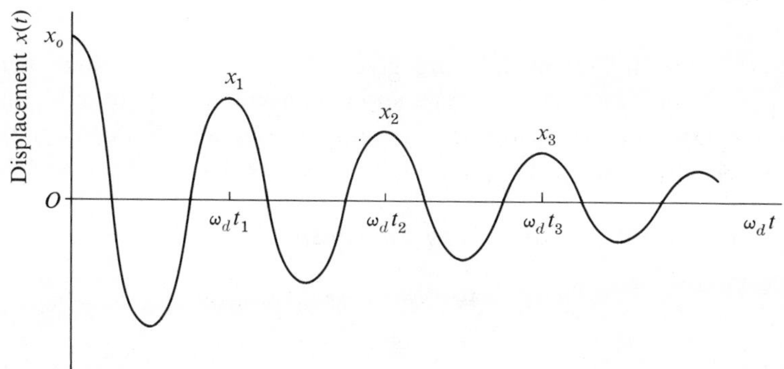

Fig. 2-22. *Free vibration with damping: initial conditions* $x(0) = x_0$ *and* $\dot{x}(0) = 0$

$(t_2 - t_1)$ is the period of oscillation, consequently $\sin(\omega_d t_1 + \psi) = \sin(\omega_d t_2 + \psi)$. The consecutive amplitude ratio is

$$\frac{x_1}{x_2} = \frac{Ae^{-\zeta\omega_n t_1}}{Ae^{-\zeta\omega_n t_2}} = e^{\zeta\omega_n(t_2 - t_1)}$$

or

$$\frac{x_1}{x_2} = e^{\zeta\omega_n\left(\frac{2\pi}{\omega_d}\right)} = e^{\frac{2\pi\zeta}{\sqrt{1-\zeta^2}}}$$

The logarithm of this ratio is called the logarithmic decrement δ.

$$\delta = \frac{2\pi\zeta}{\sqrt{1-\zeta^2}} \tag{2-62}$$

If ζ is appreciably less than unity, this logarithmic decrement can be approximated as

$$\delta \doteq 2\pi\zeta \tag{2-63}$$

The logarithmic decrement is a measure of the damping factor ζ, and therefore it affords a convenient way to measure the amount of damping in a system.

Since the rate of decay of the oscillatory motion is independent of the initial conditions imposed on the system, the logarithmic decrement must be independent of initial conditions. Furthermore, any two points on the curve (Fig. 2-22) one period apart may serve to evaluate the logarithmic decrement. The use of consecutive amplitudes, however, is more convenient.

Example 14. The following data are given for a vibratory system with viscous damping: mass $m = 5$ lb, spring constant $k = 15$ lb-in.$^{-1}$, and the amplitude decreases to 0.25 of the initial value after five consecutive cycles. Determine the damping coefficient of the damper in the system.

Solution: The amplitude ratio of any two consecutive amplitudes is

$$\frac{x_0}{x_1} = \frac{x_1}{x_2} = \frac{x_2}{x_3} = \frac{x_3}{x_4} = \frac{x_4}{x_5} = e^{\delta}$$

Hence

$$\frac{x_0}{x_5} = \frac{x_0}{x_1} \cdot \frac{x_1}{x_2} \cdot \frac{x_2}{x_3} \cdot \frac{x_3}{x_4} \cdot \frac{x_4}{x_5} = e^{5\delta} = \frac{1}{0.25}$$

or

$$\delta = \frac{1}{5} \ln 4 = 0.278 = \frac{2\pi\zeta}{\sqrt{1 - \zeta^2}}$$

The damping factor ζ and the damping coefficient c are

$$\zeta = 0.0442$$

$$c = 2\zeta\sqrt{km} = (2)(0.0442)\sqrt{(15)(5/386)} = 0.039 \text{ lb-sec-in.}^{-1}$$

Following the method given in this example, the number of cycles n required to reduce the amplitude by a factor of Q is given by the expression

$$\frac{x_0}{x_n} = Q = e^{n\delta}$$

or

$$\delta = \frac{1}{n} \ln Q \tag{2-64}$$

The relation indicated in Eq. (2-64) is plotted in Fig. 2-23.

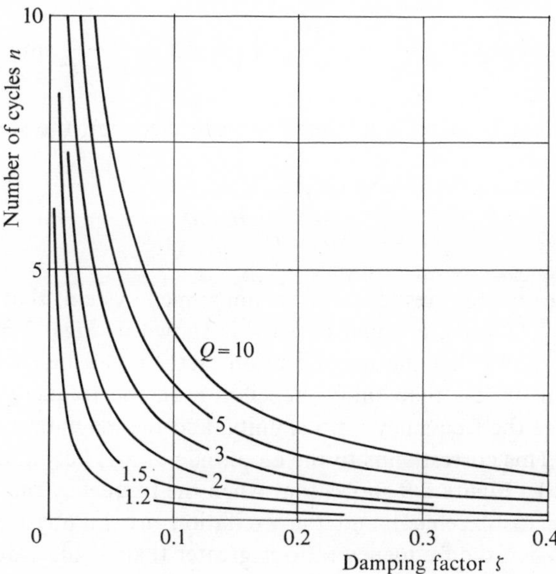

Fig. 2-23. *Number of cycles to reduce the amplitude by a factor of Q for various damping factors ζ*

2-11. UNDAMPED FORCED VIBRATION

The usual interest in the study of forced vibration with harmonic excitations is in the steady-state response of the system. Even if the system possesses very little damping, it may be tacitly assumed that the transient motion will soon die out. (See Fig. 2-23.) Hence the steady-state response of a lightly damped system can be approximated by that of an undamped system except near resonance.

Neglecting the damper in the generalized model shown in Fig. 2-6, the equation of motion, Eq. (2-18), and the corresponding displacement $x(t)$, Eq. (2-43), become

$$m\ddot{x} + kx = F \sin \omega t \tag{2-65}$$

and

$$x = A_1 \cos \omega_n t + A_2 \sin \omega_n t + \frac{F/k}{1 - r^2} \sin \omega t \tag{2-66}$$

where A_1 and A_2 are arbitrary coefficients. Assuming the initial conditions $x(0) = x_o$ and $\dot{x}(0) = \dot{x}_o$, the general solution becomes

$$x = x_o \cos \omega_n t + \left(\frac{\dot{x}_o}{\omega_n} - \frac{F/k}{1 - r^2} r\right) \sin \omega_n t + \frac{F/k}{1 - r^2} \sin \omega t \quad \textbf{(2-67)}$$

If the transient solution is assumed to have died out, the steady-state response is

$$x = \frac{F/k}{1 - r^2} \sin \omega t \quad \textbf{(2-68)}$$

The steady-state response of an undamped system, that is, when the damping factor ζ is equal to zero, is shown in Figs. 2-8 and 2-9. Figure 2-8 shows that the magnification factor κ can be considerably greater than or less than unity, depending on the frequency ratio r. At resonance the frequency ratio is unity, and the magnification factor is infinite. This corresponds to the amplitude of $x(t)$ becoming infinite in Eq. (2-68). Figure 2-9 shows that when the frequency ratio r is less than 1, the displacement and the excitation are in phase with one another. When the frequency ratio is greater than 1, they are 180 deg out of phase.

Alternatively, the behavior at resonance can be deduced from the particular integral of Eq. (2-65). Dividing this equation by m and substituting ω_n for ω in the excitation term, we obtain

$$\ddot{x} + \omega_n^2 x = \frac{F}{m} \sin \omega_n t \quad \textbf{(2-69)}$$

The particular integral of Eq. (2-69) is of the form

$$x_p = C_1 t \sin \omega_n t + C_2 t \cos \omega_n t \quad \textbf{(2-70)}$$

Substituting Eq. (2-70) in Eq. (2-69) and collecting the coefficients of the sine and cosine terms, we obtain

$$x_p = -\frac{F}{2\sqrt{km}} t \cos \omega_n t \quad \textbf{(2-71)}$$

Thus the amplitude increases proportionately with time and would theoretically become infinite. Our interest is mainly in the particular integral, because complementary functions are pure harmonic motions.

Equation (2-71) indicates that, at resonance, it takes time for the amplitude to build up. Thus, if the resonance is passed through rapidly, it is possible to bring up the speed of a piece of equipment, such as a turbine, to beyond resonance, or the *critical speed* of the equipment. Figure 2-8 shows that at frequencies considerably above resonance the magnification factor is less than unity. It may be advantageous to operate the equipment in this speed range. It should be cautioned that during shutdown the equipment would again pass through the critical speed, and excessive vibration might be encountered.

Example 15. A simple pendulum of mass m and length L is hinged at point O, as shown in Fig. 2-24(a). If the hinge point O is given a horizontal

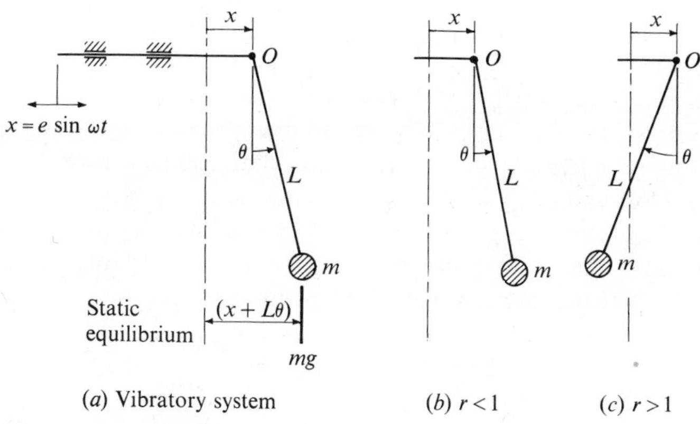

Fig. 2-24. *Simple pendulum excited at support*

motion $x(t) = e \sin \omega t$, determine (a) the angular displacement of the pendulum for frequency ratios $r \gtrless 1$, and (b) the force required to move the hinge point.

Solution: (a) Assuming small oscillations and neglecting the second-order terms, the horizontal acceleration of m is $(\ddot{x} + L\ddot{\theta})$ and the vertical acceleration is of second order. Taking moments about O, the equation of motion is

$$m(\ddot{x} + L\ddot{\theta})L = -mgL\theta$$

or

$$mL^2\ddot{\theta} + mgL\theta = -mL\ddot{x} = me\omega^2 L \sin \omega t$$

Defining $T_{\text{eq}} = me\omega^2 L$ as the amplitude of an equivalent torque, this equation has the same form as Eq. (2-65).

$$mL^2\ddot{\theta} + mgL\theta = T_{\text{eq}} \sin \omega t$$

Hence the steady-state response [see Eq. (2-68)] of this system is

$$\theta(t) = \frac{T_{\text{eq}}/mgL}{1 - r^2} \sin \omega t = \frac{\frac{e}{L} r^2}{1 - r^2} \sin \omega t$$

where $r = \dfrac{\omega}{\omega_n}$ and $\omega_n = \sqrt{g/L}$. Let us rewrite this equation in the form

$$\theta(t) = \left| \frac{\frac{e}{L} r^2}{1 - r^2} \right| \sin (\omega t - \phi)$$

where $\phi = 0$ deg when $r < 1$ and $\phi = 180$ deg when $r > 1$. Hence $\theta(t)$ and $x(t)$ are in phase with one another when $r < 1$ and 180 deg out of phase when $r > 1$. These phase relationships are illustrated in Figs. 2-24(*b*) and (*c*).

(*b*) The dynamic equilibrium of the pendulum requires that the horizontal forces at the hinge point O be equal to the horizontal component of the inertia force of the pendulum, that is,

$$F_x(t) = -m(\ddot{x} + L\ddot{\theta}) = mg\theta = \frac{me\omega^2}{1 - r^2} \sin \omega t$$

where x is positive if the motion is to the right of the static equilibrium position. This equation shows that near resonance a large force may be associated with a small amplitude e of the hinge point O. This is the principle of the bifilar-type centrifugal pendulum employed for a dynamic absorber.

Example 16. Determination of Natural Frequency

Figure 2-25 shows a schematic sketch of a control tab of an airplane elevator. The mass moment of inertia J_o of the control tab about the hinge point O is known, but the torsional spring constant k_t due to the control linkage is difficult to evaluate. The natural frequency of the control tab is $\omega_n = \sqrt{k_t/J_o}$. To determine this natural frequency experimentally, the elevator is rigidly mounted, and the tab is excited as illustrated. The excitation frequency is varied until resonance occurs. If the resonance frequency is ω_r, determine the natural frequency of the control tab.

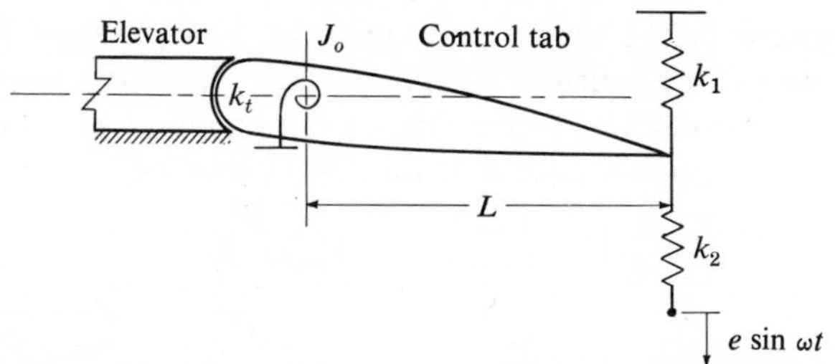

Fig. 2-25. *Determination of natural frequency of control tab*

Solution: Taking moments about the hinge point O, the equation of motion of the test system is

$$J_o\ddot{\theta} = -k_t\theta - k_1L^2\theta - k_2(L\theta - e \sin \omega t)L$$

where $(L\theta - e \sin \omega t)$ is the deformation of the spring k_2. Rearranging, this equation becomes

$$J_o\ddot{\theta} + [k_t + (k_1 + k_2)L^2]\theta = k_2eL \sin \omega t$$

At resonance,

$$\omega^2 = \omega_r^2 = \frac{k_t + (k_1 + k_2)L^2}{J_o} = \omega_n^2 + \frac{(k_1 + k_2)L^2}{J_o}$$

Hence

$$f_n = \frac{1}{2\pi}\,\omega_n = \frac{1}{2\pi}\sqrt{\omega_r^2 - \frac{(k_1 + k_2)L^2}{J_o}}$$

2-12. DAMPED FORCED VIBRATION

This section deals with the general one-degree-of-freedom systems. It is necessary to generalize the theory discussed in Secs. 2-4 and 2-5 in order that it be applicable to a larger number of problems. It was assumed that the excitation force $F \sin \omega t$ was applied to the mass and that F was constant. In some problems the excitation may be applied to the support of the system, such as vibration-measuring instruments; in others, the amplitude of the excitation may be frequency dependent, such as the excitation force owing to the unbalance of a machine component.

To generalize the theory developed, let an equivalent harmonic force $F_{eq} \sin \omega t$ be applied to the mass, where F_{eq} is the amplitude of the equivalent excitation force. Hence the equation of motion, Eq. (2-18), and the response equation, Eq. (2-43), become

$$m\ddot{x} + c\dot{x} + kx = F_{eq} \sin \omega t \tag{2-72}$$

and

$$x = Ae^{-\zeta\omega_n t} \sin (\omega_d t + \psi) + \frac{F_{eq}}{k} \kappa \sin (\omega t - \phi) \tag{2-73}$$

where

$$\phi = \tan^{-1} \frac{2\zeta r}{1 - r^2} \tag{2-74}$$

In Eq. (2-73) it is assumed that $\zeta < 1$. The first term on the right-hand side is the transient motion, and the second term is the steady-state response.

It should be noted that if F_{eq} is constant for all frequencies, then the amplitude of the steady-state response is the same as that shown in Fig. 2-8. If F_{eq} is a function of the excitation frequency, the response curves will differ from those shown in Fig. 2-8. For a given excitation $F_{eq} \sin \omega t$, however, the magnitude of F_{eq} merely changes the magnitude of the forces involved. Hence the phase relation between the steady-state response and excitation is prescribed by Eq. (2-74) as long as F_{eq} is constant for a given frequency. This phase relation is the same as that shown in Fig. 2-9.

The examples in this section are grouped together according to background and interest and are divided into seven cases. In Cases 1 to 6, the equations of motion are reduced to the form of Eq. (2-72), of which Eq. (2-73) is the general solution. Case 7 involves the use of a third-order differential equation.

Case 1. Rotating and Reciprocating Unbalance

A turbine, an electric motor, or any device with a rotor as its working part is a rotating machine. Unbalance exists in a rotating machine if the center of gravity of the rotor does not coincide with the axis of rotation. The unbalance is measured in terms of an equivalent eccentric mass m with an eccentricity e.

A rotating machine of total mass m_1 with an unbalance me is shown in Fig. 2-26. If the machine is constrained to move in the

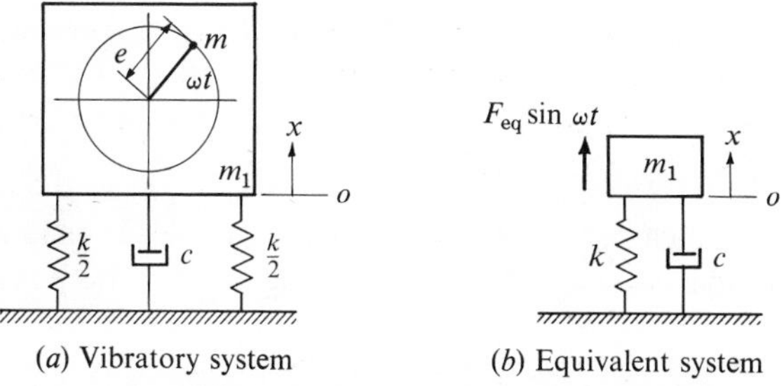

(*a*) Vibratory system (*b*) Equivalent system

Fig. 2-26. *Rotating unbalance*

vertical direction, it has one degree of freedom. If the eccentric mass m rotates with angular velocity ω, its vertical displacement is

$$x + e \sin \omega t \tag{2-75}$$

The displacement of the mass $(m_1 - m)$ is $x(t)$. Summing the vertical forces gives

$$(m_1 - m)\frac{d^2x}{dt^2} + m\frac{d^2}{dt^2}(x + e \sin \omega t) + c\frac{dx}{dt} + kx = 0 \tag{2-76}$$

which may be written as

$$m_1\ddot{x} + c\dot{x} + kx = me\omega^2 \sin \omega t \tag{2-77}$$

If we define $me\omega^2 = F_{eq}$, then this equation is of the same form as Eq. (2-72). For a given excitation frequency, $me\omega^2$ is a constant. Hence the system shown in Fig. 2-26(*b*) is the equivalent system with a harmonic excitation $F_{eq} \sin \omega t$ applied to the total mass m_1. It is noted that $me\omega^2$ is the centrifugal force of the unbalance mass m rotating about a fixed axis and $me\omega^2 \sin \omega t$ is the component of this force in the direction of the motion $x(t)$.

The steady-state response of the system is

$$x = \frac{me\omega^2}{k}\kappa \sin(\omega t - \phi) \tag{2-78}$$

where $\phi = \tan^{-1}\dfrac{2\zeta r}{1 - r^2}$ and $\omega_n = \sqrt{k/m_1}$. If $x(t)$ is multiplied and divided by m_1, this equation may be expressed as

$$x = \frac{m}{m_1}er^2\kappa \sin(\omega t - \phi) = X \sin(\omega t - \phi)$$

Equating the expressions for amplitudes and rearranging, we obtain

$$\frac{m_1 X}{me} = r^2 \kappa \tag{2-79}$$

This gives the steady-state response of the amplitude in a nondimensional form and is plotted as shown in Fig. 2-27. The phase angle of

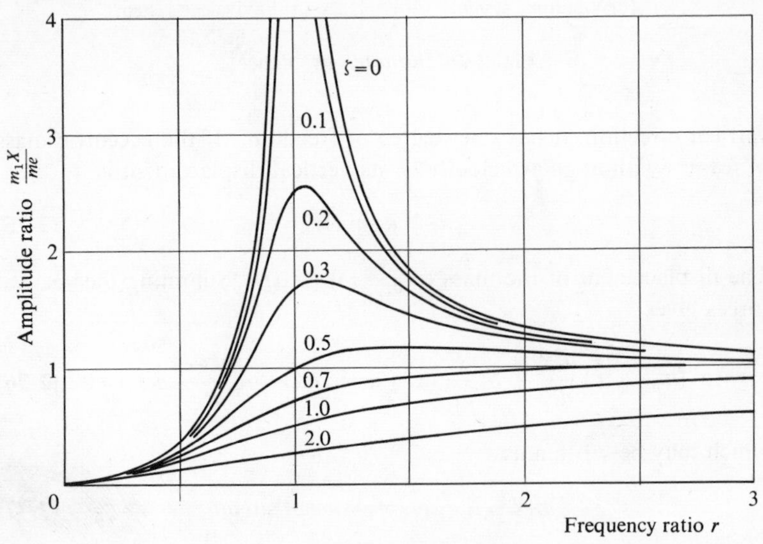

Fig. 2-27. *Steady-state response of system with inertial excitation; system shown in Fig.* 2-26

the steady-state response in Eq. (2-78) is the same as that shown in Fig. 2-9.

At low speeds, the force $me\omega^2$ is small, and the amplitude is nearly zero. At resonance, when the frequency ratio r is unity, the magnification factor κ is equal to $\dfrac{1}{2\zeta r}$ and the mass $(m_1 - m)$ has an amplitude X equal to $me/2\zeta m_1$. The amplitude is limited only by the presence of damping, and the mass $(m_1 - m)$ is 90 deg out of phase with the unbalanced mass m. For example, when $(m_1 - m)$ is moving upward and passing through its static equilibrium position, the mass m is directly above its center of rotation. When the frequency ratio is large, the mass $(m_1 - m)$ has an amplitude equal to me/m_1 and is 180 deg out of phase with the mass m. For example, when

$(m_1 - m)$ is at its topmost position, m is directly below its center of rotation.

The discussion on rotating unbalance can be extended to include reciprocating unbalance. Figure 2-28 shows a sketch of a reciprocating engine. The reciprocating mass m consists of the mass of the piston, the wrist pin, and a portion of the connecting-rod mass. The inertia force due to the reciprocating mass is approximately equal to $me\omega^2 \left(\sin \omega t + \frac{e}{L} \sin 2\omega t\right)$, where L is the length of the connecting rod and e is the crank radius.† If the ratio e/L is small, the second-harmonic term, $\frac{e}{L} \sin 2\omega t$, may be neglected. Thus the equivalent excitation force on the engine is

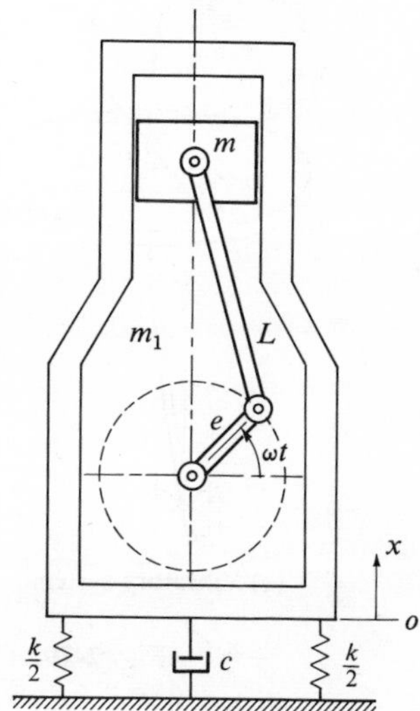

Fig. 2-28. *Reciprocating unbalance*

$$F_{eq} \sin \omega t = me\omega^2 \sin \omega t$$

which is identical to that due to rotating unbalance discussed in the previous paragraphs. Hence, neglecting the second-harmonic term, Eq. (2-77) is equally applicable to this case.

Case 2. Critical Speed of Rotating Shafts

A rotating shaft carrying an unbalanced disk at its mid-span is shown in Fig. 2-29(*a*). *Critical speed* occurs when the rotative speed of the shaft is equal to the natural frequency of lateral (beam) vibration of the shaft. Since the shaft has distributed mass and elasticity along its length, the system has more than one degree of freedom. We shall assume the shaft to be of negligible mass.

Figure 2-29(*b*) shows the top view of a general position of the rotating disk. Let G be the center of gravity of the disk, P the geometric center, and O the center of rotation. Assume that the damping force, such as air friction opposing the whirl, is proportional to the linear

† See, for example, R. T. Hinkle, *Kinematics of Machines*, 2nd ed., Prentice-Hall, Inc., Englewood Cliffs, N.J., 1960, p. 107.

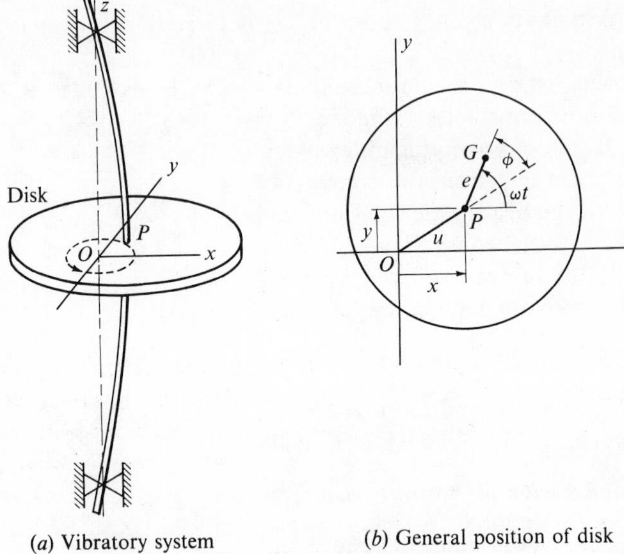

(a) Vibratory system

(b) General position of disk

Fig. 2-29. *Critical speed of rotating shaft*

speed of the geometric center P and that the bearing flexibility is negligible as compared with that of the shaft.

Resolving the forces in the x- and y-directions gives

$$m \frac{d^2}{dt^2}(x + e \cos \omega t) = -kx - c \frac{d}{dt} x$$

$$m \frac{d^2}{dt^2}(y + e \sin \omega t) = -ky - c \frac{d}{dt} y$$

Rearranging, we obtain

$$\begin{aligned} m\ddot{x} + c\dot{x} + kx &= me\omega^2 \cos \omega t \\ m\ddot{y} + c\dot{y} + ky &= me\omega^2 \sin \omega t \end{aligned} \tag{2-80}$$

which are of the same form as Eq. (2-77). Hence the steady-state responses in the x- and y-directions are

$$\begin{aligned} x &= \frac{me\omega^2}{k} \kappa \cos (\omega t - \phi) = r^2 e \kappa \cos (\omega t - \phi) \\ y &= \frac{me\omega^2}{k} \kappa \sin (\omega t - \phi) = r^2 e \kappa \sin (\omega t - \phi) \end{aligned} \tag{2-81}$$

where

$$\phi = \tan^{-1} \frac{2\zeta r}{1 - r^2}$$

The lateral deflection u of the shaft is

$$u = \sqrt{x^2 + y^2} \tag{2-82}$$

Substituting Eq. (2-81) in the above relation, we obtain

$$u/e = r^2 \kappa \tag{2-83}$$

It is noted that Eq. (2-83) is essentially the same as Eq. (2-79), with the exception that the amplitude ratio in the latter is modified by the term m_1/m. It should be remembered that, in the case of rotating unbalance, m_1 is the mass of the entire machine and m is the eccentric mass. In the case of the rotating shaft, the entire mass is causing the excitation, therefore m_1 is equal to m. Thus the frequency-response curves of Figs. 2-27 and 2-9 for rotating and reciprocating unbalance also represent the case of the rotating shaft.

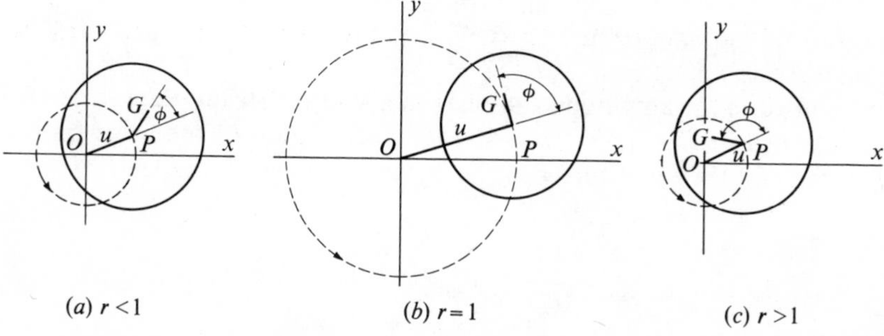

Fig. 2-30. *Phase relation of rotating shaft for various frequency ratios*

The phase relation for various operating frequencies is shown in Fig. 2-30. It is interesting to note that, when the frequency ratio r is much greater than unity, the center of gravity G tends to coincide with the center of rotation O. This can be readily demonstrated. Assume that an unbalanced rotor is rotating in a balancing machine and that a piece of chalk is moved toward the rotor until it barely touches. When the rotational speed is below critical, the chalk mark is found on the side closer to the center of gravity of the rotor. When the speed is above critical, the chalk mark is on the side away from the center of gravity.

Case 3. Vibration Isolation and Transmissibility

Machines are often mounted on springs and dampers, as shown schematically in Fig. 2-26, to minimize the transmission of forces to the foundation. The equation of motion, Eq. (2-72), and the response equation, Eq. (2-73), of this vibrating system were given in the beginning of this section. Since the machine is supported by springs and dampers, the force transmitted to the foundation is the sum of the spring force and the damping force. If the deflection of the foundation is negligible, the force transmitted F_T is

$$F_T = kx + c\dot{x} \tag{2-84}$$

The usual interest is in the force transmitted under steady-state conditions, therefore

$$\begin{aligned} F_T &= kX \sin(\omega t - \phi) + c\omega X \cos(\omega t - \phi) \\ &= X\sqrt{k^2 + c^2\omega^2} \sin(\omega t - \phi + \gamma) \\ &= F_{eq}\kappa\sqrt{1 + (2\zeta r)^2} \sin(\omega t - \phi + \gamma) \end{aligned} \tag{2-85}$$

where

$$\gamma = \tan^{-1}\frac{c\omega}{k} = \tan^{-1} 2\zeta r \tag{2-86}$$

X and ϕ are the amplitude and phase angle of the steady-state response as defined in Eqs. (2-73) and (2-74). The relation of the force transmitted and the other forces in the system is shown in Fig. 2-31.

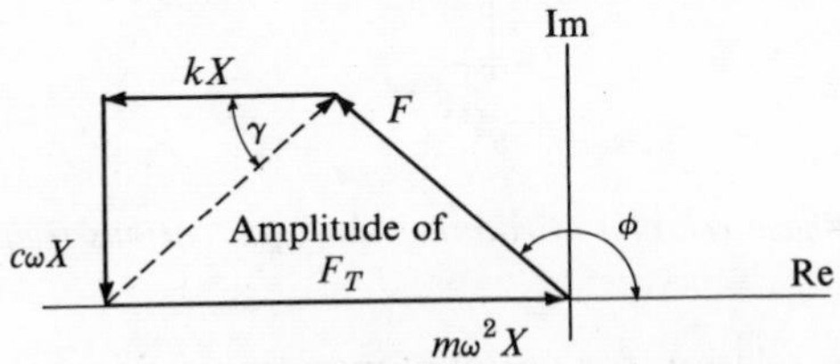

Fig. 2-31. *Relation of force transmitted and other force vectors*

The ratio of the amplitude of the force transmitted to that of the impressed force is called the *transmissibility* (TR).

$$\text{TR} = \frac{F_{eq}\kappa\sqrt{1 + (2\zeta r)^2}}{F_{eq}}$$

$$\text{TR} = \sqrt{1 + (2\zeta r)^2}\,\kappa \tag{2-87}$$

Equation (2-87) and the corresponding phase angle $\phi - \gamma$, indicated in Eq. (2-85), are plotted in Figs. 2-32 and 2-33, respectively. Since all the curves in Fig. 2-32 cross at TR $= 1$ and frequency ratio $r = \sqrt{2}$, it is apparent that the presence of damping will decrease TR when the machine is operated below this frequency ratio and will increase TR when operated above this frequency ratio. The proof of the location of this crossover point of the curves is left as an exercise.

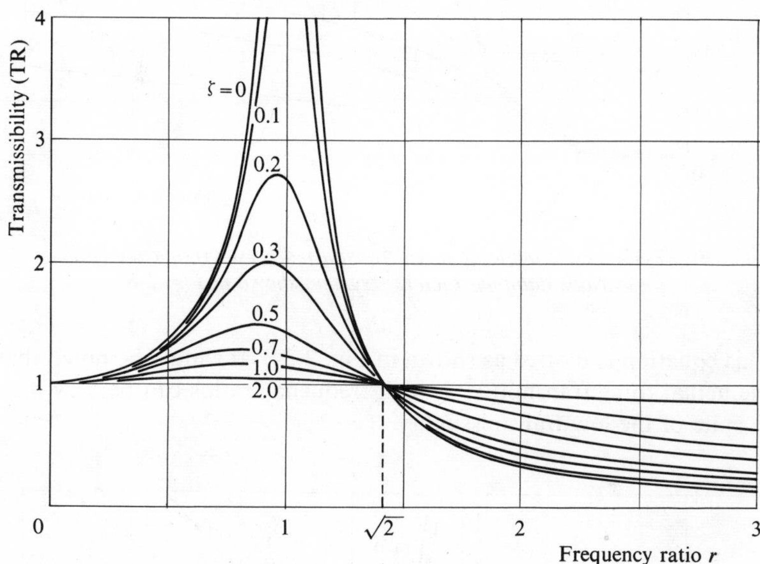

Fig. 2-32. *Transmissibility versus frequency ratio for various damping factors; system shown in Fig.* 2-6

As indicated in Fig. 2-32, the value of transmissibility is low at high frequency ratios. In the case of a constant-speed machine, the magnitude of the excitation force F_{eq} is constant. Hence the magnitude of the force transmitted is proportional to the value of transmissibility. In the case of a variable-speed machine, F_{eq}, owing to the unbalanced masses, is proportional to the speed squared. From Eq. (2-85) the magnitude of the transmitted force is

$$F_{T\max} = me\omega^2\kappa\sqrt{1 + (2\zeta r)^2}$$

where *me* is the unbalance of the system. Multiplying and dividing this equation by ω_n^2, defining $F_n = me\omega_n^2 =$ constant, and rearranging, we obtain

$$\frac{F_{T\max}}{F_n} = r^2\kappa\sqrt{1 + (2\zeta r)^2} = r^2\,(\text{TR}) \tag{2-88}$$

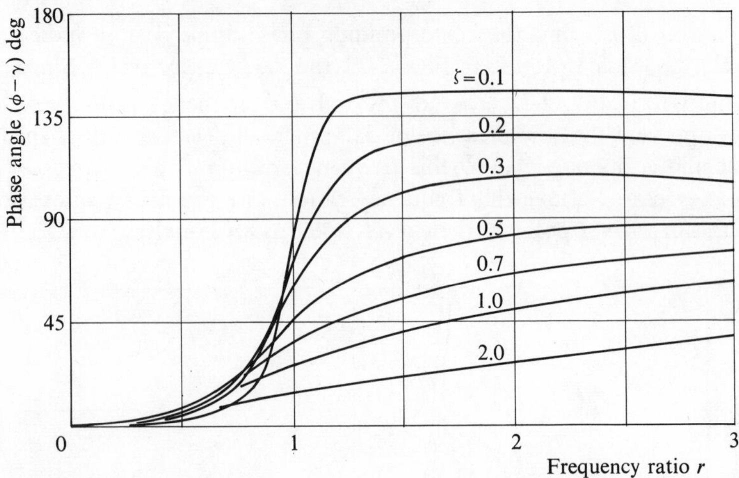

Fig. 2-33. *Phase angle of force transmitted versus frequency ratio for various damping factors; system shown in Fig.* 2-6

This equation is plotted as shown in Fig. 2-34. It should be noted that the actual force transmitted at high frequency ratios can be very high in spite of the low transmissibility.

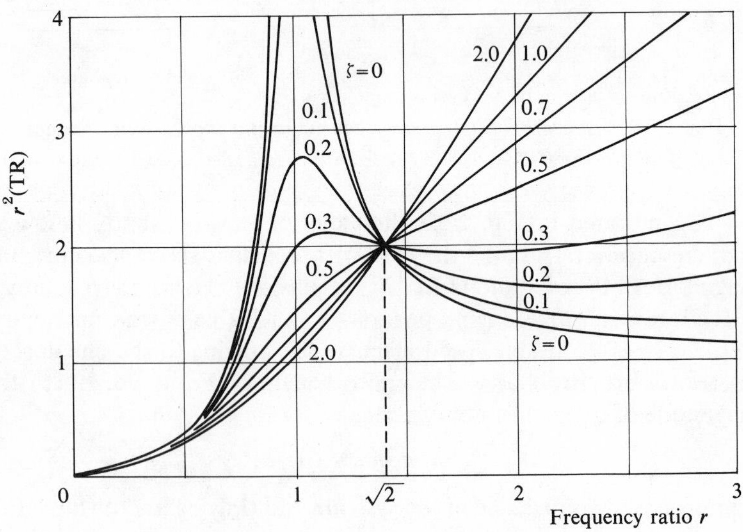

Fig. 2-34. *Force ratio versus frequency ratio for various damping factors—inertial excitation; system shown in Fig.* 2-26

Example 17. An air compressor weighs 1,000 lb and is operated at a constant speed of 1,750 rpm. The unbalanced reciprocating parts weigh 25 lb, and the rotating parts are well balanced. The crank radius is 4 in. If the dampers used for the mounting introduce a damping factor $\zeta = 0.15$, (*a*) specify the springs for the mounting such that only 20 percent of the unbalanced force is transmitted to the foundation, and (*b*) determine the amplitude of the transmitted force.

Solution: (*a*)

$$\omega = 2\pi \frac{1{,}750}{60} = 183.5 \text{ rad-sec}^{-1}$$

Since

$$\text{TR} = 0.20 = \sqrt{1 + (2\zeta r)^2}\,\kappa, \quad r = \frac{\omega}{\omega_n} = 2.76$$

$$\omega_n = \frac{183.5}{2.76} = 66.5 \text{ rad-sec}^{-1}$$

$$k = m\omega_n^2 = \frac{1{,}000}{386} \times 66.5^2 = 11{,}500 \text{ lb-in.}^{-1}$$

(*b*) Amplitude of force transmitted $= 0.2F_{eq}$

$$(0.2)(me\omega^2) = (0.2)(\tfrac{25}{386})(4)(183.5)^2 = 1{,}740 \text{ lb}$$

Case 4. Systems Attached to Moving Supports

When the excitation is applied to the support or the base of the system, as shown in Fig. 2-35, instead of applied to the mass, both the absolute motion of the mass and the relative motion between the mass and the support are of interest. First, let us examine the absolute motion $x_2(t)$ of the mass m and then discuss the relative motion in the case to follow.

The inertia force due to the mass m is $m\ddot{x}_2(t)$ where $\ddot{x}_2(t)$ is the absolute acceleration. Let the base be given a displacement $x_1(t)$. The spring force is due to the deformation of the spring and is equal to $k(x_2 - x_1)$. The damping force is proportional to the relative velocity $(\dot{x}_2 - \dot{x}_1)$ and is equal to $c(\dot{x}_2 - \dot{x}_1)$. Summing the forces gives

$$m\ddot{x}_2 + c(\dot{x}_2 - \dot{x}_1) + k(x_2 - x_1) = 0 \quad \textbf{(2-89)}$$

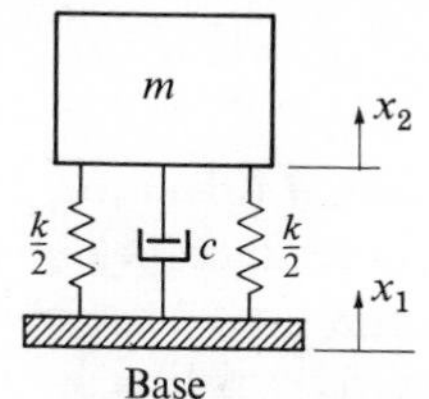

Fig. 2-35. *System attached to moving support*

Rearranging, we obtain

$$m\ddot{x}_2 + c\dot{x}_2 + kx_2 = kx_1 + c\dot{x}_1 \tag{2-90}$$

If $x_1(t) = X_1 \sin \omega t$, this equation becomes

$$\begin{aligned} m\ddot{x}_2 + c\dot{x}_2 + kx_2 &= X_1(k \sin \omega t + c\omega \cos \omega t) \\ &= X_1\sqrt{k^2 + c^2\omega^2} \sin (\omega t + \gamma) \\ &= F_{eq} \sin (\omega t + \gamma) \end{aligned} \tag{2-91}$$

where

$$\gamma = \tan^{-1} \frac{c\omega}{k} = \tan^{-1} 2\zeta r$$

Since Eq. (2-91) is of the same form as Eq. (2-72), the steady-state solution is

$$x_2 = X_1\sqrt{1 + (2\zeta r)^2}\, \kappa \sin (\omega t - \phi + \gamma) \tag{2-92}$$

where ϕ is defined by Eq. (2-74).

Example 18. Vehicle Suspension

The vehicle represents a complex system with many degrees of freedom. As a first approximation, Fig. 2-36 may be considered as a

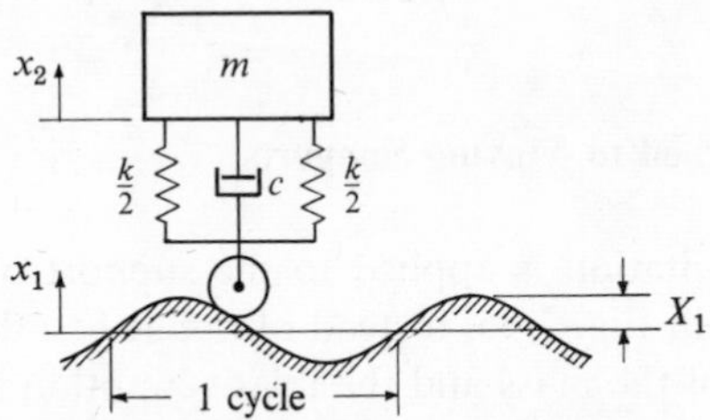

Fig. 2-36. *Schematic sketch of a vehicle moving over a rough road*

vehicle driven on a rough road. It is assumed that (1) the vehicle is constrained to one degree of freedom in the vertical direction; (2) the spring constant of the tires is infinite, that is, roughness of the road surface is directly transmitted to the suspension system of the vehicle; and (3) the tires do not leave the road surface. If a vehicle weighs 1 ton fully loaded and 1/4 ton empty, the spring constant is 2,000 lb-in.$^{-1}$, the damping factor ζ is 0.5 when the vehicle is fully loaded, vehicle speed is 60 mph, and the road surface varies sinusoidally with a period of 16 ft and an amplitude of X_1 in., determine the amplitude ratio of the vehicle when fully loaded and empty.

Solution: The excitation frequency is

$$f = \frac{(60)(5{,}280)}{3{,}600} \times \frac{1}{16} = 5.5 \text{ cps}$$

From Eq. (2-27) the damping coefficient is $c = 2\zeta\sqrt{km}$. Since the damping coefficient remains unchanged and the damping factor ζ varies inversely with the square root of the mass, ζ is equal to 1.0 when the vehicle is empty. The calculations are tabulated as follows:

Item	*Vehicle Fully Loaded*	*Vehicle Empty*
Natural frequency	$\sqrt{\frac{2{,}000g}{2{,}000}} = \sqrt{386} = \begin{cases} 19.6 \text{ rad-sec}^{-1} \\ 3.12 \text{ cps} \end{cases}$	$2 \times 19.6 = \begin{cases} 39.2 \text{ rad-sec}^{-1} \\ 6.25 \text{ cps} \end{cases}$
$(2\zeta r)^2$	$[2(0.5)(5.5/3.12)]^2 = 3.11$	$[2(1.0)(5.5/6.25)]^2 = 3.11$
$(1 - r^2)^2$	$[1 - (5.5/3.12)^2]^2 = 4.46$	0.051
$1/\kappa$	$\sqrt{4.46 + 3.11} = 2.75$	1.77
Amplitude $= X_1\sqrt{1 + (2\zeta r)^2}\kappa$	$X_1(2.03)/2.75 = 0.74X_1$	$X_1(2.03)/1.77 = 1.14X_1$

The amplitude ratio when fully loaded and empty is $0.74X_1/1.14X_1 = 1/1.56$.

Example 19. Mounting of Instruments

Figure 2-35 represents the mounting of an instrument of mass m on a vibrating body, the motion of which is $X_1 \sin \omega t$. Determine (*a*) the maximum acceleration of the instrument and (*b*) the maximum force transmitted to the instrument.

Solution: (*a*) The motion of the instrument is given by Eq. (2-92). The maximum acceleration $\ddot{x}_{2\max}$ is

$$\ddot{x}_{2\max} = \omega^2 X_2 = \omega^2 X_1\sqrt{1 + (2\zeta r)^2}\,\kappa$$

(*b*) Force is transmitted to the instrument through the spring and the damper. From Eq. (2-89), it is noted that the sum of these forces is equal to $m\ddot{x}_2$. Thus the maximum force transmitted to the instrument is equal to $m\omega^2 X_2$, which may be expressed as

$$F_{T\max} = m\omega^2 X_2 = \frac{m}{k} k\omega^2 X_2$$

$$= kr^2 X_1 \sqrt{1 + (2\zeta r)^2}\,\kappa$$

Case 5. Seismic Instruments

A seismic instrument consists essentially of a vibratory system as shown in Fig. 2-35. The support or the base of the system is attached to the body whose motion is to be measured. Figure 2-37 shows a

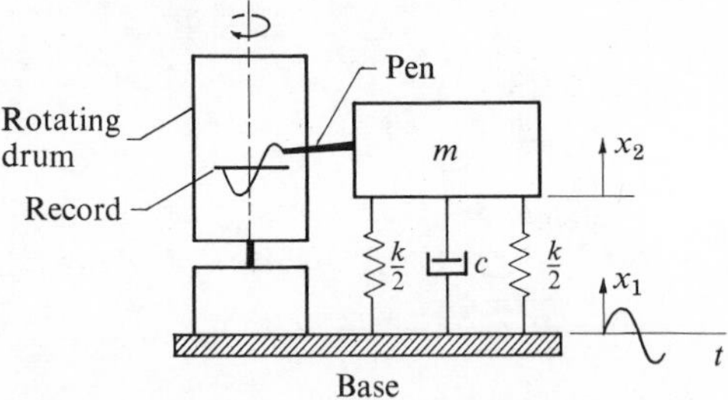

Fig. 2-37. *Schematic sketch of a seismic instrument*

schematic sketch of a seismic instrument. The motion of the base is $x_1(t)$; the motion of the mass m is $x_2(t)$; the relative motion $x_2(t) - x_1(t)$ is recorded with a pen on a rotating drum. This relative motion is used to indicate the motion of the base.

It was shown in Case 4 that the equation of motion of the system is

$$m\ddot{x}_2 + c(\dot{x}_2 - \dot{x}_1) + k(x_2 - x_1) = 0$$

Let the motion of the base be $x_1(t) = X_1 \sin \omega t$ and the relative motion between the mass and the base be $x(t) = x_2(t) - x_1(t)$. Substituting $\ddot{x}_2(t) = \ddot{x}(t) + \ddot{x}_1(t)$ in the above equation, we obtain

$$m\ddot{x} + c\dot{x} + kx = -m\ddot{x}_1 = m\omega^2 X_1 \sin \omega t \qquad \textbf{(2-93)}$$

which is of the same form as Eq. (2-72). Thus the steady-state solution is

$$x = \frac{m\omega^2 X_1}{k} \kappa \sin(\omega t - \phi)$$

$$= r^2 \kappa X_1 \sin(\omega t - \phi) = X \sin(\omega t - \phi) \quad \textbf{(2-94)}$$

where $\phi = \tan^{-1} \dfrac{2\zeta r}{1 - r^2}$. The amplitude ratio in Eq. (2-94) may be expressed as

$$\frac{X}{X_1} = r^2 \kappa \quad \textbf{(2-95)}$$

A plot of this equation is shown in Fig. 2-38.

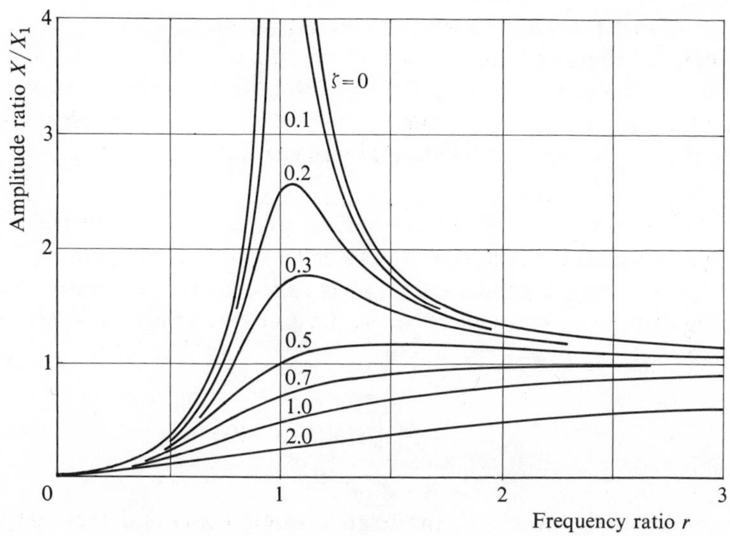

Fig. 2-38. *Steady-state response of seismic instrument; system shown in Fig.* 2-37

The construction of vibration pickups is essentially the same as that shown in Fig. 2-37. The relative motion between the mass and the base may be measured mechanically, or, through some intermediate steps, measured electrically. Figure 2-39 shows two pickups which convert the relative motions to electrical outputs.

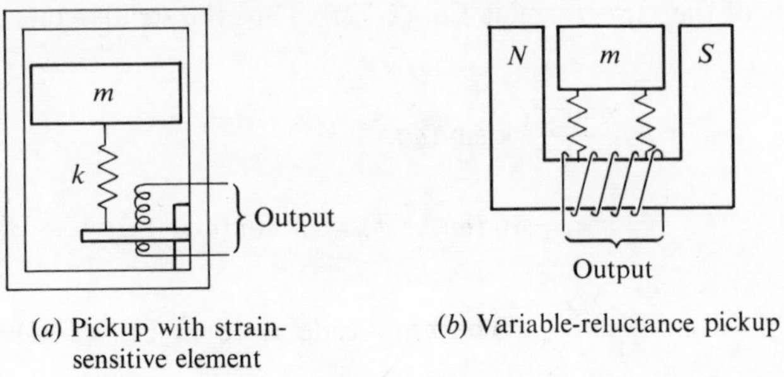

(*a*) Pickup with strain-sensitive element

(*b*) Variable-reluctance pickup

Fig. 2-39. *Vibration pickups with electrical outputs*

VIBROMETER: Figure 2-38 shows that at the high frequency range, the amplitude ratio X/X_1 approaches unity regardless of the value of ζ. Thus the relative displacement is equal in magnitude to the displacement of the base and therefore of the body to be measured. It should be noted that the phase angle of the relative displacement is nearly 180 deg out of phase with the motion of the base. An instrument for measuring displacement is called a vibrometer. Since there is no advantage in introducing damping in the system, a vibrometer is designed with damping only to minimize the transient vibrations.

ACCELEROMETER: When the natural frequency of the instrument is high compared with the vibration to be measured, the instrument can be used to measure acceleration and is called an accelerometer. Assuming that the excitation is $X_1 \sin \omega t$, its acceleration is $-\omega^2 X_1 \sin \omega t$. Rearranging Eq. (2-94), we obtain

$$x = \frac{\kappa}{\omega_n^2} \omega^2 X_1 \sin (\omega t - \phi) \qquad \textbf{(2-96)}$$

If κ is essentially constant, the relative motion $x(t)$ is a measure of acceleration.

It should be cautioned that the vibrations to be measured usually consist of a number of harmonic motions of various frequencies. *Amplitude distortion* in an accelerometer occurs if the acceleration of one harmonic is amplified more than another. From Eq. (2-96) the amplitude of the given acceleration is $\omega^2 X_1$. In order to give equal amplification to accelerations, it is desirable to have κ/ω_n^2 nearly constant for all frequencies. It is readily seen in Eq. (2-41) that $\kappa = 1$ when $r = 0$. Let us define the change in κ/ω_n^2 with respect to $r = 0$ as the amplitude distortion. The percent amplitude distortion is

$$100 \times \frac{\kappa/\omega_n^2 - 1/\omega_n^2}{1/\omega_n^2} = 100(\kappa - 1)\% \qquad \textbf{(2-97)}$$

This equation is plotted in Fig. 2-40. It is observed that an accelerometer should be built with ζ between 0.60 and 0.70 in order to minimize the amplitude distortion.

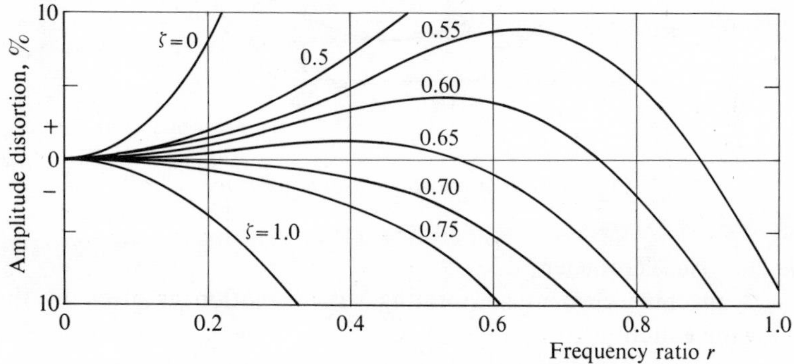

Fig. 2-40. *Amplitude distortion in accelerometer*

Phase distortion occurs if the relative phase of the harmonics recorded is different from that of the vibrations to be measured. For zero phase distortion, the phase shift ϕ should increase linearly with the frequency of the harmonic motion. The phase shift at $r = 1$ is always 90 deg. For zero phase distortion, the phase shift for $0 < r < 1$ should be r(90 deg). Hence phase distortion in an accelerometer can be defined as

$$\text{Phase distortion} = (\phi - 90r) \text{ deg} \qquad \textbf{(2-98)}$$

This expression is plotted in Fig. 2-41. Again, it is seen that appropriate damping in an accelerometer is necessary in order to minimize the phase distortion.

An accelerometer can be made quite small compared with the vibrometer. By means of integrating circuits, the electrical output of an accelerometer can be integrated to obtain the corresponding velocity or displacement.

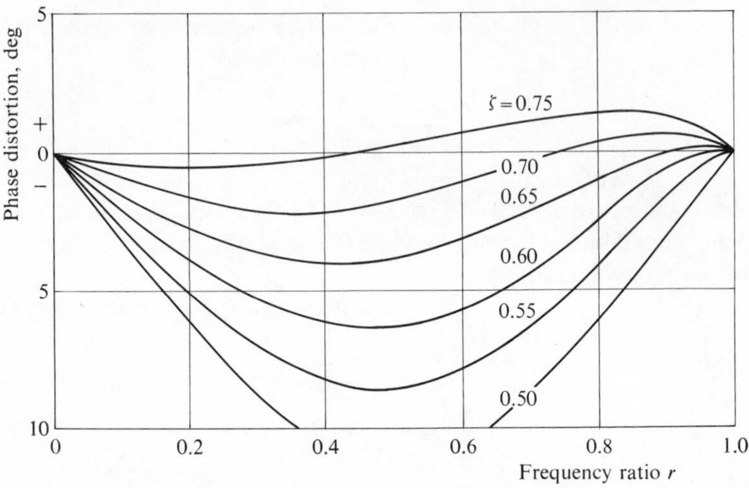

Fig. 2-41. *Phase distortion in accelerometer*

Example 20. Accelerometer

A machine element is vibrating with a motion as given by the following equation:

$$x_1 = B_1 \sin \omega_1 t + B_2 \sin \omega_2 t$$

$$x_1 = 0.10 \sin 60\pi t + 0.05 \sin 120\pi t$$

Determine the vibration record that would be obtained from a seismic instrument, as shown in Fig. 2-37, if the damping factor of the instrument is $\zeta = 0.65$ and the natural frequency of the instrument is $f_n = 1{,}500$ cycles-sec^{-1}. Compare the vibration record with the actual acceleration of the machine element.

Solution: From Eq. (2-93) the equation of motion of the instrument is

$$m\ddot{x} + c\dot{x} + kx = -m\ddot{x}_1$$

Because the equation is linear, the two components of the excitation $x_1(t)$ can be treated separately, and the results can be combined by the principle of superposition. The steady-state response of the instrument to each of the components can be determined from Eq. (2-94). The resultant steady-state response of the instrument is the sum of the individual responses, that is,

$$x = X_{(1)} \sin (\omega_1 t - \phi_1) + X_{(2)} \sin (\omega_2 t - \phi_2)$$

where

$$X_{(1)} = B_1 r_1^2 \kappa_1 \qquad X_{(2)} = B_2 r_2^2 \kappa_2$$

$$\phi_1 = \tan^{-1} \frac{2\zeta r_1}{1 - r_1^2} \qquad \phi_2 = \tan^{-1} \frac{2\zeta r_2}{1 - r_2^2}$$

For the frequency ratios considered, $r_1 = \dfrac{\omega_1}{\omega_n} = \dfrac{60\pi}{(2\pi)(1{,}500)} = \dfrac{1}{50}$ and $r_2 = \dfrac{\omega_2}{\omega_n} = \dfrac{120\pi}{(2\pi)(1{,}500)} = \dfrac{2}{50}$, κ_1 and κ_2 are practically equal to unity (see Fig. 2-40), therefore,

$$X_{(1)} = (0.10)\left(\frac{1}{50}\right)^2(1) = \frac{0.10}{2{,}500}$$

$$\phi_1 = \tan^{-1} \frac{(2)(0.65)(1/50)}{1 - (1/50)^2}$$

$$= \tan^{-1} 0.02601 = 1^\circ\ 30'$$

$$X_{(2)} = (0.05)\left(\frac{2}{50}\right)^2(1) = \frac{0.20}{2{,}500}$$

$$\phi_2 = \tan^{-1} \frac{(2)(0.65)(2/50)}{1 - (2/50)^2}$$

$$= \tan^{-1} 0.05208 = 2^\circ\ 59'$$

Thus the vibration record obtained from the accelerometer will be

$$x = \frac{1}{2{,}500}\,[0.10 \sin (60\pi t - 1^\circ\ 30') + 0.20 \sin (120\pi t - 2^\circ\ 59')]$$

The actual acceleration of the machine element is

$$\ddot{x}_1 = -(60\pi)^2(0.10 \sin 60\pi t + 0.20 \sin 120\pi t)$$

The output of an accelerometer is usually converted to an electrical signal and amplified; therefore, the recorded acceleration, when multiplied by a magnitude scale factor, is the acceleration indicated by the accelerometer. Figure 2-42 shows the displacement and acceleration curves for the machine element. The vibration record which would be obtained from the accelerometer is plotted for comparison purposes.

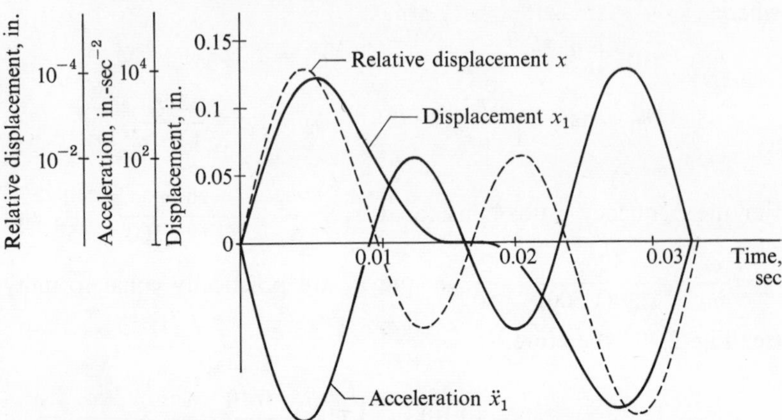

Fig. 2-42. *Displacement and corresponding acceleration as compared with relative displacement recorded by an accelerometer*

Case 6. Periodic Excitation: Fourier Series Analysis†

Forces which arise from machinery are generally periodic but seldom purely harmonic. It was illustrated in Example 20 that a periodic motion may be the sum of two harmonic functions. A function of time $F(t)$, periodic of period τ, can be separated into its harmonic components by means of the Fourier series expansion, which may be defined as

$$F(t) = \frac{a_o}{2} + \sum_{n=1}^{\infty} (a_n \cos n\,\omega t + b_n \sin n\,\omega t) \qquad \textbf{(2-99)}$$

where n is a positive integer, and a_n and b_n are the coefficients of the infinite series.

To evaluate the coefficients a_n and b_n for a given $F(t)$, we make use of the following relations:

$$\int_0^{\tau} \cos m \frac{2\pi}{\tau} t \cos n \frac{2\pi}{\tau} t\, dt = \begin{cases} 0, & m \neq n \\ \tau/2, & m = n \end{cases}$$

$$\int_0^{\tau} \sin m \frac{2\pi}{\tau} t \sin n \frac{2\pi}{\tau} t\, dt = \begin{cases} 0, & m = n \\ \tau/2, & m = n \end{cases} \qquad \textbf{(2-100)}$$

$$\int_0^{\tau} \cos m \frac{2\pi}{\tau} t \sin n \frac{2\pi}{\tau} t\, dt = 0, \text{ whatever } m \text{ and } n$$

where m and n are integers and $\tau = \dfrac{2\pi}{\omega}$ is the period of $F(t)$.

† See, for example, I. S. Sokolnikoff and R. M. Redheffer, *Mathematics of Physics and Modern Engineering*, McGraw-Hill Book Co., Inc., New York, 1958, pp. 175–196.

Rewriting the series in an expanded form, we have

$$F(t) = \left(\frac{a_o}{2} + a_1 \cos \omega t + a_2 \cos 2\omega t + \cdots\right) + (b_1 \sin \omega t + b_2 \sin 2\omega t + \cdots) \quad \textbf{(2-101)}$$

A particular coefficient a_p in the above equation can be obtained by multiplying both sides of this equation by $\cos p\omega t$ and integrating each term between the limits of 0 and τ. The set of relations in Eqs. (2-100) indicates that, except for the term containing a_p, all the integrals on the right-hand side are equal to zero. Thus

$$\int_0^\tau F(t) \cos p\omega t \, dt = 0 + \cdots + 0 + \int_0^\tau a_p \cos^2 p \frac{2\pi}{\tau} \, dt + 0 + \cdots$$

$$= a_p \frac{\tau}{2}$$

or

$$a_p = \frac{2}{\tau} \int_0^\tau F(t) \cos p\omega t \, dt$$

Similarly, a particular coefficient b_p in Eq. (2-101) can be obtained by multiplying the equation by $\sin p\omega t$ and integrating each term between the limits of 0 and τ. Thus the coefficients of the Fourier series, Eq. (2-99), are

$$a_o = \frac{2}{\tau} \int_0^\tau F(t) \, dt$$

$$a_n = \frac{2}{\tau} \int_0^\tau F(t) \cos n\omega t \, dt \quad \textbf{(2-102)}$$

$$b_n = \frac{2}{\tau} \int_0^\tau F(t) \sin n\omega t \, dt$$

If $F(t)$ of Eq. (2-99) represents a periodic excitation force applied to the mass of linear system, such as the generalized model shown in Fig. 2-6, the equation of motion is

$$m\ddot{x} + c\dot{x} + kx = \frac{a_o}{2} + \sum_{n=1}^{\infty} (a_n \cos n\omega t + b_n \sin n\omega t) \quad \textbf{(2-103)}$$

The response of the system to each of the harmonic components can be calculated. By superposition, the response of the system to a periodic excitation force is the sum of responses to the harmonic components. Hence the steady-state response of the system described by Eq. (2-103) is

$$x = \frac{a_o}{2k} + \sum_{n=1}^{\infty} \frac{a_n \cos (n\omega t - \phi_n) + b_n \sin (n\omega t - \phi_n)}{k\sqrt{(1 - n^2r^2)^2 + (2\zeta nr)^2}} \qquad \textbf{(2-104)}$$

where

$$\phi_n = \tan^{-1} \frac{2\zeta nr}{1 - n^2r^2} \qquad \textbf{(2-105)}$$

and

$$r = \frac{\omega}{\omega_n}$$

Example 21. A periodic function is as illustrated in Fig. 2-43(a). Determine the harmonic components of this function.

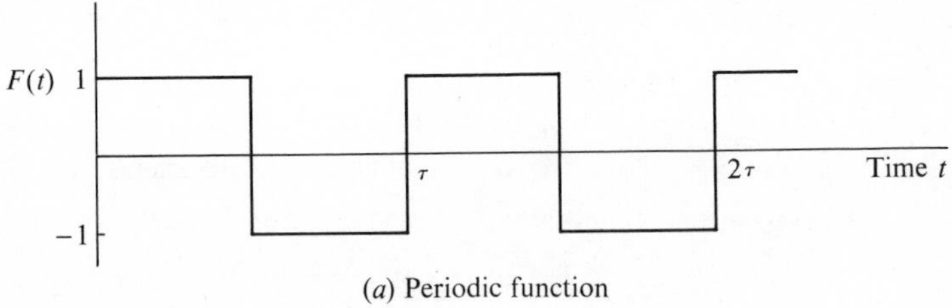

(a) Periodic function

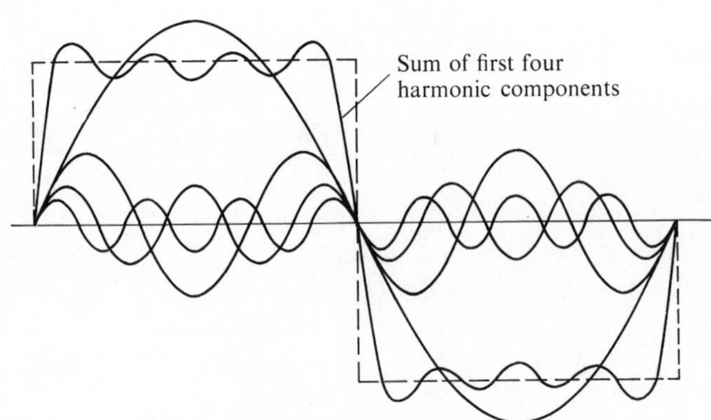

(b) Harmonic components of periodic function

Fig. 2-43. *Fourier series expansion of periodic function*

Solution: For any one cycle, the given periodic function can be expressed as

$$F(t) = \begin{cases} 1, & 0 < t < \tau/2 \\ -1, & \tau/2 < t < \tau \end{cases}$$

The coefficients of the Fourier series expansion of $F(t)$ are

$$a_o = \frac{2}{\tau}\int_0^\tau F(t)\,dt = \frac{2}{\tau}\left[\int_0^{\tau/2}(1)\,dt - \int_{\tau/2}^{\tau}(1)\,dt\right] = 0$$

$$a_n = \frac{2}{\tau}\int_0^\tau F(t)\cos\frac{2n\pi}{\tau}t\,dt$$

$$= \frac{2}{\tau}\left[\int_0^{\tau/2}\cos\frac{2n\pi}{\tau}t\,dt - \int_{\tau/2}^{\tau}\cos\frac{2n\pi}{\tau}t\,dt\right] = 0$$

$$b_n = \frac{2}{\tau}\int_0^\tau F(t)\sin\frac{2n\pi}{\tau}t\,dt$$

$$= -\frac{2}{\tau}\frac{\tau}{2n\pi}\left[\cos\frac{2n\pi}{\tau}t\Big|_0^{\tau/2} - \cos\frac{2n\pi}{\tau}t\Big|_{\tau/2}^{\tau}\right]$$

$$b_n = \begin{cases} \dfrac{4}{n\pi}, & n \text{ odd} \\ 0, & n \text{ even} \end{cases}$$

Hence the Fourier series expansion of $F(t)$ is

$$F(t) = \frac{4}{\pi}\sum_n \frac{1}{n}\sin\frac{2n\pi}{\tau}t, \; n = 1, 3, 5, \ldots$$

The first four harmonics of $F(t)$ and their sum are plotted in Fig. 2-43(*b*).

Example 22. A cam with a sawtooth motion actuates a spring-and-mass system as shown in Fig. 2-44. The total cam lift is 1 in., and the cam speed is 60 rpm. If the mass weighs 38.6 lb, $k_1 = k_2 = 20$ lb-in.$^{-1}$, and $c = 1$ lb-sec-in.$^{-1}$, determine the steady-state response of the system.

Solution: For one cycle, the sawtooth motion can be expressed as

$$x_1(t) = \frac{1}{\tau}t, \text{ for } 0 < t < \tau$$

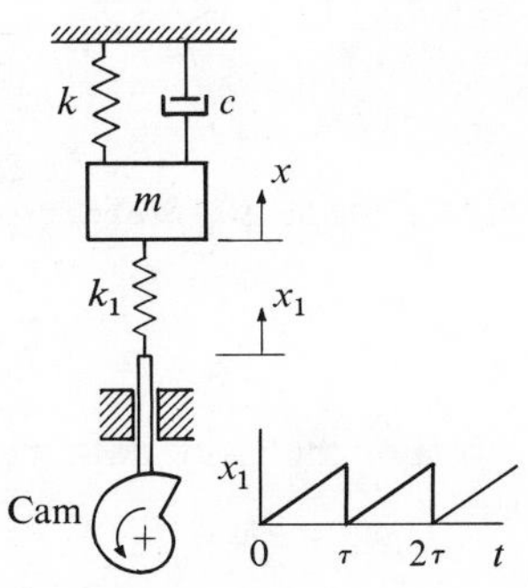

Fig. 2-44. *Nonharmonic excitation*

Since the excitation frequency is 1 cps or $\omega = 2\pi$, the period $\tau = 1$ sec-cycle^{-1}. From Eq. (2-102), the coefficients a_n and b_n are

$$a_o = \frac{2}{\tau}\int_0^{\tau} \frac{t}{\tau}\,dt = 2\int_0^1 t\,dt = 1$$

$$a_n = \frac{2}{\tau}\int_0^{\tau} \frac{t}{\tau}\cos n2\pi t\,dt = 2\int_0^1 t\cos n2\pi t\,dt = 0$$

$$b_n = \frac{2}{\tau}\int_0^{\tau} \frac{t}{\tau}\sin n2\pi t\,dt = 2\int_0^1 t\sin n2\pi t\,dt = -\frac{1}{n\pi}$$

Hence the Fourier series expansion of $x_1(t)$ is

$$x_1 = \frac{1}{2} - \frac{1}{\pi}\sum_{n=1}^{\infty}\frac{1}{n}\sin 2n\pi t$$

The equation of motion of the system is

$$m\ddot{x} + c\dot{x} + (k + k_1)x = k_1 x_1$$
$$= \frac{k_1}{2} - \frac{k_1}{\pi}\sum_{n=1}^{\infty}\frac{1}{n}\sin 2n\pi t$$

For convenience we define $\omega_n^2 = (k + k_1)/m$, $2\zeta\omega_n = c/m$, and $2n\pi/\omega_n = nr$, where r is a frequency ratio. The particular integral due to the constant excitation term $k_1/2$ is

$$x = k_1/2(k + k_1)$$

The steady-state response due to a typical harmonic excitation term $-(k_1/\pi n)\sin 2n\pi t$ is

$$x = \frac{-k_1}{(k + k_1)\pi n\sqrt{(1 - n^2r^2)^2 + (2\zeta nr)^2}}\sin(2n\pi t - \phi_n)$$

where

$$\phi_n = tan^{-1}\frac{2\zeta nr}{1 - n^2r^2}$$

Hence the steady-state response due to the total excitation is

$$x = \frac{k_1}{(k + k_1)}\left[\frac{1}{2} - \frac{1}{\pi}\sum_{n=1}^{\infty}\frac{1}{n}\frac{1}{\sqrt{(1 - n^2r^2)^2 + (2\zeta nr)^2}}\sin(2n\pi t - \phi_n)\right]$$

For the given data we have

$$\omega_n^2 = (k + k_1)/m = (20 + 20)/(38.6/386) = 400$$
$$r = \omega/\omega_n = 2\pi/20 = 0.1\pi$$
$$\zeta = c/2\sqrt{(k + k_1)m} = 1/2\sqrt{(20 + 20)(38.6/386)} = 0.25$$

Thus the steady-state response is

$$x = \frac{1}{2}\left\{\frac{1}{2} - \frac{1}{\pi}\sum_{n=1}^{\infty}\frac{1}{n}\frac{1}{\sqrt{[1-(0.1\pi n)^2]^2 + (0.05\pi n)^2}}\sin(2n\pi t - \phi_n)\right\}$$

where

$$\phi_n = \tan^{-1}\frac{0.05\pi n}{1-(0.1\pi n)^2}$$

Case 7. Elastically Supported Damper Systems

The conventional spring-damper system (see Fig. 2-26) for vibration isolation, as discussed in Case 3, can be used successfully to isolate the force transmitted to the foundation of the machine. To extend the range of operation, the damper may be elastically supported as shown in Fig. 2-45.

From the free-body sketches, the equation of motion for the mass m is

$$m\ddot{x} + c(\dot{x} - \dot{x}_1) + kx = F\sin\omega t \quad \textbf{(2-106)}$$

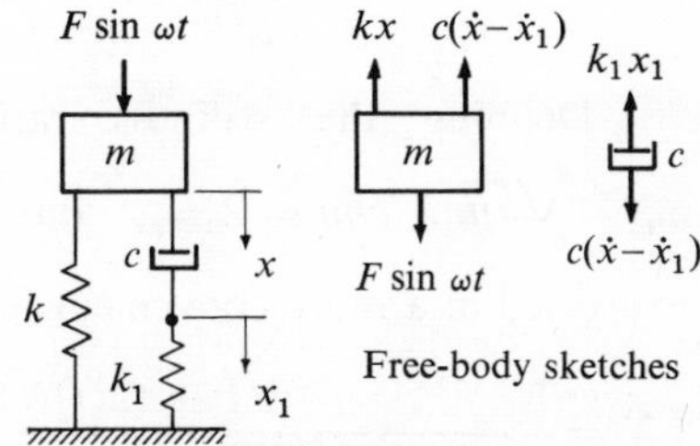

Fig. 2-45. *Elastically supported damper*

The damper c and the spring k_1 are in series; therefore, the damping force is equal to the spring force, that is,

$$c(\dot{x} - \dot{x}_1) = k_1x_1 \quad \textbf{(2-107)}$$

Substituting Eq. (2-107) in (2-106) and solving for x_1, we have

$$x_1 = \frac{1}{k_1}(F\sin\omega t - m\ddot{x} - kx) \quad \textbf{(2-108)}$$

Substituting the time derivative of Eq. (2-108) in Eq. (2-106) and rearranging, we obtain

$$\dddot{x} + \frac{k_1}{c}\ddot{x} + \frac{k_1 + k}{m}\dot{x} + \frac{k_1k}{cm}x = \frac{k_1}{cm}F\sin\omega t + \frac{\omega}{m}F\cos\omega t \quad \textbf{(2-109)}$$

Thus the motion of m is described by a third-order differential equation, the general solution of which is more difficult to determine than a second-order equation.

The steady-state solution of Eq. (2-109) can be obtained quite readily by the impedance method and solving Eqs. (2-106) and (2-107) simultaneously. Let the excitation be represented by $Fe^{j\omega t}$ and the corresponding displacements $x(t)$ and $x_1(t)$ be represented by $\bar{X}e^{j\omega t}$ and $\bar{X}_1e^{j\omega t}$, respectively, where $\bar{X}$ and $\bar{X}_1$ are complex amplitudes. Substituting these expressions in Eqs. (2-106) and (2-107) and factoring out $e^{j\omega t}$, we obtain

$$\begin{aligned}(k - m\omega^2 + jc\omega)\bar{X} - jc\omega\bar{X}_1 &= F \\ -jc\omega\bar{X} + (k_1 + jc\omega)\bar{X}_1 &= 0\end{aligned} \tag{2-110}$$

$\bar{X}$ and $\bar{X}_1$ can now be solved for by means of Cramer's rule:

$$\begin{aligned}\bar{X} &= \frac{F(k_1 + jc\omega)}{k_1(k - m\omega^2) + jc\omega(k + k_1 - m\omega^2)} \\ \bar{X}_1 &= \frac{jc\omega F}{k_1(k - m\omega^2) + jc\omega(k + k_1 - m\omega^2)}\end{aligned} \tag{2-111}$$

Defining the stiffness ratio of the springs as $N = k_1/k$, $\omega_n = \sqrt{k/m}$, $c/m = 2\zeta\omega_n$, and $r = \dfrac{\omega}{\omega_n}$, these equations can be expressed in a more convenient form as

$$\begin{aligned}\bar{X} &= \frac{F}{k}\frac{\sqrt{1 + (2\zeta r/N)^2}}{\sqrt{(1 - r^2)^2 + \left[2\zeta r\left(1 + \dfrac{1}{N} - \dfrac{r^2}{N}\right)\right]^2}}e^{-j\gamma} = Xe^{-j\gamma} \\ \bar{X}_1 &= \frac{F}{k}\frac{2\zeta r/N}{\sqrt{(1 - r^2)^2 + \left[2\zeta r\left(1 + \dfrac{1}{N} - \dfrac{r^2}{N}\right)\right]^2}}e^{-j\gamma_1} = X_1e^{-j\gamma_1}\end{aligned} \tag{2-112}$$

where

$$\gamma = \tan^{-1}\frac{2\zeta r\left(1 + \dfrac{1}{N} - \dfrac{r^2}{N}\right)}{1 - r^2} - \tan^{-1}\frac{2\zeta r}{N}$$

$$\gamma_1 = \tan^{-1}\frac{2\zeta r\left(1 + \dfrac{1}{N} - \dfrac{r^2}{N}\right)}{1 - r^2} - \frac{\pi}{2}$$

Since the excitation is a sine function, the steady-state responses are

$$\begin{aligned}x &= X\sin(\omega t - \gamma) \\ x_1 &= X_1\sin(\omega t - \gamma_1)\end{aligned} \tag{2-113}$$

where X, X_1, γ, and γ_1 are defined in Eq. (2-112).

Example 23. Vibration Isolation

A machine of mass m is mounted on a spring-damper system as shown in Fig. 2-45. (*a*) Defining $\omega_n = \sqrt{k/m}$, frequency ratio $r = \omega/\omega_n$, spring stiffness ratio $N = k_1/k$, and damping factor $\zeta = c/2\sqrt{km}$, derive the transmissibility TR equation of the system; (*b*) if $r = 0.6$, $N = 2$, and $\zeta = 0.4$, determine the transmissibility TR; (*c*) repeat part (*b*) if $r = 10$ and the other parameters remain unchanged; (*d*) compare the TR from parts (*b*) and (*c*) with that expressed in Eq. (2-87) (see Fig. 2-26).

Solution: (*a*) The force transmitted F_T to the foundation is the sum of the forces transmitted through springs k and k_1, that is,

$$F_T = kx + k_1x_1$$

The complex amplitudes of $x(t)$ and $x_1(t)$ are given in Eq. (2-111). By means of vectorial addition, the complex amplitude of F_T is

$$\begin{aligned}\bar{F}_T &= k\bar{X} + k_1\bar{X}_1 \\ &= \frac{F[k(k_1 + jc\omega) + jk_1c\omega]}{k_1(k - m\omega^2) + jc\omega(k + k_1 - m\omega^2)} \\ &= F\frac{\sqrt{1 + [2\zeta r(1 + 1/N)]^2}}{\sqrt{(1 - r^2)^2 + \left[2\zeta r\left(1 + \dfrac{1}{N} - \dfrac{r^2}{N}\right)\right]^2}}\, e^{-j\gamma_T}\end{aligned}$$

where

$$\gamma_T = \tan^{-1}\frac{2\zeta r\left(1 + \dfrac{1}{N} - \dfrac{r^2}{N}\right)}{1 - r^2} - \tan^{-1} 2\zeta r\left(1 + \frac{1}{N}\right)$$

$$\therefore \qquad \text{TR} = \sqrt{\frac{1 + [2\zeta r(1 + 1/N)]^2}{(1 - r^2)^2 + \left[2\zeta r\left(1 + \dfrac{1}{N} - \dfrac{r^2}{N}\right)\right]^2}} \qquad \textbf{(2-114)}$$

(*b*) For $r = 0.6$, $N = 2$, $\zeta = 0.4$, the transmissibility is

$$\text{TR} = \frac{1.519}{0.812} = 1.37$$

(*c*) For $r = 10$, $N = 2$, $\zeta = 0.4$, the transmissibility is

$$\text{TR} = \sqrt{\frac{145}{16 \times 10^4}} = 0.030$$

(*d*) From Eq. (2-87), the corresponding transmissibilities are

$$\text{TR} = \sqrt{\frac{1 + (2\zeta r)^2}{(1 - r^2)^2 + (2\zeta r)^2}} = \begin{cases} \sqrt{\dfrac{1.23}{0.64}} = 1.38, \text{ for } r = 0.6 \\ \sqrt{\dfrac{65}{9{,}865}} = 0.0812, \text{ for } r = 10 \end{cases}$$

Comparing parts (*b*), (*c*), and (*d*), the transmissibilities for the two isolation systems are approximately the same for $r = 0.6$. When operating at high frequency ratios the isolation system shown in Fig. 2-45 is superior to that of Fig. 2-26. It should be noted that if the excitation force is due to an unbalance in the machine, the amplitude of the excitation is proportional to the speed squared. Although the value of TR is low at high frequency ratios, the actual force transmitted from the machine to its foundation can be relatively high for rotating and reciprocating unbalances. (See Fig. 2-34.)

SUGGESTED READING

Church, A. H., *Mechanical Vibrations* (New York: John Wiley & Sons, Inc., 1957), chaps. 2, 3, and 4.

Den Hartog, J. P., *Mechanical Vibrations* (New York: McGraw-Hill Book Co., Inc., 4th ed., 1956), chap. 2.

Harris, C. M., and C. E. Crede, editors, *Shock and Vibration Handbook* (New York: McGraw-Hill Book Co., Inc., 1961), vol. 1, chaps. 10, 12, and 13; vol. 2, chap. 30.

Hetenyi, M., editor, *Handbook of Experimental Stress Analysis* (New York: John Wiley & Sons, Inc., 1950), chap. 8.

Karman, von, T., and M. A. Biot, *Mathematical Methods in Engineering* (New York: McGraw-Hill Book Co., Inc., 1940), chap. 9.

MacDuff, J. N., and J. R. Curreri, *Vibration Control* (New York: McGraw-Hill Book Co., Inc., 1958), chaps. 3 and 4.

Oldenburger, R., *Mathematical Engineering Analysis* (New York: The Macmillan Co., 1950), chaps. 1 and 2.

Scanlan, R. H., and R. Rosenbaum, *Aircraft Vibration and Flutter* (New York: The Macmillan Co., 1951), chap. 3.

Thomson, W. T., *Mechanical Vibrations* (New Jersey: Prentice-Hall, Inc., 2nd ed., 1953), chaps. 2, 3, and 4.

Timoshenko, S., and D. H. Young, *Vibration Problems in Engineering* (New York: D. Van Nostrand Co., Inc., 3rd ed., 1956), chap. 1.

Tong, K. N., *Theory of Mechanical Vibration* (New York: John Wiley & Sons, Inc., 1960), chap. 1.

The "stationary" parts of this brake lining test dynamometer constitute a two-degree-of-freedom system: the masses, springs, and damping may be varied to simulate the dynamics of an airplane landing-gear system. The dynamometer is for selecting optimum dynamic performance brake lining for an installation. (courtesy Bendix Products Aerospace Division, South Bend, Indiana)

3 SYSTEMS WITH MORE THAN ONE DEGREE OF FREEDOM

3-1. INTRODUCTION

It was shown in the previous chapter that many engineering problems can be explained by the theory of one-degree-of-freedom systems. More complex systems may possess several degrees of freedom. (See Sec. 2-2.) For example, such a system may consist of several masses connected to one another and to the foundation by springs and dampers, or it may consist of one mass which is free to vibrate with more than one mode. The simplest of these are systems with two degrees of freedom.

Since there is no basic difference in concept between systems with two or more degrees of freedom, the theory in this chapter is presented through the study of a two-degree-of-freedom system and then generalized to systems with many degrees of freedom. The equations of motion are obtained first by Newton's law of motion and in the latter part of the chapter by the method of influence coefficients. It should be remarked that the solving of the differential equations of motion becomes increasingly more laborious as the number of degrees of freedom increases. Problems involving more than three degrees of freedom are generally treated by other mathematical techniques. This chapter, however, may serve to introduce the physical concepts for the more general analysis in later chapters.

3-2. UNDAMPED FREE VIBRATION: PRINCIPAL MODES

Let us consider a two-degree-of-freedom system as shown in Fig. 3-1. The two masses are connected to one another and to the foundation by three linear springs. The displacements $x_1(t)$ and $x_2(t)$ of the masses are measured from their static equilibrium positions. This system has two degrees of freedom if the masses are constrained to move in the vertical direction. It will be shown that this system has two natural frequencies. When the entire system vibrates at one of these natural frequencies, the mode of oscillation is called a principal mode. The displacements $x_1(t)$ and $x_2(t)$ are linear combinations of the principal modes.

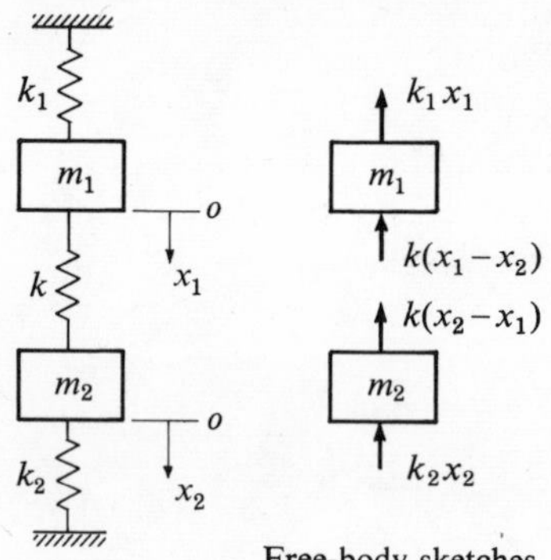

Fig. 3-1. *A two-degree-of-freedom system*

The equations of motion can be obtained from Newton's law of motion. Assuming that the masses are displaced from their static equilibrium positions, the summing of the dynamic forces indicated in the free-body sketches gives

$$m_1\ddot{x}_1 = -k_1x_1 - k(x_1 - x_2)$$
$$m_2\ddot{x}_2 = -k_2x_2 - k(x_2 - x_1)$$

Rearranging, we obtain

$$\begin{aligned} m_1\ddot{x}_1 + (k_1 + k)x_1 - kx_2 &= 0 \\ -kx_1 + m_2\ddot{x}_2 + (k_2 + k)x_2 &= 0 \end{aligned} \tag{3-1}$$

These equations are linear and homogeneous of the second order, and their solutions are of the exponential form. (See Appendix B.)

$$\begin{aligned} x_1 &= C_1 e^{st} \\ x_2 &= C_2 e^{st} \end{aligned} \tag{3-2}$$

where C_1, C_2, and s are constants. Instead of following this formal approach, let us deduce the solutions by a more direct method.

From Eqs. (3-1) and (3-2) and the discussion of differential equations in Appendix B, it can be shown that the values of s are imaginary. Using Euler's formula, $e^{\pm j\omega t} = \cos \omega t \pm j \sin \omega t$, and recalling that $x(t)$ must be real, it may be anticipated that $x_1(t)$ and $x_2(t)$ will consist of a number of harmonic functions, corresponding to the values of s. Let one of these harmonic components be

$$\begin{aligned} x_1 &= A_1 \sin (\omega t + \psi) \\ x_2 &= A_2 \sin (\omega t + \psi) \end{aligned} \tag{3-3}$$

where A_1, A_2, and ψ are arbitrary constants and ω is one of the natural frequencies of the system. Since the components are harmonic, the choice of the sine or cosine functions is arbitrary.

Substituting Eq. (3-3) in Eq. (3-1) and dividing out the factor $\sin (\omega t + \psi)$, we obtain

$$\begin{aligned} (k_1 + k - m_1\omega^2)A_1 - kA_2 &= 0 \\ -kA_1 + (k_2 + k - m_2\omega^2)A_2 &= 0 \end{aligned} \tag{3-4}$$

which are homogeneous linear algebraic equations in A_1 and A_2. This set of equations has a solution other than the trival one, $A_1 = A_2 = 0$, only if the determinant $\Delta(\omega)$ of the coefficients of A_1 and A_2 vanishes; that is,

$$\Delta(\omega) = \begin{vmatrix} k_1 + k - m_1\omega^2 & -k \\ -k & k_2 + k - m_2\omega^2 \end{vmatrix} = 0 \tag{3-5}$$

This is called the *characteristic* or the *frequency equation* of the system from which the values of ω are determined.

Expanding the determinant and dividing out the factor $m_1 m_2$, we have

$$\omega^4 - \left[\frac{k_1 + k}{m_1} + \frac{k_2 + k}{m_2}\right]\omega^2 + \frac{k_1k_2 + k_1k + k_2k}{m_1m_2} = 0 \tag{3-6}$$

which is quadratic in ω^2. Simple algebraic operations show that ω^2 may be written as

$$\omega^2 = \frac{1}{2}\left[\frac{k_1 + k}{m_1} + \frac{k_2 + k}{m_2} \mp \sqrt{\left[\frac{k_1 + k}{m_1} + \frac{k_2 + k}{m_2}\right]^2 - 4\,\frac{k_1 k_2 + k_1 k + k_2 k}{m_1 m_2}}\right] \tag{3-7}$$

or

$$\omega^2 = \frac{1}{2}\left[\frac{k_1 + k}{m_1} + \frac{k_2 + k}{m_2} \mp \sqrt{\left[\frac{k_1 + k}{m_1} - \frac{k_2 + k}{m_2}\right]^2 + 4\,\frac{k^2}{m_1 m_2}}\right] \tag{3-8}$$

Since the expression under the radical sign in Eq. (3-8) is positive, the values of ω^2 are real. Furthermore, in Eq. (3-7), the sum of the first two terms is greater than under the radical sign, therefore ω^2 is also positive. Thus we have two real and positive values for ω^2. Let them be ω_1^2 and ω_2^2. The four values of ω from Eq. (3-6) are $\pm\omega_1$ and $\pm\omega_2$. Since the solutions are sine and cosine functions, the use of the negative sign for ω merely changes the signs of the arbitrary constants of Eq. (3-3) and would not lead to new solutions. It is sufficient to consider only the frequencies with the positive sign.

It is deduced from this example that there are two natural frequencies in a two-degree-of-freedom system. The general solution of Eq. (3-1) can be expressed as

$$\begin{aligned} x_1 &= A_{11} \sin(\omega_1 t + \psi_1) + A_{12} \sin(\omega_2 t + \psi_2) \\ x_2 &= A_{21} \sin(\omega_1 t + \psi_1) + A_{22} \sin(\omega_2 t + \psi_2) \end{aligned} \tag{3-9}$$

where the A's and the ψ's are arbitrary constants.† These equations show that the vibratory motions of the masses are, in general, not harmonic but are composed of two harmonic components of frequencies ω_1 and ω_2. The lower frequency is called the *fundamental* and the higher frequency the *second harmonic.*

The amplitudes and the phase angles of the harmonic components of $x_1(t)$ and $x_2(t)$ are determined by initial conditions, but the amplitude

† ω_1, ω_2, ψ_1, and ψ_2 refer to the first and the second modes of vibration, that is, the vibrations at the fundamental and harmonic frequencies. Double subscripts are assigned to amplitudes; the first subscript refers to the coordinate and the second to the mode of vibration. For example, A_{12} is the amplitude of $x_1(t)$ due to the second mode of vibration.

ratios of the components are specified by Eq. (3-4). Substituting ω_1 and ω_2 from Eq. (3-6) in Eq. (3-4) gives

$$\begin{aligned}\frac{A_{11}}{A_{21}} &= \frac{k}{k_1 + k - m_1\omega_1^2} = \frac{k_2 + k - m_2\omega_1^2}{k} = \frac{1}{\mu_1} \\ \frac{A_{12}}{A_{22}} &= \frac{k}{k_1 + k - m_1\omega_2^2} = \frac{k_2 + k - m_2\omega_2^2}{k} = \frac{1}{\mu_2}\end{aligned} \tag{3-10}$$

where μ_1 and μ_2 are the amplitude ratios corresponding to the frequencies ω_1 and ω_2, respectively. The signs of μ_1 and μ_2 can be examined by substituting the appropriate ω from Eq. (3-8) in Eq. (3-10). Rearranging, the results are

$$\begin{aligned}k\mu_{1,2} &= k_1 + k - m_1\omega_{1,2}^2 \\ &= \frac{m_1}{2}\left[\left(\frac{k_1+k}{m_1} - \frac{k_2+k}{m_2}\right)\right. \\ &\qquad \left. \pm \sqrt{\left(\frac{k_1+k}{m_1} - \frac{k_2+k}{m_2}\right)^2 + \frac{4k^2}{m_1 m_2}}\right]\end{aligned} \tag{3-11}$$

Since $\sqrt{\left(\frac{k_1+k}{m_1} - \frac{k_2+k}{m_2}\right)^2 + \frac{4k^2}{m_1 m_2}} > \left(\frac{k_1+k}{m_1} - \frac{k_2+k}{m_2}\right)$, it is deduced that μ_1 is positive while μ_2 is negative.

Substituting the amplitude ratios μ_1 and μ_2 in Eq. (3-9), the general solutions become

$$\begin{aligned}x_1 &= A_{11}\sin(\omega_1 t + \psi_1) + A_{12}\sin(\omega_2 t + \psi_2) \\ x_2 &= \mu_1 A_{11}\sin(\omega_1 t + \psi_1) + \mu_2 A_{12}\sin(\omega_2 t + \psi_2)\end{aligned} \tag{3-12}$$

Altogether, there are four constants of integration, namely, A_{11}, A_{12}, ψ_1, and ψ_2. These constants are to be evaluated by the initial conditions $x_1(0)$, $\dot{x}_1(0)$, $x_2(0)$, and $\dot{x}_2(0)$. It should be noted that the natural frequencies ω_1 and ω_2, as obtained from Eq. (3-6), are inherent in the system parameters. Only the amplitudes A_{11} and A_{12} and the phase angles ψ_1 and ψ_2 are dependent on the initial conditions imposed on the system.

With the appropriate initial conditions, it is possible that the entire system may oscillate at one of the natural frequencies. These particular patterns of motion are called the *principal modes* of vibration. The oscillation with the lower frequency is called the *first mode* and that with the higher frequency is called the *second mode*. For example, if the initial conditions are such that the arbitrary constant A_{12} in Eq. (3-12) is made equal to zero, the motions described by the first mode are

$$\begin{aligned} x_1 &= A_{11} \sin (\omega_1 t + \psi_1) \\ x_2 &= \mu_1 A_{11} \sin (\omega_1 t + \psi_1) \end{aligned} \tag{3-13}$$

Similarly, if A_{11} is equal to zero, the second mode is

$$\begin{aligned} x_1 &= A_{12} \sin (\omega_2 t + \psi_2) \\ x_2 &= \mu_2 A_{12} \sin (\omega_2 t + \psi_2) \end{aligned} \tag{3-14}$$

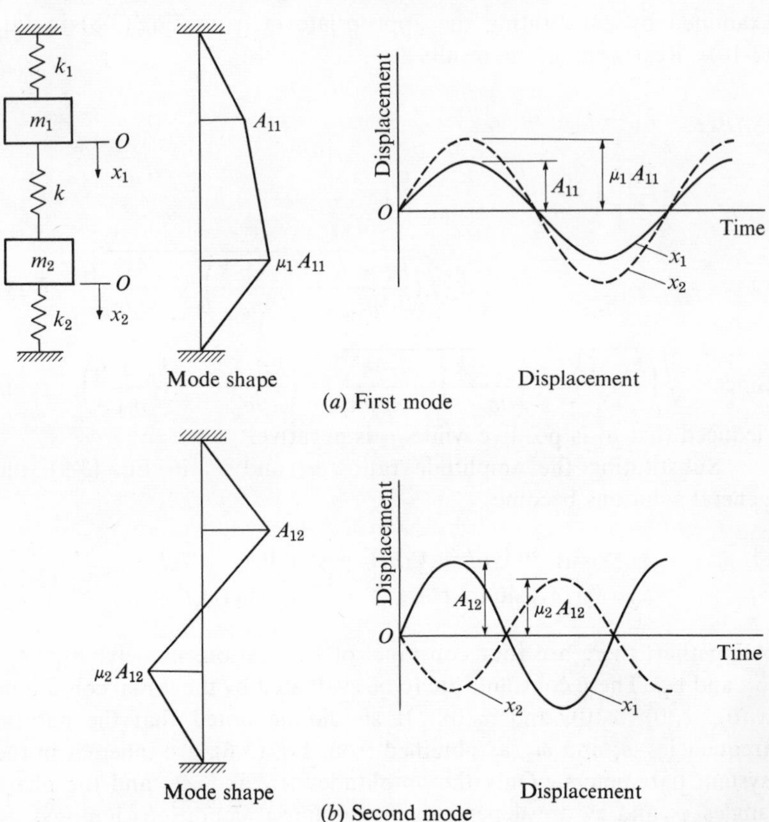

Fig. 3-2. *Principal modes of vibration; amplitude of masses plotted perpendicular to axis of motion*

The mode shapes of these principal modes of vibration are illustrated in Fig. 3-2. The amplitudes are arbitrary. It is observed in the displacement curves that, for the first mode of vibration, the displacements $x_1(t)$ and $x_2(t)$ are always in a constant ratio μ_1 and in phase with one another, corresponding to μ_1 being positive. For the second mode the displacements are always in a constant ratio μ_2 but are 180 deg out of phase, corresponding to μ_2 being negative.

Example 1. Determine the displacements $x_1(t)$ and $x_2(t)$ of the masses of the two-degrees-of-freedom system shown in Fig. 3-1. The initial conditions are $x_1(0) = 1$, $\dot{x}_1(0) = 0$, $x_2(0) = \mu_1$, $\dot{x}_2(0) = 0$.

Solution: The displacements can be obtained by a direct substitution of the given initial conditions in Eq. (3-12). From the initial conditions $x_1(0) = 1$ and $x_2(0) = \mu_1$, we obtain

$$1 = A_{11} \sin \psi_1 + A_{12} \sin \psi_2$$

$$\mu_1 = \mu_1 A_{11} \sin \psi_1 + \mu_2 A_{12} \sin \psi_2$$

Solving for A_{11} and A_{12} from these equations, we have

$$A_{11} = 1/\sin \psi_1, \text{ and } A_{12} = 0$$

From the initial conditions $\dot{x}_1(0) = \dot{x}_2(0) = 0$, we have

$$0 = \omega_1 A_{11} \cos \psi_1 + 0$$

$$0 = \mu_1 \omega_1 A_{11} \cos \psi_1 + 0$$

Equations (3-7) and (3-10) indicate that ω_1 and μ_1 are not zero. Assuming that A_{11} is not equal to zero, clearly $\cos \psi_1 = 0$, and $\psi_1 = n\pi/2$, where $n = 1, 3, 5, \cdots$ We choose $\psi_1 = \pi/2$ because the other values of ψ_1 do not lead to new solutions. Hence $A_{11} = 1/\sin \psi_1 = 1$, and

$$x_1 = A_{11} \sin (\omega_1 t + \pi/2) = \cos \omega_1 t$$

$$x_2 = \mu_1 A_{11} \sin (\omega_1 t + \pi/2) = \mu_1 \cos \omega_1 t$$

It is observed that this set of initial conditions gives the first mode of vibration. The various initial conditions necessary to obtain the first and second modes are left as an exercise.

Example 2. Determine the displacements $x_1(t)$ and $x_2(t)$ of the masses of the two-degrees-of-freedom system shown in Fig. 3-1. The initial conditions are $x_1(0) = 1$, and $\dot{x}_1(0) = x_2(0) = \dot{x}_2(0) = 0$.

Solution: The direct substitution of $x_1(0) = 1$ and $x_2(0) = 0$ in Eq. (3-12) yields

$$1 = A_{11} \sin \psi_1 + A_{12} \sin \psi_2$$

$$0 = \mu_1 A_{11} \sin \psi_1 + \mu_2 A_{12} \sin \psi_2$$

Solving for A_{11} and A_{12}, we obtain

$$A_{11} = \frac{\mu_2}{(\mu_2 - \mu_1) \sin \psi_1}, \quad \text{and} \quad A_{12} = -\frac{\mu_1}{(\mu_2 - \mu_1) \sin \psi_2}$$

From the initial conditions $\dot{x}_1(0) = \dot{x}_2(0) = 0$, we have

$$0 = \omega_1 A_{11} \cos \psi_1 + \omega_2 A_{12} \cos \psi_2$$

$$0 = \mu_1 \omega_1 A_{11} \cos \psi_1 + \mu_2 \omega_2 A_{12} \cos \psi_2$$

This set of equations in $\cos \psi_1$ and $\cos \psi_2$ has a solution other than the $\cos \psi_1 = \cos \psi_2 = 0$, only if the determinant Δ of the coefficients of $\cos \psi_1$ and $\cos \psi_2$ vanishes. The determinant is

$$\Delta = \begin{vmatrix} \omega_1 A_{11} & \omega_2 A_{12} \\ \mu_1 \omega_1 A_{11} & \mu_2 \omega_2 A_{12} \end{vmatrix} = \omega_1 \omega_2 A_{11} A_{12} (\mu_2 - \mu_1)$$

Equation (3-8) shows that neither ω_1 nor ω_2 is equal to zero. Equation (3-10) indicates that $\mu_1 \neq \mu_2$, and neither is zero. As shown in this problem, neither A_{11} nor A_{12} is zero. Hence the determinant cannot vanish, and $\cos \psi_1 = \cos \psi_2 = 0$ is the only solution; that is, $\psi_1 = n\pi/2$ and $\psi_2 = m\pi/2$, where n and m are odd integers. It can be shown that the choice of n and m other than 1 will not lead to new solutions. Substituting $\psi_1 = \psi_2 = \pi/2$ in Eq. (3-12), the displacements become

$$x_1 = A_{11} \cos \omega_1 t + A_{12} \cos \omega_2 t$$

$$x_2 = \mu_1 A_{11} \cos \omega_1 t + \mu_2 A_{12} \cos \omega_2 t$$

where $A_{11} = \dfrac{\mu_2}{\mu_2 - \mu_1}$, and $A_{12} = -\dfrac{\mu_1}{\mu_2 - \mu_1}$

Example 3. Referring to Fig. 3-1, assume $m_1 = m_2 = m$ and $k_1 = k_2 = k$. If the initial conditions are $x_1(0) = 1$, and $\dot{x}_1(0) = x_2(0) = \dot{x}_2(0) = 0$, determine the natural frequencies of the system and the displacements $x_1(t)$ and $x_2(t)$.

Solution: From Eq. (3-7) the natural frequencies are

$$\omega_1 = \sqrt{k/m}, \quad \text{and} \quad \omega_2 = \sqrt{3k/m}$$

Substituting these frequencies in Eq. (3-10), the amplitude ratios are $\mu_1 = 1$ and $\mu_2 = -1$. Using the expressions for $x_1(t)$ and $x_2(t)$ from the last example, the displacements are

$$x_1 = \frac{1}{2}\cos\sqrt{\frac{k}{m}}\,t + \frac{1}{2}\cos\sqrt{\frac{3k}{m}}\,t$$

$$x_2 = \frac{1}{2}\cos\sqrt{\frac{k}{m}}\,t - \frac{1}{2}\cos\sqrt{\frac{3k}{m}}\,t$$

These motions are plotted as shown in Fig. 3-3.

It is observed from these equations that, for the first mode of vibration, the two masses move in the same direction with equal

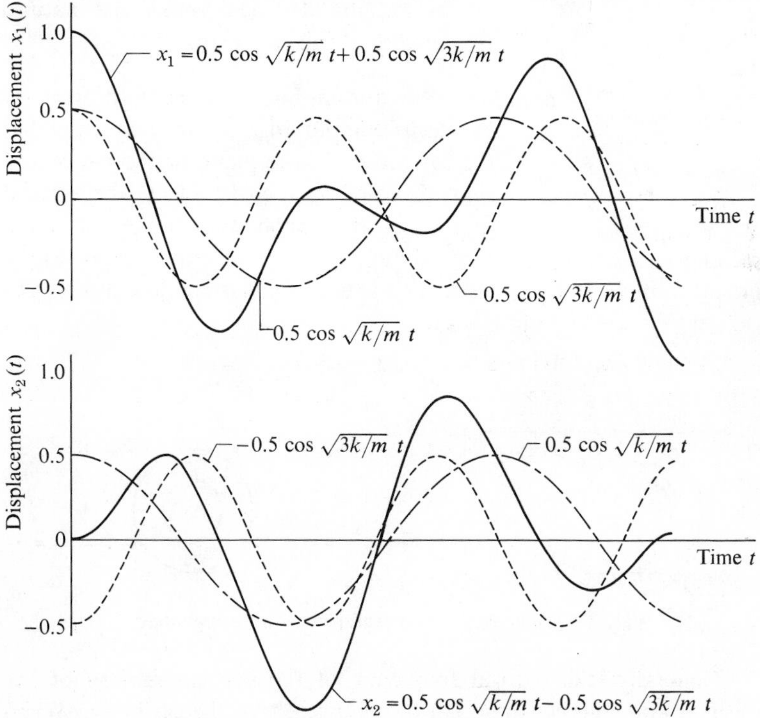

Fig. 3-3. *Curves showing the displacement of the masses of a two-degree-of-freedom system; Example 3*

amplitudes, leaving the coupling spring between them unstressed. Since the two masses behave as if they were not coupled to one another, the natural frequency of the first mode is $\sqrt{k/m}$. For the second mode, the masses move in opposite directions with equal amplitudes. Owing to symmetry, the mid-point of the coupling spring is stationary, and it is called a *node*. Each half of the system vibrates as if it were a single-degree-of-freedom system, as illustrated in Fig. 3-4. For this mode of vibration, the equivalent system consists of one of the original springs and half the coils of the coupling spring. Since only half of the number of coils of the coupling spring is considered, its spring constant is $2k$. The springs in the equivalent system are in parallel, and the natural frequency is $\sqrt{3k/m}$.

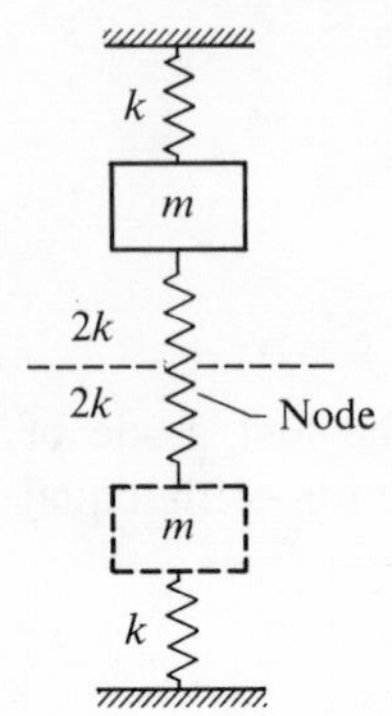

Fig. 3-4. *Second mode of vibration of a symmetrical system*

Example 4. Vehicle Suspension

Figure 3-5 is a schematic representation of an automobile. Determine the natural frequencies of the car body.

Solution: An automobile has many degrees of freedom. Simplifying, we may assume that the car moves in the plane of the paper and that the motions consist of (1) the vertical motion of the car body, (2) the rotational pitching motion of the body about its center of gravity, and (3) the vertical motion of the wheels. Even then, the system has more than two degrees of freedom.

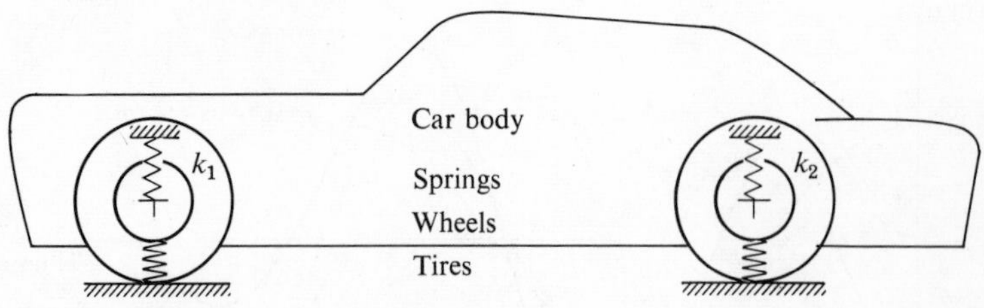

Fig. 3-5. *Schematic representation of an automobile*

Generally, the natural frequency of the vertical motion of the wheels is much higher than that of the motions of the car body. When the excitation frequency, owing to the road roughness, is high, the wheels will move up and down with great rapidity. Since this excitation frequency is much higher than the natural frequency of the car body, very little of this motion is transmitted to the car body. (See Case 4 in Sec. 2-12, and Fig. 2-8.) In other words, only the low frequency excitation is transmitted to the car body. Because of this large separation of natural frequencies between the wheels and the car body, the system representing the motions of the car body can be further simplified as shown in Fig. 3-6, which is a two-degree-of-freedom system.

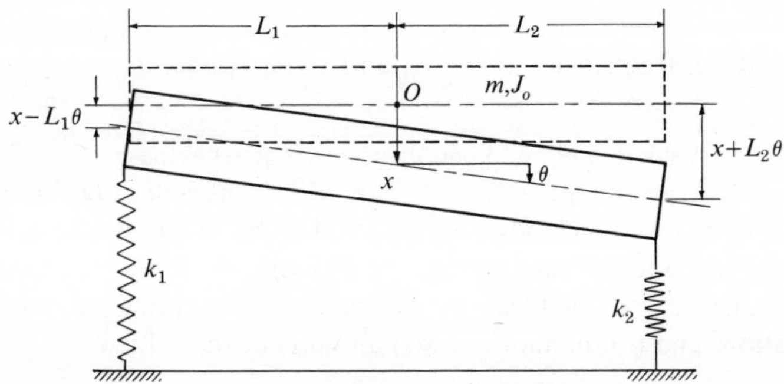

Fig. 3-6. *Simplified vibrating system of an automobile body*

Using the $x(t)$ and $\theta(t)$ coordinates to describe the motions of the car body and assuming small oscillations, the equations of motion are

$$m\ddot{x} = \sum(\text{forces in the } x\text{-direction})$$

$$m\ddot{x} = -k_1(x - L_1\theta) - k_2(x + L_2\theta)$$

and

$$J_o\ddot{\theta} = \sum(\text{moments about the center of gravity } o)$$

$$J_o\ddot{\theta} = +k_1(x - L_1\theta)L_1 - k_2(x + L_2\theta)L_2$$

Rearranging, we obtain

$$m\ddot{x} + (k_1 + k_2)x - (k_1L_1 - k_2L_2)\theta = 0$$

$$-(k_1L_1 - k_2L_2)x + J_o\ddot{\theta} + (k_1L_1^2 + k_2L_2^2)\theta = 0$$

which is of the same form as Eq. (3-1). The frequency equation is obtained by the direct application of Eq. (3-5).

$$\Delta(\omega) = \begin{vmatrix} k_1 + k_2 - m\omega^2 & k_2L_2 - k_1L_1 \\ k_2L_2 - k_1L_1 & k_1L_1^2 + k_2L_2^2 - J_o\omega^2 \end{vmatrix} = 0$$

Expanding the determinant, the roots of the frequency equation can be expressed as

$$\omega_{1,2}^2 = \frac{1}{2}\left[\frac{k_1 + k_2}{m} + \frac{k_1L_1^2 + k_2L_2^2}{J_o} \mp \sqrt{\left(\frac{k_1 + k_2}{m} + \frac{k_1L_1^2 + k_2L_2^2}{J_o}\right)^2 - \frac{4k_1k_2(L_1 + L_2)^2}{mJ_o}}\right]$$

The natural frequencies are $\frac{1}{2\pi}\omega_1$ and $\frac{1}{2\pi}\omega_2$ cps.

Example 5. A vehicle weighs 3,860 lb and has a wheelbase of 12 ft. The center of gravity (cg) is 5.25 ft from the front axle. The radius of gyration of the vehicle about the cg is 4.75 ft. If the spring constants of the front and the rear springs are 240 and 260 lb-in.$^{-1}$, respectively, determine (*a*) the natural frequencies, (*b*) the principal modes of vibration, and (*c*) the motions $x(t)$ and $\theta(t)$ of the vehicle.

Solution: (*a*) From the given data and the equations in Example 4, we have

$$\frac{k_1 + k_2}{m} = \frac{(500)(386)}{3{,}860} = 50,$$

$$\frac{k_1L_1 - k_2L_2}{m} = \frac{(15{,}130 - 21{,}050)(386)}{3{,}860} = -592$$

$$\frac{k_1L_1^2 + k_2L_2^2}{J_o} = \frac{(240)(3{,}970) + (260)(6{,}560)}{(3{,}860/386)(57)^2} = 83$$

$$\frac{4k_1k_2(L_1 + L_2)^2}{mJ_o} = \frac{4(240)(260)(144)^2}{(3{,}860/386)(3{,}860/386)(57)^2} = 15{,}900$$

$$\omega_{1,2}^2 = \tfrac{1}{2}\left[50 + 83 \mp \sqrt{(50 + 83)^2 - 15{,}900}\right] = \begin{cases} 45.3 \\ 87.6 \end{cases}$$

or

$$\omega_{1,2} = \begin{cases} 6.72 \text{ rad-sec}^{-1} = 1.07 \text{ cps} \\ 9.35 \text{ rad-sec}^{-1} = 1.49 \text{ cps} \end{cases}$$

The amplitude ratios for the two modes of vibration are

$$\frac{X}{\Theta} = \frac{(k_1L_1 - k_2L_2)/m}{(k_1 + k_2)/m - \omega_{1,2}^2} = \frac{-592}{50 - \omega_{1,2}^2} = \begin{cases} -126 \text{ in.-rad}^{-1} \\ 15.8 \text{ in.-rad}^{-1} \end{cases}$$

(*b*) The two principal modes of vibration are shown schematically in Fig. 3-7. For the first mode, the amplitude ratio is $X/\Theta = -126$ in.-rad^{-1}. This means that when $x(t)$ is positive $\theta(t)$ is negative from the assumed direction of rotation. Since $x(t)$ and $\theta(t)$ are both harmonic of the same frequency 6.72 rad-sec^{-1}, the ratio $x(t)/\theta(t)$ is also equal to -126 in.-rad^{-1}. When $x(t) = 1$ in., $\theta(t) = -\frac{1}{126}$ rad. Thus, the node is 126 in. from the center of gravity of the car body. The second mode of vibration can be interpreted similarly.

(*c*) The motions $x(t)$ and $\theta(t)$ can be expressed as

$$x = A_{11} \sin (6.72t + \psi_1) + A_{12} \sin (9.35t + \psi_2)$$

$$\theta = \frac{A_{11}}{-126} \sin (6.72t + \psi_1) + \frac{A_{12}}{15.8} \sin (9.35t + \psi_2)$$

where A_{11}, A_{12}, ψ_1, and ψ_2 are arbitrary constants to be evaluated by the initial conditions imposed on the system.

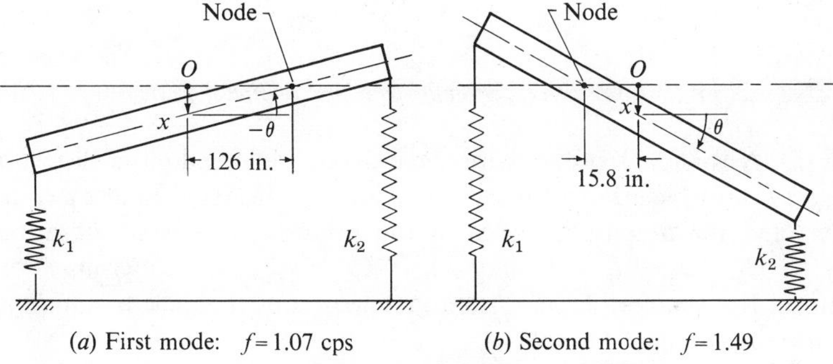

(*a*) First mode: $f = 1.07$ cps (*b*) Second mode: $f = 1.49$

Fig. 3-7. *Principal modes of vibration; Example* 4 (*diagrams not to scale*)

It should be noted that in the problems discussed in this section the motions described by the two coordinates are *elastically coupled* together. For example, in the vehicle-suspension problem, the motions in the x- and θ-directions are coupled through the quantity $(k_1L_1 - k_2L_2)$. This implies that the $x(t)$ and $\theta(t)$ motions are not independent of one another. If the system is given a displacement in the x-direction and then released with zero velocity, the car body will have a rectilinear motion in the x-direction as well as a rotational motion in the θ-direction. This is analogous to the spring-and-mass-coupled system of Fig. 3-1 in which the $x_1(t)$ and $x_2(t)$ motions are not independent of one another, although the coupling in the vehicle-suspension problem is less obvious. Coupled systems will be further discussed in Sec. 3-8.

The theory outlined in this section can easily be extended to include systems with many degrees of freedom. The procedure is to write as many differential equations of motion as there are independent coordinates and then solve the equations by the steps enumerated. It can be shown that a system has as many natural frequencies as there are degrees of freedom,† and the general motion described by each coordinate is a linear combination of its harmonic components. Correspondingly, there are as many principal modes of vibration as there are degrees of freedom. Let us illustrate the extension of the theory by means of an example with three degrees of freedom.

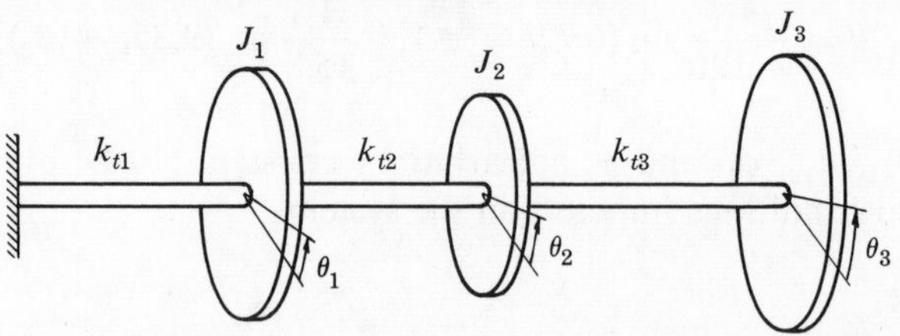

Fig. 3-8. *A three-degree-of-freedom torsional system*

Example 6. A torsional system with three degrees of freedom is shown in Fig. 3-8. (*a*) Determine the equations of motion, the frequency equation, and the amplitude ratios of the principal modes of vibration. (*b*) If $J_1 = J_2 = J_3 = J$ and $k_{t1} = k_{t2} = k_{t3} = k_t$, determine the natural frequencies of the system and write out the general solutions of the equations of motion.

Solution: (*a*) By Newton's law of motion, the equations of motion are

$$\begin{aligned} J_1\ddot{\theta}_1 &= -k_{t1}\theta_1 - k_{t2}(\theta_1 - \theta_2) \\ J_2\ddot{\theta}_2 &= -k_{t2}(\theta_2 - \theta_1) - k_{t3}(\theta_2 - \theta_3) \\ J_3\ddot{\theta}_3 &= -k_{t3}(\theta_3 - \theta_2) \end{aligned} \tag{3-15}$$

Substituting $\theta_1 = \Theta_1 \sin(\omega t + \psi)$, $\theta_2 = \Theta_2 \sin(\omega t + \psi)$, and $\theta_3 = \Theta_3 \sin(\omega t + \psi)$ in Eq. (3-15), factoring out the term $\sin(\omega t + \psi)$, and rearranging, we have

$$\begin{aligned} (k_{t1} + k_{t2} - J_1\omega^2)\Theta_1 - k_{t2}\Theta_2 &= 0 \\ -k_{t2}\Theta_1 + (k_{t2} + k_{t3} - J_2\omega^2)\Theta_2 - k_{t3}\Theta_3 &= 0 \\ -k_{t3}\Theta_2 + (k_{t3} - J_3\omega^2)\Theta_3 &= 0 \end{aligned} \tag{3-16}$$

† The frequency equation of an n-degree-of-freedom system can be obtained by following the procedure outlined in this section. Substituting $x_i = A_i \sin(\omega t + \psi)$, and equating the determinant of the coefficients of A's to zero, we obtain an nth order algebraic equation in ω^2. The results follow.

The frequency equation is obtained by equating the determinant $\Delta(\omega)$ of the coefficients of Θ_1, Θ_2, and Θ_3 to zero, that is,

$$\Delta(\omega) = \begin{vmatrix} k_{t1} + k_{t2} - J_1\omega^2 & -k_{t2} & 0 \\ -k_{t2} & k_{t2} + k_{t3} - J_2\omega^2 & -k_{t3} \\ 0 & -k_{t3} & k_{t3} - J_3\omega^2 \end{vmatrix} = 0 \quad \textbf{(3-17)}$$

The amplitude ratios of the principal modes can be obtained from Eq. (3-16) with the appropriate value of ω.

$$\frac{\Theta_1}{\Theta_2} = \frac{k_{t2}}{k_{t1} + k_{t2} - J_1\omega^2} \quad \text{and} \quad \frac{\Theta_2}{\Theta_3} = \frac{k_{t3} - J_3\omega^2}{k_{t3}} \quad \textbf{(3-18)}$$

(*b*) If $J_1 = J_2 = J_3 = J$ and $k_{t1} = k_{t2} = k_{t3} = k_t$, the frequency equation becomes

$$\omega^6 - 5\frac{k_t}{J}\omega^4 + 6\left(\frac{k_t}{J}\right)^2\omega^2 - \left(\frac{k_t}{J}\right)^3 = 0$$

This is a cubic equation in ω^2, the roots of which are $0.198k_t/J$, $1.555k_t/J$, and $3.247k_t/J$. Hence the natural frequencies are

$$f_1 = \frac{1}{2\pi}\sqrt{0.198\frac{k_t}{J}} = 0.078\sqrt{\frac{k_t}{J}}, \text{ cps}$$

$$f_2 = \frac{1}{2\pi}\sqrt{1.555\frac{k_t}{J}} = 0.198\sqrt{\frac{k_t}{J}}, \text{ cps}$$

$$f_3 = \frac{1}{2\pi}\sqrt{3.244\frac{k_t}{J}} = 0.286\sqrt{\frac{k_t}{J}}, \text{ cps}$$

From Eq. (3-18), the amplitude ratios for the first mode of vibration are

$$\frac{\Theta_{11}}{\Theta_{21}} = \frac{k_t/J}{2k_t/J - \omega_1^2} = \frac{1}{2 - 0.198} = \frac{1}{1.802}$$

$$\frac{\Theta_{21}}{\Theta_{31}} = \frac{k_t/J - \omega_1^2}{k_t/J} = \frac{1 - 0.198}{1} = \frac{1.802}{2.25}$$

Similarly, the amplitude ratios for the second and third modes are

$$\Theta_1 : \Theta_2 : \Theta_3 = \begin{cases} 1: \ \ 0.445 : -0.802, & \text{for } \omega^2 = 1.555k_t/J \\ 1: -1.247 : \ \ 0.555, & \text{for } \omega^2 = 3.244k_t/J \end{cases}$$

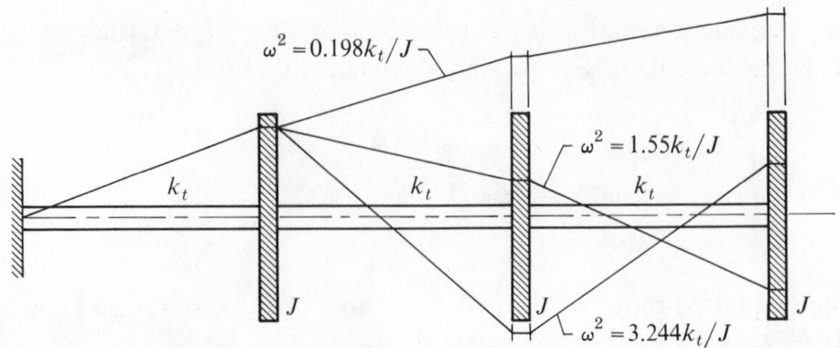

Fig. 3-9. *Principal modes of vibration; amplitude of vibration plotted perpendicular to axis of rotation*

The principal modes of vibration are illustrated in Fig. 3-9.

From the values calculated, the general solution to the equations of motion can be expressed as

$$\theta_1 = \Theta_{11} \sin(\omega_1 t + \psi_1) + \Theta_{12} \sin(\omega_2 t + \psi_2) + \Theta_{13} \sin(\omega_3 t + \psi_3)$$

$$\theta_2 = 1.802\Theta_{11} \sin(\omega_1 t + \psi_1) + 0.445\Theta_{12} \sin(\omega_2 t + \psi_2) - 1.247\Theta_{13} \sin(\omega_3 t + \psi_3)$$

$$\theta_3 = 2.25\Theta_{11} \sin(\omega_1 t + \psi_1) - 0.802\Theta_{12} \sin(\omega_3 t + \psi_2) + 0.555\Theta_{13} \sin(\omega_3 t + \psi_3)$$

where ω_1, ω_2, and ω_3 are $\sqrt{0.198k_t/J}$, $\sqrt{1.555k_t/J}$ and $\sqrt{3.244k_t/J}$, respectively. The constants Θ_{11}, Θ_{12}, Θ_{13}, ψ_1, ψ_2, and ψ_3 are to be determined by the initial conditions.

3-3. SEMIDEFINITE SYSTEMS: A SPECIAL CASE

A special case of practical importance occurs when one of the roots of the frequency equation vanishes. Thus one of the natural frequencies is equal to zero, and the system may move as a rigid body without disturbing the forces acting upon it. This class of systems represents a large group of problems and is called *semidefinite* systems.

Two semidefinite systems are shown in Fig. 3-10. The rectilinear system consists of a number of masses coupled together by springs. This system can be used to study the vibration of a train or an airplane hauling a glider. The rotational system consists of a number of disks coupled together by torsional shafts. This system has its counterpart in many rotating machines, such as a diesel engine used for marine propulsion. One of the disks may be used to represent the propeller,

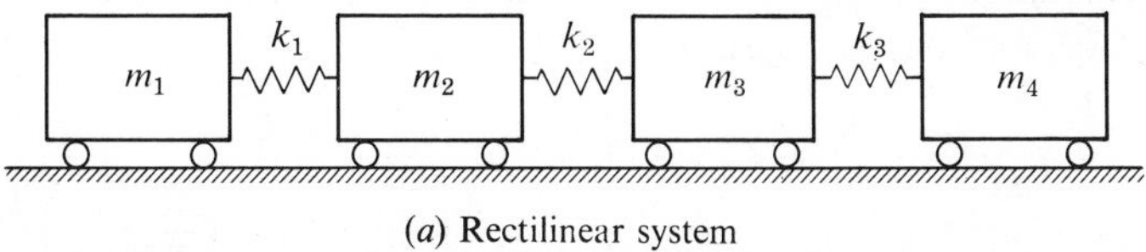

(a) Rectilinear system

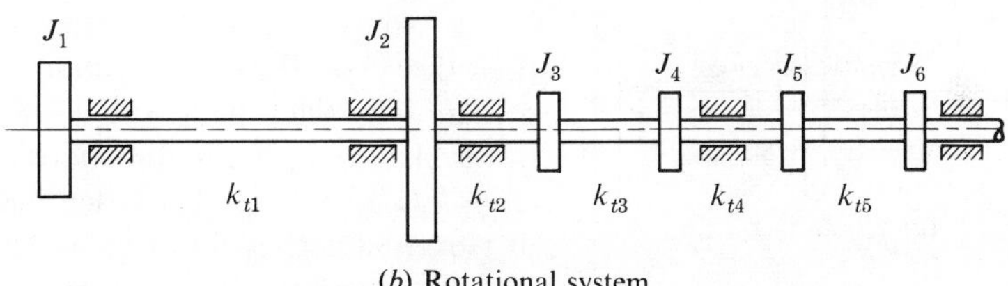

(b) Rotational system

Fig. 3-10. *Examples of semidefinite system*

another disk the flywheel, and the remaining disks may represent the rotating and the equivalent reciprocating parts of the engine.

As an example, consider a two-disk system as shown in Fig. 3-11, which may represent an electrical motor-generator set. Summing the torques about the longitudinal axis of the shafts, the equations of motion of the disks are

$$J_1\ddot{\theta}_1 = -k_t(\theta_1 - \theta_2)$$
$$J_2\ddot{\theta}_2 = -k_t(\theta_2 - \theta_1)$$

Rearranging, we obtain

$$\begin{aligned} J_1\ddot{\theta}_1 + k_t\theta_1 - k_t\theta_2 &= 0 \\ -k_t\theta_1 + J_2\ddot{\theta}_2 + k_t\theta_2 &= 0 \end{aligned} \tag{3-19}$$

This set of equations is of the same form as Eq. (3-1). Hence the frequency equation, Eq. (3-5), can be expressed as

$$\Delta(\omega) = \begin{vmatrix} k_t - J_1\omega^2 & -k_t \\ -k_t & k_t - J_2\omega^2 \end{vmatrix} = 0 \tag{3-20}$$

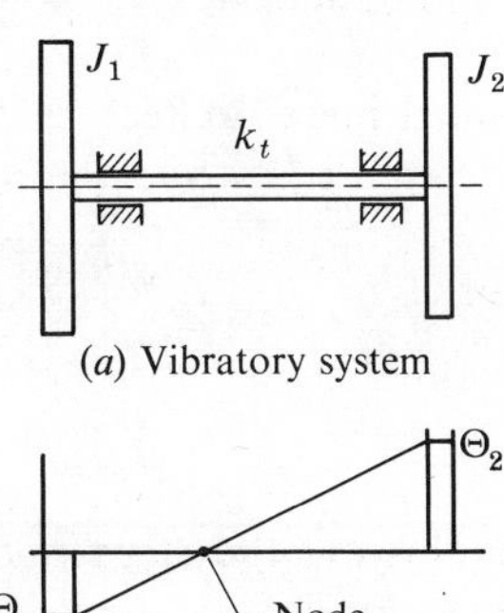

(a) Vibratory system

(b) Principal mode of vibration

Fig. 3-11. *A two-disk semidefinite system*

Expanding the determinant and dividing through by J_1J_2, we obtain

$$\omega^2[\omega^2 - (k_t/J_1 + k_t/J_2)] = 0 \tag{3-21}$$

The roots of this equation are $\omega^2 = 0$ and $\omega^2 = k_t(1/J_1 + 1/J_2)$.

The amplitude ratios of the principal modes can be obtained from Eq. (3-19) and be expressed as

$$\frac{\Theta_1}{\Theta_2} = \frac{k_t}{k_t - J_1\omega^2} = \frac{k_t - J_2\omega^2}{k_t} = \begin{cases} 1, & \text{for } \omega^2 = 0 \\ -\dfrac{J_2}{J_1}, & \text{for } \omega^2 = k_t\left(\dfrac{1}{J_1} + \dfrac{1}{J_2}\right) \end{cases} \tag{3-22}$$

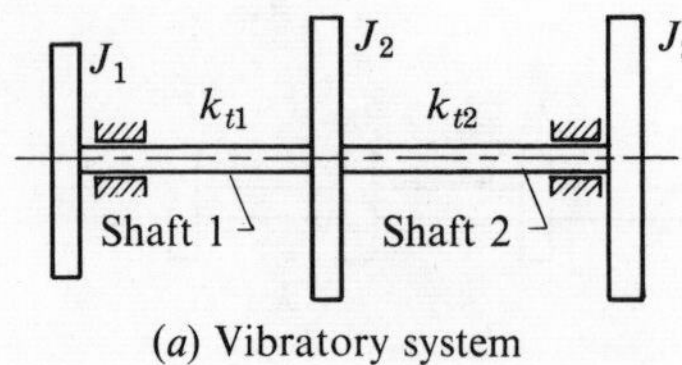

(*a*) Vibratory system

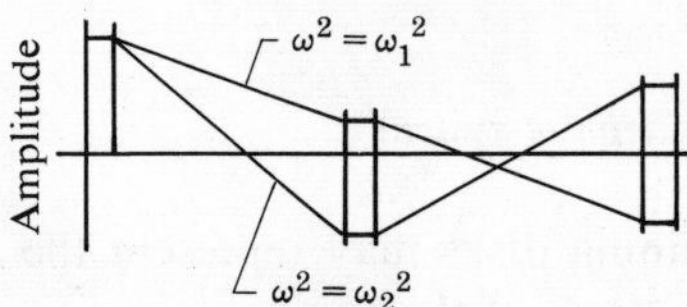

(*b*) Principal modes of vibration: $\omega_1^2 < \omega_2^2$

Fig. 3-12. *A three-disk semidefinite system*

These amplitude ratios indicate that the two disks may rotate together as a rigid body for $\omega^2 = 0$, or oscillate in opposite directions for $\omega^2 = k_t(1/J_1 + 1/J_2)$. The latter mode of vibration is shown in Fig. 3-11.

To extend the theory to systems with more than two degrees of freedom, consider the three-disk assembly of Fig. 3-12(*a*). This system could be used to represent, to the first approximation, a multidegree-of-freedom system, such as a diesel engine for marine propulsion. The disks J_1 and J_2 could represent the propeller and the flywheel, and the disk J_3 the combined inertial effect of the rotating and the reciprocating parts. From Newton's law of motion, the equations of motion of the disks are

$$\begin{aligned} J_1\ddot{\theta}_1 &= -k_{t1}(\theta_1 - \theta_2) \\ J_2\ddot{\theta}_2 &= -k_{t1}(\theta_2 - \theta_1) - k_{t2}(\theta_2 - \theta_3) \\ J_3\ddot{\theta}_3 &= -k_{t2}(\theta_3 - \theta_2) \end{aligned} \tag{3-23}$$

Following the steps outlined for the two-disk system, the frequency equation is

$$\omega^2\left\{\omega^4 - \left[k_{t1}\left(\frac{1}{J_1} + \frac{1}{J_2}\right) + k_{t2}\left(\frac{1}{J_2} + \frac{1}{J_3}\right)\right]\omega^2 + k_{t1}k_{t2}\frac{J_1 + J_2 + J_3}{J_1J_2J_3}\right\} = 0 \tag{3-24}$$

The amplitude ratios of the principal modes of vibration can be obtained from Eq. (3-23) and be expressed as

$$\frac{\Theta_1}{\Theta_2} = \frac{k_{t1}}{k_{t1} - J_1\omega^2} \quad \text{and} \quad \frac{\Theta_2}{\Theta_3} = \frac{k_{t2} - J_3\omega^2}{k_{t2}} \tag{3-25}$$

When $\omega^2 = 0$, the amplitude ratios of the disks are

$$\frac{\Theta_1}{\Theta_2} = \frac{\Theta_2}{\Theta_3} = 1$$

This indicates that the whole assembly may rotate as a rigid body. Let ω_1^2 and ω_2^2 be the other roots of the frequency equation. If ω_1^2 is the smaller root of Eq. (3-24), one of these amplitude ratios in Eq. (3-25) will be positive while the other will be negative. This indicates that two of the adjacent disks rotate in one direction while the third rotates in the opposite direction. It can be shown that if J_1/k_{t1} is less than J_3/k_{t2}, the amplitude ratio Θ_1/Θ_2 will be positive while Θ_2/Θ_3 will be negative. If ω_2^2 is the larger root, both of the amplitude ratios are negative. This means that all the adjacent disks rotate in opposition to one another. The principal modes of vibration of the system are shown in Fig. 3-12(*b*).

Example 7. A rotating machine is schematically represented by the three-disk system of Fig. 3-12(*a*). Determine the length of shaft 2 if the fundamental frequency of the system is equal to 6 cps.

Data: Shaft 1: diameter = 4 in., length = 3 ft

Shaft 2: diameter = 4 in.

Disks 1 and 2: weight = 2 tons each, radius of gyration = 2 ft

Disk 3: weight = 1 ton, radius of gyration = 2 ft

Solution:

$$J_1 = J_2 = \frac{2(2{,}000)(24)^2}{386} = 5{,}970 \text{ lb-sec}^2\text{-in.}$$

$$J_3 = 2{,}985 \text{ lb-sec}^2\text{-in.}$$

$$k_{t1} = \frac{\pi}{32}\frac{Gd^4}{L} = \frac{\pi}{32}\frac{(12 \times 10^6)(4)^4}{(3 \times 12)} = 8.38 \times 10^6 \text{ in.-lb-rad}^{-1}$$

$$\omega = 2\pi \times 6 \text{ rad-sec}^{-1}$$

Substituting these values in Eq. (3-24), we obtain

$$(12\pi)^4 - [(8.38 \times 10^6)(2/5{,}970) + k_{t2}(3/5{,}970)]\,(12\pi)^2 + k_{t2}(8.38 \times 10^6)(5/5{,}970^2) = 0$$

or

$$k_{t2} = 4.28 \times 10^6 \text{ in.-lb-rad}^{-1}$$

The corresponding length of shaft 2 is 70.1 in.

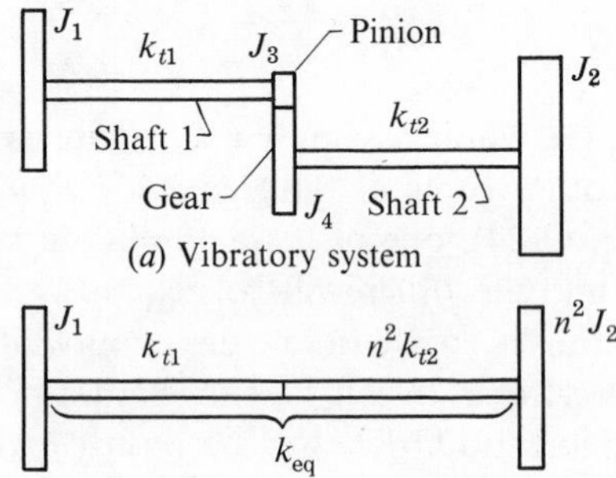

(*a*) Vibratory system

(*b*) Equivalent system, referring to shaft 1 and neglecting inertial effect of gears

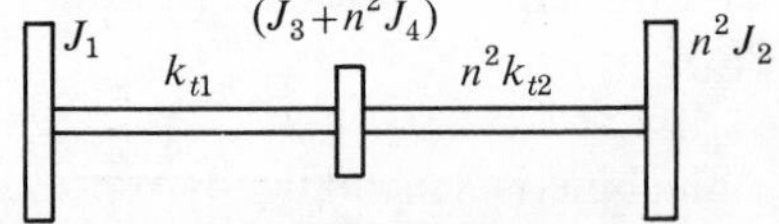

(*c*) Equivalent system, referring to shaft 1 and including inertial effect of gears

Fig. 3-13. *A two-inertial-disk system connected by shafts and gears*

Example 8. *Geared Systems*

Let a two-disk system, as shown in Fig. 3-13(*a*), be connected by shafts and a pair of gears. (*a*) Neglecting the inertial effect of the gears, determine the natural frequencies of the system. (*b*) Repeat (*a*) but include the inertial effect of the gears.

Solution: Since the two shafts are at different rotational speeds, it is more expedient to find an equivalent system in which both parts of the system refer to a common shaft. Let us choose shaft 1 as reference.

The inertia torque of disk J_2 is $T_2 = J_2\ddot{\theta}_2$; that is, $J_2 = T_2/\ddot{\theta}_2$. Let N_1 be the number of teeth on the pinion and N_2 be the number of teeth on the gear of shaft 2. Let the gear ratio $n = N_1/N_2$. Referring to shaft 1, the equivalent inertia torque of J_2 is $T_2^* = nT_2$. Similarly, the equivalent acceleration of J_2 is $\ddot{\theta}_2^* = \ddot{\theta}_2/n$. Hence the equivalent inertia of J_2 is

$$J_2^* = T_2^*/\ddot{\theta}_2^* = n^2T_2/\ddot{\theta}_2 = n^2J_2$$

Similarly, it can be shown that the equivalent spring constant k_{t2}^* of shaft 2, when referred to shaft 1, is

$$k_{t2}^* = n^2k_{t2}$$

(*a*) Neglecting the inertial effect of the gears, the equivalent system, when referred to the rotation of shaft 1, is as shown in Fig. 3-13(*b*). Since the two shafts k_{t1} and k_{t2}^* are in series, the equivalent torsional spring constant is

$$\frac{1}{k_{\text{eq}}} = \frac{1}{k_{t1}} + \frac{1}{k_{t2}^*}$$

or

$$k_{\text{eq}} = \frac{k_{t1}k_{t2}^*}{k_{t1} + k_{t2}^*} = \frac{n^2k_{t1}k_{t2}}{k_{t1} + n^2k_{t2}}$$

Since the equivalent system consists of two disks connected by a torsional shaft, the natural frequencies can be determined from Eq. (3-20). These frequencies are

$$f_n = \begin{cases} 0 \\ \dfrac{1}{2\pi}\sqrt{k_{\text{eq}}\left(\dfrac{1}{J_1} + \dfrac{1}{n^2J_2}\right)} \end{cases}$$

(*b*) Assuming that J_3 is the inertia of the pinion and J_4 that of the gear, the equivalent inertia of the pinion and the gear referred to shaft 1 is $(J_3 + n^2J_4)$. Thus the equivalent system consists of three disks and two torsional shafts as shown in Fig. 3-13(*c*). If the values of the parameters are known, the natural frequencies of the system can be calculated by the use of Eq. (3-24).

3-4. STEADY-STATE UNDAMPED FORCED VIBRATION

Consider again the system shown in Fig. 3-1 and assume that vertical excitation forces are applied to the masses. Since the system is linear, we can make use of the superposition principle. When excitation forces are applied to both masses, the resultant motions of the masses will be the linear combination of the motions due to each of the forces acting alone. Hence it is sufficient to consider that only one excitation force is being applied to the system. Let $F \sin \omega t$ be applied to the mass m_1. From Newton's law of motion, the equations of motion of the masses are

$$\begin{aligned} m_1\ddot{x}_1 &= -k_1x_1 - k(x_1 - x_2) + F \sin \omega t \\ m_2\ddot{x}_2 &= -k_2x_2 - k(x_2 - x_1) \end{aligned} \tag{3-26}$$

Rearranging these equations, we have

$$\begin{aligned} m_1\ddot{x}_1 + (k_1 + k)x_1 - kx_2 &= F \sin \omega t \\ -kx_1 + m_2\ddot{x}_2 + (k_2 + k)x_2 &= 0 \end{aligned} \tag{3-27}$$

The steady-state solution can be obtained readily by the mechanical impedance method. (See Sec. 2-6.) Let us substitute $Fe^{j\omega t}$ for $F \sin \omega t$, $\bar{X}_1e^{j\omega t}$ for $x_1(t)$, and $\bar{X}_2e^{j\omega t}$ for $x_2(t)$ in Eq. (3-27), where $\bar{X}_1$ and $\bar{X}_2$ are the complex amplitudes of $x_1(t)$ and $x_2(t)$, respectively. Rearranging and factoring out $e^{j\omega t}$, we obtain

$$\begin{aligned} (k_1 + k - m_1\omega^2)\bar{X}_1 - k\bar{X}_2 &= F \\ -k\bar{X}_1 + (k_2 + k - m_2\omega^2)\bar{X}_2 &= 0 \end{aligned} \tag{3-28}$$

$\bar{X}_1$ and $\bar{X}_2$ can now be obtained from this equation by Cramer's rule and can be expressed as

$$\begin{aligned} \bar{X}_1 &= \frac{\begin{vmatrix} F & -k \\ 0 & (k_2 + k - m_2\omega^2) \end{vmatrix}}{\Delta(\omega)} = \frac{(k_2 + k - m_2\omega^2)F}{\Delta(\omega)} = X_1e^{-j\phi_1} \\ \bar{X}_2 &= \frac{\begin{vmatrix} (k_1 + k - m_1\omega^2) & F \\ -k & 0 \end{vmatrix}}{\Delta(\omega)} = \frac{kF}{\Delta(\omega)} = X_2e^{-j\phi_2} \end{aligned} \tag{3-29}$$

where

$$\Delta(\omega) = \begin{vmatrix} k_1 + k - m_1\omega^2 & -k \\ -k & k_2 + k - m_2\omega^2 \end{vmatrix} \tag{3-30}$$

X_1, X_2, ϕ_1, and ϕ_2 are the amplitudes and phase angles of the displacements $x_1(t)$ and $x_2(t)$.

Since the excitation force is $F \sin \omega t = \text{Im}[Fe^{j\omega t}]$, the steady-state responses of the masses are

$$\begin{aligned} x_1 &= \text{Im}[\bar{X}_1e^{j\omega t}] = \text{Im}[X_1e^{-j\phi_1}e^{j\omega t}] = X_1 \sin(\omega t - \phi_1) \\ x_2 &= \text{Im}[\bar{X}_2e^{j\omega t}] = \text{Im}[X_2e^{-j\phi_2}e^{j\omega t}] = X_2 \sin(\omega t - \phi_2) \end{aligned} \tag{3-31}$$

It is noted that all the quantities in Eqs. (3-29) and (3-30) are real. Hence the phase angles ϕ_1 and ϕ_2 may assume only the values of either zero or 180 deg. In other words, $x_1(t)$ and $x_2(t)$ may be either in phase or 180 deg out of phase with the excitation. For a single-degree-of-freedom system, it was shown in Chap. 2 that the phase angle of undamped forced vibrations can be only zero or 180 deg. It is seen here that this is also true for multidegree-of-freedom systems.

The phase angles, as indicated in Eq. (3-29), may be examined by expressing the equations as

$$\bar{X}_1 = \frac{\left(\dfrac{k_2 + k}{m_2} - \omega^2\right) F}{m_1(\omega^2 - \omega_1^2)(\omega^2 - \omega_2^2)} = X_1 e^{-j\phi_1}$$
$$\bar{X}_2 = \frac{kF}{m_1 m_2(\omega^2 - \omega_1^2)(\omega^2 - \omega_2^2)} = X_2 e^{-j\phi_2} \qquad \textbf{(3-32)}$$

where ω_1^2 and ω_2^2 are the roots of the characteristic equation $\Delta(\omega) = 0$. For the system considered, these roots are indicated in Eq. (3-7). Hence ω_1 and ω_2 are also the natural frequencies of the system. It can be shown by simple algebraic operations that $\omega_1^2 < (k_2 + k)/m_2 < \omega_2^2$. Hence, if $\omega^2 < \omega_1^2$, both of the ratios in Eq. (3-32) are positive. Consequently, both ϕ_1 and ϕ_2 are zero. If $\omega_1^2 < \omega^2 < (k_2 + k)/m_2$, both of these ratios are negative. Consequently, both ϕ_1 and ϕ_2 are 180 deg. The remaining phase relations of $x_1(t)$ and $x_2(t)$ for other frequency ranges can be deduced in the same manner.

When the excitation frequency is equal to one of the natural frequencies of the system, resonance occurs, and the amplitudes of $x_1(t)$ and $x_2(t)$ become infinite. Since this system has two natural frequencies, there are two excitation frequencies at which the amplitudes of both $x_1(t)$ and $x_2(t)$ become infinite.

It is noted in Eq. (3-32) that when $\omega^2 = (k_2 + k)/m_2$, the amplitude of $x_1(t)$ is zero. The amplitude of $x_2(t)$, however, is never zero for a finite value of ω. When numerical values are substituted in Eq. (3-31), curves similar to those shown in Fig. 3-14 are obtained. The dashed portions of the curves indicate that the displacements of the masses are 180 deg out of phase with the excitation force.

Example 9. *Dynamic Absorber*

Figure 3-15(*a*) is a schematic sketch of a spring-mounted machine whose operating frequency and the natural frequency of the system are nearly equal, and the force transmitted to the foundation is excessive. (*a*) Show that a mass m_2 attached to the base of the machine by spring

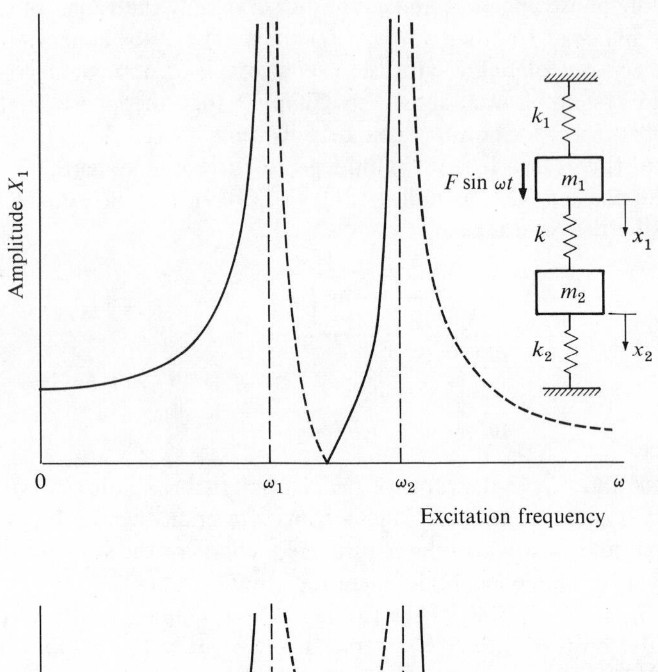

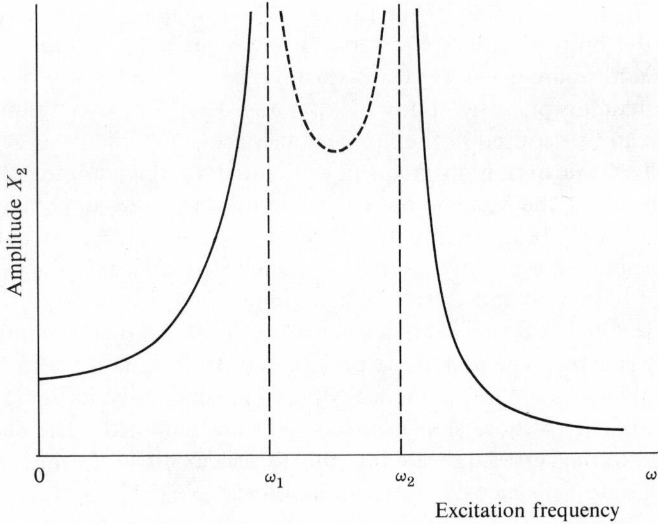

Fig. 3-14. *Steady-state response of a two-degree-of-freedom system*

k_2 will act as a dynamic absorber to remedy the situation. (*b*) Indicate the effect of the size of the absorber mass m_2.

Solution: (*a*) The equivalent system with the dynamic absorber mounted is as shown in Fig. 3-15(*b*). Summing the forces in the vertical direction, we obtain

$$m_1\ddot{x}_1 = -k_1x_1 - k_2(x_1 - x_2) + F\sin\omega t$$
$$m_2\ddot{x}_2 = -k_2(x_2 - x_1)$$

Rearranging yields

$$m_1\ddot{x}_1 + (k_1 + k_2)x_1 - k_2x_2 = F\sin\omega t$$
$$-k_2x_1 + m_2\ddot{x}_2 + k_2x_2 = 0$$

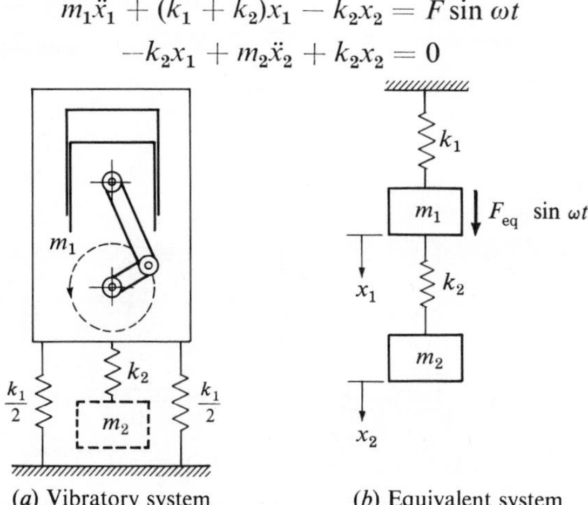

(*a*) Vibratory system (*b*) Equivalent system

Fig. 3-15. *Dynamic absorber*

These equations of motion are of the same form as Eq. (3-27). Hence the procedure outlined previously can be used to solve this problem. The steady-state responses of the masses are

$$x_1 = X_1 \sin(\omega t - \phi_1)$$
$$x_2 = X_2 \sin(\omega t - \phi_2)$$

where

$$X_1 = |\bar{X}_1| = \left|\frac{(k_2 - m_2\omega^2)F}{\Delta(\omega)}\right|, \quad X_2 = |\bar{X}_2| = \left|\frac{k_2F}{\Delta(\omega)}\right|,$$

and

$$\Delta(\omega) = (k_1 + k_2 - m_1\omega^2)(k_2 - m_2\omega^2) - k_2^2$$

The phase angles ϕ_1 and ϕ_2 are either zero or 180 deg.

These equations indicate that when $(k_2 - m_2\omega^2) = 0$, $\bar{X}_1 = 0$ and $\bar{X}_2 = -F/k_2$. Clearly, when $x_1(t) = 0$, the force transmitted to the foundation is nil. The force transmitted through the spring k_2 to the base of the machine is $k_2x_2 = -F \sin \omega t$. Physically, this means that the motion of m_2 is 180 deg out of phase with the excitation force, and the force due to the deformation of the spring k_2 is equal and opposite to the excitation force.

(*b*) The purpose of the dynamic absorber is to minimize the vibrations of the original system when operating at a frequency nearly

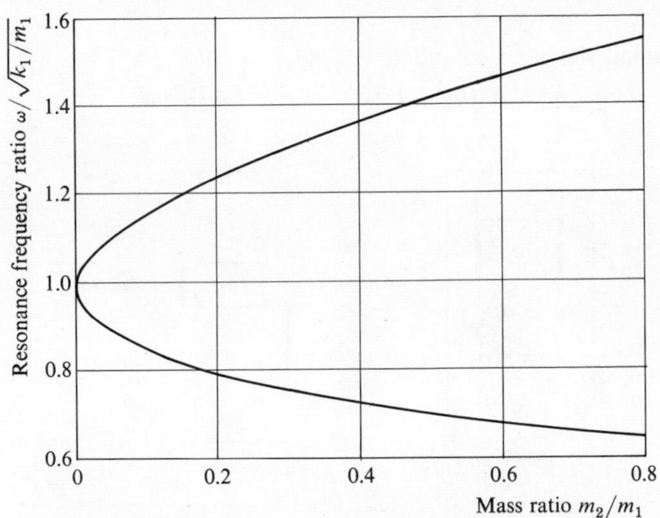

Fig. 3-16. *Effect of size of absorber mass m_2 on resonance; system shown in Fig. 3-15. Curve is plotted for $k_1/m_1 = k_2/m_2$*

equal to $\sqrt{k_1/m_1}$. It was shown that when the excitation frequency ω is equal to $\sqrt{k_2/m_2}$, the amplitude of $x_1(t)$ is zero. Hence an undamped dynamic absorber is generally tuned so that $k_1/m_1 = k_2/m_2$.†

The frequency equation of the system is

$$(k_1 + k_2 - m_1\omega^2)(k_2 - m_2\omega^2) - k_2^2 = 0$$

which can be rearranged as

$$\frac{m_1 m_2}{k_1 k_2}\,\omega^4 - \left[\left(1 + \frac{k_2}{k_1}\right)\frac{m_2}{k_2} + \frac{m_1}{k_1}\right]\omega^2 + 1 = 0$$

Assume that the dynamic absorber is "tuned" for $k_1/m_1 = k_2/m_2$. Defining

$$\omega/\sqrt{k_1/m_1} = \omega/\sqrt{k_2/m_2} = r$$

† This applies only to an undamped dynamic absorber.

and equating

$$k_2/k_1 = m_2/m_1$$

we obtain

$$r^4 - \left(2 + \frac{m_2}{m_1}\right) r^2 + 1 = 0$$

The resonance frequencies of the "tuned" system can be determined from the roots of this equation with the mass ratio m_2/m_1 as a parameter. Since there are two resonance frequencies, the steady-state response of the dynamic absorber is similar to that shown in Fig. 3-14. The frequency ratio r versus the mass ratio m_2/m_1 is plotted as shown in Fig. 3-16.

The curve in Fig. 3-16 shows that the effect of the size of the absorber mass m_2 is to change the range of the resonance frequencies. When m_2/m_1 is very small, the absorber mass has very little effect, and the resonance frequencies are close to those of the original system. When m_2/m_1 is appreciable, the resonance frequencies are separated. For example, when $m_2/m_1 = 0.4$, the resonance frequency ratios are, 0.73 and 1.36; that is, resonance occurs at frequencies 0.732 and 1.36 times those of the original system.

To extend the theory presented in this section to systems with more than two degrees of freedom, let us consider the three-mass system of Fig. 3-17 in which the masses are attached to a taut string. It is assumed that (1) the string is of negligible mass, (2) the tension S of the string is constant, and (3) the masses are constrained to vibrate vertically in the plane of the paper.

Resolving the forces in the vertical direction and assuming small oscillations, the equations of motion are

$$m_1\ddot{x}_1 = -S\frac{x_1}{L_1} - S\frac{x_1 - x_2}{L_2} + F\sin\omega t$$

$$m_2\ddot{x}_2 = -S\frac{x_2 - x_1}{L_2} - S\frac{x_2 - x_3}{L_3}$$

$$m_3\ddot{x}_3 = -S\frac{x_3 - x_2}{L_3} - S\frac{x_3}{L_4}$$

Rearranging these equations, we obtain

$$m_1\ddot{x}_1 + S\left(\frac{1}{L_1} + \frac{1}{L_2}\right)x_1 - S\left(\frac{1}{L_2}\right)x_2 = F\sin\omega t$$

$$-S\left(\frac{1}{L_2}\right)x_1 + m_2\ddot{x}_2 + S\left(\frac{1}{L_2} + \frac{1}{L_3}\right)x_2 - S\left(\frac{1}{L_3}\right)x_3 = 0 \qquad \textbf{(3-33)}$$

$$-S\left(\frac{1}{L_3}\right)x_2 + m_3\ddot{x}_3 + S\left(\frac{1}{L_3} + \frac{1}{L_4}\right)x_3 = 0$$

Using the mechanical impedance method to solve for the steady-state response, we substitute $Fe^{j\omega t}$ for $F \sin \omega t$ and $\bar{X}e^{j\omega t}$ for the displacement $x(t)$, where $\bar{X}$ is a complex amplitude. Rearranging and factoring out the term $e^{j\omega t}$, we obtain

$$\begin{aligned} [S(1/L_1 + 1/L_2) - m_1\omega^2]\bar{X}_1 - S(1/L_2)\bar{X}_2 &= F \\ -S(1/L_2)\bar{X}_1 + [S(1/L_2 + 1/L_3) - m_2\omega^2]\bar{X}_2 - S(1/L_3)\bar{X}_3 &= 0 \\ -S(1/L_3)\bar{X}_2 + [S(1/L_3 + 1/L_4) - m_3\omega^2]\bar{X}_3 &= 0 \end{aligned} \qquad (3\text{-}34)$$

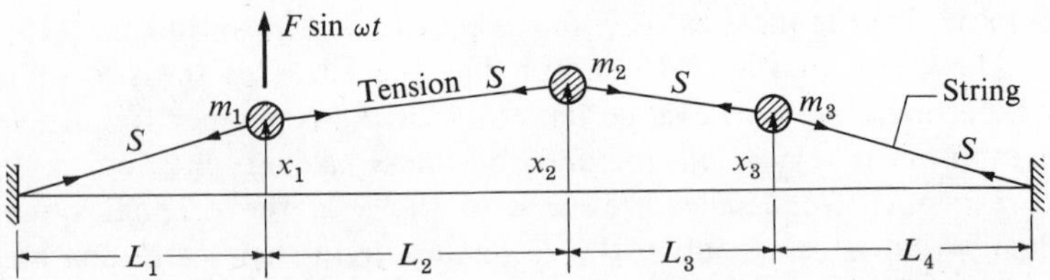

Fig. 3-17. *Forced vibration of a three-degree-of-freedom system*

The complex amplitudes $\bar{X}_1$, $\bar{X}_2$, and $\bar{X}_3$ can now be obtained by Cramer's rule, as indicated in Eq. (3-29). The frequency equation of the system is

$$\Delta(\omega) = \begin{vmatrix} [S(1/L_1 + 1/L_2) - m_1\omega^2] & -S(1/L_2) & 0 \\ -S(1/L_2) & [S(1/L_2 + 1/L_3) - m_2\omega^2] & -S(1/L_3) \\ 0 & -S(1/L_3) & [S(1/L_3 + 1/L_4) - m_3\omega^2] \end{vmatrix} = 0 \qquad (3\text{-}35)$$

Since the excitation force is $F \sin \omega t = \text{Im}[Fe^{j\omega t}]$, the steady-state responses of the masses are

$$\begin{aligned} x_1 &= \text{Im}[\bar{X}_1 e^{j\omega t}] = \text{Im}[X_1 e^{-j\phi_1} e^{j\omega t}] = X_1 \sin(\omega t - \phi_1) \\ x_2 &= \text{Im}[\bar{X}_2 e^{j\omega t}] = \text{Im}[X_2 e^{-j\phi_2} e^{j\omega t}] = X_2 \sin(\omega t - \phi_2) \\ x_3 &= \text{Im}[\bar{X}_3 e^{j\omega t}] = \text{Im}[X_3 e^{-j\phi_3} e^{j\omega t}] = X_3 \sin(\omega t - \phi_3) \end{aligned} \qquad (3\text{-}36)$$

The system has no damping, and the phase angles ϕ_1, ϕ_2, and ϕ_3 can be either zero or 180 deg out of phase with the excitation.

Example 10. Consider the three-degree-of-freedom system, as shown in Fig. 3-17, in which $m_1 = m_2 = m_3 = m$ and $L_1 = L_2 = L_3 = L_4 = L$. (*a*) Determine the natural frequencies of the system and the principal modes of vibration. (*b*) If an excitation force $F \sin \omega t$ is applied to the mass m_1, determine the steady-state response of the masses and sketch the response curves for a range of excitation frequency.

Solution: (*a*) Define $b^2 = S/Lm$ and $r = \omega/b$. It can be shown that r is nondimensional and may be regarded as a frequency ratio. Using these notations, the frequency equation, Eq. (3-35), can be expressed as

$$r^6 - 6r^4 + 10r^2 - 4 = 0$$

The roots of this cubic equation in r^2 are 0.59, 2.0, and 3.41. Hence the natural frequencies of the system are

$$f_n = \frac{1}{2\pi}\sqrt{\frac{0.59S}{Lm}}, \quad \frac{1}{2\pi}\sqrt{\frac{2.0S}{Lm}}, \quad \text{and} \quad \frac{1}{2\pi}\sqrt{\frac{3.41S}{Lm}}$$

The amplitude ratios of the principal modes of vibration can be obtained from Eq. (3-34) by assuming that $F = 0$. Expressing these ratios in terms of the frequency ratio r and permitting the positive and negative signs to account for the phase relation, we have

$$\frac{X_1}{X_2} = \frac{1}{2 - r^2}, \quad \text{and} \quad \frac{X_2}{X_3} = \frac{2 - r^2}{1}$$

Substituting the appropriate values of r^2 in these relations gives

$$\frac{X_1}{X_2} = \frac{1}{2 - 0.59} = \frac{1}{1.41}, \quad \text{and} \quad \frac{X_2}{X_3} = \frac{1.41}{1}, \quad \text{for } r^2 = 0.59$$

$$\frac{X_1}{X_2} = \frac{1}{2 - 2.0} = \frac{1}{0}, \quad \text{and} \quad \frac{X_2}{X_3} = \frac{0}{1}, \quad \text{for } r^2 = 2.0$$

$$\frac{X_1}{X_2} = \frac{1}{2 - 3.41} = \frac{1}{-1.41}, \quad \text{and} \quad \frac{X_2}{X_3} = \frac{-1.41}{1}, \quad \text{for } r^2 = 3.41$$

The principal modes of vibration corresponding to these amplitude ratios are shown in Fig. 3-18.

(*b*) From Eq. (3-36), the steady-state response of the masses, which are due to an excitation force $F \sin \omega t$ applied to m_1, can be expressed as

$$x_1 = X_1 \sin(\omega t - \phi_1)$$

$$x_2 = X_2 \sin(\omega t - \phi_2)$$

$$x_3 = X_3 \sin(\omega t - \phi_3)$$

The amplitudes of the steady-state response can be calculated from Eq. (3-34) as

$$X_1 = \left| \frac{(-FL/S)(r^4 - 4r^2 + 3)}{r^6 - 6r^4 + 10r^2 - 4} \right|$$

$$X_2 = \left| \frac{(FL/S)(r^2 - 2)}{r^6 - 6r^4 + 10r^2 - 4} \right|$$

$$X_3 = \left| \frac{-FL/S}{r^6 - 6r^4 + 10r^2 - 4} \right|$$

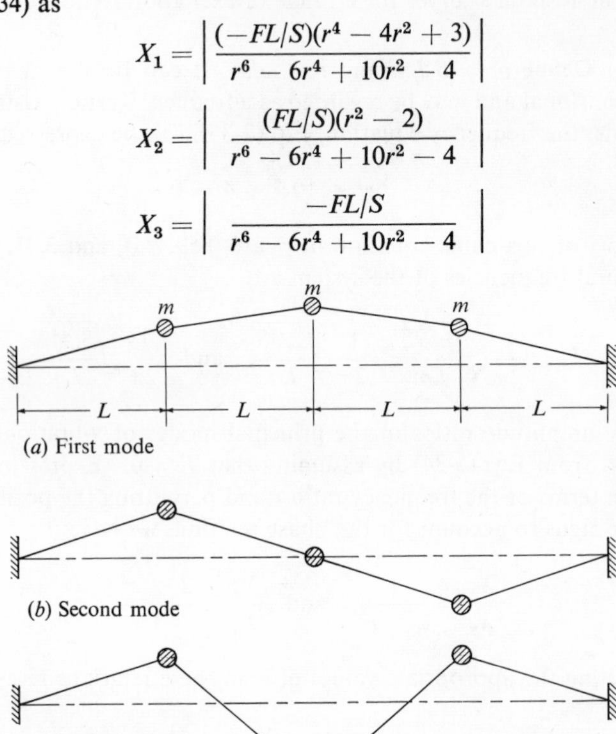

Fig. 3-18. *Principal modes of vibration of a three-degree-of-freedom system*

The phase angles ϕ_1, ϕ_2, and ϕ_3 can be either 0 or 180 deg, depending on whether the quantities within the absolute value signs are positive or negative.

The steady-state response curves are plotted in Fig. 3-19. The dashed portions of the curves indicate that the displacements of the masses are 180 deg out of phase with the excitation force. It is noted in these curves that, at frequency ratios of $r = 1$ and $\sqrt{3}$, the amplitude of m_1 is zero. This may be interpreted as the dynamic absorber effect discussed in the previous example. The system has three natural frequencies, and the amplitudes of the masses m_1 and m_3 tend to become infinite at these resonance frequencies. The amplitude of m_2, however, is finite at the second resonance frequency.

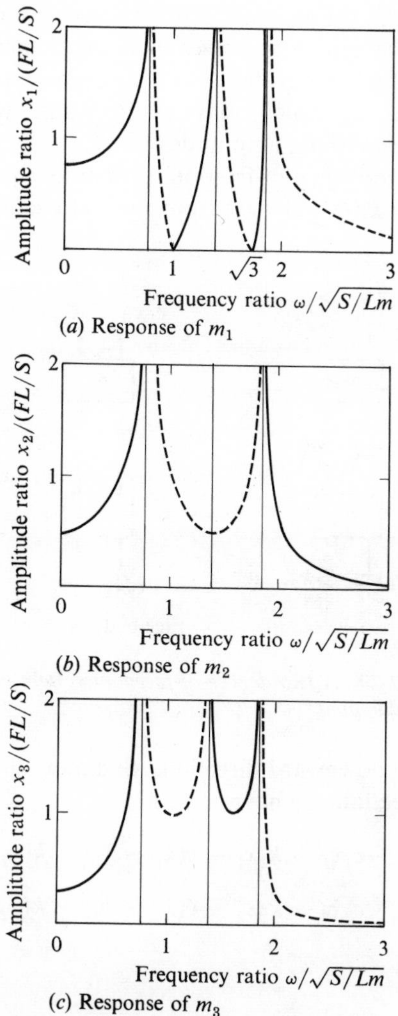

(*a*) Response of m_1

(*b*) Response of m_2

(*c*) Response of m_3

Fig. 3-19. *Steady-state response of a three-degree-of-freedom system; system shown in Fig.* 3-17

3-5. DAMPED FREE VIBRATION

Undamped free vibrations of systems with several degrees of freedom were discussed in Sec. 3-2. It was shown that the motion of a mass in the system is the resultant of the superposition of its harmonic components. If the system possesses damping, it may be anticipated that the motion of a mass is due to the superposition of a number of components, some of which may be aperiodic while others are oscillatory with diminishing amplitudes.

Consider a two-degree-of-freedom system as shown in Fig. 3-20. The displacements $x_1(t)$ and $x_2(t)$ from the static equilibrium position

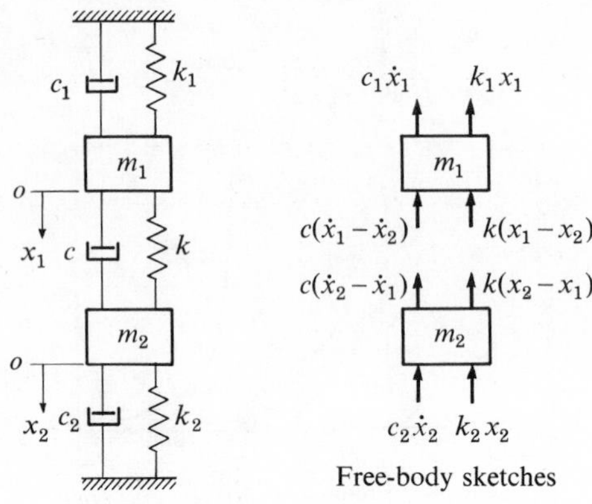

Fig. 3-20. *A two-degree-of-freedom system with damping*

are positive in the downward direction. Summing the dynamic forces in the vertical direction, we have

$$m_1\ddot{x}_1 = -c_1\dot{x}_1 - k_1x_1 - c(\dot{x}_1 - \dot{x}_2) - k(x_1 - x_2)$$

$$m_2\ddot{x}_2 = -c_2\dot{x}_2 - k_2x_2 - c(\dot{x}_2 - \dot{x}_1) - k(x_2 - x_1)$$

Rearranging gives

$$\begin{aligned} m_1\ddot{x}_1 + (c_1 + c)\dot{x}_1 + (k_1 + k)x_1 - c\dot{x}_2 - kx_2 &= 0 \\ -c\dot{x}_1 - kx_1 + m_2\ddot{x}_2 + (c_2 + c)\dot{x}_2 + (k_2 + k)x_2 &= 0 \end{aligned} \tag{3-37}$$

These are linear differential equations with constant coefficients, the solutions of which are of the form (see Appendix B)

$$\begin{aligned} x_1 &= C_1e^{st} \\ x_2 &= C_2e^{st} \end{aligned} \tag{3-38}$$

where C_1, C_2, and s are constants.

Substituting Eq. (3-38) in Eq. (3-37) and factoring out e^{st}, we have

$$\begin{aligned} [m_1 s^2 + (c_1 + c)s + (k_1 + k)]C_1 - (cs + k)C_2 &= 0 \\ -(cs + k)C_1 + [m_2 s^2 + (c_2 + c)s + (k_2 + k)]C_2 &= 0 \end{aligned} \tag{3-39}$$

which are homogeneous linear algebraic equations in C_1 and C_2. This set of equations has a solution other than the trival one, $C_1 = C_2 = 0$, only if the determinant $\Delta(s)$ of the coefficients of C_1 and C_2 vanishes; that is,

$$\Delta(s) = \begin{vmatrix} [m_1 s^2 + (c_1 + c)s + k_1 + k] & -(cs + k) \\ -(cs + k) & [m_2 s^2 + (c_2 + c)s + k_2 + k] \end{vmatrix} = 0 \tag{3-40}$$

Expanding the determinant, we get

$$[m_1 s^2 + (c_1 + c)s + k_1 + k][m_2 s^2 + (c_2 + c)s + k_2 + k] - (cs + k)^2 = 0 \tag{3-41}$$

This equation is called the *characteristic equation* of the system, from which the values of s are determined. The values of s, or the roots of the characteristic equation, may be real or complex.

Since Eq. (3-41) is quartic in s, there are four values of s for which Eq. (3-38) are solutions of the equations of motion. Hence the general solution can be expressed as

$$\begin{aligned} x_1 &= C_{11}e^{s_1 t} + C_{12}e^{s_2 t} + C_{13}e^{s_3 t} + C_{14}e^{s_4 t} \\ x_2 &= C_{21}e^{s_1 t} + C_{22}e^{s_2 t} + C_{23}e^{s_3 t} + C_{24}e^{s_4 t} \end{aligned} \tag{3-42}$$

The coefficients in this equation are related by Eq. (3-39) and can be expressed as

$$\begin{aligned} \frac{C_{1i}}{C_{2i}} &= \frac{cs_i + k}{m_1 s_i^2 + (c_1 + c)s_i + k_1 + k} \\ &= \frac{m_2 s_i^2 + (c_2 + c)s_i + k_2 + k}{cs_i + k} = \frac{1}{\mu_i} \end{aligned} \tag{3-43}$$

where i may be 1, 2, 3, or 4, and μ_i may be real or complex.

Since damping exists in the system, the motions $x_1(t)$ and $x_2(t)$ must diminish with time. Thus the roots of the characteristic equation, Eq. (3-41), must be either real and negative or complex with negative real parts. (See Sec. 2-5.) If all the roots are real and negative, the motions expressed by Eq. (3-42) are aperiodic, and no oscillation can be expected of the system for free vibration. If one of the roots is complex, there must be a conjugate complex root. Thus the motions may be expressed as

$$\begin{aligned} x_1 &= C_{11}e^{-(b+j\omega_d)t} + C_{12}e^{-(b-j\omega_d)t} + C_{13}e^{s_3t} + C_{14}e^{s_4t} \\ x_2 &= \mu_1 C_{11}e^{-(b+j\omega_d)t} + \mu_2 C_{12}e^{-(b-j\omega_d)t} + \mu_3 C_{13}e^{s_3t} + \mu_4 C_{14}e^{s_4t} \end{aligned} \tag{3-44}$$

where $-(b \pm j\omega_d)$ are assumed to be the complex roots with negative real parts. Since all physical quantities are real, the motions $x_1(t)$ and $x_2(t)$ must be real. If C_{11} is complex, C_{12} must be its complex conjugate.

It can be shown that, if one of the roots of the characteristic equation is complex, the motions $x_1(t)$ and $x_2(t)$ can be expressed as†

$$\begin{aligned} x_1 &= A_{11}e^{-bt}\sin(\omega_d t + \psi_1) + C_{13}e^{s_3t} + C_{14}{}^{s_4t} \\ x_2 &= A_{21}e^{-bt}\sin(\omega_d t + \psi_2) + \mu_3 C_{13}e^{s_3t} + \mu_4 C_{14}e^{s_4t} \end{aligned} \tag{3-45}$$

where A_{11} and ψ_1 are arbitrary, and A_{21} and ψ_2 can be expressed in terms of these constants and the system parameters. The last two terms of $x_1(t)$ and $x_2(t)$ in Eq. (3-45) are assumed to give aperiodic motions. The type of motion described by Eq. (3-45) is composed of motions which are aperiodic and oscillatory with diminishing amplitude.

The theory presented can easily be extended to include systems with several degrees of freedom. The mathematical manipulation, however, can be very tedious. Since damped free vibration is usually of less interest than forced vibration, we shall not pursue this subject any further.

† If C_{11} is complex and C_{12} is its complex conjugate, the first two terms of $x_1(t)$ in Eq. (3-44) can be expressed as

$$\begin{aligned} x_1 &= e^{-bt}[(C_{11} + C_{12})\cos\omega_d t - j(C_{11} - C_{12})\sin\omega_d t] \\ &= e^{-bt}(A_1\cos\omega_d t + A_2\sin\omega_d t) \\ &= A_{11}e^{-bt}\sin(\omega_d t + \psi_1) \end{aligned}$$

where A_1 and A_2 are real, $A_{11} = \sqrt{A_1^2 + A_2^2}$, and $\psi_1 = \tan^{-1}\dfrac{A_1}{A_2}$.

Let μ_1 and μ_2 from Eq. (3-43) be the complex conjugates $(p \pm jq)$. Therefore, the first two terms of $x_2(t)$ in Eq. (3-44) can be expressed as

$$\begin{aligned} x_2 &= (p + jq)C_{11}e^{-(b+j\omega_d)t} + (p - jq)C_{12}e^{-(b-j\omega_d)t} \\ &= e^{-bt}\{[p(C_{11} + C_{12}) + jq(C_{11} - C_{12})]\cos\omega_d t \\ &\qquad + [-jp(C_{11} - C_{12}) + q(C_{11} + C_{12})]\sin\omega_d t\} \\ &= e^{-bt}[(pA_1 - qA_2)\cos\omega_d t + (pA_2 + qA_1)\sin\omega_d t] \\ &= e^{-bt}\sqrt{p^2 + q^2}\,A_{11}\sin(\omega_d t + \psi_2) \end{aligned}$$

where

$$\psi_2 = \tan^{-1}\frac{pA_1 - qA_2}{pA_2 + qA_1}$$

3-6. STEADY-STATE FORCED VIBRATION WITH DAMPING

Consider again the two-degree-of-freedom system of Fig. 3-20 and assume that an excitation force $F \sin \omega t$ is applied to the mass m_1. From Newton's law of motion, the equations of motion of the masses are

$$\begin{aligned} m_1\ddot{x}_1 + (c_1 + c)\dot{x}_1 + (k_1 + k)x_1 - c\dot{x}_2 - kx_2 &= F \sin \omega t \\ -c\dot{x}_1 - kx_1 + m_2\ddot{x}_2 + (c_2 + c)\dot{x}_2 + (k_2 + k)x_2 &= 0 \end{aligned} \tag{3-46}$$

Using the mechanical impedance method to solve for the steady-state solutions, we substitute $Fe^{j\omega t}$ for $F \sin \omega t$, $\bar{X}_1 e^{j\omega t}$ for $x_1(t)$, and $\bar{X}_2 e^{j\omega t}$ for $x_2(t)$ in Eq. (3-46), where $\bar{X}_1$ and $\bar{X}_2$ are the complex amplitudes of $x_1(t)$ and $x_2(t)$, respectively. Rearranging and factoring out $e^{j\omega t}$, we obtain

$$\begin{aligned} [(k_1 + k) - m_1\omega^2 + j(c_1 + c)\omega]\bar{X}_1 - (k + jc\omega)\bar{X}_2 &= F \\ -(k + jc\omega)\bar{X}_1 + [(k_2 + k) - m_2\omega^2 + j(c_2 + c)\omega]\bar{X}_2 &= 0 \end{aligned} \tag{3-47}$$

$\bar{X}_1$ and $\bar{X}_2$ can be obtained from these equations by Cramer's rule and be expressed as

$$\begin{aligned} \bar{X}_1 &= \frac{\begin{vmatrix} F & -(k + jc\omega) \\ 0 & k_2 + k - m_2\omega^2 + j(c_2 + c)\omega \end{vmatrix}}{\Delta(\omega)} = X_1 e^{-j\phi_1} \\ \bar{X}_2 &= \frac{\begin{vmatrix} k_1 + k - m_1\omega^2 + j(c_1 + c)\omega & F \\ -(k + jc\omega) & 0 \end{vmatrix}}{\Delta(\omega)} = X_2 e^{-j\phi_2} \end{aligned} \tag{3-48}$$

where

$$\Delta(\omega) = \begin{vmatrix} k_1 + k - m_1\omega^2 + j(c_1 + c)\omega & -(k + jc\omega) \\ -(k + jc\omega) & k_2 + k - m_2\omega^2 + j(c_2 + c)\omega \end{vmatrix} \tag{3-49}$$

and ϕ_1 and ϕ_2 are the phase angles of the complex amplitudes $\bar{X}_1$ and $\bar{X}_2$, respectively.

Corresponding to the excitation force $F \sin \omega t$, the steady-state responses of the masses are

$$\begin{aligned} x_1 &= X_1 \sin (\omega t - \phi_1) \\ x_2 &= X_2 \sin (\omega t - \phi_2) \end{aligned} \tag{3-50}$$

where X_1, X_2, ϕ_1, and ϕ_2 are as defined in Eq. (3-48).

Following the steps outlined, the theory can easily be extended to systems with several degrees of freedom. It is noted that the procedure is relatively straightforward, but the algebraic manipulation can be very tedious.

Example 11. Dynamic Absorber with Damping

Consider the dynamic absorber of Example 9 in which a damper c is installed in parallel with the spring k_2. Briefly discuss the effect of the damper c on the motion of the mass m_1.

Solution: The equations of motion of the masses are

$$m_1\ddot{x}_1 + c\dot{x}_1 + (k_1 + k_2)x_1 - c\dot{x}_2 - k_2x_2 = F \sin \omega t$$
$$-c\dot{x}_1 - k_2x_1 + m_2\ddot{x}_2 + c\dot{x}_2 + k_2x_2 = 0$$

Analogous to Eq. (3-48), the complex amplitudes of $x_1(t)$ and $x_2(t)$ are

$$\bar{X}_1 = \frac{\begin{vmatrix} F & -(k_2 + jc\omega) \\ 0 & k_2 - m_2\omega^2 + jc\omega \end{vmatrix}}{\Delta(\omega)} = X_1 e^{-j\phi_1}$$

$$\bar{X}_2 = \frac{\begin{vmatrix} k_1 + k_2 - m_1\omega^2 + jc\omega & F \\ -(k_2 + jc\omega) & 0 \end{vmatrix}}{\Delta(\omega)} = X_2 e^{-j\phi_2}$$

where

$$\Delta(\omega) = \begin{vmatrix} k_1 + k_2 - m_1\omega^2 + jc\omega & -(k_2 + jc\omega) \\ -(k_2 + jc\omega) & k_2 - m_2\omega^2 + jc\omega \end{vmatrix}$$

The steady-state responses of the masses are

$$x_1 = X_1 \sin (\omega t - \phi_1)$$
$$x_2 = X_2 \sin (\omega t - \phi_2)$$

The effect of the damper c for a given set of parameters can be examined through these equations.

Alternatively, let us examine the effect of the damper qualitatively by assuming two extreme values of c. If $c = 0$, the system is that of an undamped dynamic absorber, the steady-state response of which was discussed in Example 9. If $c = \infty$, that is, the mass m_2 is securely attached to the mass m_1, the response of the system is that of a single-degree-of-freedom system. The vibrating mass is equal to $(m_1 + m_2)$, and the spring constant is equal to k_1. The response of this system to a sinusoidal excitation was discussed in Chap. 2. Clearly, for

$0 < c < \infty$, the steady-state response of m_1 must be intermediate between these two extreme conditions. The steady-state response curves of m_1 for $0 < c < \infty$ are shown in Fig. 3-21. Curve 1 is that of an undamped system, and curve 2 corresponds to $c = \infty$. Where these two curves intersect, the damping can range from zero to infinity.

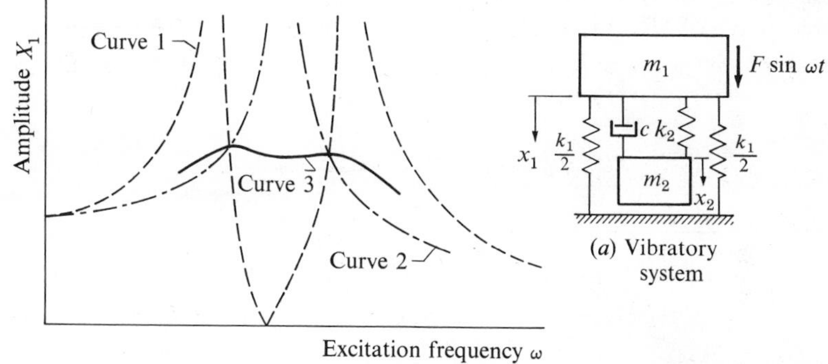

Fig. 3-21. *Dynamic absorber with damping*

Hence curve 3 for intermediate damping must pass through these intersections. As shown, curve 3 is that of a properly tuned dynamic absorber with the appropriate damping.†

3-7. INFLUENCE COEFFICIENTS

In the preceding sections Newton's law of motion was used to derive the equations of motion. Another approach is by the method of influence coefficients. This method, together with more advanced mathematics and digital computers, is widely used for the analysis of complex structures such as airplane wings.

An influence coefficient δ_{ij} is defined as the static deflection of the system at station i owing to a unit force applied at station j of the system. Hence the influence coefficient is a measure of the elastic properties of a system. As an example, consider the simply supported beam of Fig. 3-22 in which two vertical forces F_1 and F_2 are applied at stations 1 and 2. In this illustration the influence coefficients are

† J. P. Den Hartog, *Mechanical Vibrations*, 4th ed., McGraw-Hill Book Co., Inc., New York, 1956, pp. 93–102.

δ_{11}, δ_{12}, δ_{22}, and δ_{21}. For example, the deflection at station 1 owing to the force F_2 applied at station 2 is $F_2\delta_{12}$.

Consider again the system of Fig. 3-22 and assume that the procedure of loading is separated into two steps. First, assume that the force F_1 is applied to station 1 and then the force F_2 is applied to station 2. When F_1 is applied alone, the potential energy in the beam, by virtue of its deformation, is equal to $\frac{1}{2}F_1^2\delta_{11}$. Now, when F_2 is applied, the additional deflection at station 1 owing to the force F_2 is

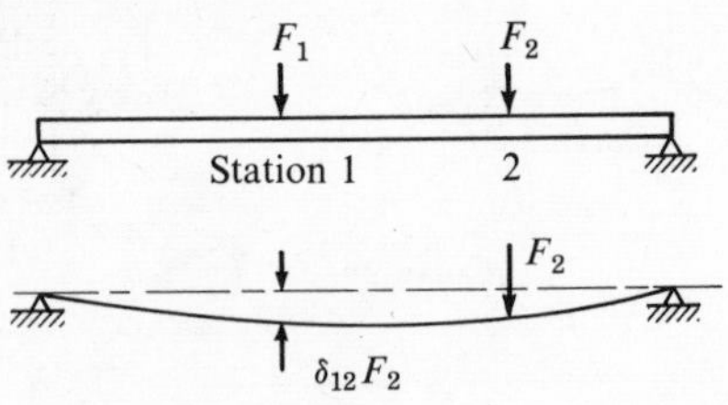

Fig. 3-22. *A simply supported beam with two loads*

$F_2\delta_{12}$. The work done by F_1 corresponding to this deflection is $F_1(F_2\delta_{12})$. Hence the total potential energy in the system is

$$U = \tfrac{1}{2}F_1^2\delta_{11} + F_1(F_2\delta_{12}) + \tfrac{1}{2}F_2^2\delta_{22}$$

The last two terms of this equation represent the additional potential energy which is due to the application of F_2.

Second, assume that the force F_2 is applied to station 2 and then the force F_1 is applied to station 1. The total potential energy of the system is

$$U = \tfrac{1}{2}F_2^2\delta_{22} + F_2(F_1\delta_{21}) + \tfrac{1}{2}F_1^2\delta_{11}$$

The last two terms of this equation are due to the application of F_1.

Since the final states of the system are identical for the two methods of loading, by the law of conservation of energy, the potential energies expressed in the two cases are the same. Thus it is deduced that $\delta_{12} = \delta_{21}$. This proves the reciprocity theorem for the case of two loads. The generalization of the proof for several loads is analogous. The important relationship

$$\delta_{ij} = \delta_{ji} \tag{3-51}$$

is sometimes called Maxwell's reciprocity theorem, and it holds for any linear system.

In a more complex system, both rectilinear and rotational motions may be encountered. Hence the force F can be generalized to mean a force or a moment, and the influence coefficient δ_{ij} can be interpreted to mean a rectilinear or an angular deflection.

We defined the influence coefficient as the static elastic property of a system in order to neglect the inertial effect of its components. When the method of influence coefficients is applied to a dynamic system, the inertia forces may be substituted for the assumed static forces. The total deflection at a station of a system is the sum of the product of the inertia forces and the appropriate influence coefficients.

Example 12. Undamped Free Vibration

A dynamic system consists of three masses attached to a taut string as shown in Fig. 3-23. (*a*) Assuming small deflections and that the tension S in the string is constant, determine the influence coefficients.

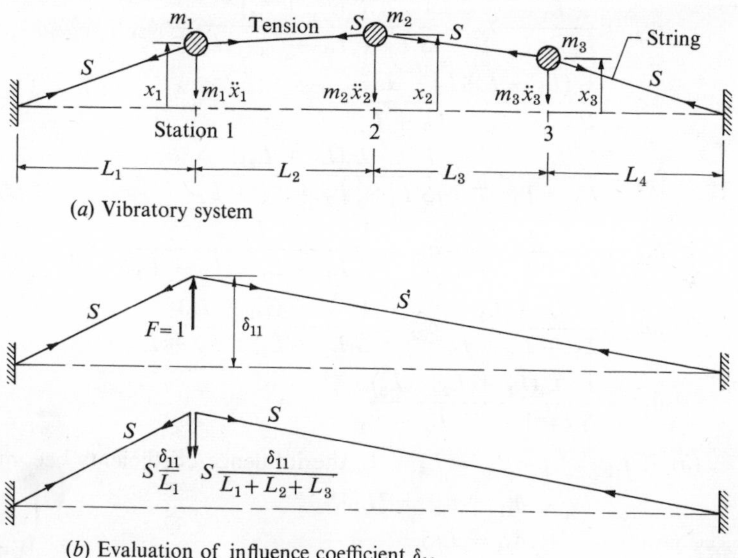

Fig. 3-23. *Taut spring carrying three mass particles*

(*b*) If $m_1 = m_2 = m_3 = m$ and $L_1 = L_2 = L_3 = L$, determine the frequency equation of the system and the principal modes of vibration.

Solution: (*a*) Assume that a unit vertical force is applied at station 1 as shown in Fig. 3-23(*b*). The string is deflected by an amount δ_{11} at this location. For equilibrium, the vertical reaction due to the tension S of the string is

$$S\left(\frac{\delta_{11}}{L_1}\right) + S\left(\frac{\delta_{11}}{L_2 + L_3 + L_4}\right) = 1$$

or

$$\delta_{11} = \frac{1}{S}\frac{L_1(L_2 + L_3 + L_4)}{L_1 + L_2 + L_3 + L_4}$$

It is noted in this figure that by similar triangles

$$\delta_{21} = \frac{L_3 + L_4}{L_2 + L_3 + L_4}\delta_{11} = \frac{1}{S}\frac{L_1(L_3 + L_4)}{L_1 + L_2 + L_3 + L_4}$$

$$\delta_{31} = \frac{L_4}{L_2 + L_3 + L_4}\delta_{11} = \frac{1}{S}\frac{L_1 L_4}{L_1 + L_2 + L_3 + L_4}$$

The other influence coefficients are determined in like manner. They are listed as follows:

$$\delta_{12} = \frac{L_1}{L_1 + L_2}\delta_{22} = \frac{1}{S}\frac{L_1(L_3 + L_4)}{L_1 + L_2 + L_3 + L_4} = \delta_{21}$$

$$\delta_{22} = \frac{1}{S}\frac{(L_1 + L_2)(L_3 + L_4)}{L_1 + L_2 + L_3 + L_4}$$

$$\delta_{32} = \frac{L_4}{L_3 + L_4}\delta_{22} = \frac{1}{S}\frac{L_4(L_1 + L_2)}{L_1 + L_2 + L_3 + L_4}$$

$$\delta_{13} = \frac{L_1}{L_1 + L_2 + L_3}\delta_{33} = \frac{1}{S}\frac{L_1 L_4}{L_1 + L_2 + L_3 + L_4} = \delta_{31}$$

$$\delta_{23} = \frac{L_1 + L_2}{L_1 + L_2 + L_3}\delta_{33} = \frac{1}{S}\frac{L_4(L_1 + L_2)}{L_1 + L_2 + L_3 + L_4} = \delta_{32}$$

$$\delta_{33} = \frac{1}{S}\frac{L_4(L_1 + L_2 + L_3)}{L_1 + L_2 + L_3 + L_4}$$

(*b*) If $L_1 = L_2 = L_3 = L_4 = L$, the influence coefficients become

$$\begin{aligned} \delta_{11} &= \delta_{33} = 3L/4S \\ \delta_{22} &= L/S \\ \delta_{12} &= \delta_{21} = \delta_{23} = \delta_{32} = L/2S \\ \delta_{13} &= \delta_{31} = L/4S \end{aligned} \tag{3-52}$$

The deflections $x_1(t)$, $x_2(t)$, and $x_3(t)$ due to the inertia forces $m_1\ddot{x}_1$, $m_2\ddot{x}_2$, and $m_3\ddot{x}_3$ are

$$\begin{aligned} x_1 &= -m_1\ddot{x}_1\delta_{11} - m_2\ddot{x}_2\delta_{12} - m_3\ddot{x}_3\delta_{13} \\ x_2 &= -m_1\ddot{x}_1\delta_{21} - m_2\ddot{x}_2\delta_{22} - m_3\ddot{x}_3\delta_{23} \\ x_3 &= -m_1\ddot{x}_1\delta_{31} - m_2\ddot{x}_2\delta_{32} - m_3\ddot{x}_3\delta_{33} \end{aligned} \tag{3-53}$$

Following the method outlined in Sec. 3-2, let the solutions of Eq. (3-53) be $x_i = X_i \sin \omega t$, where $i = 1$, 2, and 3. Substituting these relations in Eq. (3-53), factoring out the term $\sin \omega t$, and rearranging, we obtain

$$\begin{aligned}(1 - m_1\omega^2\delta_{11})X_1 - m_2\omega^2\delta_{12}X_2 - m_3\omega^2\delta_{13}X_3 &= 0\\ -m_1\omega^2\delta_{21}X_1 + (1 - m_2\omega^2\delta_{22})X_2 - m_3\omega^2\delta_{23}X_3 &= 0\\ -m_1\omega^2\delta_{31}X_1 - m_2\omega^2\delta_{32}X_2 + (1 - m_3\omega^2\delta_{33})X_3 &= 0\end{aligned} \tag{3-54}$$

These are linear homogeneous algebraic equations in X_1, X_2, and X_3. The frequency equation can be obtained by equating the determinant $\Delta(\omega)$ of the coefficients to zero, that is,

$$\Delta(\omega) = \begin{vmatrix} 1 - m_1\omega^2\delta_{11} & -m_2\omega^2\delta_{12} & -m_3\omega^2\delta_{13} \\ -m_1\omega^2\delta_{21} & 1 - m_2\omega^2\delta_{22} & -m_3\omega^2\delta_{23} \\ -m_1\omega^2\delta_{31} & -m_2\omega^2\delta_{32} & 1 - m_3\omega^2\delta_{33} \end{vmatrix} = 0 \tag{3-55}$$

The natural frequencies are obtained by substituting the appropriate values of the influence coefficients in this equation.

Substituting the values of the influence coefficients from Eq. (3-52) in this frequency equation and defining $r^2 = \omega^2/(S/Lm)$, simple algebraic operations show that the frequency equation can be expressed as

$$\Delta(r) = \begin{vmatrix} 4 - 3r^2 & -2r^2 & -r^2 \\ -2r^2 & 4(1 - r^2) & -2r^2 \\ -r^2 & -2r^2 & 4 - 3r^2 \end{vmatrix} = 0$$

Expanding the determinant and simplifying, we obtain

$$r^6 - 6r^4 + 10r^2 - 4 = 0 \tag{3-56}$$

which is identical to the frequency equation of the same system calculated in Example 10. Hence the natural frequencies determined by the two methods are identical. This must be so because the natural frequencies are inherent in the system and must be independent of the method of analysis.

The amplitude ratios of the principal modes can be obtained from Eq. (3-54). Using the values of the influence coefficients from Eq. (3-52), defining $r^2 = \omega^2/(S/Lm)$, and rearranging, we obtain

$$\begin{aligned}(4 - 3r^2)X_1 \quad -2r^2X_2 \quad -r^2X_3 &= 0\\ -2r^2X_1 + 4(1 - r^2)X_2 \quad -2r^2X_3 &= 0\\ -r^2X_1 \quad -2r^2X_2 + (4 - 3r^2)X_3 &= 0\end{aligned}$$

Using the first two equations and eliminating X_3 yields

$$\frac{X_1}{X_2} = \frac{1}{2 - r^2}$$

Similarly, using the second and third of these equations and eliminating X_1 gives

$$\frac{X_2}{X_3} = \frac{2 - r^2}{1}$$

These expressions for the amplitude ratios are identical to those derived in Example 10. Hence the principal modes of vibration are the same as those shown in Fig. 3-18.

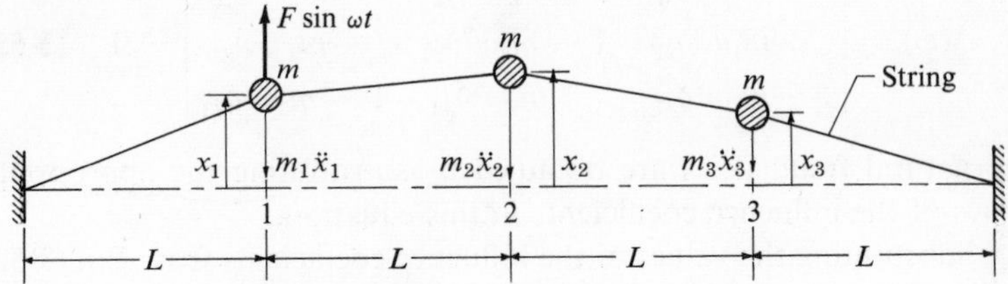

Fig. 3-24. *Harmonic excitation force applied to system*

Example 13. Undamped Forced Vibration

Figure 3-24 shows an excitation force $F \sin \omega t$ applied to the mass m_1 of the system discussed in Example 12. Determine the amplitudes of the steady-state response of the masses of the system.

Solution: Since $F \sin \omega t$ is applied at station 1 of the system, the deflections due to the excitation at stations 1, 2, and 3 are $\delta_{11} F \sin \omega t$, $\delta_{21} F \sin \omega t$, and $\delta_{31} F \sin \omega t$, respectively. Together with inertia forces, the deflections can be expressed as

$$\begin{aligned} x_1 &= (F \sin \omega t - m_1\ddot{x}_1)\delta_{11} - m_2\ddot{x}_2\delta_{12} - m_3\ddot{x}_3\delta_{13} \\ x_2 &= (F \sin \omega t - m_1\ddot{x}_1)\delta_{21} - m_2\ddot{x}_2\delta_{22} - m_3\ddot{x}_3\delta_{23} \\ x_3 &= (F \sin \omega t - m_1\ddot{x}_1)\delta_{31} - m_2\ddot{x}_2\delta_{32} - m_3\ddot{x}_3\delta_{33} \end{aligned} \tag{3-57}$$

Since the excitation is sinusoidal, the steady-state response must be sinusoidal and of the same frequency. Let $x_1 = X_1 \sin \omega t$, $x_2 = X_2 \sin \omega t$, and $x_3 = X_3 \sin \omega t$ be the steady response of the masses. Since the system does not possess damping, the phase angle of the steady-state response can be only zero or 180 deg. Let the phase angle be accounted for by the positive and negative signs of X_1, X_2, and X_3. Substituting these relations in Eq. (3-57), factoring out $\sin \omega t$, and rearranging, we obtain

$$
\begin{aligned}
(1 - m_1\omega^2\delta_{11})X_1 - m_2\omega^2\delta_{12}X_2 - m_3\omega^2\delta_{13}X_3 &= F\delta_{11} \\
-m_1\omega^2\delta_{21}X_1 + (1 - m_2\omega^2\delta_{22})X_2 - m_3\omega^2\delta_{33}X_3 &= F\delta_{21} \qquad \textbf{(3-58)} \\
-m_1\omega^2\delta_{31}X_1 - m_2\omega^2\delta_{32}X_2 + (1 - m_3\omega^2\delta_{33})X_3 &= F\delta_{31}
\end{aligned}
$$

The values of X_1, X_2, and X_3 can be obtained from this equation by Cramer's rule.

For the system considered, the values of the influence coefficients are given in Eq. (3-52). Substituting these in Eq. (3-58) and using the notation $r^2 = \omega^2/(S/Lm)$, we obtain

$$X_1 = -\frac{FL}{S}\frac{r^4 - 4r^2 + 3}{r^6 - 6r^4 + 10r^2 - 4}$$

$$X_2 = \frac{FL}{S}\frac{r^2 - 2}{r^6 - 6r^4 + 10r^2 - 4}$$

$$X_3 = -\frac{FL}{S}\frac{1}{r^6 - 6r^4 + 10r^2 - 4}$$

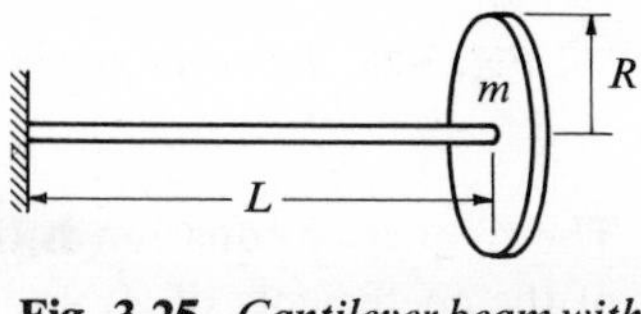

Fig. 3-25. *Cantilever beam with mass at one end*

Again, these values are the same as those determined in Example 10.

Example 14. Determine the natural frequencies of the system shown in Fig. 3-25. Assume (1) that the flexural stiffness of the shaft is EI, (2) that the inertial effect of the shaft is negligible, (3) that the radius of the disk at the end of the shaft is $R = L/4$, and (4) that the system is shown in its static equilibrium position.

Solution: The inertia forces of the system are as shown in Fig. 3-26(*a*), and the influence coefficients are defined in Fig. 3-26(*b*). From elementary beam theory it can be shown that the influence coefficients are

$$\delta_{11} = L^3/3EI, \quad \delta_{22} = L/EI, \quad \text{and} \quad \delta_{12} = \delta_{21} = L^2/2EI$$

The deflections $x(t)$ and $\theta(t)$ are

$$
\begin{aligned}
x &= -m\ddot{x}\delta_{11} - J_2\ddot{\theta}\delta_{12} = -(L^2/6EI)(2Lm\ddot{x} + 3J_2\ddot{\theta}) \\
\theta &= -m\ddot{x}\delta_{21} - J_2\ddot{\theta}\delta_{22} = -(L/2EI)(Lm\ddot{x} + 2J_2\ddot{\theta})
\end{aligned}
$$

Assume that $x = A_1 \sin(\omega t + \psi)$ and $L\theta = A_2 \sin(\omega t + \psi)$. Substituting these in the above equations and factoring out $\sin(\omega t + \psi)$, we obtain

$$
\begin{aligned}
(6EI - 2L^3m\omega^2)A_1 - 3LJ_2\omega^2A_2 &= 0 \\
-L^3m\omega^2A_1 + (2EI - 2LJ_2\omega^2)A_2 &= 0
\end{aligned}
$$

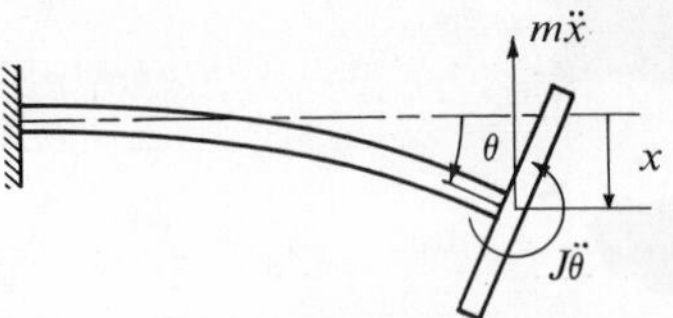

(a) Vibratory system

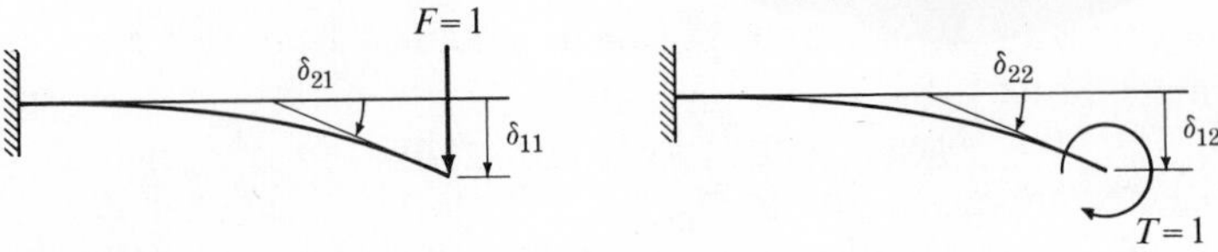

(b) Determination of influence coefficients

Fig. 3-26. *Influence coefficients due to force and moment applied to a cantilever*

The frequency equation is obtained by equating the determinant $\Delta(\omega)$ of the coefficients of A_1 and A_2 to zero, that is,

$$\Delta(\omega) = \begin{vmatrix} 6EI - 2L^3m\omega^2 & -3LJ_2\omega^2 \\ -L^3m\omega^2 & 2EI - 2LJ_2\omega^2 \end{vmatrix} = 0$$

Substituting $J_2 = mR^2/4$ and $R = L/4$ and simplifying, we obtain

$$\omega^4 - 268(EI/mL^3)\omega^2 + 768(EI/mL^3)^2 = 0$$

giving

$$\omega = \sqrt{3.85EI/mL^3} \quad \text{and} \quad \sqrt{264EI/mL^3}$$

3-8. GENERALIZED COORDINATES AND COORDINATE COUPLING

If a system has n degrees of freedom, its configuration can be defined by n independent spatial coordinates. In writing the differential equations of motion, we often use the displacements from the static equilibrium positions of the masses and the rotations about the centers of gravity as variables. This choice of coordinates is convenient but, nonetheless, arbitrary. It is possible that a configuration can be specified by more than one set of independent geometric quantities, which may be lengths, angles, or their combinations. Any set of such quantities is called the *generalized coordinates*. Thus generalized coordinates may be defined as a set of n independent geometric quantities necessary to specify the configuration of an n-degree-of-freedom system.

Consider the two-degree-of-freedom system shown in Fig. 3-27. Any pair of the coordinates (x_1,θ), (x_2,θ), and (x_3,θ) can be used to specify its configuration.

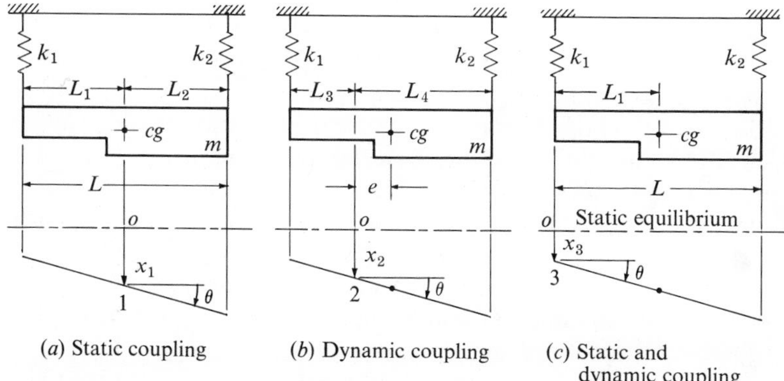

(a) Static coupling (b) Dynamic coupling (c) Static and dynamic coupling

Fig. 3-27. *The configuration of a two-degree-of-freedom system as specified by the (x_1,θ), (x_2,θ), and (x_3,θ) coordinates*

Referring to Fig. 3-27(*a*) with the (x_1,θ) coordinates and assuming small oscillations, the equations of motion are

$$m\ddot{x}_1 = -k_1(x_1 - L_1\theta) - k_2(x_1 + L_2\theta)$$
$$J_1\ddot{\theta} = +k_1(x_1 - L_1\theta)L_1 - k_2(x_1 + L_2\theta)L_2$$

Rearranging, we obtain

$$\begin{aligned} m\ddot{x}_1 + (k_1 + k_2)x_1 - (k_1L_1 - k_2L_2)\theta = 0 \\ -(k_1L_1 - k_2L_2)x_1 + J_1\ddot{\theta} + (k_1L_1^2 + k_2L_2^2)\theta = 0 \end{aligned} \tag{3-59}$$

Since each of these equations is expressed in terms of both $x_1(t)$ and $\theta(t)$, the equations are interdependent on each other. The motions of the mass are said to be coupled, and $(k_1L_1 - k_2L_2)$ is the coupling term. If the system is given an initial displacement in the x_1-direction and then released with zero initial velocity, the resultant motion of the body will consist of rectilinear and rotational components. (See Examples 4 and 5.) If a static force is applied through the center of gravity in the x_1-direction, the body will rotate as well as translate in the x_1-direction. Conversely, if a torque is applied through the center of gravity, the body will translate in the x_1-direction as well as rotate.

The system, as described by the (x_1,θ) coordinates, is said to be *elastically*, or *statically, coupled.* The rectilinear and rotational motions of the body are uncoupled only if the center of gravity is located so that $k_1L_1 = k_2L_2$.

Now consider the same system but define the configuration by the coordinates (x_2,θ), as shown in Fig. 3-27(*b*). Let e be the distance between the new origin and the center of gravity of the body. The origin does not coincide with the center of gravity of the system but is selected so that $k_1L_3 = k_2L_4$. Thus if a static force is applied to the body through point 2 to cause a static displacement x_2, there is no tendency for the body to rotate, and there is no static coupling. When the body vibrates in the x_2-direction, however, the inertia force $m\ddot{x}_2$, which goes through the center of gravity, will create a moment $m\ddot{x}_2e$ tending to rotate the body in the θ direction. Conversely, a motion in the θ-direction will create a force $me\ddot{\theta}$ in the x_2-direction. Thus dynamic coupling is anticipated with the (x_2,θ) description of the system.

The equation of motion can be obtained from Eq. (3-59) using the relations $x_2 = x_1 - e\theta, J_2 = J_1 + me^2, L_3 = L_1 - e$, and $L_4 = L_2 + e$. Alternatively, it is expedient to consider the motion to consist of two parts, a rectilinear displacement x_2 and a rotation θ. Thus the equations of motion are

$$-k_1(x_2 - L_3\theta) - k_2(x_2 + L_4\theta) - m\ddot{x}_2 - me\ddot{\theta} = 0$$

$$+k_1(x_2 - L_3\theta)L_3 - k_2(x_2 + L_4\theta)L_4 - J_2\ddot{\theta} - m\ddot{x}_2e = 0$$

where $me\ddot{\theta}$ is the inertia force in the x_2-direction due to the rotation θ and $m\ddot{x}_2e$ is the inertia torque due to x_2. Rearranging and canceling k_1L_3 with k_2L_4, we obtain

$$\begin{aligned} m\ddot{x}_2 + (k_1 + k_2)x_2 + me\ddot{\theta} &= 0 \\ me\ddot{x}_2 + J_2\ddot{\theta} + (k_1L_3^2 + k_2L_4^2)\theta &= 0 \end{aligned} \tag{3-60}$$

It should be noted that, with the (x_2,θ) description of the system, we have replaced the static coupling term with a *dynamic*, or *inertia, coupling* term.

Lastly, consider the same system as described by the (x_3,θ) coordinates in Fig. 3-27(*c*). It can be shown that the equations of motion are

$$\begin{aligned} m\ddot{x}_3 + (k_1 + k_2)x_3 + mL_1\ddot{\theta} + k_2L\theta &= 0 \\ mL_1\ddot{x}_3 + k_2Lx_3 + J_3\ddot{\theta} + k_2L^2\theta &= 0 \end{aligned} \tag{3-61}$$

With this description of the system, the motions of the body are both elastically and dynamically coupled.

It is seen that the configuration of the system shown in Fig. 3-27 can be adequately defined by (x_1,θ), (x_2,θ), or (x_3,θ) coordinates, each of which is a set of generalized coordinates. If (q_1,q_2) are the generalized coordinates of a two-degree-of-freedom system, the equations of motion for undamped free vibration can be expressed as

$$\begin{aligned} m_{11}\ddot{q}_1 + m_{12}\ddot{q}_2 + k_{11}q_1 + k_{12}q_2 &= 0 \\ m_{21}\ddot{q}_1 + m_{22}\ddot{q}_2 + k_{21}q_1 + k_{22}q_2 &= 0 \end{aligned} \tag{3-62}$$

Comparing with Eq. (3-61), m_{12} and m_{21} are the dynamic coupling terms, and k_{12} and k_{21} are the static coupling terms.

Following the procedure outlined in Sec. 3-2, the frequency equation can be obtained by substituting $q_1 = A_1 \sin(\omega t + \psi)$ and $q_2 = A_2 \sin(\omega t + \psi)$ in Eq. (3-62), factoring out $\sin(\omega t + \psi)$, and setting the determinant of the coefficients of A_1 and A_2 to zero; that is,

$$\Delta(\omega) = \begin{vmatrix} k_{11} - m_{11}\omega^2 & k_{12} - m_{12}\omega^2 \\ k_{21} - m_{21}\omega^2 & k_{22} - m_{22}\omega^2 \end{vmatrix} = 0 \tag{3-63}$$

From Eqs. (3-59), (3-60), (3-61) the frequency equation of the system shown in Fig. 3-27 can be expressed in this form. It is obvious that the natural frequencies of a system are inherent in the system parameters and are independent of the coordinates used to specify its configuration. Furthermore, these equations are mutually convertible from one to the other by a change of coordinates. The verification of these statements is left as an exercise for the reader.

The generalization of Eq. (3-62) to systems with several degrees of freedom is immediate. For example, the equations of motion of a three-degree-of-freedom system for undamped free vibration can be expressed as

$$\begin{aligned} m_{11}\ddot{q}_1 + m_{12}\ddot{q}_2 + m_{13}\ddot{q}_3 + k_{11}q_1 + k_{12}q_2 + k_{13}q_3 &= 0 \\ m_{21}\ddot{q}_1 + m_{22}\ddot{q}_2 + m_{23}\ddot{q}_3 + k_{21}q_1 + k_{22}q_2 + k_{23}q_3 &= 0 \\ m_{31}\ddot{q}_1 + m_{32}\ddot{q}_2 + m_{33}\ddot{q}_3 + k_{31}q_1 + k_{32}q_2 + k_{33}q_3 &= 0 \end{aligned} \tag{3-64}$$

3-9. PRINCIPAL COORDINATES

It was shown in the previous section that the configuration of a system can be specified by more than one set of coordinates. By a proper selection, it is always possible to choose a set of coordinates such that there is neither static nor dynamic coupling in the equations of motion. In other words, each equation contains only one dependent variable. Since these equations can be solved independent of one another, their solutions are harmonic functions, each of which has its own amplitude, frequency, and phase angle. Coordinates satisfying this condition are called the *principal coordinates.* Principal coordinates will be discussed in greater detail in Chap. 7. At present, let us show the existence of such coordinates.

Consider a two-degree-of-freedom system. The general solution [see Eq. (3-12)] for undamped free vibration can be expressed as

$$\begin{aligned} x_1 &= A_{11} \sin (\omega_1 t + \psi_1) + A_{12} \sin (\omega_2 t + \psi_2) \\ x_2 &= \mu_1 A_{11} \sin (\omega_1 t + \psi_1) + \mu_2 A_{12} \sin (\omega_2 t + \psi_2) \end{aligned} \tag{3-65}$$

where ω_1 and ω_2 are the natural frequencies of the system; A_{11}, A_{12}, ψ_1, and ψ_2 are arbitrary; and μ_1 and μ_2 are the amplitude ratios of the principal modes of vibration.

Let us choose a set of generalized coordinates (q_1,q_2) that satisfies the linear transformation

$$\begin{aligned} x_1 &= a_1 q_1 + a_2 q_2 \\ x_2 &= b_1 q_1 + b_2 q_2 \end{aligned} \tag{3-66}$$

Choosing $a_1 = a_2 = 1$, $b_1 = \mu_1$, and $b_2 = \mu_2$, we obtain

$$x_1 = q_1 + q_2 \qquad x_2 = \mu_1 q_1 + \mu_2 q_2$$

and (3-67)

$$q_1 = \frac{\mu_2 x_1 - x_2}{\mu_2 - \mu_1} \qquad q_2 = \frac{x_2 - \mu_1 x_1}{\mu_2 - \mu_1}$$

Substituting Eq. (3-65) in Eq. (3-67), we have

$$\begin{aligned} q_1 &= A_{11} \sin (\omega_1 t + \psi_1) \\ q_2 &= A_{12} \sin (\omega_2 t + \psi_2) \end{aligned} \tag{3-68}$$

Hence $q_1(t)$ and $q_2(t)$ are harmonic motions. Their corresponding differential equations of motion would be of the form

$$\begin{aligned} m_{11}\ddot{q}_1 + k_{11} q_1 &= 0 \\ m_{22}\ddot{q}_2 + k_{22} q_2 &= 0 \end{aligned} \tag{3-69}$$

Since there are neither static nor dynamic coupling terms in Eq. (3-69), the coordinates (q_1,q_2) are the principal coordinates.

Example 15. Determine the principal coordinates for the system shown in Fig 3-1 if $m_1 = m_2 = m$ and $k_1 = k_2 = k$.

Solution: From Eq. (3-1), the equations of motion of the masses are

$$m\ddot{x}_1 + (k + k)x_1 - kx_2 = 0$$
$$m\ddot{x}_2 + (k + k)x_2 - kx_1 = 0$$

Adding and subtracting the two equations, we obtain

$$m(\ddot{x}_1 + \ddot{x}_2) + k(x_1 + x_2) = 0$$
$$m(\ddot{x}_1 - \ddot{x}_2) + 3k(x_1 - x_2) = 0$$

which are of the same form as Eq. (3-69). Thus the principal coordinates are

$$q_1 = x_1 + x_2$$
$$q_2 = x_1 - x_2$$

Alternatively, from Example 3, the natural frequencies and the amplitude ratios are $\omega_1 = \sqrt{k/m}$, $\omega_2 = \sqrt{3k/m}$, $\mu_1 = 1$, and $\mu_2 = -1$. Hence the motions of the masses can be expressed as

$$x_1 = A_{11} \sin (\sqrt{k/m}\, t + \psi_1) + A_{12} \sin (\sqrt{3k/m}\, t + \psi_2)$$
$$x_2 = +A_{11} \sin (\sqrt{k/m}\, t + \psi_1) - A_{12} \sin (\sqrt{3k/m}\, t + \psi_2)$$

From Eq. (3-67), we may choose the principal coordinates to be

$$q_1 = \tfrac{1}{2}(x_1 + x_2)$$
$$q_2 = \tfrac{1}{2}(x_1 - x_2)$$

Since the amplitude of vibration is arbitrary, the factor of 2 between the two choices of principal coordinates is secondary.

3-10. ORTHOGONALITY OF THE PRINCIPAL MODES OF VIBRATION

The orthogonality principle is a fundamental property of vibrating systems having two or more degrees of freedom. It is important for the practical calculation of natural frequencies and for the treatment of forced vibrations. Let us develop this principle for a two-degree-of-freedom system.

As an example, consider the system of Fig. 3-28 in which the springs are in the x–y-plane and the mass m is constrained to move in this plane. We assume that the mass is at its static equilibrium position at the origin of the x–y-axes.

If the mass m is given a displacement x, the deformation of the spring k_1 is $x \cos \alpha_1$, and the corresponding spring force along the axis

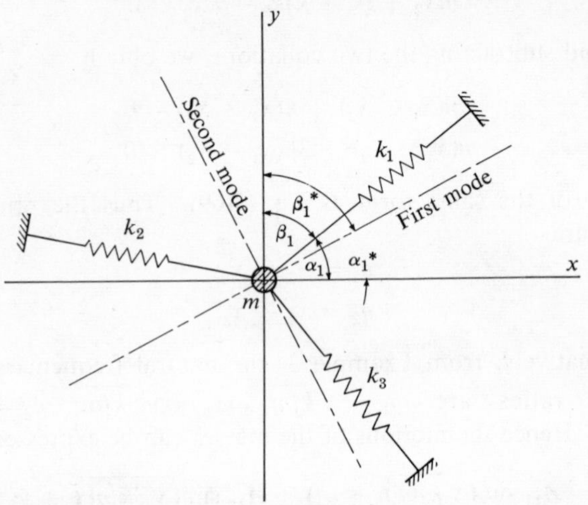

Fig. 3-28. *Orthogonality of principal modes of vibration —a two-degree-of-freedom system*

of the spring is $k_1 x \cos \alpha_1$. The x-component of this spring force is $k_1 x \cos^2 \alpha_1$. Considering all three springs, the total spring force in the x-direction is

$$(k_1 \cos^2 \alpha_1 + k_2 \cos^2 \alpha_2 + k_3 \cos^2 \alpha_3)x = k_{11} x$$

where k_{11}, defined as the quantity in the parentheses, is an equivalent spring constant.

If the mass is given a displacement y, the deformation of spring k_1 is $y \cos \beta_1$. The corresponding spring force along the axis of the spring is $k_1 y \cos \beta_1$. The x-component of the spring force is $k_1 y \cos \beta_1 \cos \alpha_1$. Hence the total spring force in the x-direction due to the three springs is

$$(k_1 \cos \alpha_1 \cos \beta_1 + k_2 \cos \alpha_2 \cos \beta_2 + k_2 \cos \alpha_3 \cos \beta_3)y = k_{12} y$$

where k_{12}, as defined in this equation, is an equivalent spring constant and a static coupling term.

Similarly, k_{22} and k_{21} may be defined as

$$k_{22} = k_1 \cos^2 \beta_1 + k_2 \cos^2 \beta_2 + k_3 \cos^2 \beta_3$$
$$k_{21} = k_1 \cos \alpha_1 \cos \beta_1 + k_2 \cos \alpha_2 \cos \beta_2 + k_3 \cos \alpha_3 \cos \beta_3$$

Evidently, $k_{12} = k_{21}$.

Having defined these equivalent spring constants, the equations of motion of the mass m due to a general x–y displacement are

$$\begin{aligned} m\ddot{x} &= -k_{11}x - k_{12}y \\ m\ddot{y} &= -k_{21}x - k_{22}y \end{aligned} \tag{3-70}$$

This equation is of the same form as Eq. (3-1).

Following the procedure outlined in Sec. 3-2, the frequency equation [see Eq. (3-5)] of this system can be expressed as

$$\Delta(\omega) = \begin{vmatrix} k_{11} - m\omega^2 & k_{12} \\ k_{21} & k_{22} - m\omega^2 \end{vmatrix} = 0 \tag{3-71}$$

Assuming that ω_1 and ω_2 are the natural frequencies calculated from this equation, the amplitude ratios of the principal modes of vibration [see Eq. (3-10)] are

$$\begin{aligned} \frac{A_{11}}{A_{21}} &= \frac{k_{12}}{k_{11} - m\omega_1^2} = \frac{k_{22} - m\omega_1^2}{k_{21}} = \frac{1}{\mu_1} \\ \frac{A_{12}}{A_{22}} &= \frac{k_{12}}{k_{11} - m\omega_2^2} = \frac{k_{22} - m\omega_2^2}{k_{21}} = \frac{1}{\mu_2} \end{aligned} \tag{3-72}$$

The general motion [see Eq. (3-12)] of the masses can be expressed as

$$\begin{aligned} x &= A_{11} \sin (\omega_1 t + \psi_1) + A_{12} \sin (\omega_2 t + \psi_2) \\ y &= \mu_1 A_{11} \sin (\omega_1 t + \psi_1) + \mu_2 A_{12} \sin (\omega_2 t + \psi_2) \end{aligned} \tag{3-73}$$

where A_{11}, A_{12}, ψ_1, and ψ_2 are arbitrary. The principal modes of vibration are

$$\begin{cases} x = A_{11} \sin (\omega_1 t + \psi_1) \\ y = \mu_1 A_{11} \sin (\omega_1 t + \psi_1) \end{cases} \quad \text{and} \quad \begin{cases} x = A_{12} \sin (\omega_2 t + \psi_2) \\ y = \mu_2 A_{12} \sin (\omega_2 t + \psi_2) \end{cases} \tag{3-74}$$

When the mass is vibrating in one of these principal modes, the $x(t)$ and $y(t)$ motions are always in the same ratio. Therefore, the motion of the mass is along a straight line through the origin of the x–y-coordinates, as shown in Fig. 3-28. The amplitudes of the principal modes of vibration are arbitrary. For simplicity, let the amplitudes be unity. Rewriting Eq. (3-74), the principal modes can be expressed as

$$\begin{cases} x = A_{11}^* \sin (\omega_1 t + \psi_1) \\ y = A_{21}^* \sin (\omega_1 t + \psi_1) \end{cases} \quad \text{and} \quad \begin{cases} x = A_{12}^* \sin (\omega_2 t + \psi_2) \\ y = A_{22}^* \sin (\omega_2 t + \psi_2) \end{cases} \tag{3-75}$$

where

$$\begin{aligned} A_{11}^{*\,2} + A_{21}^{*\,2} &= 1 \\ A_{12}^{*\,2} + A_{22}^{*\,2} &= 1 \end{aligned} \tag{3-76}$$

With the amplitude of vibration equal to 1, A_{11}^*, A_{21}^*, A_{12}^*, and A_{22}^* are now direction cosines. They can be expressed as

$$\begin{cases} A_{11}^* = \cos \alpha_1^* \\ A_{21}^* = \cos \beta_1^* \end{cases} \quad \text{and} \quad \begin{cases} A_{12}^* = \cos \alpha_2^* \\ A_{22}^* = \cos \beta_2^* \end{cases} \tag{3-77}$$

Substituting the principal modes of vibration from Eq. (3-75) in Eq. (3-70) and factoring out the sine terms, we obtain

$$\begin{aligned} m\omega_1^2 A_{11}^* &= k_{11}A_{11}^* + k_{12}A_{21}^* \\ m\omega_1^2 A_{21}^* &= k_{21}A_{11}^* + k_{22}A_{21}^* \\ m\omega_2^2 A_{12}^* &= k_{11}A_{12}^* + k_{12}A_{22}^* \\ m\omega_2^2 A_{22}^* &= k_{21}A_{12}^* + k_{22}A_{22}^* \end{aligned} \tag{3-78}$$

We multiply the first of these equations by $-A_{12}$, the second by $-A_{22}$, the third by A_{11}, and the fourth by A_{21} and add. Remembering that $k_{12} = k_{21}$, the resultant equation becomes

$$m(A_{11}^*A_{12}^* + A_{21}^*A_{22}^*)(\omega_2^2 - \omega_1^2) = 0$$

Since we assume that $\omega_2^2 \neq \omega_1^2$, it is deduced that

$$A_{11}^*A_{12}^* + A_{21}^*A_{22}^* = 0 \tag{3-79}$$

From Eq. (3-77) this equation may be expressed as

$$\cos \alpha_1^* \cos \alpha_2^* + \cos \beta_1^* \cos \beta_2^* = 0 \tag{3-80}$$

which is the cosine of the angle between the two paths along which the principal modes vibrate. The cosine of this angle is zero, and the two paths must be perpendicular to each other; that is, the principal modes are orthogonal.

Returning to Eq. (3-75) in which the amplitudes of the principal modes are unity, these modes are called the *normal modes* of vibration. In other words, the amplitude of vibration of a principal mode is arbitrary, and that of a normal mode is normalized. In this case the amplitude is unity, as indicated by Eq. (3-76).

Generalizing the discussion to a vibrating system with n degrees of freedom, the orthogonality relation of Eq. (3-79) can be expressed as

$$\sum_{i=1}^{n} m_i A_{ir} A_{is} = 0, \quad \text{for } r \neq s \tag{3-81}$$

The asterisk is omitted from this equation for ease of writing. Equation (3-76) can be generalized as

$$\sum_{i=1}^{n} m_i A_{ir}^2 = M, \quad \text{a constant} \tag{3-82}$$

where M may be defined as

$$M = \frac{1}{n} \sum_{n=1}^{n} m_i \tag{3-83}$$

The normal modes of vibration of an n-degree-of-freedom system may not be literally perpendicular to one another. The orthogonality relation is as specified in Eq. (3-81).

Example 16. Determine the normal modes of vibration of the three-degrees-of-freedom torsional system as shown in Fig. 3-9. (This system was discussed in Example 6.)

Solution: Let the normal modes of vibration be

$$\begin{aligned} \theta_1 &= \Theta_{1i} \sin (\omega_i t + \psi_i) \\ \theta_2 &= \Theta_{2i} \sin (\omega_i t + \psi_i) \\ \theta_3 &= \Theta_{3i} \sin (\omega_i t + \psi_i) \end{aligned} \tag{3-84}$$

where $i = 1, 2,$ or 3 corresponding to the three modes of vibration.

It is given $J_1 = J_2 = J_3 = J$ and $k_{t1} = k_{t2} = k_{t3} = k_t$. Let us define a frequency ratio $r_i = \omega_i / \sqrt{k_t/J}$. Substituting this relation in Eq. (3-16) gives

$$\begin{aligned} (2 - r_i^2)\Theta_{1i} \quad - \Theta_{2i} \quad &= 0 \\ -\Theta_{1i} + (2 - r_i^2)\Theta_{2i} \quad - \Theta_{3i} &= 0 \\ -\Theta_{2i} + (1 - r_i^2)\Theta_{3i} &= 0 \end{aligned} \tag{3-85}$$

The frequency ratios, as determined from Example 6, are $r_1^2 = 0.198$, $r_2^2 = 1.555$, and $r_3^2 = 3.247$.

From Eq. (3-83) we have

$$M = \tfrac{1}{3}(3J) = J$$

From Eq. (3-82) we obtain

$$\Theta_{1i}^2 + \Theta_{2i}^2 + \Theta_{3i}^2 = 1 \tag{3-86}$$

Solving Eqs. (3-85) and (3-86) gives

$$\begin{aligned} &\Theta_{11} = 0.328, \Theta_{21} = 0.591, \Theta_{31} = 0.736, \quad \text{for } r_1^2 = 0.198 \\ &\Theta_{12} = 0.736, \Theta_{22} = 0.328, \Theta_{32} = -0.591, \quad \text{for } r_2^2 = 1.555 \\ &\Theta_{13} = 0.591, \Theta_{32} = -0.736, \Theta_{33} = 0.328, \quad \text{for } r_3^2 = 3.247 \end{aligned}$$

which are the coefficients of the normal modes of vibration.

The orthogonality relation of Eq. (3-81) may be satisfied by these coefficients. For example, using the first and third modes, and substituting A for Θ, we have

$$\sum_{i=1}^{3} A_{i1}A_{i3} = A_{11}A_{13} + A_{21}A_{23} + A_{31}A_{33}$$
$$= (0.328)(0.591) + (0.591)(-0.736) + (0.736)(0.328) = 0$$

SUGGESTED READING

Den Hartog, J. P., *Mechanical Vibrations* (New York: McGraw-Hill Book Co., Inc., 4th ed., 1956), chaps. 3 and 4.

Harris, C. M., and C. E. Crede, editors, *Shock and Vibration Handbook* (New York: McGraw-Hill Book Co., Inc., 1961), vol. 1, chaps. 6 and 10.

Karman, von, T., and M. A. Biot, *Mathematical Methods in Engineering* (New York: McGraw-Hill Book Co., Inc., 1940), chaps. 5 and 6.

Rayleigh, Lord, *Theory of Sound* (New York: Dover Publications, Inc., 1945), vol. 1.

Thomson, W. T., *Mechanical Vibrations* (New Jersey: Prentice-Hall, Inc., 2nd ed., 1953), chap. 5.

Timoshenko, S., and D. H. Young, *Vibration Problems in Engineering* (New York: D. Van Nostrand Co., Inc., 3rd ed., 1956), chaps. 3 and 4.

Tong, K. N., *Theory of Mechanical Vibration* (New York: John Wiley & Sons, Inc., 1960), chap. 2.

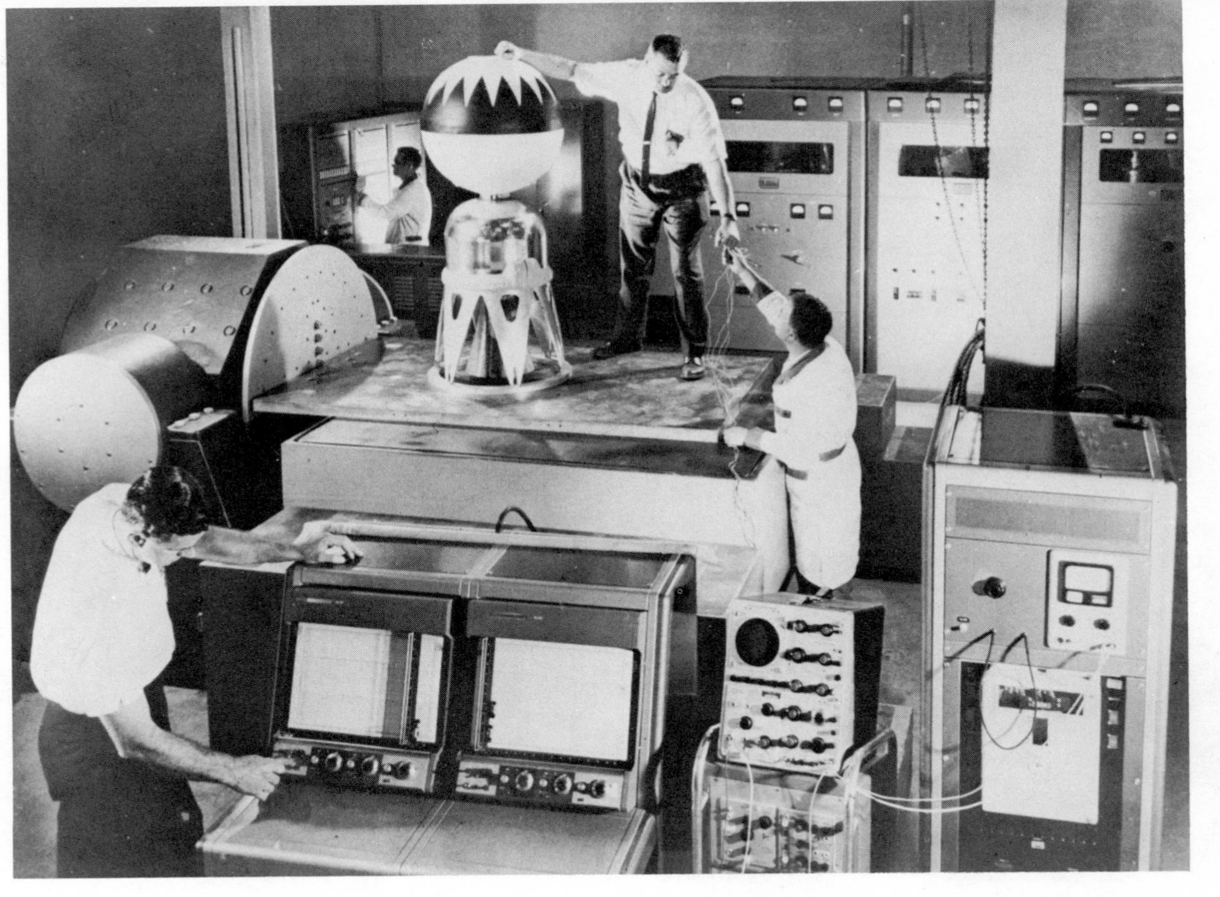

Lunar capsule vibration test simulates bumpy journey which will be encountered by the Ranger spacecraft as it travels through our atmosphere, and as it will slow down to land on the moon. (courtesy Ford Motor Company, Aeronutronic Division)

4 THE LAGRANGE EQUATIONS

4-1. INTRODUCTION

It was shown in the previous chapter that if a system is holonomic and has n degrees of freedom, its geometric configuration can be specified by a set of n generalized coordinates. The choice of coordinates depends on the nature of the problem. Often, it is convenient to specify the configuration by one set of coordinates, such as the Cartesian coordinates. The transformation of the equations of motion from one set of coordinates, together with the conditions of constraint, to the desired generalized coordinates can, however, be very involved. Lagrange's equations permit the equations of motion to be written directly in terms of any generalized coordinates.

The Lagrange method allows the equations of motion to be written when certain basic energy expressions of the system are known. Since energies are scalar quantities, they can be expressed in any convenient coordinates with little regard to the conditions of constraint or the kinematics of the system. Thus this method may be regarded as a generalization of the energy method discussed in Chap. 2. When applied to a dynamic system, Lagrange's equations yield the same equations of motion as obtained by Newton's laws. Although this method offers nothing that is basically new, it is a powerful tool in the analysis of complex dynamic systems.

In this chapter we shall first illustrate the Lagrange method with two simple examples and then derive Lagrange's equations from the concepts of virtual displacement and d'Alembert's principle. Although Lagrange's equations can also be derived from a more general principle, known as Hamilton's principle, the discussion of this topic is beyond the scope of this text.

4-2. SIMPLE EXPOSITION

Consider the simple mass-spring system as shown in Fig. 4-1. It was shown in Chap. 2 (see Sec. 2-3) that the kinetic energy of the system is $T = \frac{1}{2}m\dot{x}^2$, and the potential energy is $U = \frac{1}{2}kx^2$. The sum of the kinetic and potential energies is the total mechanical energy of the system. Since the system is conservative, once the system is set into motion its total energy is constant. The equation of motion can be derived by equating the time derivative of this total energy to zero, that is,

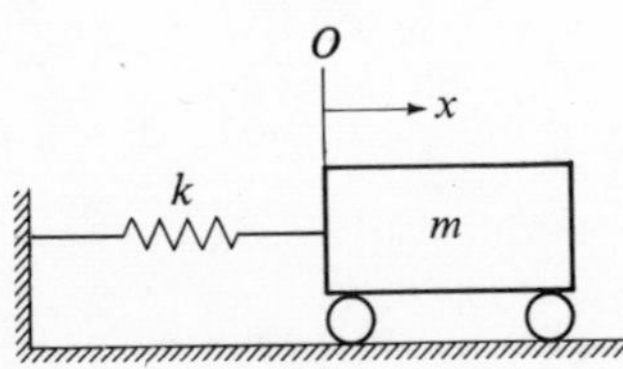

Fig. 4-1. *Simple spring-mass system*

$$\frac{d}{dt}(T + U) = \frac{d}{dt}(\tfrac{1}{2}m\dot{x}^2 + \tfrac{1}{2}kx^2) = 0 \qquad \textbf{(4-1)}$$

or

$$m\ddot{x} + kx = 0 \qquad \textbf{(4-2)}$$

Now let us write Newton's law of motion as

$$\frac{d}{dt}(m\dot{x}) = (\text{force in the } x\text{-direction}) \qquad \textbf{(4-3)}$$

It is noted that the quantity $m\dot{x}$, or the momentum of the mass, can be obtained by differentiating the kinetic energy T with respect to the velocity $\dot{x}$. Furthermore, the spring force $-kx$, which is the only force acting on the mass in the x-direction, can be obtained by differentiating the potential energy U with respect to the displacement x. The negative sign indicates that the spring force tends to restore the mass to its equilibrium position. Substituting these derivatives into Eq. (4-3) gives

$$\frac{d}{dt}\left(\frac{dT}{d\dot{x}}\right) = Q \tag{4-4}$$

$$\frac{d}{dt}\left(\frac{dT}{d\dot{x}}\right) + \frac{dU}{dx} = 0 \tag{4-5}$$

where Q is the force acting on the mass in the x-direction. If the mass m is assumed to be constant, it can readily be shown that the equation of motion, Eq. (4-2), can be obtained from Eq. (4-5). Hence we have derived the equation of motion in terms of the energy functions.

If a system has n degrees of freedom with the generalized coordinates $(q_1, q_2, \cdots, q_n)$ it may be postulated that Eqs. (4-4) and (4-5) can be written as

$$\frac{d}{dt}\left(\frac{\partial T}{\partial \dot{q}_i}\right) = Q_i \tag{4-6}$$

$$\frac{d}{dt}\left(\frac{\partial T}{\partial \dot{q}_i}\right) + \frac{\partial U}{\partial q_i} = 0 \tag{4-7}$$

where $i = 1, 2, \cdots, n$, and Q_i is a generalized force associated with the generalized coordinate q_i. We shall prove these equations in a more general form in a later section. Equation (4-6) consists of a set of n equations known as Lagrange's equations of motion. It is applicable to cases of small oscillations. Equation (4-7) is Lagrange's equation for a conservative system with small oscillations.

Example 1. Determine the equations of motion of the double pendulum as shown in Fig. 4-2 for small oscillations.

Solution: Let (θ_1, θ_2) be the generalized coordinates. For convenience, let the kinetic energy of the system be expressed as

$$T = \tfrac{1}{2}m_1(\dot{x}_1^2 + \dot{y}_1^2) + \tfrac{1}{2}m_2(\dot{x}_2^2 + \dot{y}_2^2)$$

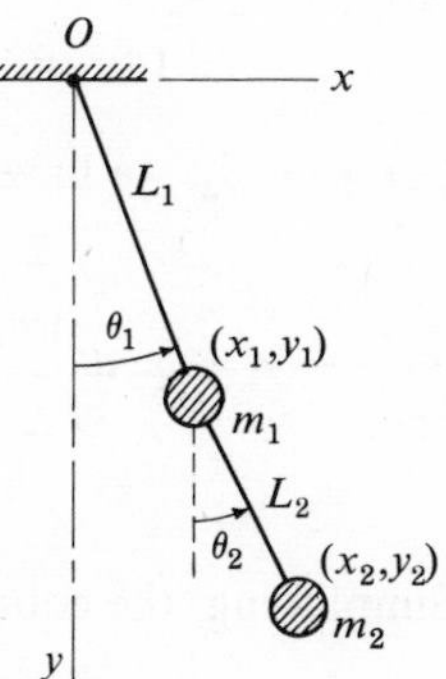

Fig. 4-2. *Double pendulum*

The Cartesian coordinates can be expressed in terms of the generalized coordinates as

$$\begin{cases} x_1 = L_1 \sin \theta_1 \\ y_1 = L_1 \cos \theta_1 \end{cases} \quad \text{and} \quad \begin{cases} x_2 = L_1 \sin \theta_1 + L_2 \sin \theta_2 \\ y_2 = L_1 \cos \theta_1 + L_2 \cos \theta_2 \end{cases}$$

Hence the kinetic energy of the system becomes

$$\begin{aligned} T &= \tfrac{1}{2}m_1 L_1^2\dot{\theta}_1^2 \\ &\qquad + \tfrac{1}{2}m_2[L_1^2\dot{\theta}_1^2 + 2L_1L_2(\cos\theta_1\cos\theta_2 + \sin\theta_1\sin\theta_2)\dot{\theta}_1\dot{\theta}_2 + L_2^2\dot{\theta}_2^2] \\ &= \tfrac{1}{2}m_1L_1^2\dot{\theta}_1^2 + \tfrac{1}{2}m_2[L_1^2\dot{\theta}_1^2 + 2L_1L_2\cos(\theta_1 - \theta_2)\dot{\theta}_1\dot{\theta}_2 + L_2^2\dot{\theta}_2^2] \\ &\doteq \tfrac{1}{2}m_1L_1^2\dot{\theta}_1^2 + \tfrac{1}{2}m_2(L_1^2\dot{\theta}_1^2 + 2L_1L_2\dot{\theta}_1\dot{\theta}_2 + L_2^2\dot{\theta}_2^2) \end{aligned}$$

Letting the potential energy at static equilibrium equal zero, the potential energy of the system is due to the elevation of m_1 and m_2 above their static equilibrium positions. The potential energy of the system is

$$\begin{aligned} U &= m_1gL_1(1 - \cos\theta_1) + m_2g[L_1(1 - \cos\theta_1) + L_2(1 - \cos\theta_2)] \\ &\doteq m_1gL_1\theta_1^2/2 + m_2g(L_1\theta_1^2/2 + L_2\theta_2^2/2) \\ &\doteq \tfrac{1}{2}(m_1 + m_2)gL_1\theta_1^2 + \tfrac{1}{2}m_2gL_2\theta_2^2 \end{aligned}$$

The equations of motion are obtained by substituting these energy functions in Eq. (4-7). For $q_1 = \theta_1$, we have

$$\frac{d}{dt}(m_1L_1^2\dot{\theta}_1 + m_2L_1^2\dot{\theta}_1 + m_2L_1L_2\dot{\theta}_2) + (m_1 + m_2)gL_1\theta_1 = 0$$

or

$$(m_1 + m_2)L_1^2\ddot{\theta}_1 + m_2L_1L_2\ddot{\theta}_2 + (m_1 + m_2)gL_1\theta_1 = 0$$

For $q_2 = \theta_2$, we have

$$\frac{d}{dt}(m_2L_1L_2\dot{\theta}_1 + m_2L_2^2\dot{\theta}_2) + m_2gL_2\theta_2 = 0$$

or

$$m_2L_1L_2\ddot{\theta}_1 + m_2L_2^2\ddot{\theta}_2 + m_2gL_2\theta_2 = 0$$

Simplifying, the equations of motion become

$$\begin{aligned} (m_1 + m_2)L_1\ddot{\theta}_1 + m_2L_2\ddot{\theta}_2 + (m_1 + m_2)g\theta_1 &= 0 \\ L_1\ddot{\theta}_1 + L_2\ddot{\theta}_2 + g\theta_2 &= 0 \end{aligned}$$

Example 2. Consider the pendulum as shown in Fig. 4-3, in which the length of the cord suspending the mass m is varied according to a prescribed function of time. Derive the equation of motion of the system.

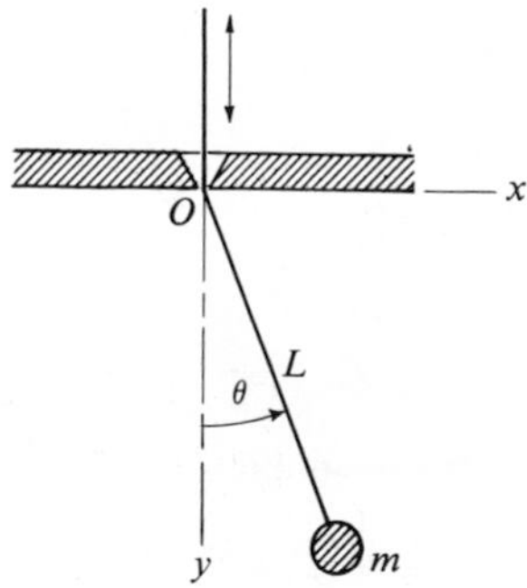

Fig. 4-3. *Pendulum with time-varying constraint*

Solution: The position of the mass m is defined by the angle θ and the length L. Since L is given as a function of time, we have a one-degree-of-freedom system. Let θ be the generalized coordinate. For convenience, let the kinetic energy be expressed in terms of the Cartesian coordinates as

$$T = \tfrac{1}{2}m(\dot{x}^2 + \dot{y}^2)$$

The two coordinate systems are related by the equations

$$x = L\sin\theta \quad \text{and} \quad y = L\cos\theta$$

Hence, in terms of the generalized coordinates, the kinetic energy is

$$\begin{aligned} T &= \tfrac{1}{2}m[(\dot{L}\sin\theta + L\cos\theta\dot{\theta})^2 + (\dot{L}\cos\theta - L\sin\theta\dot{\theta})] \\ &= \tfrac{1}{2}m(\dot{L}^2 + L^2\dot{\theta}^2) \end{aligned}$$

The potential energy of the system is

$$U = mgL(1 - \cos\theta)$$

Substituting these energy functions in Eq. (4-7) gives

$$\frac{d}{dt}(mL^2\dot{\theta}) + mgL\sin\theta = 0$$

Assuming small oscillations, performing the time derivative, and simplifying, we obtain

$$\ddot{\theta} + \frac{2\dot{L}}{L}\dot{\theta} + \frac{g}{L}\theta = 0$$

4-3. VIRTUAL DISPLACEMENT

The method of virtual displacement is essentially a static analysis in which the inertia forces are neglected. The concept is extended to the study of dynamic systems in the sections to follow.

A virtual displacement may be defined as a small arbitrary displacement that is compatible with the geometric constraints of the system. Consider a rigid bar pivoting about a frictionless pin C, as shown in Fig. 4-4. The bar is constrained to rotate about the longitudinal axis of the pin. Assume that the forces X_1 and X_2 of constant magnitude are applied to the bar and the bar is in equilibrium. Let the bar be rotated clockwise by a small angular displacement $\delta\theta$. The virtual displacements associated with the forces X_1 and X_2 are δx_1 and δx_2, respectively. The work done by the applied forces is

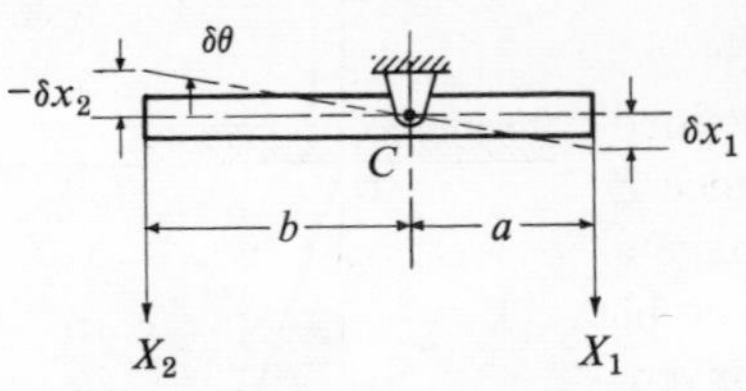

Fig. 4-4. *Rigid bar hinged at C*

$$\delta W_v = X_1\,\delta x_1 - X_2\,\delta x_2 = (X_1 a - X_2 b)\,\delta\theta$$

The bar is acted upon by the applied forces and the reactive force at the pin. Since the pin is frictionless, the work done by the reactive force is zero and need not be considered. Since the bar is assumed to be at equilibrium, the work done by the applied forces is zero.

In this simple illustration we assume that (1) the reaction is frictionless (such constraints are called workless constraints), (2) the system is in equilibrium, and (3) the virtual displacements are compatible with the geometric constraints of the system. In this case the virtual displacements are related by the equation $(b\,\delta x_1 - a\,\delta x_2) = 0$. With these assumptions, it is deduced that the work done by the applied forces, or the *virtual work*, is equal to zero.

The same assumptions and deduction may be applied to a general system. The principle of virtual work, sometimes known as the principle of virtual displacements, may be stated as follows: If a system with workless constraint is in equilibrium, the total virtual work done by the applied forces on all of the virtual displacements must equal zero. The reactive forces at the constraints are not considered, because if the constraints are frictionless, the reactive force must be perpendicular to the virtual displacement. For example, if a particle is constrained to move on a smooth surface, the virtual displacement satisfying the constraint must be tangential to the surface at the point of contact. In the absence of friction, the reactive force at the point of contact must be perpendicular to the surface. Hence, for workless constraints, the contribution of the reactive force to the virtual work is nil.

Consider a system with p mass points with r geometric constraints of the form

$$f_j(x_1,y_1,z_1, \ldots, x_p,y_p,z_p) = 0 \tag{4-8}$$

where $j = 1,2, \ldots, r$ and (x,y,z) are Cartesian coordinates. The virtual displacements x_i,y_i,z_i $(i = 1,2, \ldots, p)$ are arbitrary but not independent. They must satisfy the r relations

$$\sum_{i=1}^{p} \left(\frac{\partial f_j}{\partial x_i} \delta x_i + \frac{\partial f_j}{\partial y_i} \delta y_i + \frac{\partial f_j}{\partial z_i} \delta z_i \right) = 0 \tag{4-9}$$

Let the components of the applied forces to the particle m_i be X_i, Y_i, and Z_i. From the principle of virtual work, the virtual work is

$$\delta W_v = \sum_{i=1}^{p} (X_i\, \delta x_i + Y_i\, \delta y_i + Z_i\, \delta z_i) = 0 \tag{4-10}$$

For example, the geometric constraint for the spherical pendulum, as shown in Fig. 4-5, is $x^2 + y^2 + z^2 - L^2 = 0$. Assuming that the length L is constant, the virtual displacements must satisfy the relation

$$x\, \delta x + y\, \delta y + z\, \delta z = 0$$

For a virtual displacement of the spherical pendulum, the virtual work is

$$\delta W_v = X\, \delta x + Y\, \delta y + Z\, \delta z = 0$$

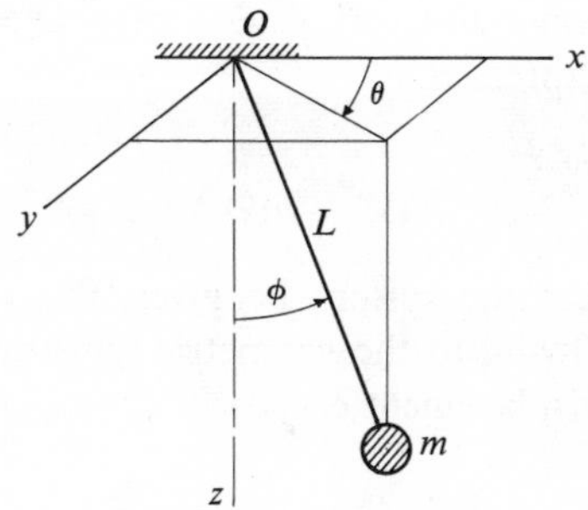

Fig. 4-5. *Spherical pendulum*

4-4. D'ALEMBERT'S PRINCIPLE

Combining the principle of virtual displacement and d'Alembert's principle, the discussion in the previous section can be extended to dynamic systems. D'Alembert's principle may be stated as follows: Every state of motion may be considered at any instant as a state of equilibrium if the inertia forces are taken into consideration. The inertia force is defined as the product of the mass and the negative acceleration. Including the inertia forces as part of the force system, Eq. (4-10) becomes

$$\delta W_v = \sum_{i=1}^{p} [(X_i - m_i\ddot{x}_i)\, \delta x_i + (Y_i - m_i\ddot{y}_i)\, \delta y_i + (Z_i - m_i\ddot{z}_i)\, \delta z_i] = 0 \tag{4-11}$$

This equation leads to a set of differential equations with time as the independent variable. It is analogous to Eq. (4-10), in the respect that the same assumptions are made for both of the equations; namely, the reactions are frictionless, the system is in equilibrium, and the displacements are compatible with the geometric constraints of the system.

Example 3. Use the method of virtual work to find the equations of motion of the spherical pendulum as shown in Fig. 4-5.

Solution: It is more expedient to use the generalized coordinates θ and ϕ to solve this problem. We shall first write the equations of motion in terms of the Cartesian coordinates, to illustrate the application of Eq. (4-11), and then express the resultant equations in the generalized coordinates.

Since the system has only one mass point and one equation of constraint, $x^2 + y^2 + z^2 - L^2 = 0$, Eqs. (4-9) and (4-11) can be expressed as

$$x\,\delta x + y\,\delta y + z\,\delta z = 0 \qquad \textbf{(4-12)}$$

$$(X - m\ddot{x})\,\delta x + (Y - m\ddot{y})\,\delta y + (Z - m\ddot{z})\,\delta z = 0 \qquad \textbf{(4-13)}$$

Let the system be given the virtual displacements δx, δy, and δz. Owing to the geometric constraint, if δy and δz are independent, the displacement δx is

$$\delta x = -\frac{y}{x}\,\delta y - \frac{z}{x}\,\delta z \qquad \textbf{(4-14)}$$

Substituting Eq. (4-14) in Eq. (4-13) and rearranging gives

$$\left[(Y - m\ddot{y}) - (X - m\ddot{x})\frac{y}{x}\right]\delta y + \left[(Z - m\ddot{z}) - (X - m\ddot{x})\frac{z}{x}\right]\delta z = 0 \qquad \textbf{(4-15)}$$

Since δy and δz are arbitrary and independent of one another, Eq. (4-15) is satisfied if each of the forces associated with the displacements δy and δz is equal to zero; that is,

$$\begin{aligned} (Y - m\ddot{y}) - (X - m\ddot{x})\frac{y}{x} &= 0 \\ (Z - m\ddot{z}) - (X - m\ddot{x})\frac{z}{x} &= 0 \end{aligned} \qquad \textbf{(4-16)}$$

or

$$\begin{aligned} m(\ddot{x}y - \ddot{y}x) &= Xy - Yx \\ m(\ddot{x}z - \ddot{z}x) &= Xz - Zx \end{aligned} \qquad \textbf{(4-17)}$$

For the free vibration of the spherical pendulum, the applied forces $X = Y = 0$ and $Z = mg$. Hence the equations of motion become

$$\begin{aligned} m(\ddot{x}y - \ddot{y}x) &= 0 \\ m(\ddot{x}z - \ddot{z}x) &= -mgx \end{aligned} \tag{4-18}$$

Now let us express these equations in terms of the generalized coordinates by the relations

$$\begin{aligned} x &= L \sin \phi \cos \theta \\ y &= L \sin \phi \sin \theta \\ z &= L \cos \phi \end{aligned} \tag{4-19}$$

Before substituting these relations in Eq. (4-18), it is convenient to rewrite the equations of motion in the form

$$m \frac{d}{dt} (\dot{x}y - \dot{y}x) = 0 \tag{4-20}$$

$$m \frac{d}{dt} (\dot{x}z - \dot{z}x) = -mgx \tag{4-21}$$

Substituting Eq. (4-19) in Eq. (4-20) and simplifying, we have

$$m \frac{d}{dt} L^2(-\sin^2 \phi\dot{\theta}) = 0$$

or

$$\sin \phi\ddot{\theta} + 2 \cos \phi\dot{\phi}\dot{\theta} = 0 \tag{4-22}$$

Substituting Eq. (4-19) in Eq. (4-21) and simplifying gives

$$m \frac{d}{dt} L^2(-\sin \phi \cos \phi \sin \theta\dot{\theta} + \cos \theta\dot{\phi}) = -mgL \sin \phi \cos \theta$$

Performing the differentiation indicated in this equation, using the relation in Eq. (4-22), and simplifying, we obtain

$$\ddot{\phi} - \sin \phi \cos \phi\dot{\theta}^2 + \frac{g}{L} \sin \phi = 0 \tag{4-23}$$

Eqs. (4-22) and (4-23) are the equations of motion in terms of the generalized coordinates ϕ and θ.

4-5. LAGRANGE'S EQUATIONS

It was shown in the last section that the equations of motion of a dynamic system could be obtained from the principle of virtual work.

The procedure, however, is rather cumbersome. The Lagrange method makes use of the generalized coordinates and gives directly the equations of motion. We shall use the principle of virtual displacement to derive Lagrange's equations for a system of particles; the proof can easily be extended to a system of rigid bodies.

Consider a dynamic system of n degrees of freedom with p mass points. Let m_i be the mass and x_i, y_i, and z_i $(i = 1,2, \cdots, p)$ be the Cartesian coordinates of the ith particle and let the system be described by a set of generalized coordinates $(q_1, q_2, \ldots, q_n)$. Let the Cartesian and the generalized coordinates be related as

$$\begin{aligned} x_i &= x_i(q_1, q_2, \ldots, q_n) \\ y_i &= y_i(q_1, q_2, \ldots, q_n) \\ z_i &= z_i(q_1, q_2, \ldots, q_n) \end{aligned} \tag{4-24}$$

where $i = 1,2, \ldots, p$. Lagrange's equations are derived by expressing Eq. (4-11) in terms of the generalized coordinates and introducing the kinetic energy function into the equation.

Let us first rewrite Eq. (4-11) in the form

$$\sum_{i=1}^{p}(X_i\,\delta x_i + Y_i\,\delta y_i + Z_i\,\delta z_i) - \sum_{i=1}^{p} m_i(\ddot{x}_i\,\delta x_i + \ddot{y}_i\,\delta y_i + \ddot{z}_i\,\delta z_i) = 0 \tag{4-25}$$

The first term of this equation represents the work done by the applied forces expressed in terms of the Cartesian coordinates. For infinitesimal variations, δx_i, δy_i, and δz_i, the corresponding variations in the generalized coordinates are

$$\delta x_i = \sum_{j=1}^{n} \frac{\partial x_i}{\partial q_j}\,\delta q_j, \quad \delta y_i = \sum_{j=1}^{n} \frac{\partial y_i}{\partial q_j}\,\delta q_j, \quad \text{and} \quad \delta z_i = \sum_{j=1}^{n} \frac{\partial z_i}{\partial q_j}\,\delta q_j \tag{4-26}$$

Hence the work done δW by the applied forces can be expressed as

$$\delta W = \sum_{j=1}^{n}\left[\sum_{i=1}^{p}\left(X_i\frac{\partial x_i}{\partial q_j} + Y_i\frac{\partial y_i}{\partial q_j} + Z_i\frac{\partial z_i}{\partial q_j}\right)\right]\delta q_j \tag{4-27}$$

or

$$\delta W = \sum_{j=1}^{n} Q_j\,\delta q_j \tag{4-28}$$

where Q_j, as defined by these equations, is called the *generalized force* corresponding to the coordinate q_j. The product of Q_j and δq_j has the dimension of work. When q_j is a length, Q_j has the dimension of a force. When q_j is an angular displacement, Q_j has the dimension of a moment.

Before expressing the second term of Eq. (4-25) in terms of the generalized coordinates, let us introduce the kinetic energy function T as

$$T = \tfrac{1}{2}\sum_{i=1}^{p} m_i(\dot{x}_i^2 + \dot{y}_i^2 + \dot{z}_i^2) \tag{4-29}$$

Since

$$\frac{\partial T}{\partial \dot{x}_i} = m_i\dot{x}_i, \quad \frac{\partial T}{\partial \dot{y}_i} = m_i\dot{y}_i, \quad \text{and} \quad \frac{\partial T}{\partial \dot{z}_i} = m_i\dot{z}_i$$

and

$$\frac{d}{dt}\frac{\partial T}{\partial \dot{x}_i} = m_i\ddot{x}_i, \quad \frac{d}{dt}\frac{\partial T}{\partial \dot{y}_i} = m_i\ddot{y}_i, \quad \text{and} \quad \frac{d}{dt}\frac{\partial T}{\partial \dot{z}_i} = m_i\ddot{z}_i$$

together with Eq. (4-26) the second term in Eq. (4-25) can be expressed as

$$\delta W = \sum_{j=1}^{n}\sum_{i=1}^{p}\left[\frac{d}{dt}\left(\frac{\partial T}{\partial \dot{x}_i}\right)\frac{\partial x_i}{\partial q_j} + \frac{d}{dt}\left(\frac{\partial T}{\partial \dot{y}_i}\right)\frac{\partial y_i}{\partial q_j} + \frac{d}{dt}\left(\frac{\partial T}{\partial \dot{z}_i}\right)\frac{\partial z_i}{\partial q_j}\right]\delta q_j \tag{4-30}$$

To simplify Eq. (4-30), we make use of the relation

$$\frac{d}{dt}\left(\frac{\partial T}{\partial \dot{x}_i}\frac{\partial x_i}{\partial q_j}\right) = \frac{d}{dt}\left(\frac{\partial T}{\partial \dot{x}_i}\right)\frac{\partial x_i}{\partial q_j} + \frac{\partial T}{\partial \dot{x}_i}\frac{\partial \dot{x}_i}{\partial q_j}$$

or

$$\frac{d}{dt}\left(\frac{\partial T}{\partial \dot{x}_i}\right)\frac{\partial x_i}{\partial q_j} = \frac{d}{dt}\left(\frac{\partial T}{\partial \dot{x}_i}\frac{\partial x_i}{\partial q_j}\right) - \frac{\partial T}{\partial \dot{x}_i}\frac{\partial \dot{x}_i}{\partial q_j} \tag{4-31}$$

Expressions similar to Eq. (4-31) can be written for y_i and z_i. Substituting these expressions in Eq. (4-30) leads to

$$\delta W = \sum_{j=1}^{n}\sum_{i=1}^{p}\left[\frac{d}{dt}\left(\frac{\partial T}{\partial \dot{x}_i}\frac{\partial x_i}{\partial q_j} + \frac{\partial T}{\partial \dot{y}_i}\frac{\partial y_i}{\partial q_j} + \frac{\partial T}{\partial \dot{z}_i}\frac{\partial z_i}{\partial q_j}\right) - \left(\frac{\partial T}{\partial \dot{x}_i}\frac{\partial \dot{x}_i}{\partial q_j} + \frac{\partial T}{\partial \dot{y}_i}\frac{\partial \dot{y}_i}{\partial q_j} + \frac{\partial T}{\partial \dot{z}_i}\frac{\partial \dot{z}_i}{\partial q_j}\right)\right]\delta q_j \tag{4-32}$$

To simplify further this equation, it is noted that, from Eq. (4-24), the original velocity components $\dot{x}_i$, $\dot{y}_i$, and $\dot{z}_i$ can be expressed in terms of the generalized coordinates as

$$\dot{x}_i = \sum_{j=1}^{n}\frac{\partial x_i}{\partial q_j}\dot{q}_j, \quad \dot{y}_i = \sum_{j=1}^{n}\frac{\partial y_i}{\partial q_j}\dot{q}_j, \quad \text{and} \quad \dot{z}_i = \sum_{j=1}^{n}\frac{\partial z_i}{\partial q_j}\dot{q}_j$$

Since these are linear functions, we obtain the relations

$$\frac{\partial \dot{x}_i}{\partial \dot{q}_j} = \frac{\partial x_i}{\partial q_j}, \quad \frac{\partial \dot{y}_i}{\partial \dot{q}_j} = \frac{\partial y_i}{\partial q_j}, \quad \text{and} \quad \frac{\partial \dot{z}_i}{\partial \dot{q}_j} = \frac{\partial z_i}{\partial q_j} \tag{4-33}$$

Substituting Eq. (4-33) in Eq. (4-32) gives

$$\delta W = \sum_{j=1}^{n} \sum_{i=1}^{p} \left[\frac{d}{dt} \left(\frac{\partial T}{\partial \dot{x}_i} \frac{\partial \dot{x}_i}{\partial \dot{q}_j} + \frac{\partial T}{\partial \dot{y}_i} \frac{\partial \dot{y}_i}{\partial \dot{q}_j} + \frac{\partial T}{\partial \dot{z}_i} \frac{\partial \dot{z}_i}{\partial \dot{q}_j} \right) - \left(\frac{\partial T}{\partial \dot{x}_i} \frac{\partial \dot{x}_i}{\partial q_j} + \frac{\partial T}{\partial \dot{y}_i} \frac{\partial \dot{y}_i}{\partial q_j} + \frac{\partial T}{\partial \dot{z}_i} \frac{\partial \dot{z}_i}{\partial q_j} \right) \right] \delta q_j \tag{4-34}$$

It is observed that the kinetic energy function T in the Cartesian coordinates is a function of the velocities $\dot{x}_i$, $\dot{y}_i$, and $\dot{z}_i$ only, and it is independent of x_i, y_i, and z_i. T in terms of the generalized coordinates, however, may be functions of the q_j and the $\dot{q}_j$. Differentiating T with respect to $\dot{q}_j$ and q_j, we obtain

$$\frac{\partial T}{\partial \dot{q}_j} = \sum_{i=1}^{p} \left(\frac{\partial T}{\partial \dot{x}_i} \frac{\partial \dot{x}_i}{\partial \dot{q}_j} + \frac{\partial T}{\partial \dot{y}_i} \frac{\partial \dot{y}_i}{\partial \dot{q}_j} + \frac{\partial T}{\partial \dot{z}_i} \frac{\partial \dot{z}_i}{\partial \dot{q}_j} \right) \tag{4-35}$$

$$\frac{\partial T}{\partial q_j} = \sum_{i=1}^{p} \left(\frac{\partial T}{\partial \dot{x}_i} \frac{\partial \dot{x}_i}{\partial q_j} + \frac{\partial T}{\partial \dot{y}_i} \frac{\partial \dot{y}_i}{\partial q_j} + \frac{\partial T}{\partial \dot{z}_i} \frac{\partial \dot{z}_i}{\partial q_j} \right) \tag{4-36}$$

Comparing Eqs. (4-35), (4-36), and (4-34), it is noted that Eq. (4-34) can be written as

$$\delta W = \sum_{j=1}^{n} \left[\frac{d}{dt} \left(\frac{\partial T}{\partial \dot{q}_j} \right) - \frac{\partial T}{\partial q_j} \right] \delta q_j \tag{4-37}$$

Now we have completed the transformation of coordinates for each of the terms in Eq. (4-25). Substituting Eqs. (4-28) and (4-37) in Eq. (4-25) gives

$$\sum_{j=1}^{n} \left(\frac{d}{dt} \frac{\partial T}{\partial \dot{q}_j} - \frac{\partial T}{\partial q_j} - Q_j \right) \delta q_j = 0 \tag{4-38}$$

Since $(q_1, q_2, \ldots, q_n)$ are independent coordinates, and since δq_j is arbitrary, the quantity in the parentheses must be zero. Rearranging this quantity gives

$$\frac{d}{dt} \frac{\partial T}{\partial \dot{q}_j} - \frac{\partial T}{\partial q_j} = Q_j \tag{4-39}$$

This is called Lagrange's equations of motion. Since we assumed that the given system has n degrees of freedom, there are n such equations corresponding to the number of generalized coordinates.

4-6. GENERALIZED FORCES

Lagrange's equations express the generalized force as a function of the kinetic energy of the system. The generalized force, as defined in Eq. (4-28), may consist of three distinct forces: (1) the spring force due to the change of potential energy, (2) the damping force due to the energy dissipation in the equivalent damper, and (3) the applied force externally applied to the system.

The spring force is due to the change of potential energy of the system. Let $q_1, q_2, \ldots, q_n$ be the generalized coordinates describing its geometric configuration. The coordinates are chosen such that, when the system is at its stable equilibrium position, $q_1 = q_2 = \ldots = q_n = 0$. A stable equilibrium position is a configuration about which a system oscillates for all times if it is set in motion with sufficiently small initial disturbances. The potential energy of the system is a function of coordinates only. Hence it can be expanded by Maclaurin's series about the stable equilibrium position as

$$U(q_1,q_2, \ldots, q_n) = U_o + \sum_{i=1}^{n} \left(\frac{\partial U}{\partial q_i}\right)_o q_i + \sum_{i=1}^{n} \sum_{j=1}^{n} \frac{1}{2} \left(\frac{\partial^2 U}{\partial q_i \, \partial q_j}\right)_o q_i q_j + \ldots \quad \textbf{(4-40)}$$

where the subscript o denotes the values at the equilibrium position.

If the potential energy of the system is measured from the equilibrium position, the value U_o can be defined equal to zero. Since the spring force is conservative, the total spring force acting on the system is in equilibrium at $q_1 = q_2 = \ldots = q_n = 0$, and the component of the total force along any direction is zero; that is,

$$\left(\frac{\partial U}{\partial q_i}\right)_o = 0$$

Alternatively, it may be said that if the potential energy is a minimum at equilibrium, its first derivative must vanish. Assuming small oscillations and neglecting all terms that are higher than the second order, the potential energy of a system can be expressed as

$$U = \frac{1}{2} \sum_{i=1}^{n} \sum_{j=1}^{n} \left(\frac{\partial^2 U}{\partial q_i \, \partial q_j}\right)_o q_i q_j \quad \textbf{(4-41)}$$

or

$$U = \frac{1}{2} \sum_{i=1}^{n} \sum_{j=1}^{n} k_{ij} q_i q_j \quad \textbf{(4-42)}$$

where $k_{ij} = k_{ji}$ by definition is the equivalent spring constant to be evaluated at the equilibrium position as indicated in Eq. (4-41). The spring force Q_j associated with the coordinate q_j is

$$Q_j = -\frac{\partial U}{\partial q_j} = -\sum_{i=1}^{n} k_{ij} q_i \tag{4-43}$$

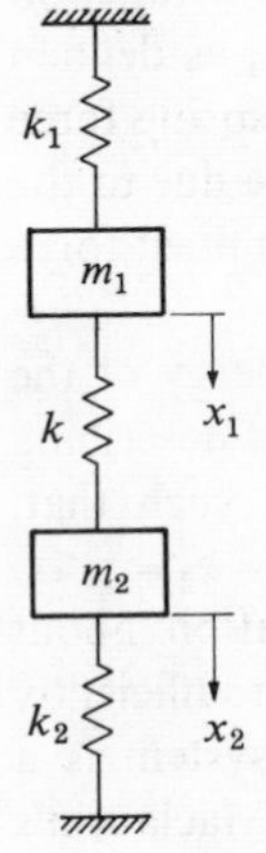

Fig. 4-6. *Two-degree-of-freedom system*

To illustrate the application of Eq. (4-43), consider the mass-spring system as shown in Fig. 4-6. The potential energy due to the deformation of the springs is

$$U = \tfrac{1}{2}k_1x_1^2 + k(x_1 - x_2)^2 + k_2x_2^2$$

This equation can be written as

$$\begin{aligned} U &= \tfrac{1}{2}[(k_1 + k)x_1x_1 - kx_1x_2 - kx_2x_1 + (k_2 + k)x_2x_2] \\ &= \tfrac{1}{2}(k_{11}x_1x_1 + k_{12}x_1x_2 + k_{21}x_2x_1 + k_{22}x_2x_2) \\ &= \tfrac{1}{2}\sum_{i=1}^{2}\sum_{j=1}^{2} k_{ij}x_ix_j \end{aligned}$$

The spring forces in the x_1 and x_2 directions are

$$Q_1 = -\frac{\partial U}{\partial x_1} = -(k_1 + k)x_1 + kx_2 = -k_{11}x_1 - k_{12}x_2 = -\sum_{j=1}^{2} k_{1j}x_j$$

$$Q_2 = -\frac{\partial U}{\partial x_2} = +kx_1 - (k_2 + k)x_2 = -k_{21}x_1 - k_{22}x_2 = -\sum_{j=1}^{2} k_{2j}x_j$$

As indicated in these equations, each of the spring forces acting on m_1 and m_2 consists of two parts. For example, the $k_{11}x_1$ term in Q_1 represents the spring force on m_1 due to the displacement x_1, and the elastic coupling term $k_{12}x_2$ represents the spring force on m_1 due to the displacement x_2. The terms in Q_2 can be explained in like manner. It should be noted that the potential energy stored in a spring, owing to its deformation, is a function of the relative displacement between its two ends and not a direct function of x_1 and x_2 which are absolute displacements. This illustration shows, however, that the spring force can be expressed as a linear function of the coordinates, as shown in Eq. (4-43), with the appropriate values of k_{ij}.

By analogy with the potential energy function, let us define a *dissipation function D*, which, when differentiated with respect to the velocity, yields the damping, or the dissipative, force. The dissipation function is defined as

$$D = \tfrac{1}{2}\sum_{i=1}^{n}\sum_{j=1}^{n} c_{ij}\dot{q}_i\dot{q}_j \tag{4-44}$$

where $c_{ij} = c_{ji}$. Analogous to Eq. (4-43), the damping force Q_j associated with the velocity $\dot{q}_j$ is

$$Q_j = -\frac{\partial D}{\partial \dot{q}_j} = -\sum_{i=1}^{n} c_{ij}\dot{q}_i \tag{4-45}$$

It should be noted that the damping force of a dashpot is proportional to the relative velocity between its two ends and is not a linear function of the velocities. On the basis of the discussion on the spring force, however, Eq. (4-45) can be assumed without the loss of generality. It is observed that the dissipation function, as defined in Eq. (4-44), is equal to half the rate of energy dissipation in the damper.

Combining the spring force, the damping force, and the applied force to form the generalized force, Lagrange's equations can be written as

$$\frac{d}{dt}\frac{\partial T}{\partial \dot{q}_j} - \frac{\partial T}{\partial q_j} + \frac{\partial D}{\partial \dot{q}_j} + \frac{\partial U}{\partial q_j} = Q_j \tag{4-46}$$

where $j = 1,2, \ldots, n$ corresponding to the number of generalized coordinates of the system, and Q_j denotes the applied force corresponding to q_j.

If the system is conservative, Eq. (4-46) reduces to the form of

$$\frac{d}{dt}\frac{\partial T}{\partial \dot{q}_j} - \frac{\partial T}{\partial q_j} + \frac{\partial U}{\partial q_j} = 0 \tag{4-47}$$

Since the potential energy U is a function of the coordinates only, this equation can be further simplified by introducing the Lagrangian function $L = T - U$. Hence Eq. (4-47) can be expressed as

$$\frac{d}{dt}\frac{\partial L}{\partial \dot{q}_j} - \frac{\partial L}{\partial q_j} = 0 \tag{4-48}$$

which is Lagrange's equation of conservative systems.

Example 4. Derive the equations of motion of the spherical pendulum as shown in Fig. 4-5, and compare the resultant equations with those derived in Example 3.

Solution: The kinetic energy of the system is

$$T = \tfrac{1}{2}m(\dot{x}^2 + \dot{y}^2 + \dot{z}^2)$$

Let θ and ϕ be the generalized coordinates. The kinetic energy can be expressed in terms of the generalized coordinates by the relations

$$x = L \sin \phi \cos \theta$$

$$y = L \sin \phi \sin \theta$$

$$z = L \cos \phi$$

Substituting these relations in the kinetic energy function gives

$$T = \tfrac{1}{2}mL^2(\sin^2 \phi\dot{\theta}^2 + \dot{\phi}^2)$$

Alternatively, if the kinetic energy function is written in terms of the velocity components of the mass in spherical polar coordinates, we have

$$T = \tfrac{1}{2}m[(L \sin \phi\dot{\theta})^2 + (L\dot{\phi})^2]$$

It is recognized that these two expressions of the kinetic energy are identical. Hence the energy functions can be written in any convenient coordinates, with little regard to the constraints of the system. The potential energy of the system is

$$U = mgL(1 - \cos \phi)$$

The equations of motion can be obtained by the direct substitution of the kinetic and potential energy functions in Eq. (4-47). For $q_1 = \theta$, we have

$$\frac{d}{dt}(mL^2)(\sin^2 \phi\dot{\theta}) + 0 + 0 = 0$$

or

$$\sin \phi\ddot{\theta} + 2 \cos \phi\dot{\phi}\dot{\theta} = 0$$

For $q_2 = \phi$, we have

$$\frac{d}{dt}(mL^2\dot{\phi}) - mL^2(\sin \phi \cos \phi\dot{\theta}^2) + mgL \sin \phi = 0$$

or

$$\ddot{\phi} - \sin \phi \cos \phi\dot{\theta}^2 + \frac{g}{L} \sin \phi = 0$$

The two equations of motion are identical to those obtained in Example 3.

Example 5. A simple mass-spring system is as shown in Fig. 4-7. The mass is allowed to swing from side to side as well as to oscillate along the axis of the spring. Determine the equations of motion of the system.

Solution: The kinetic energy of the system is

$$T = \tfrac{1}{2}m(\dot{x}^2 + \dot{y}^2)$$

This energy function can be expressed in terms of the generalized coordinates (θ,η) by the relations

$$x = (L + \eta)\sin\theta$$
$$y = (L + \eta)\cos\theta$$

Performing the transformation of coordinates and simplifying, we obtain

$$T = \tfrac{1}{2}m[(L + \eta)^2\dot{\theta}^2 + \dot{\eta}^2]$$

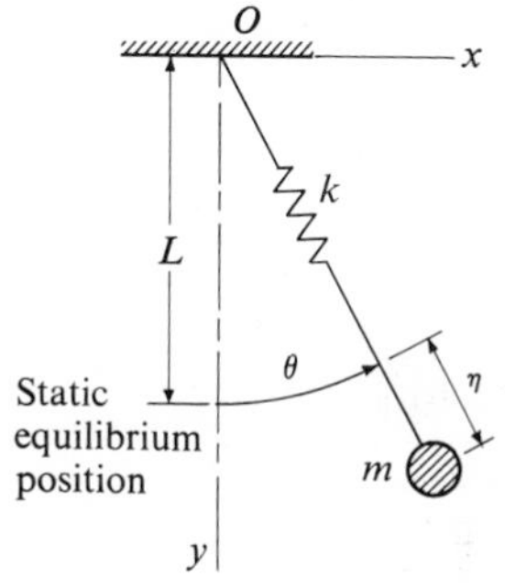

Fig. 4-7. *Simple spring-mass system with two degrees of freedom*

The change in potential energy due to a displacement from the static equilibrium position is equal to the sum of the strain energy in the spring k and the potential energy change due to the change in elevation of the mass m.

$$U = \int_0^{\eta} (k\eta + mg)d\eta + mg[L - (L + \eta)\cos\theta]$$
$$U = \tfrac{1}{2}k\eta^2 + mg(L + \eta)(1 - \cos\theta)$$

The equations of motion can be obtained by the direct substitution of these energy functions in Eq. (4-47). For $q_1 = \eta$, we have

$$m\ddot{\eta} - m(L + \eta)\dot{\theta}^2 + k\eta + mg(1 - \cos\theta) = 0$$

For $q_2 = \theta$, we obtain

$$\frac{d}{dt}[m(L + \eta)^2\dot{\theta}] - 0 + mg(L + \eta)\sin\theta = 0$$

or

$$m(L + \eta)^2\ddot{\theta} + 2m(L + \eta)\dot{\eta}\dot{\theta} + mg(L + \eta)\sin\theta = 0$$

Example 6. A centrifugal pendulum dynamic absorber is as shown in Fig. 4-8. The disk rotating at constant angular velocity Ω has an oscillating torque $T_o \sin\omega t$ acting on it. Assume (1) that the pendulums are identical, (2) that the pendulums oscillate only in the plane of rotation, (3) that viscous damping exists at the pin joints of the pendulum with a coefficient of c_t in.-lb-sec-rad^{-1}, and (4) that the gravitational field is small compared with the centrifugal field. (*a*) Determine the equations of motion of the system for small oscillations of the pendulums. (*b*) Neglecting the friction at the pin joints, find the condition under which the pendulums will act as a dynamic absorber.

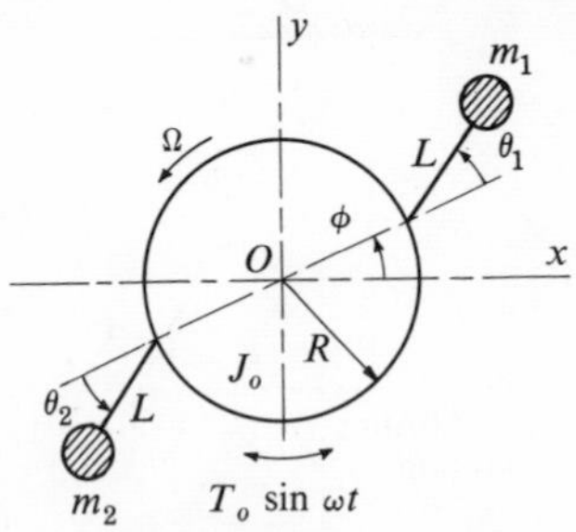

Fig. 4-8. *Pendulum dynamic absorber*

Solution: (*a*) Let the generalized coordinates describing the system be (ϕ,θ_1,θ_2). The energy functions pertaining to this system are

$$T = \tfrac{1}{2}J_o\dot{\phi}^2 + \tfrac{1}{2}m_1(\dot{x}_1^2 + \dot{y}_1^2) + \tfrac{1}{2}m_2(\dot{x}_2^2 + \dot{y}_2^2)$$
$$D = \tfrac{1}{2}c_t(\dot{\theta}_1^2 + \dot{\theta}_2^2)$$
$$U = 0 \text{ (neglecting the gravitational effect)}$$

The (x,y)-coordinates can be transformed into the generalized coordinates by the relations

$$x_1 = R\cos\phi + L\cos(\phi + \theta_1)$$
$$y_1 = R\sin\phi + L\sin(\phi + \theta_1)$$
$$x_2 = -R\cos\phi - L\cos(\phi + \theta_2)$$
$$y_2 = -R\sin\phi - L\sin(\phi + \theta_1)$$

It can be shown, through simple algebraic operations, that the kinetic energy function can be expressed as

$$T = \tfrac{1}{2}J_o\dot{\phi}^2 + \tfrac{1}{2}m[2R^2\dot{\phi}^2 + L^2(\dot{\phi} + \dot{\theta}_1)^2 + L^2(\dot{\phi} + \dot{\theta}_2)^2 + 2RL\dot{\phi}(\dot{\phi} + \dot{\theta}_1)\cos\theta_1 + 2RL\dot{\phi}(\dot{\phi} + \dot{\theta}_2)\cos\theta_2]$$

The equations of motion can be obtained by the direct substitution of the energy functions enumerated in Eq. (4-46). The intermediate steps are shown as follows:

$$\frac{\partial T}{\partial \dot{\phi}} = J_o\dot{\phi} + 2m[R^2 + L^2 + RL(\cos\theta_1 + \cos\theta_2)]\dot{\phi} + mL(L + R\cos\theta_1)\dot{\theta}_1 + mL(L + R\cos\theta_2)\dot{\theta}_2$$

$$\frac{\partial T}{\partial \dot{\theta}_1} = mL(L + R\cos\theta_1)\dot{\phi} + mL^2\dot{\theta}_1$$

$$\frac{\partial T}{\partial \dot{\theta}_2} = mL(L + R\cos\theta_2)\dot{\phi} + mL^2\dot{\theta}_2$$

$$-\frac{\partial T}{\partial \phi} = 0$$

$$-\frac{\partial T}{\partial \theta_1} = mRL\dot{\phi}(\dot{\phi} + \dot{\theta}_1)\sin\theta_1$$

$$-\frac{\partial T}{\partial \theta_2} = mRL\dot{\phi}(\dot{\phi} + \dot{\theta}_2)\sin\theta_2$$

$$\frac{\partial D}{\partial \phi} = 0, \quad \frac{\partial D}{\partial \dot{\theta}_1} = c_t\dot{\theta}_1, \quad \text{and} \quad \frac{\partial D}{\partial \dot{\theta}_2} = c_t\dot{\theta}_2$$

$$\frac{\partial U}{\partial \phi} = \frac{\partial U}{\partial \theta_1} = \frac{\partial U}{\partial \theta_2} = 0$$

$$Q_\phi = T_o \sin\omega t, \quad Q_{\theta_1} = Q_{\theta_2} = 0$$

Substituting these expressions in Eq. (4-46) and simplifying, we obtain

$$\begin{aligned}&\{J_o + 2m[R^2 + L^2 + RL(\cos\theta_1 + \cos\theta_2)]\}\ddot{\phi}\\ &\quad + mL(L + R\cos\theta_1)\ddot{\theta}_1 + mL(L + R\cos\theta_2)\ddot{\theta}_2\\ &\quad - mRL(2\dot{\phi} + \dot{\theta}_1)\dot{\theta}_1\sin\theta_1 - mRL(2\dot{\phi} + \dot{\theta}_2)\dot{\theta}_2\sin\theta_2 = T_o\sin\omega t\end{aligned}$$

$$mL(L + R\cos\theta_1)\ddot{\phi} + mL^2\ddot{\theta}_1 + mRL\dot{\phi}^2\sin\theta_1 + c_t\dot{\theta}_1 = 0$$

$$mL(L + R\cos\theta_2)\ddot{\phi} + mL^2\ddot{\theta}_2 + mRL\dot{\phi}^2\sin\theta_2 + c_t\dot{\theta}_2 = 0$$

Since the disk is acted on by a torque $T_o \sin\omega t$, let us examine the oscillation of the disk about its dynamic equilibrium position. The equilibrium position of ϕ is Ωt. Let

$$\phi = \Omega t + \theta$$

where θ is a small angle. Substituting ϕ and its derivatives in the equations of motion gives

$$\begin{aligned}&\{J_o + 2m[R^2 + L^2 + RL(\cos\theta_1 + \cos\theta_2)]\}\ddot{\theta} + mL(L + R\cos\theta_1)\ddot{\theta}_1\\ &\quad + mL(L + R\cos\theta_2)\ddot{\theta}_2 - mRL(2\Omega + 2\dot{\theta} + \dot{\theta}_1)\dot{\theta}_1\sin\theta_1\\ &\quad - mRL(2\Omega + 2\dot{\theta} + \dot{\theta}_2)\dot{\theta}_2\sin\theta_2 = T_o\sin\omega t\end{aligned}$$

$$mL(L + R\cos\theta_1)\ddot{\theta} + mL^2\ddot{\theta}_1 + mRL(\Omega + \dot{\theta})^2\sin\theta_1 + c_t\dot{\theta}_1 = 0$$

$$mL(L + R\cos\theta)\ddot{\theta} + mL^2\ddot{\theta}_2 + mRL(\Omega + \dot{\theta})^2\sin\theta_2 + c_t\dot{\theta}_2 = 0$$

These nonlinear equations can be linearized by assuming $\sin\theta \doteq \theta$, $\cos\theta \doteq 1$, etc., and neglecting the second- and higher-order terms $\dot\theta\dot\theta_1$, $\dot\theta\dot\theta_2$, etc. The resultant linearized equations are

$$[J_o + 2m(R^2 + L^2 + 2RL)]\ddot\theta + mL(L+R)\ddot\theta_1 + mL(L+R)\ddot\theta_2 = T_o \sin\omega t$$
$$mL(L+R)\ddot\theta + mL^2\ddot\theta_1 + mRL\Omega^2\theta_1 + c_t\dot\theta_1 = 0$$
$$mL(L+R)\ddot\theta + mL^2\ddot\theta_2 + mRL\Omega^2\theta_2 + c_t\dot\theta_2 = 0$$

(*b*) The steady-state solution of these equations can be obtained readily by the mechanical impedance method. (See Sec. 3-4, Chap. 3.) Let us substitute $T_o e^{j\omega t}$ for $T_o \sin\omega t$, $\bar\Theta_1 e^{j\omega t}$ for θ_1, etc., where $\bar\Theta_1$ is the complex amplitude of θ_1. Neglecting the friction term, factoring out $e^{j\omega t}$, and rearranging, we obtain

$$-\omega^2[J_o + 2m(R^2 + L^2 + 2RL)]\bar\Theta - \omega^2 mL(L+R)\bar\Theta_1 - \omega^2 mL(L+R)\bar\Theta_2 = T_o$$
$$-\omega^2 mL(L+R)\bar\Theta + (mRL\Omega^2 - \omega^2 mL^2)\bar\Theta_1 + 0 = 0$$
$$-\omega^2 mL(L+R)\bar\Theta + 0 + (mRL\Omega^2 - \omega^2 mL^2)\bar\Theta = 0$$

Solving for $\bar\Theta$ from this set of algebraic equations, we have

$$\bar\Theta = \frac{(R\Omega^2 - \omega^2 L)T_o}{\omega^2\{\omega^2 J_o L - R\Omega^2[J_o + 2m(R^2 + L^2 + 2RL)]\}}$$

Hence the amplitude of θ can be made equal to zero if the pendulums are tuned so that

$$R/L = \omega^2/\Omega^2$$

SUGGESTED READING

Karman, von, T., and M. A. Biot, *Mathematical Methods in Engineering* (New York: McGraw-Hill Book Co., Inc., 1940), chaps. 3, 5, and 6.

Keller, E. G., *Mathematics of Modern Engineering* (New York: Dover Publications, Inc., 1961), vol. 2, chap. 1.

Scanlan, R. H., and R. Rosenbaum, *Aircraft Vibration and Flutter* (New York: The Macmillan Co., 1951), chap. 2.

Whittaker, E. T., *Analytical Dynamics of Particles and Rigid Bodies* (London: Cambridge University Press, 4th ed., 1937), chap. 2.

Complete jet engine compressor being balanced in a vertical compensator balancing machine (courtesy Tinius Olsen Testing Machine Company)

5 APPLICATIONS

5-1. INTRODUCTION

The basic theory of vibratory systems was developed in the preceding chapters. It was shown that the equations developed could readily be applied to the types of problems considered. Although this is true in many instances, for a large number of physical situations the formulation of the problem and the determination of the equivalent mass, spring stiffness, and damping coefficient may require considerable skill and judgment. The topics discussed in this chapter may serve to illustrate that our previous discussions may have to be modified in order to apply to a physical problem. The object is to point out the problem rather than to go into the ramifications. For example, balancing is introduced, but field balancing, which is often necessary for large high-speed rotors, is omitted; branched geared systems are discussed, but the effect of backlash in gears is not considered. After studying some of these problems, the reader of this book will be aware of other considerations when faced with a physical problem.

5-2. EQUIVALENT VISCOUS DAMPING

In our discussion of linear systems, we assumed (1) that the mass is a rigid body, (2) that the spring is linear and that Hooke's law is obeyed, and (3) that the damper is viscous, that is, the damping force is proportional to the relative velocity between the two ends of the damper. For the types of problems considered, the mass of a system does not change and is sufficiently rigid; within limits, a spring can be made linear. Thus the first two assumptions are reasonably valid. Viscous damping, however, can only be obtained under idealized conditions, and the major nonlinearity originates in damping. In this section we shall discuss a useful method for approximating various types of damping.

Damping in a system is a complex phenomenon. It may originate from (1) the rubbing of two surfaces, dry or lubricated; (2) the moving of a body through a fluid; (3) the forcing of a fluid through an orifice, circular or irregular; (4) the stressing of an imperfect elastic material; and (5) the passing of an electric conductor through a magnetic field. In general, a damping force exists whenever there is energy dissipation in the system. It is probable that more than one type of damping may exist at the same time. Furthermore, damping may be dependent on the operating conditions of a system. It is well known that the viscosity of a fluid is sensitive to temperature changes.

We shall consider Coulomb damping, velocity-squared damping, and solid damping. It should be remembered that the cases considered are simplified mathematical descriptions of a rather complex problem. The hypothesis is that if the damping in a system is small, these damping effects can be represented by that of an equivalent viscous damper.

Consider the equation of motion of a one-degree-of-freedom system

$$m\ddot{x} + c\dot{x} + kx = F_o \sin \omega t \tag{5-1}$$

If the steady-state response of the system is $x = X \sin(\omega t - \phi)$, the net work per cycle ΔE at steady state is

$$\begin{aligned}
\Delta E &= \int (\text{force})\, dx = \int_0^{\tau} (F_o \sin \omega t)\dot{x}\, dt \\
&= \omega F_o X \int_0^{\tau} \sin \omega t \cos(\omega t - \phi)\, dt = F_o X \int_0^{2\pi} \sin \omega t \cos(\omega t - \phi)\, d(\omega t) \\
&= F_o X \pi \sin \phi
\end{aligned} \tag{5-2}$$

Equation (5-2) indicates that if the phase angle ϕ between the force and the displacement is zero, the net work per cycle is zero, since the spring and the mass are conservative elements. When $\phi = 90°$, ΔE is a maximum. Thus the harmonic force $F_o \sin \omega t$ can be considered to be composed of two components—one in phase, or 180° out of phase, with the displacement, and the other in phase with the velocity. The net work is due to the force in phase with the velocity; the damping force is opposed to this force component.

The energy dissipation per cycle ΔE due to a harmonic damping force is

$$\Delta E = \int (\text{damping force})\, dx = \int_0^T (\text{damping force})\, \dot{x}\, dt \tag{5-3}$$

In a system with viscous damping, this energy dissipation is

$$\begin{aligned} \Delta E &= \int_0^T (c\dot{x})\dot{x}\, dt = \int_0^T c\omega^2 X^2 \cos^2(\omega t - \phi)\, dt \\ &= c\omega\pi X^2 \end{aligned} \tag{5-4}$$

Since our interest is in the magnitude of energy dissipation, the negative sign associated with $c\dot{x}$ is neglected. If the damping is nonviscous, an equivalent viscous damping coefficient c_{eq} can be used to describe the damping. Corresponding to Eq. (5-4), we have

$$\Delta E = c_{\text{eq}}\omega\pi X^2 \tag{5-5}$$

The value of c_{eq} is determined by (1) assuming that the steady-state response to a harmonic excitation is also harmonic, (2) evaluating the integral indicated in Eq. (5-3), and then (3) equating Eqs. (5-3) and (5-5). Thus the effect of a nonviscous damper can be approximated.

Since the steady-state motion with nonviscous damping may not be harmonic, the assumption of harmonic motion is reasonable only if the damping is not large enough to change its wave form appreciably. In many practical problems the damping in a system is small, and its effect may be neglected except near resonance. At resonance it is the damping that governs the amplitude of the motion. Hence, equivalent damping is often used to determine the resonance amplitude.

ase 1. Coulomb Damping

The Coulomb (dry friction) damping force is generally assumed to be proportional to the normal force between two sliding bodies. Hence it is independent of the displacement and its derivatives, and, for a given sliding body, the frictional force is of constant magnitude. It should be remembered that in a physical system the force required to start the motion is usually greater than that required to maintain the motion. Furthermore, the frictional coefficient is not necessarily constant, depending somewhat on the surface roughness of the sliding bodies.

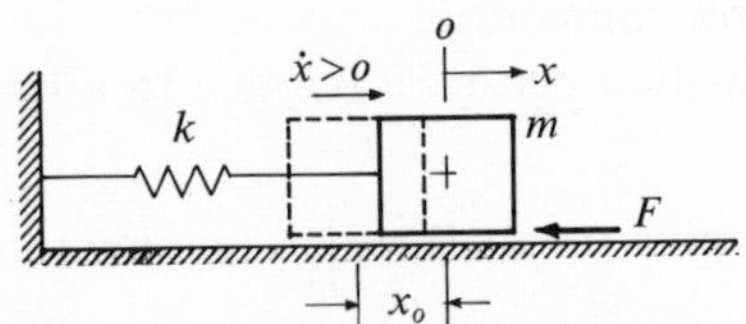

Fig. 5-1. *Free vibration with Coulomb damping*

The free vibration of a system with Coulomb friction is as shown in Fig. 5-1. Assume the spring is unstressed at $x = 0$. The equation of motion of the mass is

$$m\ddot{x} + kx = -F \quad \text{for } \dot{x} > 0 \tag{5-6}$$

$$m\ddot{x} + kx = F \quad \text{for } \dot{x} < 0 \tag{5-7}$$

where the frictional force F is assumed to be constant. Since frictional force always opposes the motion, if the motion is from left to right ($\dot{x} > 0$) the frictional force is $-F$, and vice versa. The solutions of Eqs. (5-6) and (5-7) are

$$x = C_1 \cos \omega_n t + C_2 \sin \omega_n t \mp F/k \tag{5-8}$$

where $\omega_n = \sqrt{k/m}$. The oscillatory motion, described by the circular functions in Eq. (5-8), is harmonic and of frequency ω_n; that is, as if there is no damping in the system. The values of C_1 and C_2 are determined by the initial conditions appropriate to the portion of the cycle considered.

Assuming that at $t = 0$ the mass is displaced by an amount $-x_o$ and released with zero initial velocity, the motion from left to right is

$$x = (F/k - x_o) \cos \omega_n t - F/k \tag{5-9}$$

This equation is valid for $\dot{x} > 0$ corresponding to $0 \leq \omega_n t \leq \pi$. For one half cycle ($\omega_n t = \pi$), the displacement is

$$x = x_o - 2F/k \tag{5-10}$$

and the velocity is $\dot{x} = 0$. Hence the decrease in amplitude is $2F/k$ per half cycle. Now, if the motion is from right to left ($\dot{x} < 0$) with the initial conditions $x(0) = (x_o - 2F/k)$ and $\dot{x}(0) = 0$, it can be shown that the decrease in amplitude is again $2F/k$ per half cycle. Thus, for free vibration, the total decrease in amplitude is $4F/k$ per cycle, as indicated in Fig. 5-2. The corresponding analysis by the energy method is left as an exercise.

If a harmonic force is applied to the mass, the steady-state energy dissipation per cycle can be obtained from Eq. (5-3). Alternatively, if

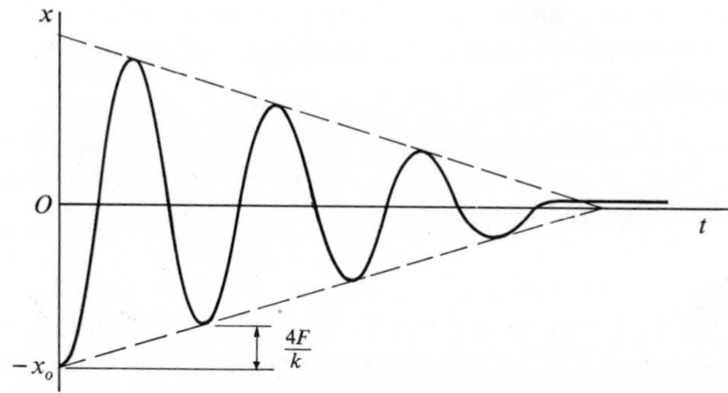

Fig. 5-2. *Rate of amplitude decay for free vibration with Coulomb damping*

the damping force is of constant magnitude F and the total displacement per cycle is $4X$, then

$$\Delta E = 4FX \tag{5-11}$$

The equivalent viscous damping coefficient c_{eq} is obtained by equating ΔE from Eqs. (5-5) and (5-11); that is,

$$c_{\text{eq}} = 4F/\omega\pi X \tag{5-12}$$

It may be noted that c_{eq} is not a constant, as is generally assumed for viscous damping. It is dependent on the excitation frequency ω, the amplitude of oscillation X, as well as the magnitude of the damping force F.

The amplitude of the steady-state response is obtained from the solution of Eq. (5-1) and substituting c_{eq} from Eq. (5-12) for c. From Eq. (2-39) in Chap. 2, this amplitude is

$$X = \frac{F_o/k}{\sqrt{(1 - r^2)^2 + (c_{\text{eq}}\omega/k)^2}} \tag{5-13}$$

where $r = \omega/\omega_n$. Substituting Eq. (5-12) in Eq. (5-13) and simplifying yields

$$X = \frac{F_o}{k}\frac{\sqrt{1 - (4F/\pi F_o)^2}}{1 - r^2} \tag{5-14}$$

This equation gives a real value of X only if $4F/\pi F_o < 1$. It is noted in Eq. (5-14) that the amplitude at resonance is always theoretically infinite. The resonance amplitude can also be viewed from energy considerations. From Eq. (5-2), the phase angle ϕ at resonance is 90°, and the energy input is $F_0 X\pi$. From Eq. (5-11) the energy dissipation per cycle is $4FX$. Hence, if $4F/\pi F_0 < 1$, the energy input exceeds the energy dissipation, and the excess energy is used to build up the amplitude of oscillation. If the damping in a system is heavy, the motion cannot be assumed to be harmonic, and a more exact analysis must be used.

Case 2. Velocity-Squared Damping

When a mass vibrates in a fluid or a fluid is forced through an orifice, the fluid friction is generally assumed to be proportional to the square of the velocity, although the exact value of the velocity exponent is dependent on many factors. Consider a damping force $a\dot{x}^2$ which is proportional to the square of the velocity $\dot{x}$. The frictional force is negative when the velocity $\dot{x}$ is positive, and vice versa. If the motion is harmonic, $x = X \sin \omega t$, the energy dissipation per cycle can be obtained from Eq. (5-3).

$$\Delta E = 2 \int_{-X}^{X} a\dot{x}^2\, dx = 2X^3 \int_{-\pi/2}^{\pi/2} a\omega^2 \cos^3 \omega t\, d(\omega t)$$

$$= \frac{8}{3} a\omega^2 X^3 \qquad \textbf{(5-15)}$$

Equating ΔE in Eqs. (5-5) and (5-15), the equivalent viscous damping coefficient is

$$c_{eq} = 8a\omega X/3\pi \qquad \textbf{(5-16)}$$

Again, c_{eq} is not a constant. It is dependent on the excitation frequency and the amplitude of oscillation.

The amplitude of the steady-state response can be obtained by substituting c_{eq} from Eq. (5-16) in Eq. (5-13). Through simple algebraic operations, it can be shown that this amplitude is

$$X = \frac{3\pi m}{8ar^2} \sqrt{-\frac{(1-r^2)^2}{2} + \sqrt{\frac{(1-r^2)^4}{4} + \left(\frac{8ar^2F_o}{3\pi km}\right)^2}} \qquad \textbf{(5-17)}$$

Case 3. Solid Damping

The spring is considered to be a conservative element in a vibratory system. Solid damping exists if the elastic material is imperfect. For example, when a spring is subjected to a cyclic load reversal, the stress has higher value for the same strain when it is increasing than when it is decreasing. A plot of the stress-strain relation forms a closed loop, and the energy dissipation per cycle is proportional to the area enclosed by the loop. Usually, the magnitude of this type of damping is quite small. Solid damping is variously called hysteresis damping, structural damping, and displacement damping.

Consider a damping force that is proportional to the displacement and independent of frequency.

$$\text{Damping force} = bx = hkx \tag{5-18}$$

where b and h are constants. Since b has the dimension of k, a spring constant, h is defined as a nondimensional coefficient, and $hk = b$.

It was observed in Eq. (5-2) that the damping force is opposite in phase to the velocity. In order to have this force proportional to the displacement, it is expedient to write the equation of motion, Eq. (5-1), in the exponential form and then solve for the steady-state amplitude X by the impedance method. (See Sec. 2-6, Chap. 2.)

$$m\ddot{x} + k(1 + jh)x = F_o e^{j\omega t} \tag{5-19}$$

$$X = \frac{F_o}{|k - m\omega^2 + jhk|} = \frac{F_o/k}{\sqrt{(1 - r^2)^2 + h^2}} \tag{5-20}$$

Comparing Eqs. (5-13) and (5-20), the equivalent viscous damping coefficient is

$$c_{\text{eq}} = hk/\omega = b/\omega \tag{5-21}$$

Thus, from Eq. (5-5), the energy dissipation per cycle is

$$\Delta E = b\pi X^2 \tag{5-22}$$

In solid damping it is generally assumed that the energy dissipation per cycle is independent of frequency and proportional to the square of the strain amplitude, as indicated in the last equation. With this assumption, and using c_{eq} in Eq. (5-21), the resultant equations of motion become linear equations. The nature of structural damping is rather complex, however, and it is not yet well understood. For mild steel, the energy dissipation is found to be proportional to $X^{2.3}$. For other cases, the value of the amplitude exponent may range from 2 to 3.

5-3. BALANCING OF MACHINES

The motions of a large majority of machine elements are of the three following classes: plane rotation, rectilinear translation, and a combination of these two. Plane rotation is the simplest and will be treated first.

MACHINE ELEMENTS WITH PLANE ROTATION

Two disks on a shaft that passes through fixed bearings are shown in Fig. 5-3. If one of the disks is not balanced, as indicated by a weight at *A*, the system will rotate until *A* is at its lowest position. The system is said to be statically and dynamically unbalanced. It can be statically balanced by putting an appropriate weight at *B* such that, statically, the

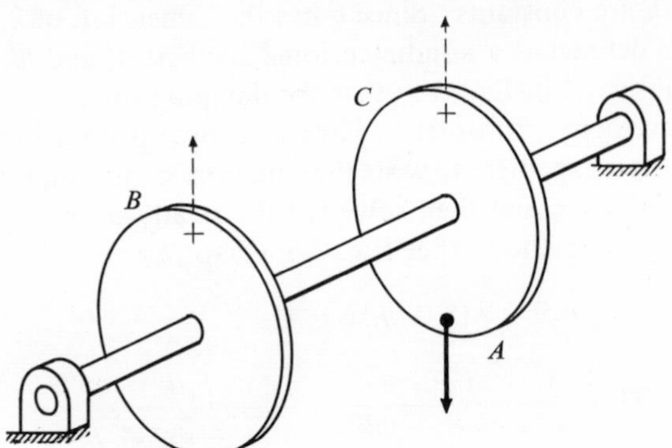

Fig. 5-3. *Static and dynamic balancing; unbalance at A is statically balanced by placing weight at B, or dynamically balanced by placing weight at C*

system is at neutral equilibrium. When the system is in rotation, the weights at *A* and *B* have equal angular velocity, and hence the centrifugal forces produced by the weights are equal. These forces, however, also produce a moment which results in bearing pressure, and the system is not dynamically balanced. If the balancing weight is placed at *C* instead of *B*, the system is then in static and dynamic balance, since there are neither unbalanced forces nor unbalanced moments when the system is in rotation. Any system that is dynamically balanced is also statically balanced, but the converse may not be true.

The general case of balancing a rotating member is shown in Fig. 5-4. In this example the unbalance results from three weights attached to the disks at locations A, B, and C. Let the corresponding centrifugal forces for a given speed of rotation be of magnitudes A, B, and C, respectively. Dynamic balance can be achieved by adding two weights in arbitrarily chosen planes that are perpendicular to the axis of rotation. Let planes 1 and 2 be chosen for the locations of the

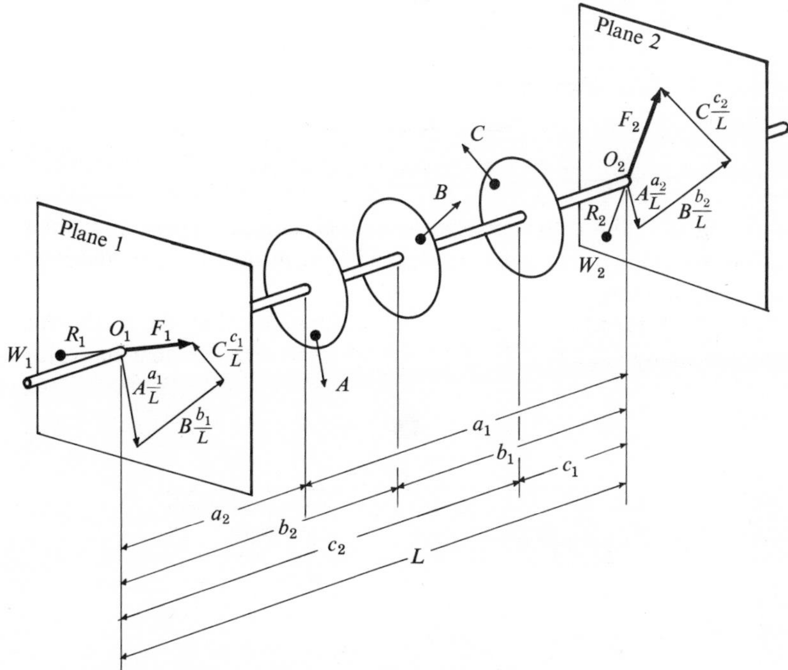

Fig. 5-4. *Dynamic balancing; correction weights can be placed in two arbitrarily chosen planes*

balancing weights. Using plane 2 as reference, the moments of these forces about O_2 are Aa_1, Bb_1, and Cc_1. It should be noted that these moments are vectorial quantities. Let the vectorial sum of these moments be F_1L. The corresponding force in plane 1 at a distance L from plane 2 is F_1. The relation of these forces is as shown in Fig. 5-4. Hence, if a balancing weight W_1 is introduced in plane 1 such that the centrifugal force produced is equal and opposite to F_1, the sum of the moments about O_2 is zero. Similarly, a weight W_2 can be placed in plane 2. Thus the dynamic forces and moments are in balance. Since the entire system rotates at one speed, if the shaft is rigid, it is necessary only to balance the system at an arbitrary speed of rotation.

A practical application of this theory is the balancing of rotors where it is convenient to place a balancing weight at each end, as shown in Fig. 5-5. Rotors are manufactured with reasonable care, but many of them, when they leave the assembly line, are not sufficiently balanced to operate smoothly. There are a number of commercially available

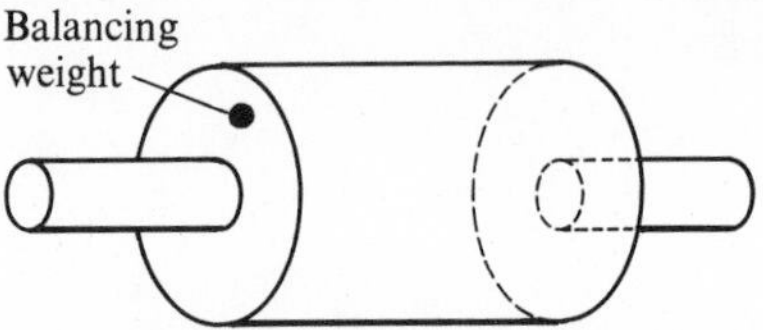

Fig. 5-5. *Method of balancing rotor*

balancing machines to perform the final check. These machines quickly determine the magnitudes and locations of the required correction weights.

In the foregoing discussion, it was assumed that the shaft was rigid. In many cases the flexibility of the shaft requires that the rotor

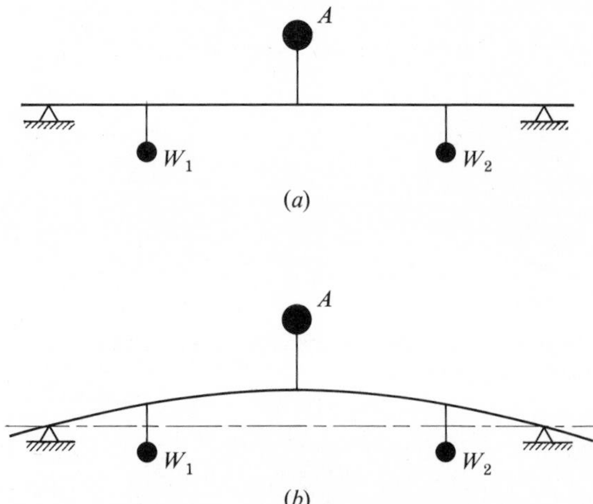

Fig. 5-6. *A rotating member balanced at one speed may not be balanced at other speeds*

be balanced for a particular operating speed. For example, in Fig. 5-6(*a*), let the weights A, W_1, and W_2 be balanced at a low speed of rotation. At a higher speed, if the shaft is deflected as shown in Fig. 5-6(*b*), the system is no longer in balance. Large turbine rotors must be balanced in the field where the rotor is in its own bearings and the housing has been permanently mounted. To achieve a satisfactory balance, it is sometimes necessary to place correction weights in three or more planes.

MACHINE ELEMENTS WITH ROTATION AND TRANSLATION

The balancing of members that do not move with plane rotation is much more difficult. A typical four-bar mechanism, the most common of all mechanisms, is shown in Fig. 5-7. Crank 2 rotates with a constant angular velocity, rocker 4 oscillates, and coupler 3

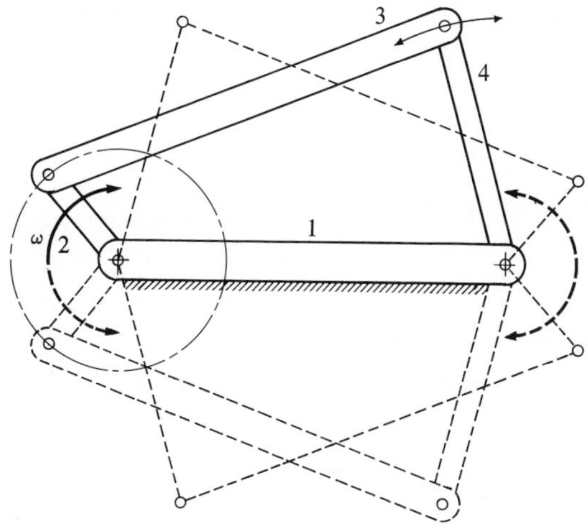

Fig. 5-7. *Four-bar linkage with mirror-image balancing linkage*

moves with a combination of rotation and translation. The system can be balanced by introducing mechanisms that produce the opposite effects. A mechanism that is the mirror image of the original, but moves in the opposite sense, can be introduced to balance the vertical shaking forces and the moments due to the angular accelerations of links 3 and 4. The motion of this combination of linkages would give unbalanced horizontal shaking forces, which can be balanced by a mechanism that is a mirror image of this combination but moves in the opposite sense. Although this method of balancing can be used in most cases, it is usually impractical.

Partial balance of the four-bar mechanism can easily be obtained as shown in Fig. 5-8. First, it is necessary to consider equivalent links. Two members are dynamically equivalent if they have the same total weight, the same center of gravity, and the same moment of inertia. This will be treated in detail in the discussion of the slider-crank mechanism. In this case only the shaking forces will be balanced, hence we do not need to fulfill the requirement of equal moments of inertia. The equivalent coupler shown in Fig. 5-8 satisfies the first two con-

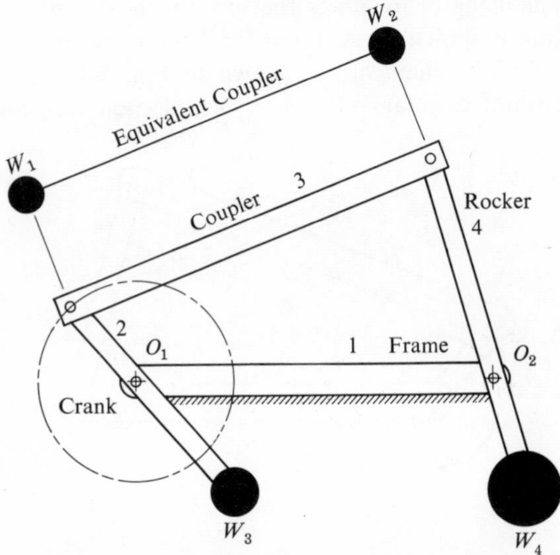

Fig. 5-8. *Partial balance of four-bar linkage*

ditions and is to replace the original coupler. Link 2 and weight W_1 can be balanced by W_3 to bring the center of gravity to the center of rotation O_1. Similarly, link 4 and W_2 can be balanced by W_4. This will eliminate all vertical and horizontal shaking forces, but there will remain an unbalanced and variable torque on the frame owing to the angular accelerations of links 3 and 4.

In many machines, as in the foregoing example, it is neither practical nor necessary to obtain complete balance. Reasonable steps are taken to reduce the unbalance, and then other means are used to control the remainder. Some of these are increasing the damping, avoiding resonance, using dynamic absorbers, and employing proper mounting.

SLIDER-CRANK MECHANISM

The slider-crank mechanism, as shown in Fig. 5-9, is a special case of the four-bar linkage. This mechanism is widely used, so it will be treated in more detail. We shall first consider the motions of the components of the system before discussing the forces involved.

Figure 5-9(a) shows a general position of the slider crank in which R is the crank radius and L the length of the connecting rod. Assume that the crank rotates counterclockwise with constant velocity ω. The displacement of point P at the wrist pin is

$$y = (R + L) - (R \cos \theta + L \cos \beta) \tag{5-23}$$

It can be shown that the approximate relation†

$$\cos \beta = 1 - \frac{R^2}{2L^2} \sin^2 \theta \tag{5-24}$$

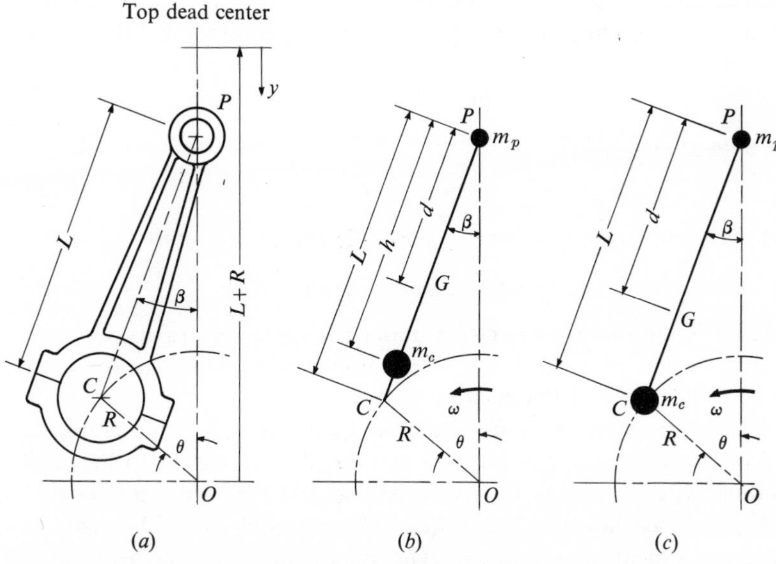

Fig. 5-9. *Dynamic analysis of a slider-crank mechanism*

† From triangle OCP we have the identity

$$R/\sin \beta = L/\sin \theta$$

Using the identity $\sin^2 \beta + \cos^2 \beta = 1$ and the previous one, we obtain

$$\cos \beta = \left(1 - \frac{R^2}{L^2} \sin^2 \theta\right)^{1/2}$$

This can be expanded by the binomial series as

$$\left(1 - \frac{R^2}{L^2} \sin^2 \theta\right)^{1/2} = 1 - \frac{1}{2}\frac{R^2}{L^2} \sin^2 \theta - \frac{1}{8}\frac{R^4}{L^4} \sin^4 \theta - \frac{1}{16}\frac{R^6}{L^6} \sin^6 \theta - \cdots$$

If all terms except the first two on the right side of this equation are neglected, Eq. (5-24) follows. For example, when $R/L = 1/4$, it can be shown that the maximum error introduced in the acceleration calculation is less than 0.6 percent.

is sufficiently accurate for all practical purposes. Substituting Eq. (5-24) in Eq. (5-23) and rearranging gives

$$y = R\left[1 + \frac{L}{R} - \cos\theta - \left(\frac{L}{R} - \frac{R}{2L}\sin^2\theta\right)\right] \tag{5-25}$$

Recalling that $\theta = \omega t$ and using the identity $\cos 2\theta = \cos^2\theta - \sin^2\theta$, we obtain

$$\ddot{y} = R\omega^2\left(\cos\theta + \frac{R}{L}\cos 2\theta\right) \tag{5-26}$$

The motion of the connecting rod at the wrist pin is reciprocating while that at the crankpin is rotating. To calculate its inertia forces, it is desirable to replace the connecting rod with an equivalent two-mass system. The systems in Fig. 5-9(*a*) and (*b*) are equivalent if they have the same mass

$$m = m_p + m_c \tag{5-27}$$

the same center of gravity (consider moments about point P)

$$md = m_c h \tag{5-28}$$

and the same moment of inertia about the center of gravity

$$I_G = mk^2 = m_p d^2 + m_c(h - d)^2 \tag{5-29}$$

where k is the radius of gyration about the center of gravity. It is of interest to note that the center of percussion with respect to the wrist pin P is the center of the mass m_c.

Since the mass m_c is usually close to the center of the crank C, and in many designs it is brought to C by adding weight to the connecting-rod cap, it is customary to assume that the masses are located at P and C, as shown in Fig. 5-9(*c*). This assumption will be made in the development that follows. The requirements of the equivalent system, as shown in Fig. 5-9(*c*), are

$$m = m_p + m_c \tag{5-30}$$

$$md = m_c L \tag{5-31}$$

$$I_G = mk^2 = m_p d^2 + m_c(L - d)^2 \tag{5-32}$$

The equivalent masses, as obtained from these equations, are

$$m_c = md/L$$

$$m_p = m(1 - d/L)$$

The motion of the mass m_c is purely rotational, hence m_c can be balanced by adding weights to the crankshaft. The mass m_p has rectilinear motion, which will produce a vertical force. The acceleration of point P at the wrist pin is prescribed by Eq. (5-26). Since all the reciprocating parts attached to the wrist pin have the same motion, the total vertical inertia force is

$$F_y = -m'_p\ddot{y} = -m'_p R\omega^2 \left(\cos\theta + \frac{R}{L}\cos 2\theta\right) \tag{5-33}$$

where $m'_p = m_p +$ (mass of piston assembly). The force $m'_p R\omega^2 \cos\theta$ is called the primary unbalance or shaking force, and $m'_p(R^2/L)\omega^2 \cos 2\theta$ is the secondary unbalance or shaking force. Thus the primary unbalance has the same frequency as the crankshaft, but the secondary unbalance has twice this frequency.

It is not possible to balance a reciprocating force for all positions of the crank by adding weights to the crankshaft. It can be minimized

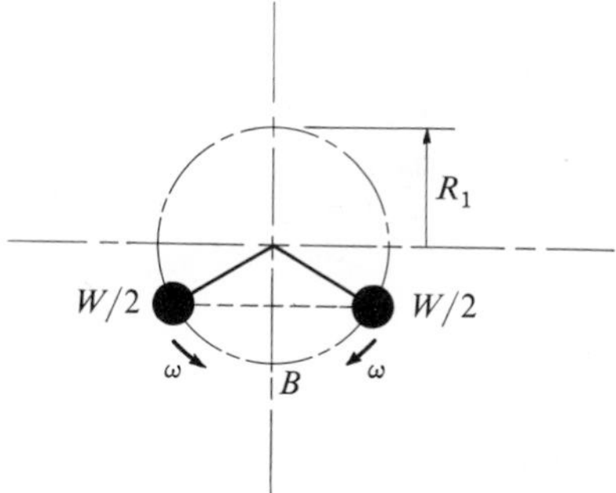

Fig. 5-10. *Balancing of a reciprocating force*

by adding balancing weights to the crankshaft. The shaking forces, as indicated in Eq. (5-33), can be balanced by (1) introducing an opposite effect into the system, using dummy pistons or some form of opposed-piston arrangement for a multicylinder engine, and (2) using counter-rotating weights to obtain a reciprocating force.

Figure 5-10 shows two equal weights rotating at the same radius R_1 with the same angular speed ω but opposite in direction. Since the system is symmetrical about the vertical diameter of the circle, the horizontal components of the centrifugal forces due to the weights cancel one another, and the resultant force is the sum of the vertical components of the centrifugal forces. Thus the resultant force is harmonic along the vertical diameter with a frequency of ω rad-sec^{-1}. If $(W/g)R_1 = m'_p R$ and the weights are at position B when the piston is at top center (Fig. 5-9), then these counterrotating weights will completely balance the primary shaking force if the forces are in the same plane. The secondary shaking force can be balanced in like manner by using counterrotating weights revolving at twice the crankshaft speed. The counterrotating motion is generally obtained by a gear system.

It is evident, from Fig. 5-10, that if only one of the balancing weights $W/2$ is used, half of the primary shaking force is balanced. Owing to the horizontal component of the centrifugal force of $W/2$, a corresponding horizontal unbalance is introduced in the system. The maximum unbalance in either case, however, is only half of that of the original value, and thus the unbalance is minimized. In this case the balancing weight can be attached directly to the crankshaft. The weight can be selected to obtain the degree of unbalance in the vertical and the horizontal directions. For example, in a single-cylinder motorcycle, it is usually desirable to have more unbalanced force in the fore-and-aft direction rather than up and down.

In multicylinder engines the attempt is made to arrange the crank angles so that the unbalance of the various cylinders will nullify each other. Even when this is achieved, there is usually a rocking moment in the vertical plane through the crankshaft. The theory for the standard arrangements of the various types of multicylinder engines is included in most books on dynamics of machinery.

5-4. APPLICATIONS

It was stated in the introduction that our discussion in the preceding chapters may have to be modified in order to apply to a physical situation. For example, the gyroscopic effect of a rotating disk may not be negligible in determining the critical speed of shafts. We shall illustrate these considerations with a few examples.

Example 1. Gyroscopic Effect

The critical speed of a shaft with a disk at mid-span was considered in Chap. 2. If the disk vibrates or whirls in its own plane, the centrifugal forces of the mass particles of the disk lie in one plane. Since these forces cannot introduce additional moments to deflect the shaft, the critical speed could be calculated as the resonant frequency of the system in transverse vibration. The rotation merely furnishes the excitation due to the eccentricity of the disk. Hence the critical speed can also be determined by applying to the system a transverse harmonic force of varying frequency.

Now let us consider a nonsymmetrical system such as that illustrated in Fig. 5-11. We assume that the disk is in complete balance, that the disk and shaft have the same rotational speed in the same direction, that the mass of the shaft is negligible, that the system is undamped, and that it is slightly displaced, as shown in the figure. The whirling motion of the disk is such that every point of the disk moves in a circle in a plane perpendicular to the undistorted center line of the shaft.

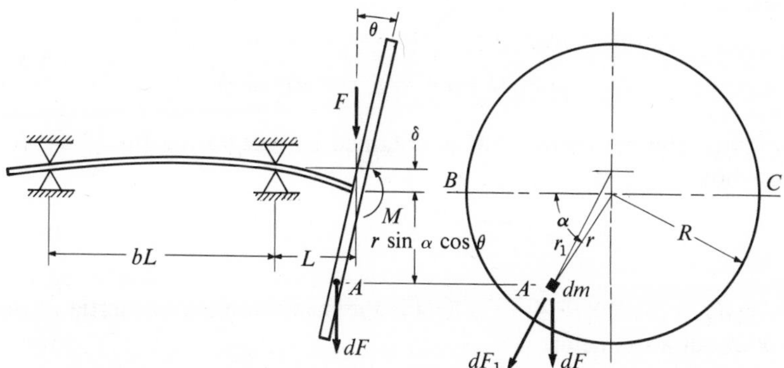

Fig. 5-11. *Centrifugal forces on a disk mounted on a flexible shaft*

It is evident that the centrifugal forces of the mass particles of the disk do not lie in one plane. Thus, in addition to the net centrifugal force tending to bend the shaft, a moment is introduced tending to straighten the shaft. This is called the gyroscopic effect.

The theory can be developed either with the equations of the gyroscope, or, more simply, by a force analysis. The latter will be used here. Furthermore, the analysis can be simplified by considering the rotational system as equivalent to a bent shaft acted upon by a force and a moment due to the centrifugal forces, as shown in Fig. 5-11.

The centrifugal force dF_1 due to a mass dm at point A of the disk is

$$dF_1 \doteq \omega^2 r_1 \, dm$$

Since a small deflection is assumed, we have $\cos\theta \doteq 1$ and $\sin\theta \doteq \theta$. It should be remembered that this force is perpendicular to the undistorted center line of the shaft. The vertical component dF of dF_1 is

$$\begin{aligned} dF &\doteq \omega^2(\delta + r\sin\alpha)\,dm = \omega^2(\delta + r\sin\alpha)(\rho r\,dr\,d\alpha) \\ &= \rho\omega^2(r\delta + r^2\sin\alpha)\,dr\,d\alpha \end{aligned} \tag{5-34}$$

where ρ is mass per unit surface area of the disk. The horizontal components of dF_1 need not be considered, since, for the entire disk, these components cancel one another. Integrating Eq. (5-34) for $0 \leq r \leq R$ and $0 \leq \alpha \leq 2\pi$ to obtain the net centrifugal force F in the vertical direction, we obtain

$$F = \rho\pi R^2\omega^2\delta = m\omega^2\delta \tag{5-35}$$

where m is the mass of the entire disk. This equation shows that the net centrifugal force equals that of a concentrated mass.

The moment dM due to dF, Eq. (5-34), about the horizontal diameter BC is

$$\begin{aligned} dM &\doteq \rho\omega^2(r\delta + r^2 \sin \alpha)(r \sin \alpha \sin \theta)\, dr\, d\alpha \\ &\doteq \rho\omega^2(r^2\delta \sin \alpha + r^3 \sin^2 \alpha)\theta\, dr\, d\alpha \end{aligned} \tag{5-36}$$

Similarly, the net moment M is obtained by integrating Eq. (5-36) for the whole disk; that is,

$$M = \rho \frac{R^4}{4} \pi\omega^2\theta = I_d\omega^2\theta \tag{5-37}$$

where $I_d = (\rho\pi R^2)(R^2/4) = mR^2/4$ is the mass moment of inertia of the disk about a diameter.

Equations (5-35) and (5-37) give the equivalent force F and moment M, respectively, for the analysis of the system shown in Fig. 5-11. The deflections δ and θ, caused by a force F and moment M, can be related as

$$\begin{aligned} \delta &= F\delta_F + M\delta_\theta \\ \theta &= F\theta_F + M\theta_M \end{aligned} \tag{5-38}$$

The influence coefficients (see Sec. 3-7, Chap. 3) are defined as

$\delta_F =$ deflection of shaft per unit force applied

$\quad = (L^3/3EI)(1 + b) \equiv L^3C_1$

$\delta_M =$ deflection of shaft per unit moment applied

$\quad = (L^2/6EI)(3 + 2b) \equiv L^2C_2$

$\theta_F =$ slope of shaft per unit force applied

$\quad = (L^2/6EI)(3 + 2b) \equiv L^2C_3 = L^2C_2$ (Maxwell's reciprocal relation)

$\theta_M =$ slope of shaft per unit moment applied

$\quad = (L/3EI)(3 + b) \equiv LC_4$

The deflection due to F is taken positive and that due to M, which tends to straighten the bent shaft, is negative. Hence the counterclockwise moment here is considered negative. Substituting Eqs. (5-35) and (5-37) in Eq. (5-38) gives

$$\begin{aligned} \delta &= (m\omega^2\delta)\delta_F + (-I_d\omega^2\theta)\delta_M \\ \theta &= (m\omega^2\delta)\theta_F + (-I_d\omega^2\theta)\theta_M \end{aligned} \tag{5-39}$$

The frequency equation of the system can be determined from Eq. (5-39). Upon rearranging, this equation gives

$$\begin{aligned} (1 - m\omega^2\delta_F)\delta + (I_d\omega^2\delta_M)\theta &= 0 \\ (m\omega^2\theta_F)\delta - (1 + I_d\omega^2\theta_M)\theta &= 0 \end{aligned} \tag{5-40}$$

The frequency equation is obtained by equating the determinant of the coefficients of δ and θ to zero. (See Sec. 3-2, Chap. 3.)

$$\Delta(\omega) = \begin{vmatrix} 1 - m\omega^2\delta_F & I_d\omega^2\delta_M \\ m\omega^2\theta_F & -1 - I_d\omega^2\theta_M \end{vmatrix} = 0 \tag{5-41}$$

or

$$mI_d(\delta_F\theta_M - \delta_M\theta_F)\omega^4 - (I_d\theta_M - m\delta_F)\omega^2 - 1 = 0 \tag{5-42}$$

Substituting the influence coefficients defined for the system in Eq. (5-42) yields

$$mI_dL^4(C_1C_4 - C_2^2)\omega^4 - L(I_dC_4 - mL^2C_1)\omega^2 - 1 = 0 \tag{5-43}$$

or

$$\omega^2 = \frac{L(I_dC_4 - mL^2C_1) \pm \sqrt{L^2(I_dC_4 - mL^2C_1)^2 + 4mI_dL^4(C_1C_4 - C_2^2)}}{2mI_dL^4(C_1C_4 - C_2^2)} \tag{5-44}$$

It can be shown that $(C_1C_4 - C_2^2) = (1/6EI)^2(3 + 4b)$, which is a positive quantity. Thus, if ω is real, ω^2 has to be positive, and only the positive sign before the radical in Eq. (5-44) needs to be considered. It is deduced that the system, as shown in Fig. 5-11, has only one natural frequency. It can be shown that this frequency is greater than that of a corresponding system with a concentrated mass at the end of the shaft. The proof of this statement is left as an exercise.

As a second illustration, consider the system shown in Fig. 5-12. The influence coefficients of the system are

$$\begin{aligned} &\delta_F = (L^3/3EI)(ab)^2 \equiv L^3C_1, \quad &&\delta_M = (L^2/3EI)ab(b - a) \equiv L^2C_2 \\ &\theta_F = \delta_M = L^2C_2, &&\theta_M = (L/3EI)(a^2 + b^2 - ab) \equiv LC_4 \end{aligned} \tag{5-45}$$

Again, the counterclockwise moment in this system tends to straighten the shaft and to decrease θ, and M is considered negative. Hence, except that different values are assigned for C_1, C_2, C_3, and C_4, Eq.

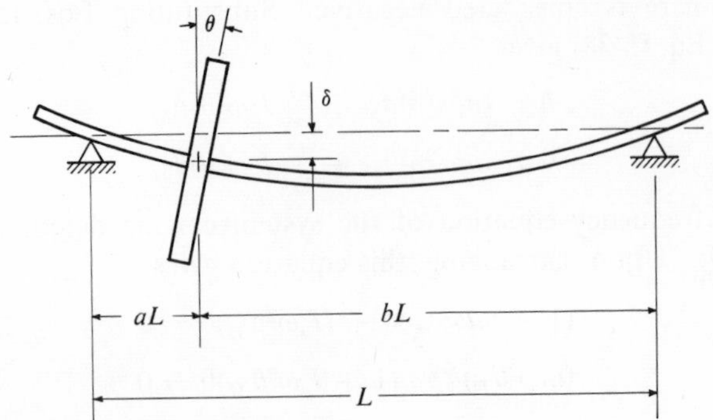

Fig. 5-12. *Disk on flexible shaft*

(5-44) is the frequency equation of this system. It can be shown that the quantity $(C_1C_4 - C_2^2)$ is also a positive quantity.

Let us further illustrate the systems enumerated with specific values.

(*a*) Referring to Fig. 5-11, if $b = 0$, the system becomes a fixed-end cantilever with a rotating disk. From Eq. (5-44) we obtain

$$\omega^2 = \frac{2EI}{mL^3}\left[\left(3 - \frac{mL^2}{I_d}\right) + \sqrt{\left(3 - \frac{mL^2}{I_d}\right)^2 + \frac{3mL^2}{I_d}}\right] \tag{5-46}$$

(*b*) Referring to Fig. 5-11, if $b = 2$, the direct application of Eq. (5-44) gives

$$\omega^2 = \frac{18EI}{11mL^3}\left[\left(\frac{5}{3} - \frac{mL^2}{I_d}\right) + \sqrt{\left(\frac{5}{3} - \frac{mL^2}{I_d}\right)^2 + \frac{11mL^2}{9I_d}}\right] \tag{5-47}$$

(*c*) Referring to Fig. 5-12, if $a = 1/4$ and $b = 3/4$, the direct application of Eqs. (5-44) and (5-45) yields

$$\omega^2 = \frac{8EI}{mL^3}\left[\left(\frac{112}{9} - \frac{mL^2}{I_d}\right) + \sqrt{\left(\frac{112}{9} - \frac{mL^2}{I_d}\right)^2 + \frac{64mL^2}{3I_d}}\right] \tag{5-48}$$

Equations (5-46), (5-47), and (5-48) are plotted in Fig. 5-13 for $\sqrt{I_d/mL^2}$ versus the ratio of the frequencies with disks and with concentrated masses. The quantity $\sqrt{I_d/mL^2}$ is the ratio of the radius of gyration of the disk about its diameter to the beam length as defined in Figs. 5-11 and 5-12; hence it is a measure of the disk effect.

Figure 5-13 shows that the ratio of the frequencies is always greater than unity. In the systems considered, the effect of the rotating disk is to straighten the shaft. If the shaft is stiffened, it is to be expected

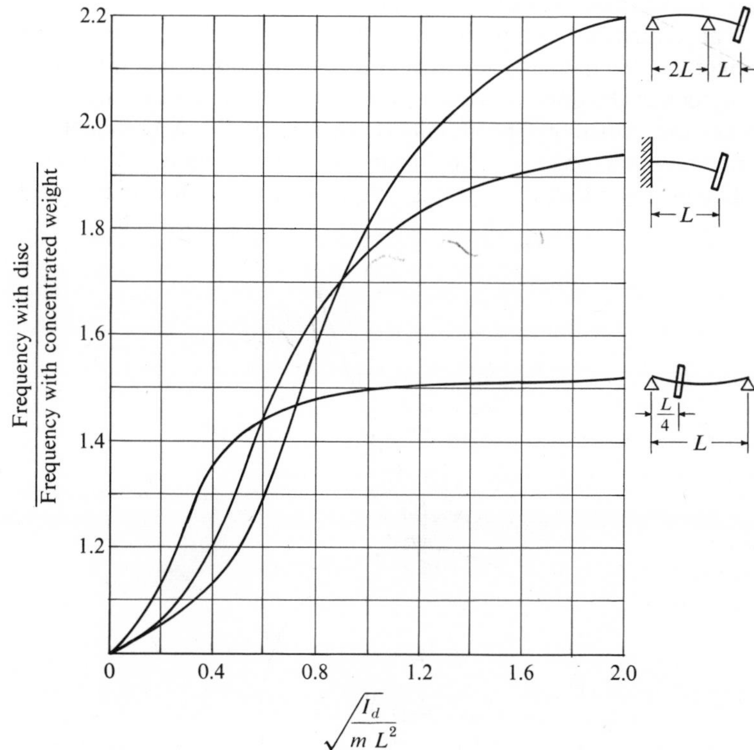

Fig. 5-13. *Change in the natural frequency of a rotating shaft caused by the gyroscope effect of a disk*

that the natural frequency of the system is increased. This is the usual case when the shaft and the disk rotate in the same direction and at the same speed.

It has been observed that the shaft may whirl in one direction and the disk rotate in the opposite direction at the same speed. It can be shown that the corresponding gyroscopic effect is to lower the natural frequency of the system by increasing the shaft deflection. This is an unusual condition and probably is of little practical importance.† The theory for this and for multiple-disk systems is contained in Stodola.‡

† J. P. Den Hartog, *Mechanical Vibrations*, 4th ed., McGraw-Hill Book Co., Inc., New York, 1956, pp. 260–265.

‡ A. Stodola, *Steam and Gas Turbines*, McGraw-Hill Book Co., Inc., New York, 1927, Vol. I, Chap. 5.

Example 2. Centrifugal Pendulum Dynamic Absorber

The centrifugal pendulum, as shown in Fig. 5-14, is somewhat analogous to the dynamic absorber, Fig. 3-15(*a*), discussed in Chap. 3. The dynamic absorber is used to nullify a vertical disturbing force, whereas the centrifugal pendulum is used to nullify a torsional disturbing moment on a rotating member. The dynamic absorber is effective at a particular disturbing frequency for which it is designed,

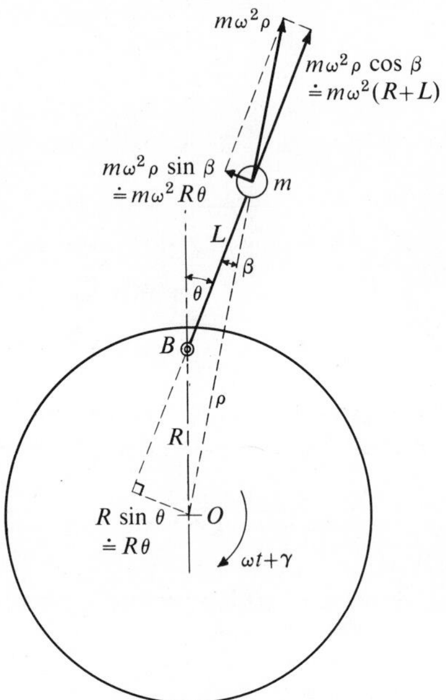

Fig. 5-14. *Centrifugal pendulum*

but it also introduces two resonance frequencies, which may greatly limit the range of operating speed of the system. In a rotating member the number of disturbing cycles per revolution is generally fixed, as in the case of an internal-combustion engine. When a centrifugal pendulum is properly "tuned" for this condition, it is effective for all speeds of operation.

In Fig. 5-14 assume that the disk rotates at an average speed ω with a superimposed oscillation $\gamma = \Gamma \sin n\omega t$, where n is the number of disturbing cycles per revolution. The pendulum is free to oscillate about B, and it is under the influence of a centrifugal field when the system is in rotation. It is assumed that the gravitational field is negligible compared with the centrifugal field at moderate and high speeds. As shown in the figure, the pendulum bob m is subjected to a centrifugal force $m\omega^2\rho$. The tangential component of this force normal to L is $m\omega^2\rho \sin \beta$. From triangle OBm, we have

$$\frac{R}{\sin \beta} = \frac{\rho}{\sin (180 - \theta)} = \frac{\rho}{\sin \theta}$$

or

$$\sin \beta = \frac{R}{\rho} \sin \theta \doteq \frac{R}{\rho} \theta$$

For small angular displacements the tangential acceleration of m is $L\ddot{\theta} + (R + L)\ddot{\gamma}$. Taking moments about B and using d'Alembert's principle, the equation of motion of the pendulum can be expressed as

$$m[L\ddot{\theta} + (R + L)\ddot{\gamma}]L + m\omega^2 RL\theta = 0$$

Substituting $\ddot{\gamma} = -\Gamma(n\omega)^2 \sin n\omega t$ in this equation and rearranging gives

$$\ddot{\theta} + \left(\omega^2 \frac{R}{L}\right) \theta = \frac{R + L}{L} n^2\omega^2\Gamma \sin n\omega t \tag{5-49}$$

This equation is essentially that of the forced vibration of a one-degree-of-freedom system discussed in Chap. 2. Its solution is of the form

$$\theta = \Theta \sin (n\omega t - \phi) \tag{5-50}$$

Substituting Eq. (5-50) in Eq. (5-49), the ratio of the amplitudes can be obtained.

$$\frac{\Gamma}{\Theta} = \frac{\omega^2 R/L - n^2\omega^2}{n^2\omega^2(R + L)/L} \tag{5-51}$$

This equation shows that if $n = \sqrt{R/L}$, the amplitude ratio $\Gamma/\Theta = 0$. Physically, this means that a finite value of Θ is possible for an arbitrarily small value of Γ. If the superimposed oscillation $\gamma = \Gamma \sin n\omega t$ is due to a disturbing torque of the same frequency, then, for a disturbing torque of finite magnitude, the resultant oscillation γ of the disk can be very small and yet Θ will be of finite value. In the limit, if the centrifugal pendulum is "tuned" such that $n = \sqrt{R/L}$, we have $\Gamma = 0$, and the superimposed torque or moment is balanced by the inertia torque of the pendulum. Hence, if the disturbance per cycle n is constant, the pendulum acts as a dynamic absorber for all speeds of rotation of the disk. From Fig. 5-14 the magnitudes of the disturbing torque T_o and Θ are related as

$$T_o = m\omega^2(R + L)R\Theta \tag{5-52}$$

We shall illustrate the use of these equations as follows:

An engine operating at 1,800 rpm has a disturbing torque of magnitude $T_o = 4{,}600$ lb-in. with $n = 4$. The most convenient length for R is 3.75 in. Assuming that the maximum amplitude of oscillation of the pendulum is 10°, determine the length L and the weight of the pendulum.

Solution: The length of a properly tuned pendulum is

$$L = R/n^2 = 3.75/16 = 0.234 \text{ in.}$$

From Eq. (5-52), the required weight of the pendulum is

$$\text{Weight} = mg = \frac{T_o g}{\omega^2(R + L)R\Theta}$$

$$= \frac{(4{,}600)(386)}{(60\pi)^2(3.75 + 0.234)(3.75)(\pi/18)} = 19.2 \text{ lb}$$

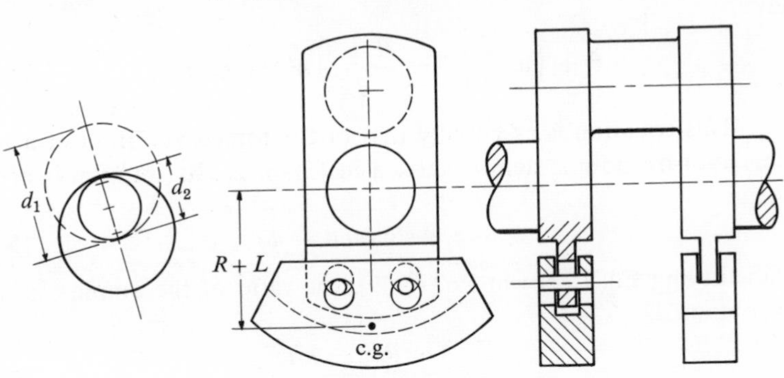

Fig. 5-15. *Bifilar type of centrifugal-pendulum absorber*

The length of the pendulum is usually small, in this example being 0.234 in. This poses a design problem in providing sufficient mass with such a small length. The problem is solved by using the bifilar type of centrifugal pendulum shown in Fig. 5-15. Dividing the pendulum weight in two equal parts, each is mounted on two loosely fitted pins of diameter d_2 through holes of equal diameter d_1 in the weights and the crank. Each weight can then move with curvilinear translation; each point on the weight moves in the arc of a circle of radius $(d_1 - d_2)$. Thus the length of the pendulum is $L = (d_1 - d_2)$. When used on a radial internal-combustion engine, the pendulum is usually attached to the crank web. Not only is this convenient, but the opposing moment is close to the disturbing torque. In addition, the pendulum can serve as a counterweight for the rotating mass of the crank and connecting rod.

The centrifugal pendulum is a dynamic absorber in that it minimizes vibrations due to a disturbing torque by creating an equal and opposite torque. Except for the unavoidable friction in the system, it is not a friction-type damper.

Example 3. Lancaster Damper

The Lancaster torsional vibration damper is a friction damper. Unlike the dynamic absorber discussed in the last example, this damper limits the amplitude of oscillation through energy dissipation due to friction and relative motion between the parts of the damper. Since its action depends upon relative motion, it cannot completely eliminate vibrations, and it is most effective when placed in a position at which the largest amplitude of oscillation is anticipated. The Lancaster damper has found application in torsional systems, such as diesel engines, which must operate over a wide speed range, including the critical speeds. The damper is effective in limiting the amplitudes of vibration at critical speeds.

A dry-friction type of Lancaster damper is shown in Fig. 5-16(*a*). The damper consists of a hub and disk keyed to a shaft that is subjected to torsional vibration, and two spring-loaded flywheels that can rotate relative to the disk. The flywheels are driven by friction existing between themselves and the disk. Obviously, if the spring load is very high, the flywheels will not slip but will oscillate with the shaft, and there will be no energy dissipation. If the spring load is zero, the disk will oscillate freely between the flywheels, and again there will be no energy dissipation. Optimum energy dissipation, and thereby damping action, must lie somewhere between these two extremes.

The mathematical treatment of the Lancaster damper is rather involved,† and we shall discuss it qualitatively. With proper adjustment, the flywheels will oscillate with the shaft for small oscillations. For larger oscillations the flywheels will slip during part of each cycle, as shown in Fig. 5-16(*b*). With Coulomb friction the flywheels are driven with a constant torque while slipping; hence the flywheels move with constant acceleration, as indicated by the straight-line segments in the figure. The amount of slipping increases as the amplitude of oscillation increases until the flywheels slip continuously, as shown in Fig. 5-16(*c*). The flywheels thus produce a continuous drag on the disk. Energy dissipation in the damper is equal to the integral of the Coulomb frictional torque times the relative angular displacement $d\theta$ between the

† J. P. Den Hartog and J. Ormondroyd, "Torsional-Vibration Dampers," *Trans. ASME*, APM-52-13, September–December, 1930, pp. 133–152.

disk and the flywheels. Since the torque is constant and $d\theta = (d\theta/dt)\,dt$, energy dissipation is proportional to the shaded areas between the curves, as shown in Fig. 5-16(*b*) and (*c*).

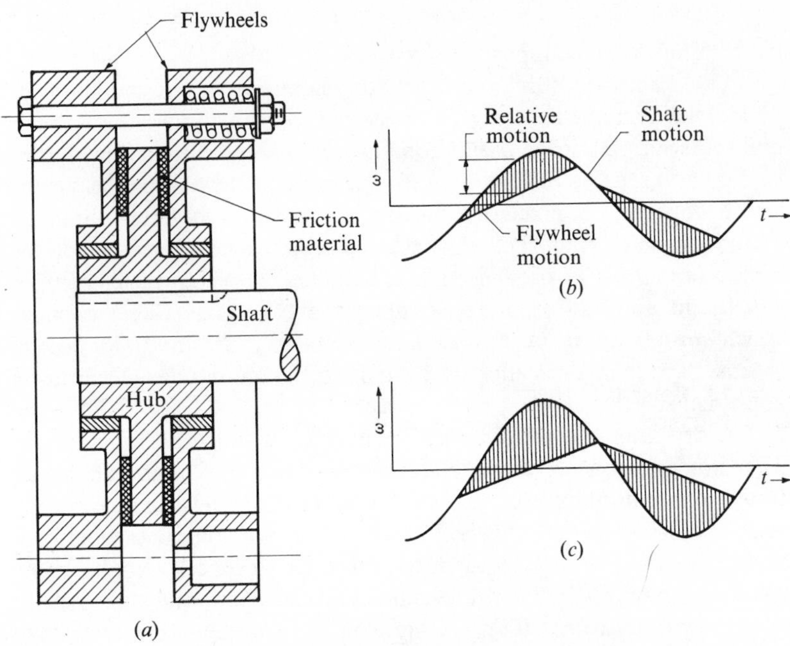

Fig. 5-16. *Lancaster damper*

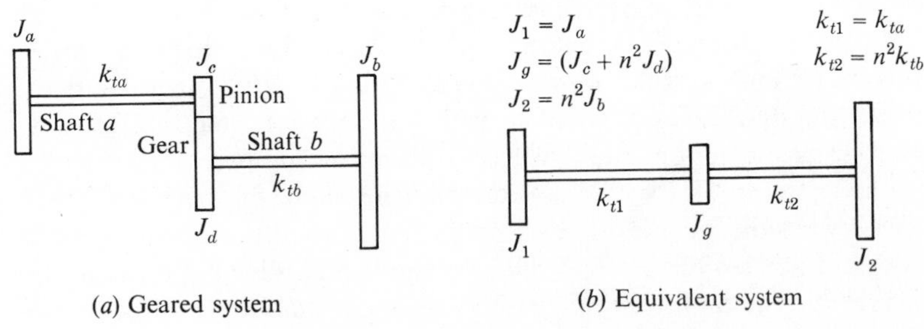

Fig. 5-17. *A two-shaft geared system*

Example 4. Branched Geared Systems

Geared systems are commonly encountered in engineering. In an automobile the engine torque is transmitted to the rear wheels by gearing. In marine propulsion two prime movers may be connected by gearing to a common transmission shaft. A two-shaft geared system was considered in Chap. 3, and we shall briefly discuss multishaft branched geared systems in this example.

It was shown in Chap. 3 (see Sec. 3-3) that the natural frequencies of a geared system can be determined by reducing the geared system to an equivalent system such that all gears and shafts have the same speed. For the two-shaft system shown in Fig. 5-17(*a*), if N_1 is the number of teeth of the pinion and N_2 that of the gear and $n = N_1/N_2$, the equivalent system referring to shaft *a* is as shown in Fig. 5-17(*b*).

The equations of motion of the equivalent system, Fig. 5-17(*b*), are

$$\begin{aligned} J_1\ddot{\theta}_1 &= -k_{t1}(\theta_1 - \theta_g) \\ J_g\ddot{\theta}_g &= -k_{t1}(\theta_g - \theta_1) - k_{t2}(\theta_g - \theta_2) \\ J_2\ddot{\theta}_2 &= -k_{t2}(\theta_2 - \theta_g) \end{aligned} \qquad \textbf{(5-53)}$$

The natural frequencies of the system can be determined by the procedure prescribed in Chap. 3. Assuming that $\theta_i = \Theta_i \sin(\omega t + \psi)$, $i = 1, 2$, and g, substituting this expression in Eq. (5-53), factoring out the sine term, and rearranging, we obtain

$$\begin{aligned} (k_{t1} - \omega^2 J_1)\Theta_1 - k_{t1}\Theta_g &= 0 \\ -k_{t1}\Theta_1 + (k_{t1} + k_{t2} - \omega^2 J_g)\Theta_g - k_{t2}\Theta_2 &= 0 \\ -k_{t2}\Theta_g + (k_{t2} - \omega^2 J_2)\Theta_2 &= 0 \end{aligned} \qquad \textbf{(5-54)}$$

The frequency equation is obtained by equating the determinant of the coefficients of Θ_i to zero; that is,

$$\Delta(\omega) = \begin{vmatrix} k_{t1} - \omega^2 J_1 & -k_{t1} & 0 \\ -k_{t1} & k_{t1} + k_{t2} - \omega^2 J_g & -k_{t2} \\ 0 & -k_{t2} & k_{t2} - \omega^2 J_2 \end{vmatrix} = 0 \qquad \textbf{(5-55)}$$

The natural frequencies are found by expanding the determinant and solving the resultant equation for ω^2.

For the type of problem considered, it is expedient to express the frequency equation in a modified form. Defining $\omega_1^2 = k_{t1}/J_1$ and $\omega_2^2 = k_{t2}/J_2$ and using simple algebraic operations, Eq. (5-55) can be expressed as

$$J_g = \frac{k_{t1}}{\omega^2 - \omega_1^2} + \frac{k_{t2}}{\omega^2 - \omega_2^2} \qquad \textbf{(5-56)}$$

The derivation of Eq. (5-56) from Eq. (5-55) is left as an exercise. Since the system considered is semidefinite and has three disks, it has three natural frequencies, one of which is zero. The zero frequency can be obtained from Eq. (5-55), but it is not indicated in Eq. (5-56).

Alternatively, Eq. (5-56) can be derived directly from Eq. (5-54). Summing the three equations of the set of equations gives

$$\omega^2(J_1\Theta_1 + J_g\Theta_g + J_2\Theta_2) = 0 \tag{5-57}$$

This shows that one of the natural frequencies is zero, corresponding to the zero mode of a semidefinite system. The other frequencies are determined from

$$J_1\Theta_1 + J_g\Theta_g + J_2\Theta_2 = 0 \tag{5-58}$$

Since the amplitude of the inertia torque of disk J_i is $J_i\omega^2\Theta_i$, Eq. (5-58) indicates that the sum of the inertia torque at a principal mode of vibration is zero. This is necessarily true of a semidefinite system. In fact, from Eq. (5-53) we also have

$$J_1\ddot{\theta}_1 + J_g\ddot{\theta}_g + J_2\ddot{\theta}_2 = 0 \tag{5-59}$$

This shows that the sum of the inertia torque must vanish at all times.

Using Eq. (5-58) and the first and third equations of Eq. (5-54), we have

$$J_g\Theta_g + J_1\Theta_1 + J_2\Theta_2 = \left(J_g - \frac{k_{t1}}{\omega^2 - \omega_1^2} - \frac{k_{t2}}{\omega^2 - \omega_2^2}\right)\Theta_g = 0 \tag{5-60}$$

If Θ_g is not zero, clearly Eq. (5-56) can be obtained from Eq. (5-60). Thus two natural frequencies other than $\omega^2 = 0$ can be obtained. If Θ_g is zero, physically this means that J_g is located at a node. From Eq. (5-58) we have $J_1\Theta_1 + J_2\Theta_2 = 0$. This implies that, at this particular principal mode, the inertia torques on the right and left side of J_g must balance each other. Since the system is at a principal mode with $\Theta_g = 0$, one of the natural frequencies is

$$\omega = \omega_1 = \sqrt{k_{t1}/J_1} = \omega_2 = \sqrt{k_{t2}/J_2} \tag{5-61}$$

If $\omega_1 = \omega_2 \equiv \Omega$, Eq. (5-56) becomes

$$J_g = \frac{k_{t1} + k_{t2}}{\omega^2 - \Omega^2} = \frac{\Omega^2(J_1 + J_2)}{\omega^2 - \Omega^2} \tag{5-62}$$

Rearranging this equation gives

$$\omega^2 = \Omega^2(J_1 + J_2 + J_g)/J_g \tag{5-63}$$

Hence, in a two-shaft system, the nonzero natural frequencies can be obtained from Eq. (5-56). If $\omega_1 = \omega_2$, the nonzero frequencies are specified in Eqs. (5-61) and (5-63). This discussion can easily be extended to multishaft systems.

Consider the equivalent system of a three-shaft geared system shown in Fig. 5-18(*b*). Analogous to Eqs. (5-54) and (5-58), we have

$$
\begin{aligned}
(k_{t1} - \omega^2 J_1)\Theta_1 - k_{t1}\Theta_g &= 0 \\
-k_{t1}\Theta_1 + (k_{t1} + k_{t2} + k_{t3} - \omega^2 J_g)\Theta_g - k_{t2}\Theta_2 - k_{t3}\Theta_3 &= 0 \\
-k_{t2}\Theta_g + (k_{t2} - \omega^2 J_2)\Theta_2 &= 0 \\
-k_{t3}\Theta_g + (k_{t3} - \omega^2 J_3)\Theta_3 &= 0
\end{aligned}
\tag{5-64}
$$

and

$$J_1\Theta_1 + J_g\Theta_g + J_2\Theta_2 + J_3\Theta_3 = 0 \tag{5-65}$$

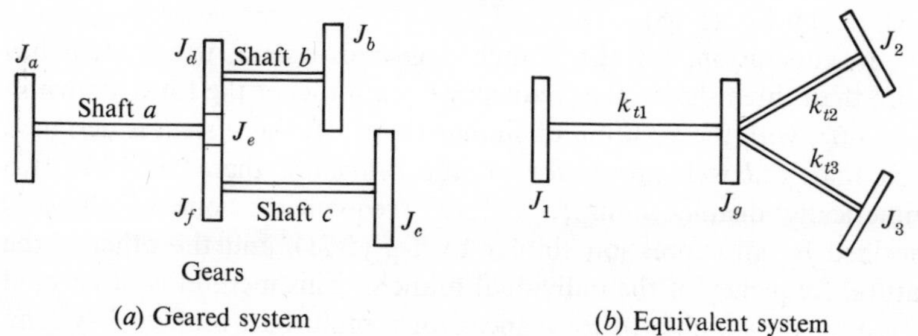

(*a*) Geared system (*b*) Equivalent system

Fig. 5-18. *A three-shaft geared system*

Using Eq. (5-65) and the first, third, and fourth equations of Eq. (5-64), we have

$$\left(J_g - \frac{k_{t1}}{\omega^2 - \omega_1^2} - \frac{k_{t2}}{\omega^2 - \omega_2^2} - \frac{k_{t3}}{\omega^2 - \omega_3^2}\right)\Theta_g = 0 \tag{5-66}$$

where $\omega_i^2 = k_{ti}/J_i$, $i = 1$, 2, and 3. If Θ_g is not zero, the nonzero natural frequencies can be obtained from the equation

$$J_g = \frac{k_{t1}}{\omega^2 - \omega_1^2} + \frac{k_{t2}}{\omega^2 - \omega_2^2} + \frac{k_{t3}}{\omega^2 - \omega_3^2} \tag{5-67}$$

This equation shows that a three-shaft system has three nonzero natural frequencies. If Θ_g is zero, J_g is located at a node. From Eq. (5-65) we have

$$J_1\Theta_1 + J_2\Theta_2 + J_3\Theta_3 = 0 \tag{5-68}$$

Hence the inertia torques due to J_1, J_2, and J_3 must be in balance. If one of the disks is not oscillating, the remaining two branches must have the same natural frequency.

If the system is designed such that

$$\omega_1 = \sqrt{k_{t1}/J_1} = \omega_2 = \sqrt{k_{t2}/J_2} = \omega_3 = \sqrt{k_{t3}/J_3} \equiv \Omega \tag{5-69}$$

then Eq. (5-67) becomes

$$J_g = \frac{k_{t1} + k_{t2} + k_{t3}}{\omega^2 - \Omega^2} = \frac{\Omega^2(J_1 + J_2 + J_3)}{\omega^2 - \Omega^2} \tag{5-70}$$

Rearranging this equation gives

$$\omega^2 = \Omega^2(J_1 + J_2 + J_3 + J_g)/J_g \tag{5-71}$$

It should be noted that a three-shaft system has three nonzero natural frequencies, but only two are indicated in Eqs. (5-69) and (5-71). It is deduced that if k_t/J is identical for all the branches, two of the natural frequencies are equal to Ω, Eq. (5-69), and the third is as specified in Eq. (5-71). The modes, corresponding to the equal frequencies, are specified by Eq. (5-68).

In conclusion, for the branched geared system, if a system has more than three shafts, the frequency equation is of the form shown in Eq. (5-67) with the addition of similar terms. If the system is designed such that k_t/J is identical for all the branches, there are only two numerically distinct nonzero natural frequencies; one of these is specified by an expression similar to Eq. (5-71), and the other is the natural frequency of the individual branch. This method is often used to reduce the number of resonances for a multishaft branched system. It is called the *nodal drive.*

Example 5. Elasticity of Bearings and Supports

Perfectly rigid bearings were assumed in the calculation of the critical speed of shafts in Chap. 2. Figure 5-19 shows a pulley assembly

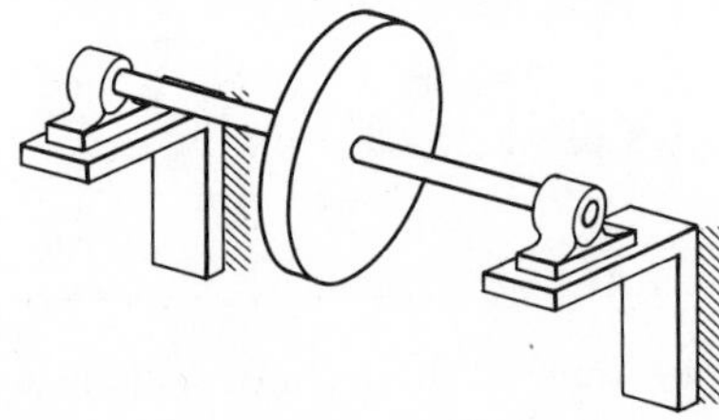

Fig. 5-19. *Pulley assembly with flexible bearing supports*

in which the supporting brackets can be deflected more easily up and down than sideways. It may be necessary to include the elasticity of the supports in the critical-speed calculation. In general, the elasticity of the bearings and supports is difficult to determine, and it is not necessarily the same in the vertical and lateral directions. The effect of the elasticity is to render the system more flexible and therefore to lower the critical speed. The lowering of the critical speed can be greater than 25 percent in some installations.

Consider the system, as shown in Fig. 5-20, in which the elasticity of the bearings and supports is represented by springs mounted in rigid

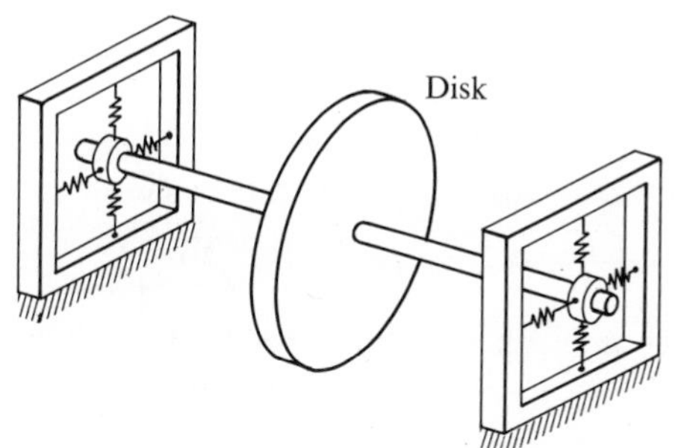

Fig. 5-20. *System with elastic bearings and supports*

frames. Thus the equivalent spring constants k_x and k_y in the x- and y-directions, respectively, are due partly to the stiffness of the shaft (see Sec. 2-12, Chap. 2) and partly to that of the bearings and their

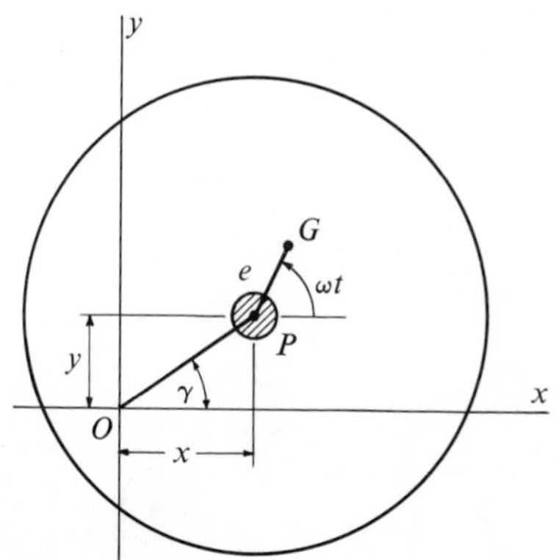

Fig. 5-21. *General position of rotating disk*

supports. A general position of the rotating disk in this system is shown in Fig. 5-21. The disk is keyed to the shaft, and it is rotating with angular speed ω about point P. P is the geometric center and G the mass center of the disk. O is the center of rotation of the system corresponding to the static equilibrium position of the shaft axis.

Neglecting the gravitational and gyroscopic effects, assuming that the system is undamped, and resolving the forces in the horizontal and vertical directions, we obtain

$$m\frac{d^2}{dt^2}(x + e\cos\omega t) + k_x x = 0$$
$$m\frac{d^2}{dt^2}(y + e\sin\omega t) + k_y y = 0 \tag{5-72}$$

or

$$m\ddot{x} + k_x x = me\omega^2\cos\omega t$$
$$m\ddot{y} + k_y y = me\omega^2\sin\omega t \tag{5-73}$$

These equations indicate that the system has two natural frequencies and therefore two critical speeds. Defining $\omega_{nx} = \sqrt{k_x/m}$, $\omega_{ny} = \sqrt{k_y/m}$, $r_x = \omega/\omega_{nx}$, and $r_y = \omega/\omega_{ny}$, from Eq. (2-81), Chap. 2, the steady-state solutions to Eq. (5-73) are

$$x = \frac{r_x^2 e}{1 - r_x^2}\cos\omega t = X\cos\omega t$$
$$y = \frac{r_y^2 e}{1 - r_y^2}\sin\omega t = Y\sin\omega t \tag{5-74}$$

Hence the critical-speed ratios are $r_x = 1$ and $r_y = 1$; that is, $\omega = \omega_{nx}$ and $\omega = \omega_{ny}$. The angle γ indicated in Fig. 5-21 is

$$\gamma = \tan^{-1}(y/x) \tag{5-75}$$

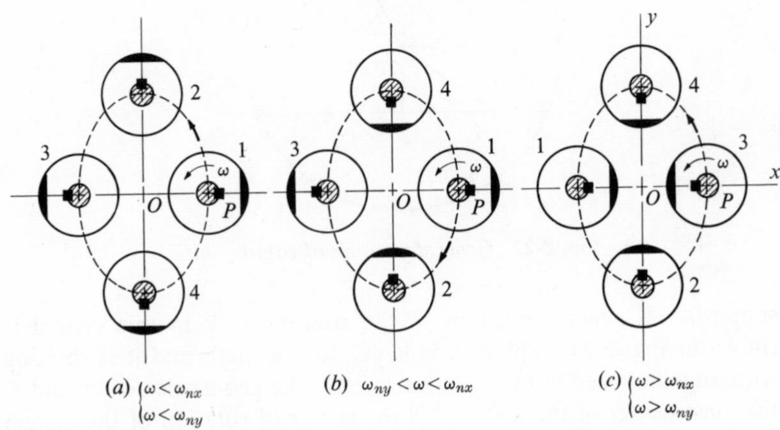

Fig. 5-22. *Rotation of disk about O for various frequencies*

The x- and y-motions in Eq. (5-74) can be combined to give

$$\frac{x^2}{X^2} + \frac{y^2}{Y^2} = 1 \tag{5-76}$$

Hence point P moves in an ellipse, as illustrated in Fig. 5-22. The shaded segment of the disk indicates the heavy side corresponding to the location of the mass center. Let the disk be rotating about point P with angular speed ω, and let P be rotating about point O. When $\omega < \omega_{nx}$ and $\omega < \omega_{ny}$, Fig. 5-22(*a*), both the disk and P rotate in the same direction with the same speed. Assuming that $\omega_{nx} > \omega_{ny}$, when $\omega_{ny} < \omega < \omega_{nx}$, Fig. 5-22(*b*), the disk and P rotate in opposite directions with the same speed. When ω is greater than the critical speeds, again the disk and P rotate in the same direction with the same speed, as indicated in Fig. 5-22(*c*). It is interesting to note that when the excitation is above or below the critical speeds, there is no reversal in stresses in the shaft; that is, the compression side of the shaft remains in compression while it is revolving, and the tension side remains in tension. When the excitation is between the two critical speeds, the shaft undergoes two reversals in stress per revolution.

SUGGESTED READING

Den Hartog, J. P., *Mechanical Vibrations* (New York: McGraw-Hill Book Co., Inc., 4th ed., 1956), chaps. 5, 6, and 8.

Harris, C. M., and C. E. Crede, editors, *Shock and Vibration Handbook* (New York: McGraw-Hill Book Co., Inc., 1961), vol. 1, chap. 6; vol. 2, chaps. 35, 36, and 37; vol. 3, chaps. 38 and 39.

Holowehko, A. R., *Dynamics of Machinery* (New York: John Wiley & Sons, Inc., 1955), chaps. 17–20.

MacDuff, J. N., and J. R. Curreri, *Vibration Control* (New York: McGraw-Hill Book Co., Inc., 1958), chap. 4.

Myklestad, N. O., *Fundamentals of Vibration Analysis* (New York: McGraw-Hill Book Co., Inc., 1956), chap. 4.

Stodola, A., translated by L. C. Loewenstein, *Steam and Gas Turbines* (New York: McGraw-Hill Book Co., Inc., 1927), chap. 5.

Wilson, W. K., *Practical Solution of Torsional Vibration Problems* (New York: John Wiley & Sons, Inc., 2nd ed., 1948), chap. 5.

Ling vibration system consists of 30,000 pound-force shaker (foreground), control console, and 175 KVA amplifier. System installed at Itek Corporation's Lexington, Massachusetts laboratory facilities (courtesy Itek Laboratories)

6 METHODS OF DETERMINING NATURAL FREQUENCIES

6-1. INTRODUCTION

The natural frequencies of multidegree-of-freedom systems can be determined by the method outlined in Chap. 3. If the number of degrees of freedom is large, however, the solving of the determinant of large order becomes increasingly laborious. These problems are generally solved by machines or other methods, such as the Rayleigh, Rayleigh-Ritz, Holzer, Stodola, Myklestad, and matrix methods. In this chapter we shall discuss the Rayleigh and the Holzer methods, which are commonly used and may be considered as representative of this type of analysis. The matrix method is discussed in Chap. 7. It should be mentioned that for a system with complex geometric configuration, such as a beam of irregular shape, an analytical approach may prove to be extremely difficult. Hence, natural frequencies are often determined experimentally by exciting the system with a variable frequency exciter.

6-2. RAYLEIGH METHOD

If a conservative system is vibrating at one of its principal modes, all the masses execute simple harmonic motion of the same frequency. The Rayleigh method is based on the assumption that, at a principal mode, the maximum kinetic energy equals the maximum potential energy. This technique was applied to a one-degree-of-freedom system in Chap. 2. (See Example 3.)

To evaluate the energy quantities, it is necessary to estimate the deflection of the system under dynamic conditions. For example, to find the fundamental frequency of a beam, a "reasonable" deflection curve corresponding to the first mode is assumed. If the assumed curve is not the exact deflection curve under dynamic conditions, it is equivalent to the application of an additional constraint to the vibratory system. Hence the calculated frequency is higher than the true value.† If the assumed curve does not deviate from the exact curve, the frequency calculated is also exact. If a system has many masses, there are as many deflection curves corresponding to the modes. Each possible deflection curve will give a frequency based on equating the energies. Owing to difficulties in assuming deflection curves for the higher modes, the Rayleigh method is used mainly for estimating the fundamental frequency. If the assumed curve is a reasonable approximation of the exact curve for the first mode, the calculated frequency will be very close to, but somewhat higher than, the actual fundamental frequency.

The basic equation of this method can be derived by considering a beam of negligible mass supporting several lumped masses along its span. Figure 6-1(*a*) illustrates such a system with three masses. Figure 6-1(*b*) represents the static deflection curve of the system, and Fig. 6-1(*c*) shows a deflection curve that may be used to approximate the fundamental mode shape. The static deflection curve, or its modification, is often used to approximate the dynamic curve of the system.

The potential energy of the system is equal to the strain energy in the beam. Assuming that the dynamic deflection curve of the beam is the same as that due to the masses acting as static loads, the maximum potential energy is also equal to the work done by the static loads in attaining their respective deflections, that is,

$$U_{\max} = \tfrac{1}{2}(W_1y_1 + W_2y_2 + W_3y_3) \tag{6-1}$$

† J. P. Den Hartog, *Mechanical Vibrations*, 4th ed., McGraw-Hill Book Co., Inc., New York, 1956, p. 161.

The kinetic energy of the system is the sum of the kinetic energy of the masses. By virtue of the assumed simple harmonic motion, the maximum velocity of each mass is equal to its displacement amplitude times the circular frequency. Thus the maximum kinetic energy of the system is

$$T_{\max} = \frac{\omega^2}{2g}(W_1y_1^2 + W_2y_2^2 + W_3y_3^2) \tag{6-2}$$

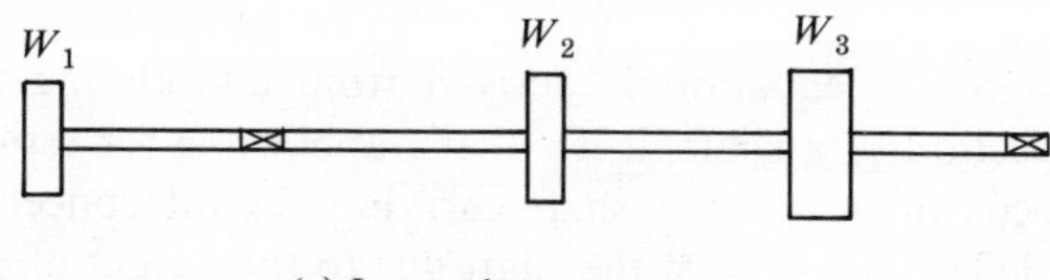

(*a*) Lumped-mass system

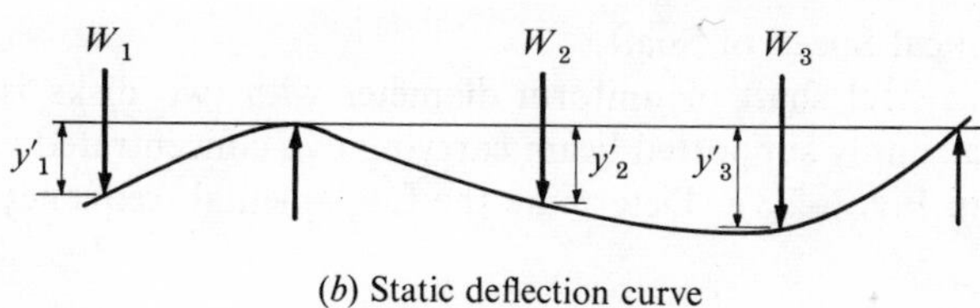

(*b*) Static deflection curve

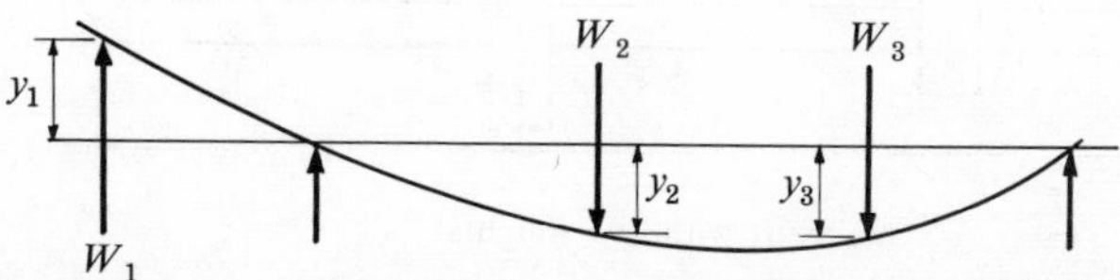

(*c*) Assumed deflection curve

Fig. 6-1. *Fundamental frequency determination by the Rayleigh method*

Equating the maximum potential and kinetic energies and simplifying, we obtain

$$\omega^2 = \frac{g(W_1y_1 + W_2y_2 + W_3y_3)}{(W_1y_1^2 + W_2y_2^2 + W_3y_3^2)} \tag{6-3}$$

The fundamental frequency, as determined in Eq. (6-3), will generally be within 5 percent of the correct value, because a considerable change in the loadings is required in order to change appreciably the deflection curve. Equation (6-3) can be generalized for a system of n masses as

$$\omega^2 = \frac{g \sum_{i=1}^{n} W_i y_i}{\sum_{i=1}^{n} W_i y_i^2} \tag{6-4}$$

Although this equation is derived from considering the lateral (beam) deflection of a shaft, it is equally applicable for estimating the critical speeds of a rotating shaft carrying several concentric disks. The static deflection curve of the shaft due to the weight of the disks is assumed to be the mode shape of the shaft under rotating or whirling conditions. The transverse vibration frequency of the system is the critical speed at which whirling may take place. (See Secs. 2-12 and 5-4.)

Example 1. Critical Speed of Shaft.

A solid steel shaft of uniform diameter with two disks is represented by a simply supported beam carrying two concentrated weights, as shown in Fig. 6-2(a). Determine the fundamental frequency of the system.

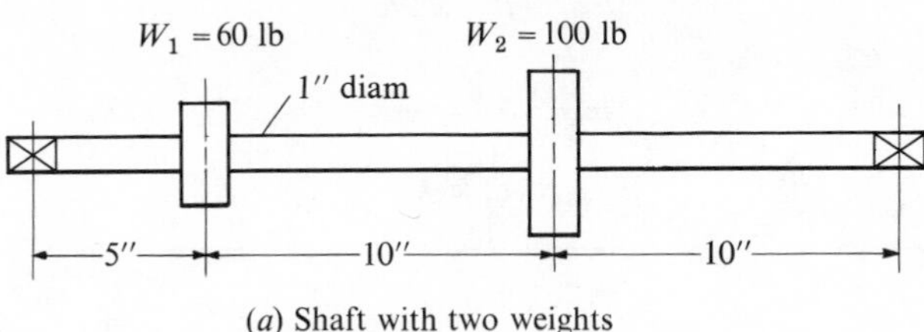

(a) Shaft with two weights

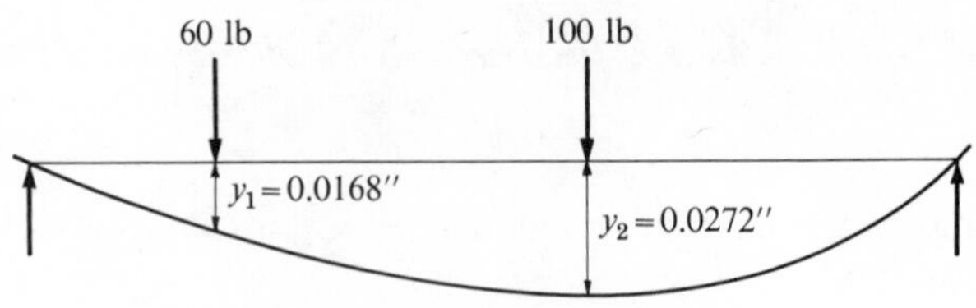

(b) Static deflection curve

Fig. 6-2. *Critical speed of a shaft*

Solution: Considering the weights as concentrated static loads, the resultant deflections at each weight are as shown in Fig. 6-2(*b*). Direct substitution of these values in Eq. (6-4) gives

$$\omega^2 = (386)\,\frac{60(0.0168) + 100(0.0272)}{60(0.0168)^2 + 100(0.0272)^2} = 15{,}800$$

$$\omega = 126 \text{ rad-sec}^{-1}, \text{ or } 1{,}200 \text{ rpm}$$

It may be remarked that Eq. (6-4) can be further generalized to include systems with distributed mass. This affords an easy method for estimating the fundamental frequency of beams and shafts if a reasonable deflection curve can be assumed.

Consider a uniform beam of w lb per unit length. The potential energy of a beam of length L in bending is

$$U = \frac{1}{2}\int_o^L M\,d\theta \tag{6-5}$$

where M is the bending moment and $d\theta$ is the change of slope of the beam owing to the applied moment M. From simple beam theory we have

$$\frac{d\theta}{dx} = \frac{M}{EI} \quad \text{and} \quad \frac{d^2y}{dx^2} = \frac{M}{EI}$$

Substituting these relations in Eq. (6-5) and simplifying gives

$$U = \frac{EI}{2}\int_o^L \left(\frac{d^2y}{dx^2}\right)^2 dx \tag{6-6}$$

The kinetic energy due to the mass of the beam is

$$T = \frac{1}{2g}\int_o^L w\omega^2 y^2\,dx \tag{6-7}$$

The maximum potential and kinetic energy of the beam are as expressed by the last two equations. Equating the energy functions yields

$$\omega^2 = \frac{gEI}{w}\,\frac{\displaystyle\int_o^L \left(\frac{d^2y}{dx^2}\right)^2 dx}{\displaystyle\int_o^L y^2\,dx} \tag{6-8}$$

Example 2. Determine the fundamental frequency of a cantilever beam of length L and w lb per unit length.

Solution: The static deflection curve of a uniform cantilever beam is

$$y = \frac{w}{24EI}(x^4 - 4L^3x + 3L^4)$$

where x is measured from the free end of the beam. The fundamental frequency can be obtained by the direct substitution of this expression in Eq. (6-8). For the purpose of integration, it is convenient to define a new variable $z = x/L$ and obtain the expressions

$$y = y_o(z^4 - 4z + 3)/3 \quad \text{and} \quad \frac{d^2y}{dx^2} = \frac{1}{L^2}\frac{d^2y}{dz^2} = \frac{4y_oz^2}{L^2}$$

Substituting the last two expressions in Eq. (6-8) and simplifying gives

$$\omega^2 = \frac{144EIg}{wL^4} \frac{\int_o^1 (y_oz^2)^2\,dz}{\int_o^1 [y_o(z^4 - 4z + 3)]^2\,dz} = 12.46\frac{EIg}{wL^4}$$

$$\omega = 3.53\sqrt{gEI/wL^4}$$

The exact solution is $\omega = 3.52\sqrt{EIg/wL^4}$.

6-3. RAYLEIGH METHOD: GRAPHICAL TECHNIQUE

In the preceding section the Rayleigh method was applied to two simple cases in which (1) the inertial effect of the shaft is negligible, and (2) this effect is considered, but the deflection curve can be easily estimated. If the effects of the distributed shaft weight, the variations in the moment of inertia and the modulus of elasticity of the shaft, and the effects of shear forces on the resulting deflections are considered, the deflection curve becomes more difficult to obtain. Graphical integration is one of the techniques commonly used for the analysis of such beam-and-shaft problems.

In this section we shall first outline the steps for graphical integration, verify the procedure, and establish the scale factor, and then apply the technique to a shaft problem.

Consider a given curve as shown in Fig. 6-3(a). The steps for the graphical integration of the curve are as follows:

1. Divide the given curve into conveniently spaced intervals by choosing the points B, C, D, etc., along the x-axis. The accuracy of

the graphical integration process is improved by selecting these points at close intervals.

2. Locate the points p_1, p_2, p_3, etc., on the curve. These points represent the mean value of the ordinate (y) within the selected intervals.

3. Project these points horizontally to a line parallel to the y-axis, determining the points labeled 1, 2, 3, etc.

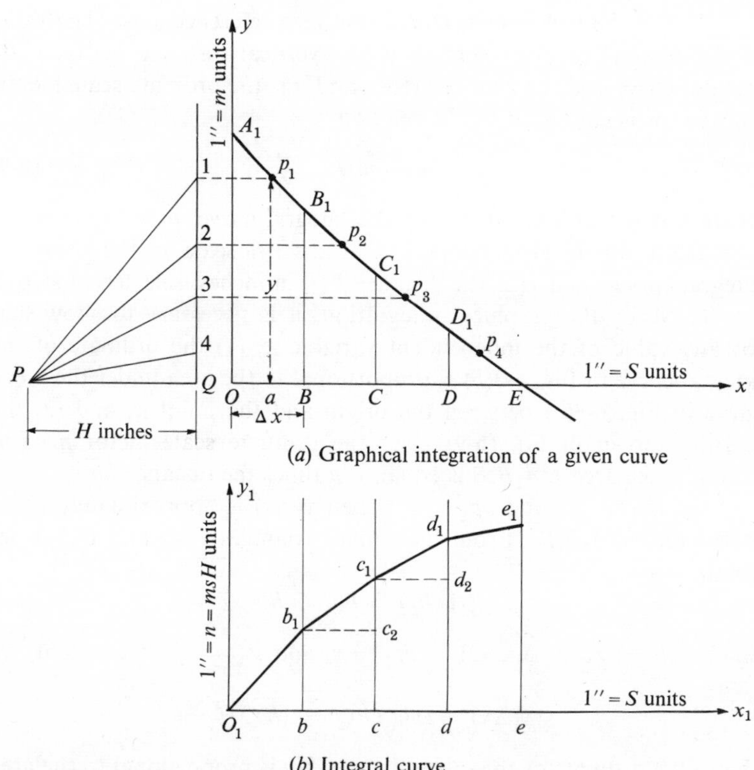

(*a*) Graphical integration of a given curve

(*b*) Integral curve

Fig. 6-3. ***Graphical integration***

4. Choose a convenient distance $H = PQ$, thereby locating the pole P on an extension of the x-axis.

5. Construct the lines $P1$, $P2$, $P3$, etc.

6. Directly below the given curve, construct a new set of axes as shown in Fig. 6-3(b).

7. Divide the x_1-axis into the same intervals as those used in step 1.

8. Construct the line segments O_1b_1 parallel to $P1$, b_1c_1 parallel to $P2$, etc., to obtain the integral curve.

The true origin of the zero reference axis for the integral curve is not, in general, located at the point O_1, because the constant of integration has not been evaluated by this procedure. Therefore, any ordinate value of the integral curve will differ from the true value of the integral by a constant amount. A method for determining the constant will be illustrated in Example 3.

If O_1 is the true origin, O_1x_1 is the zero reference axis. The value of the integral is y_1n where y_1 is the vertical distance between the integral curve and the zero reference and n is the ordinate scale factor. This factor is expressed by the relation

$$n = msH \tag{6-9}$$

where n is the ordinate scale for the integral curve, m is the ordinate scale factor for the given curve, s is the abscissa scale for the given and integral curves, and H is the distance PQ, in inches, selected in step 4.

To verify the graphical integration, it is necessary to show that for any value of the independent variable x, (1) the ordinate of the integral curve in Fig. 6-3(b) is proportional to the area under the given curve in Fig. 6-3(a) between the origin and the point x, and (2) the relation shown in Eq. (6-9) gives the ordinate scale factor n. For example, the area OA_1B_1B is equal to n times the distance bb_1.

Let $OB = \Delta x$ and $ap_1 = y$. Then $y(\Delta x)$ is approximately equal to the area OA_1B_1B. From the similar triangles $P1Q$ and O_1b_1b, we obtain

$$Q1\!:\!PQ = bb_1\!:\!O_1b$$

Since $O_1b = OB = \Delta x$, $Q1 = ap_1 = y$, and $PQ = H$, a constant, we have

$$y(\Delta x) = (bb_1)(PQ) = (bb_1)H$$

Hence the ordinate of the integral curve bb_1 is proportional to the area OA_1B_1B. Similarly, using the triangles $P2Q$ and $b_1c_1c_2$, it can be shown that the distance c_1c_2 is proportional to the area BB_1C_1C; that is, the increase in ordinate of the integral curve is proportional to the increase in area under the given curve.

To establish Eq. (6-9) it is noted that

$$\text{Area } OA_1B_1B = m(ap_1)s(OB)$$

It is desired to have the relationship

$$\text{Area } OA_1B_1B = n(bb_1)$$

Equating the last two equations gives

$$n = ms(ap_1)(OB)/(bb_1) \tag{6-10}$$

From the similar triangles $P1Q$ and O_1b_1b we obtain

$$H = PQ = (Q1)(O_1b)/(bb_1) = (ap_1)(OB)/(bb_1) \tag{6-11}$$

It is evident that Eq. (6-9) can be established from Eqs. (6-10) and (6-11).

Graphical integration can be used to determine the deflection curve of a beam. The beam curvature d^2y/dx^2 can be expressed as†

$$\frac{d^2y}{dx^2} = \frac{M}{EI} + K\frac{W}{GAL} \tag{6-12}$$

where M is the bending moment, E the modulus of elasticity, G the shear modulus, I the moment of inertia, A the cross-sectional area, W the load on the beam (including its own weight) over the length L, and K the ratio of the maximum transverse shear stress to the average shear stress. The second term on the right side of Eq. (6-12) can be omitted if the effects of shear are negligible. If the right side of Eq. (6-12) can be established, its first integration with respect to x gives the slope curve dy/dx, and the second integration yields the deflection curve. We shall illustrate this process with an example.

Example 3. Determine the static deflection curve and the critical speed of the system as shown in Fig. 6-4(a).

Solution: The physical dimensions of the steel shaft with the weights and locations of all attached members, such as gears, pulleys, and impellers, are as indicated in Fig. 6-4(a). The shaft is supported by two bearings which are assumed to be simple beam supports. The effects of the shaft weight and shear forces on the resulting deflection and critical speed will be considered here.

Although the weight of the shaft constitutes a distributed load, it is considered as a series of concentrated loads. The shaft is treated as a series of short sections, with the load of each section concentrated at its midpoint. Figure 6-4(b) shows the assumed loading and the corresponding sections. The letters $a, b, c, \ldots$ serve to identify the stations at the end points of the sections for purposes of calculation.

The beam curvature d^2y/dx^2 is computed at each of the assigned stations $a, b, c, \ldots$. These values are shown in Table 6-1 and plotted in Fig. 6-4(c).

† See, for example, R. T. Hinkle, *Design of Machines*, Prentice-Hall, Inc., Englewood Cliffs, N.J., 1957, p. 131.

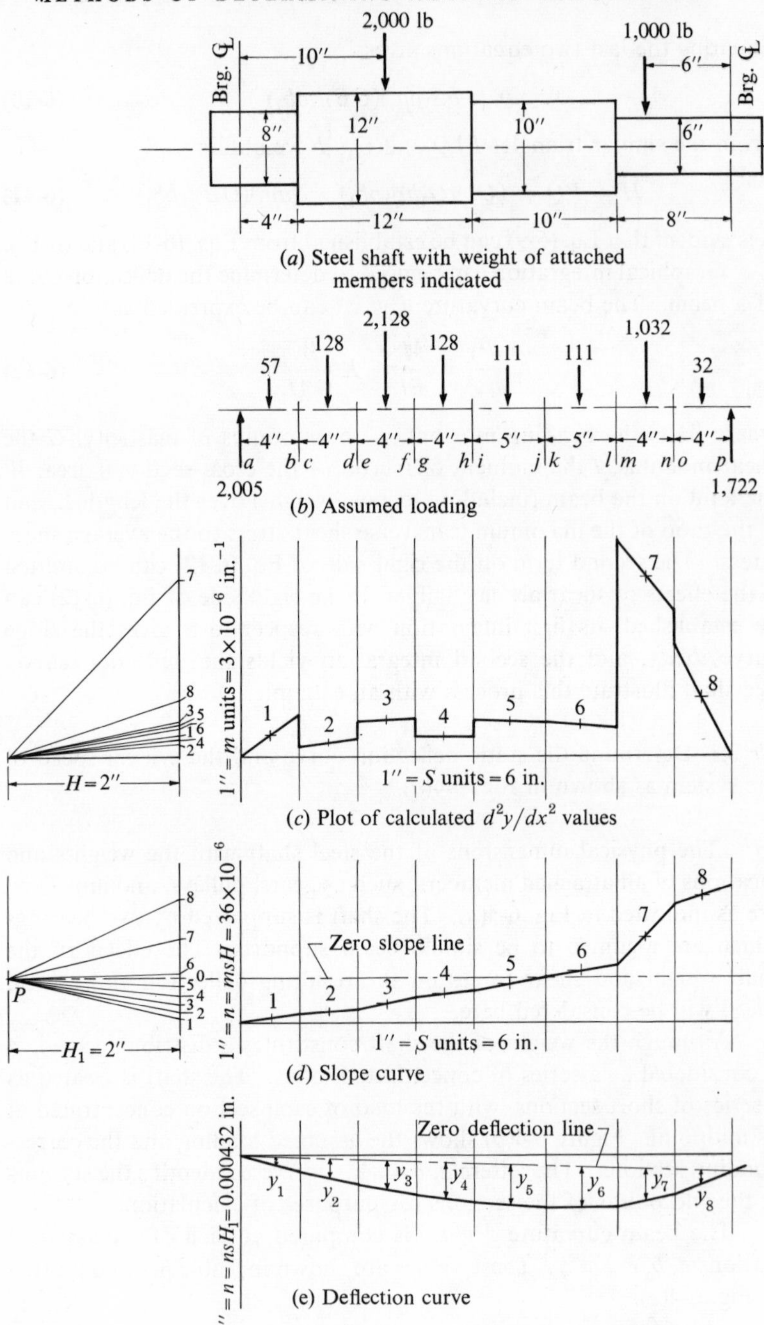

(a) Steel shaft with weight of attached members indicated

(b) Assumed loading

(c) Plot of calculated d^2y/dx^2 values

(d) Slope curve

(e) Deflection curve

Fig. 6-4. *Deflection curve by graphical integration*

TABLE 6-1

CALCULATED DATA FOR EXAMPLE 3

STATION	A (in.^{2})	I (in.^{4})	W (lb)	L (in.)	M (in.-lb)	$\frac{M}{EI} \times 10^6$ (in.$^{-1}$)	$K\frac{W}{GAL} \times 10^6$ (in.$^{-1}$)	$\frac{d^2y}{dX^2} \times 10^6$ (in.$^{-1}$)
a	40.24	201	57	4	0	0	0.041	0.041
b	40.24	01	57	4	7,906	1.310	0.041	1.351
c	113.04	1,018	128	4	7,906	0.264	0.036	0.300
d	113.04	1,018	128	4	15,440	0.506	0.036	0.542
e	113.04	1,018	2,128	4	15,440	0.506	0.600	1.106
f	113.04	1,018	2,128	4	18,470	0.606	0.600	1.206
g	113.04	1,018	128	4	18,470	0.606	0.036	0.642
h	13. 4	1,018	128	4	16,980	0.556	0.036	0.592
i	79.50	491	111	5	16,980	1.150	0.032	1.182
j	79.50	491	111	5	14,520	0.989	0.032	1.021
k	79.50	491	111	5	14,520	0.989	0.032	1.021
l	79.50	491	111	5	11,520	0.782	0.032	0.814
m	28.26	63.6	1,032	4	11,520	6.040	1.060	7.100
n	28.26	63.6	1,032	4	6,820	3.570	1.060	4.630
o	28.26	63.6	32	4	6,820	3.570	0.033	3.603
p	28.26	63.6	32	4	0	0	0.033	0.033

$E = 30 \times 10^6$ psi; $G = 11.5 \times 10^6$ psi.
$K = \frac{4}{3}$ (solid circular section).

Following the graphical integration procedure and using the lengths selected as subdivisions for integration, Fig. 6-4(*c*) is integrated to give the slope curve dy/dx as shown in Fig. 6-4(*d*). From Eq. (6-9) the corresponding ordinate scale factor is $(3 \times 10^{-6})(6)(2) = 36 \times 10^{-6}$. It is unnecessary to establish a zero reference, or the zero slope line, in order to proceed with the second integration. Hence, choosing a convenient location for the pole P in Fig. 6-4(*d*), the integration is repeated, and the deflection curve is obtained as shown in Fig. 6-4(*e*). The corresponding scale factor is $(36 \times 10^{-6})(6)(2) = 0.000432$.

The zero deflection curve, or the zero reference line of Fig. 6-4(*e*), is obtained by assuming that the shaft bearings are not deflected. Thus a straight line joining the end points of the deflection curve gives the zero deflection line. This is equivalent to applying the boundary conditions to the integration process for evaluating the constants of integration. The shaft deflection is equal to the scale factor, 0.000432, times the vertical distance between the deflection curve and the zero deflection line. For example, the shaft deflection at the 2,128 lb load is $(n)(y_3) = (0.000432)(0.32) = 0.000138$ in. Similarly, other values for the shaft deflections can be obtained.

The zero slope line is obtained by assuming that the bearing supports remain horizontal. Hence the zero deflection line will have zero slope. This line is transferred back to the integrating polygon of Fig. 6-4(*d*) to give the line labeled *PO*, and the point *O* is projected horizontally to establish the zero slope line.

TABLE 6-2

DEFLECTION VALUES FROM FIG. 6-4(*e*)

	1	2	3	4	5	6	7	8	
W (lb)	57	128	2,128	128	111	111	1,032	32	
$y \times 10^4$ (in.)	0.432	0.950	1.382	1.728	1.944	1.944	1.555	0.648	Totals
$Wy \times 10^3$	2.46	12.15	294.0	22.2	21.6	21.6	161.0	2.07	537.08
$Wy^2 \times 10^8$	10.62	115.2	4,060	384	420	420	2,500	13.4	7,923

The deflections y at the appropriate stations and the values of Wy and Wy^2 are tabulated as shown in Table 6-2. By the direct application of Eq. (6-4), the fundamental frequency of the given system is

$$\omega^2 = (386)(537 \times 10^{-3})/(7{,}923 \times 10^{-8}) = 2.62 \times 10^6$$
$$\omega = 1{,}620 \text{ rad-sec}^{-1}, \text{ or } 17{,}000 \text{ rpm}$$

6-4. HOLZER METHOD

The Holzer method is a tabular method for the analysis of multi-mass lumped-parameter systems. It is applicable for the study of free and forced vibrations, systems with or without damping,† semidefinite

† J. P. Den Hartog and J. P. Li, "Forced Torsional Vibrations with Damping: An Extension of Holzer's Method," *J. Appl. Mechanics*, December, 1946, pp. A-276–280.

or systems with fixed ends, and systems with angular or rectilinear motion. We shall discuss the Holzer method mainly in connection with the free vibration of conservative semidefinite torsional systems to which the method is commonly applied.

The Holzer method is a trial-and-error method. It can be used to find the natural frequencies, and the higher modes can be determined independent of the lower modes. In addition to the natural frequencies, this method also gives the amplitude ratios of the masses and the nodes in a system at its principal modes of vibration.

Applied to a semidefinite system, this method is based on the assumption that no external torque is required to maintain a conservative system to vibrate at its principal modes. The tabular procedure

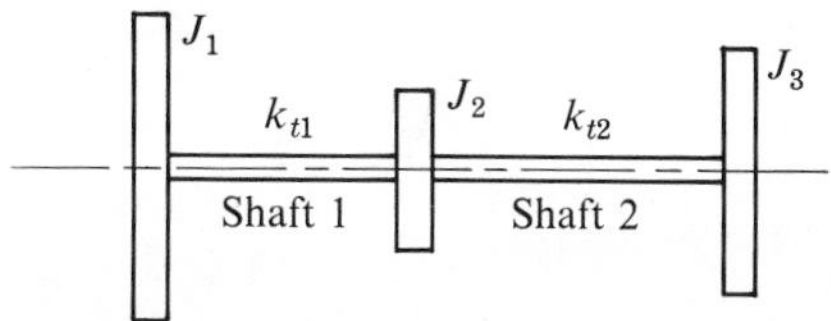

Fig. 6-5. *A three-disk torsional system*

essentially follows the way in which the torque of the shaft changes from one section to the next. A trial value for a natural frequency is first assumed. Correspondingly, the inertia torques are calculated. If the assumed frequency is a natural frequency, then the sum of the inertia torques is zero. We shall illustrate the method with a three-disk and two-shaft system and then generalize the equations to an n-disk and $(n - 1)$-shaft system.

Consider the system shown in Fig. 6-5. The equations of motion are

$$\begin{aligned} J_1\ddot{\theta}_1 &= -k_{t1}(\theta_1 - \theta_2) \\ J_2\ddot{\theta}_2 &= -k_{t1}(\theta_2 - \theta_1) - k_{t2}(\theta_2 - \theta_3) \\ J_3\ddot{\theta}_3 &= -k_{t2}(\theta_3 - \theta_2) \end{aligned} \tag{6-13}$$

Summing the three equations in Eq. (6-13) gives

$$J_1\ddot{\theta}_1 + J_2\ddot{\theta}_2 + J_3\ddot{\theta}_3 = \sum J_i\ddot{\theta}_i = 0 \tag{6-14}$$

Hence the sum of the inertia torques must equal zero at all times. If the system is vibrating at one of its principal modes, the motions of all the disks are simple harmonic. Thus Eq. (6-14) becomes

$$\sum J_i\Theta_i\omega^2 = 0 \tag{6-15}$$

where Θ_i is the amplitude of the displacement of J_i, and ω is a natural frequency. If the system considered has n disks, analogous to Eq. (6-15), we have

$$\sum_{i=1}^{n} J_i\Theta_i\omega^2 = 0 \tag{6-16}$$

Referring to Fig. 6-5, if Θ_1 is the displacement of J_1, then $J_1\Theta_1\omega^2$ is the inertia torque of J_1, which is also equal to the torque of shaft 1; that is, $-J_1\Theta_1\omega^2 = -k_{t1}(\Theta_1 - \Theta_2)$. The angular twist of shaft 1 is $(\Theta_1 - \Theta_2)$. Hence the angular displacement of J_2 is

$$\Theta_2 = \Theta_1 - J_1\Theta_1\omega^2/k_{t1} \tag{6-17}$$

The torque on shaft 2 is equal to the sum of the inertia torques of J_1 and J_2; that is, $-(J_1\Theta_1 + J_2\Theta_2)\omega^2 = -k_{t2}(\Theta_2 - \Theta_3)$. Hence the displacement of J_3 is

$$\Theta_3 = \Theta_2 - \omega^2(J_1\Theta_1 + J_2\Theta_2)/k_{t2} \tag{6-18}$$

If the system considered has n disks and $(n - 1)$ shafts, Eq. (6-18) can be generalized as

$$\Theta_j = \Theta_{j-1} - \frac{\omega^2}{k_{t(j-1)}}\sum_{i=1}^{j-1} J_i\Theta_i; \quad j = 2, 3, \ldots, n \tag{6-19}$$

Equations (6-16) and (6-19) provide the basis for the Holzer method. Assuming a trial value for a natural frequency, the process is carried out by letting the angular displacement of the first disk arbitrarily be 1 radian. The angular displacement of each disk of the system is found by applying Eq. (6-19) in succeeding order. If the algebraic sum of the inertia torques is zero, that is, if Eq. (6-16) is satisfied, then the assumed frequency is a natural frequency. If Eq. (6-16) is not satisfied, a new value of ω is assumed, and the process is repeated.

The difficulties often encountered with the Holzer method are (1) in estimating the initial trial value for ω^2, and (2) in selecting a second trial value for ω^2 if the initial trial value does not satisfy Eq. (6-16). The estimation of the initial trial value may depend on judgment and experience, but reasonable trial values for the lower modes can be obtained by reducing the multidisk system to a two- or three-disk system. Thus the natural frequencies can be estimated from the equations developed in Chap. 3. There are no general rules telling how to accomplish this reduction in degrees of freedom. Grouping together disks that have shafts with relatively high stiffness between them or neglecting disks with inertias that are small compared to the other disks, however, will give fair approximations.

If the trial value of frequency is not one of the natural frequencies, the sum of the inertia torques, Eq. (6-16), will be different from zero. This torque remainder from the calculation represents an applied torque at the last disk at the assumed frequency to cause the system to oscillate with the calculated displacement amplitudes. Thus we have essentially a condition of steady-state undamped forced vibration with the excitation applied to the last disk. The displacement amplitudes of the disks will be proportional to the magnitude of the torque applied to the system. Thus the torque remainder may be considered as an "external" torque, and a representative plot of the external torque versus ω^2 is as shown in Fig. 6-6. It can be seen that the external

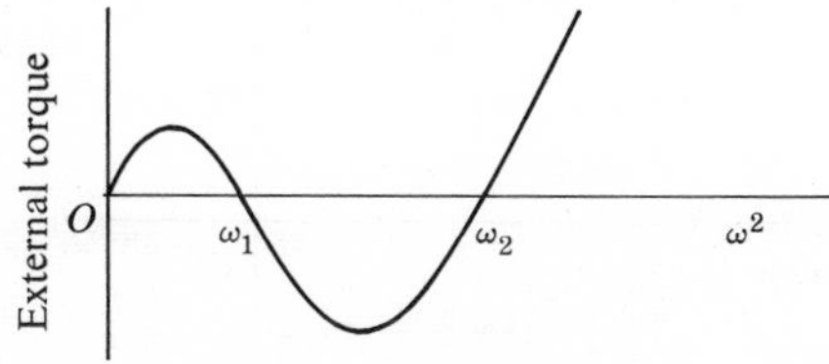

Fig. 6-6. *A plot of external torque versus* ω^2

torque has a positive value as it approaches the fundamental frequency and a negative value as it approaches the second harmonic. Similar statements can be made for the higher modes.

It should be noted that the graph of the external torque versus ω^2, Fig. 6-6, is essentially a plot of the characteristic function of the system versus ω^2. The proof of this statement is left as an exercise. Hence the curve in Fig. 6-6 must be of the form indicated. If the characteristic function is equated to zero, we obtain the characteristic or the frequency equation. Hence the natural frequencies must correspond to the zero values of the plot in Fig. 6-6.

It may be recalled that there is one node in the system when it vibrates at its first mode, there are two nodes at the second mode, and so on. (See Fig. 6-7.) A node is located between any two disks which have opposite signs for their displacement amplitudes. Hence the mode of vibration can be determined from the changes in sign of Θ. For example, if there are two sign changes in Θ the frequency considered must correspond to the second mode.

Example 4. For the torsional system shown in Fig. 6-7(a), determine the natural frequencies and locate the relative positions of the nodes.

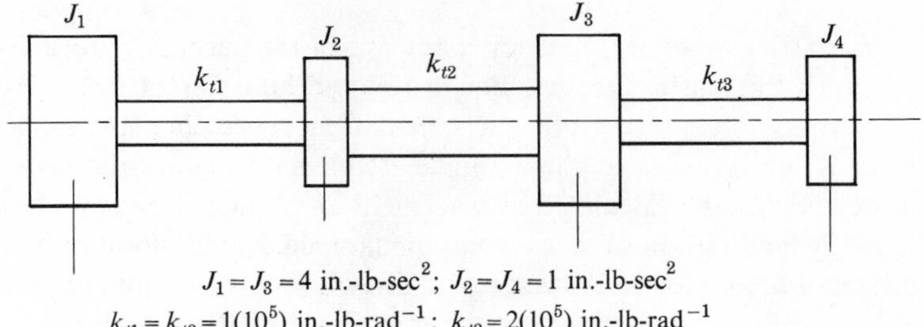

$J_1 = J_3 = 4$ in.-lb-sec^2; $J_2 = J_4 = 1$ in.-lb-sec^2

$k_{t1} = k_{t3} = 1(10^5)$ in.-lb-rad^{-1}; $k_{t2} = 2(10^5)$ in.-lb-rad^{-1}

(*a*) Four-disk torsional system

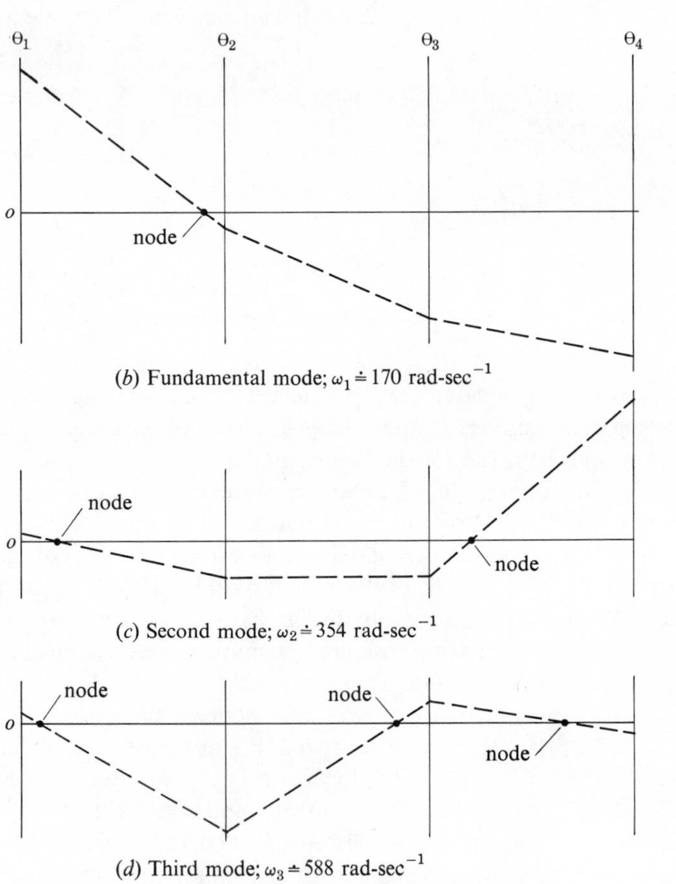

Fig. 6-7. *Natural frequencies of a torsional system*

Solution: A trial value for the fundamental frequency is determined by considering a two-disk system with disk J_1 at one end and $(J_3 + J_4)$ at the opposite end of a shaft with stiffness equal to the combination of k_{t1} and k_{t2} in series. Thus the equivalent spring constant is

$$k_t = \frac{1}{1/k_{t1} + 1/k_{t2}} = \frac{1}{1/10^5 + 1/(2)(10^5)}$$
$$= (2/3)(10^5) \text{ in.-lb-rad}^{-1}$$

and the trial frequency is given by

$$\omega = \sqrt{k_t(J_1 + J_3 + J_4)/J_1(J_3 + J_4)}$$
$$= \sqrt{(2/3)(10^5)(4 + 4 + 1)/(4)(4 + 1)} = 173 \text{ rad-sec}^{-1}$$

With this trial frequency the Holzer tabular method is applied to the original system, and the computations are shown in Table 6-3(*a*). First, the values of J and k_t as specified are entered in columns 1 and 5, respectively. The remaining values are determined as follows:

Row 1: Column 2: Assume that $\Theta_1 = 1$ radian.

Column 3: Compute $J_1\Theta_1\omega^2 = (4)(1)(3)(10^4) = 12(10^4)$.

Column 4: Sum values in column 3 $= 12(10^4)$.

Column 6: Divide column 4 by column 5 $= 12(10^4)/10^5 = 1.2$ rad (twist in shaft k_{t1}).

Row 2: Column 2: Compute $\Theta_2 = \Theta_1 -$ (twist in shaft k_{t1}) $= 1 - 1.2 = -0.2$ rad.

Column 3: Compute $J_2\Theta_2\omega^2 = (1)(-0.2)(3)(10^4) = -0.6(10^4)$.

Column 4: Sum values in column 3 $= (12 - 0.6)(10^4) = 11.4(10^4)$.

Column 6: Divide column 4 by column 5 $= 11.4(10^4)/2(10^5) = 0.57$ rad (twist in shaft k_{t2}).

The values in rows 3 and 4 are determined in like manner. The sum of the inertia torques indicated in row 4 column 4 is $-0.8(10^4)$, which is not zero. Hence $\omega = 173$ rad-sec^{-1} is not the fundamental frequency. Since this remainder torque is negative, the assumed frequency is too high.

A frequency $\omega = 165$ rad-sec^{-1} is used for the second trial, and the results are shown in Table 6-3(*b*). The remainder torque is $1.14(10^4)$, hence the assumed frequency is too low. Linear interpolation between these two trial values gives $\omega = 170$ rad-sec^{-1}. The values of

TABLE 6-3

Holzer's Method Tabulations for Example 4

(*a*) First trial for fundamental frequency ($\omega_1 = 173$ rad-sec^{-1})

	1	2	3	4	5	6
	J (in.-lb-sec^2)	Θ (radian)	$J\Theta\omega^2$ (in.-lb)	$\sum J\Theta\omega^2$ (in.-lb)	k_t (in.-lb-rad^{-1})	$\frac{1}{k_t}\sum J\Theta\omega^2$ (radian)
1	4	1.000	12 (10^4)	12 (10^4)	1 (10^5)	1.200
2	1	−0.200	−0.6 (10^4)	11.4 (10^4)	2 (10^5)	0.570
3	4	−0.770	−9.24 (10^4)	2.16 (10^4)	1 (10^5)	0.216
4	1	−0.986	−2.96 (10^4)	−0.8 (10^4)	= external torque	

(*b*) Second trial for fundamental frequency ($\omega_1 = 165$ rad-sec^{-1})

	1	2	3	4	5	6
1	4	1.000	10.89 (10^4)	10.89 (10^4)	1 (10^5)	1.089
2	1	−0.089	−0.24 (10^4)	10.65 (10^4)	2 (10^5)	0.532
3	4	−0.621	−6.76 (10^4)	3.89 (10^4)	1 (10^5)	0.389
4	1	−1.010	−2.75 (10^4)	1.14 (10^4)	= external torque	

(*c*) Final trial for fundamental frequency ($\omega_1 = 170$ rad-sec^{-1})

	1	2	3	4	5	6
1	4	1.000	11.56 (10^4)	11.56 (10^4)	1 (10^5)	1.156
2	1	−0.156	−0.45 (10^4)	11.11 (10^4)	2 (10^5)	0.556
3	4	−0.712	−8.23 (10^4)	2.88 (10^4)	1 (10^5)	0.288
4	1	−1.000	−2.89 (10^4)	−0.01 (10^4)	= external torque	

(*d*) Final trial for second-mode frequency ($\omega_2 = 354$ rad-sec^{-1})

	1	2	3	4	5	6
1	4	1.000	50 (10^4)	50 (10^4)	1 (10^5)	5.000
2	1	−4.000	−50 (10^4)	0	2 (10^5)	0
3	4	−4.000	−200 (10^4)	−200 (10^4)	1 (10^5)	−20.000
4	1	16.000	200 (10^4)	0	= external torque	

a third trial with this frequency are shown in Table 6-3(c). Although the remainder torque is not exactly zero, it can be assumed that the fundamental frequency is approximately 170 rad-sec^{-1}. The displacement amplitudes from Table 6-3(c) are plotted in Fig. 6-7(b) to indicate the relative angular displacements of the disks and to locate the node.

The second mode is determined by carrying out another series of trial solutions. The values of the final trial for this mode are given in Table 6-3(d). Figure 6-7(c) illustrates the mode shape and locates the two nodes for this natural frequency.

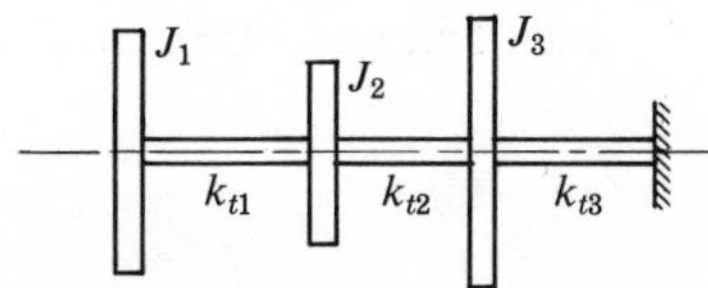

Fig. 6-8. *Torsional system with fixed end*

The frequency for the third mode is approximately 588 rad-sec^{-1}. The corresponding mode shape and the locations of the three nodes are shown in Fig. 6-7(d).

Our discussion of the Holzer method has been confined to semidefinite systems. To illustrate another application of this method, let us consider a torsional system, as shown in Fig. 6-8, in which the shaft at one end is fixed. The corresponding boundary condition is that the angular displacement at the fixed end is zero. The computation procedure used in the last example can be modified to apply to this system. Assuming a trial frequency and a unit angular displacement for the disk at the free end, the procedure is essentially the same as before; it traces the way in which the torque changes from one section to the next. Instead of requiring the sum of the inertia torques to be zero, however, we now demand that the angular displacement at the fixed end must be zero. The trial frequency is a natural frequency if this boundary condition is satisfied. A plot of the remainder angular displacement versus ω^2 is of the form shown in Fig. 6-9.

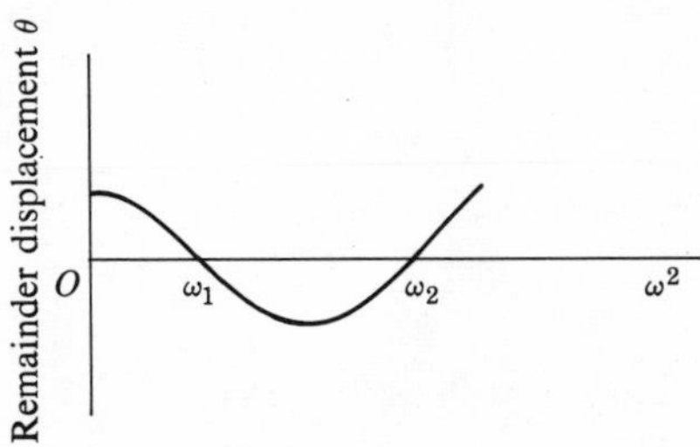

Fig. 6-9. *Natural frequency of torsional system with one fixed end*

SUGGESTED READING

Den Hartog, J. P., *Mechanical Vibrations* (New York: McGraw-Hill Book Co., Inc., 4th ed., 1956), chap. 4.

Scanlan, R. H., and R. Rosenbaum, *Aircraft Vibration and Flutter* (New York: The Macmillan Co., 1951), chap. 7.

Thomson, W. T., *Mechanical Vibrations* (New Jersey: Prentice-Hall, Inc., 2nd ed., 1953), chap. 6.

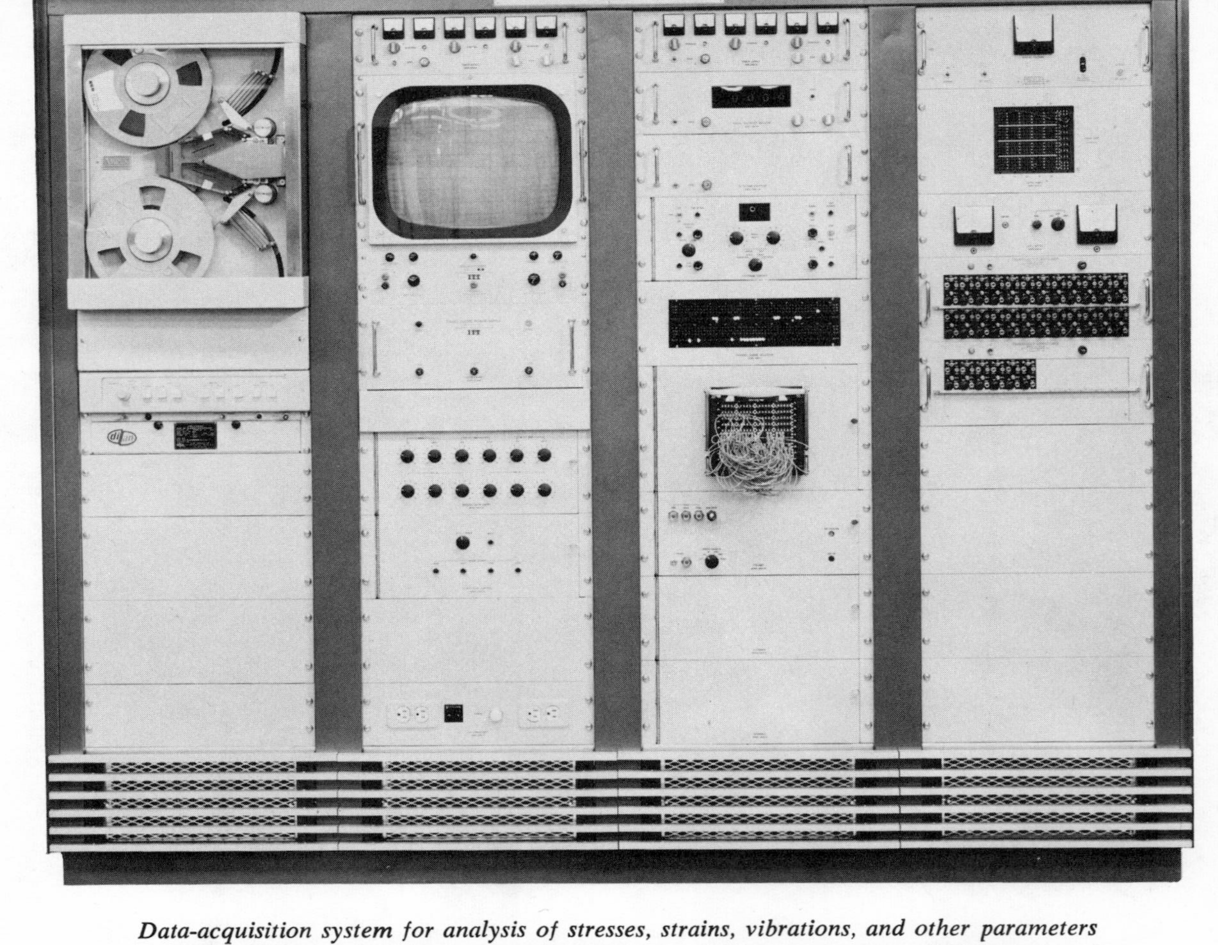

Data-acquisition system for analysis of stresses, strains, vibrations, and other parameters affecting models of aircraft in wind-tunnel tests. This system scans 100 inputs and samples at a rate of 100 samples per second. (courtesy Systron-Donner Corporation)

7 MULTI-DEGREE-OF-FREEDOM SYSTEM—MATRIX METHOD

7-1. INTRODUCTION

It was shown in Chap. 3 that the equations of motion for the free vibration of an undamped three-degree-of-freedom system are [see Eq. (3-64)]

$$\begin{aligned} m_{11}\ddot{q}_1 + m_{12}\ddot{q}_2 + m_{13}\ddot{q}_3 + k_{11}q_1 + k_{12}q_2 + k_{13}q_3 &= 0 \\ m_{21}\ddot{q}_1 + m_{22}\ddot{q}_2 + m_{23}\ddot{q}_3 + k_{21}q_1 + k_{22}q_2 + k_{23}q_3 &= 0 \\ m_{31}\ddot{q}_1 + m_{32}\ddot{q}_2 + m_{33}\ddot{q}_3 + k_{31}q_1 + k_{32}q_2 + k_{33}q_3 &= 0 \end{aligned} \tag{7-1}$$

These equations can be written more concisely in matrix notation as

$$M\{\ddot{q}\} + K\{q\} = \{0\} \tag{7-2}$$

where

$$M = \begin{bmatrix} m_{11} & m_{12} & m_{13} \\ m_{21} & m_{22} & m_{23} \\ m_{31} & m_{32} & m_{33} \end{bmatrix}, \quad K = \begin{bmatrix} k_{11} & k_{12} & k_{13} \\ k_{21} & k_{22} & k_{23} \\ k_{31} & k_{32} & k_{33} \end{bmatrix}, \quad \{\ddot{q}\} = \begin{bmatrix} \ddot{q}_1 \\ \ddot{q}_2 \\ \ddot{q}_3 \end{bmatrix},$$

and
$$\{q\} = \begin{bmatrix} q_1 \\ q_2 \\ q_3 \end{bmatrix}$$

$\{q\}$ is called the *coordinate matrix*. $M = [m_{ij}]$ is called the *inertia matrix*, and m_{ij} are the inertia coefficients. $K = [k_{ij}]$ is called the *stiffness matrix*, and k_{ij} are the stiffness coefficients. It is apparent that Eq. (7-1) can be easily extended to cover a linear system with multidegree of freedom.

More generally, a linear system with applied excitation can be described by a matrix equation of the form

$$M\{\ddot{q}\} + C\{\dot{q}\} + K\{q\} = \{Q\} \tag{7-3}$$

where $\{Q\}$ are the generalized forces corresponding to the generalized coordinates $\{q\}$, and C is the viscous *damping matrix* of the same order as M and K. In this chapter we are concerned with small oscillations of systems described by equations of this type.

The use of matrix notation and matrix method is a saving of effort, both physical and mental, in the study of multidegree-of-freedom-systems. Furthermore, matrix notation has become the accepted language in vibrations and is used by writers in this field. A review of the mathematical techniques useful for this chapter is given in Appendix C.

7-2. EQUATIONS OF MOTION: SMALL OSCILLATIONS OF CONSERVATIVE SYSTEMS

The equations of motion of a conservative system can be obtained from Lagrange's equation [see Eq. (4-48)]

$$\frac{d}{dt}\frac{\partial L}{\partial \dot{q}_j} - \frac{\partial L}{\partial q_j} = 0 \tag{7-4}$$

where $L = T - U$ is the Lagrangian function, T and U are the kinetic and potential energies of the system, and q_j are the generalized coordinates. We shall examine these energy functions in further detail in order to derive the equations of motion for small oscillations of conservative systems.

The potential energy function can be expressed as [see Eqs. (4-41) and (4-42)]

$$U = \frac{1}{2}\sum_{i=1}^{n}\sum_{j=1}^{n}\frac{\partial^2 U}{\partial q_i\,\partial q_j}q_iq_j = \frac{1}{2}\sum_{i=1}^{n}\sum_{j=1}^{n}k_{ij}q_iq_j \tag{7-5}$$

where $k_{ij} = k_{ji}$. This energy function was derived by considering small oscillations of a system about its equilibrium position. The configuration $q_1 = q_2 = \cdots = q_n = 0$ is one of stable, neutral, or unstable equilibrium, depending on whether U increases, remains unchanged, or decreases in numerical value when any or all of the generalized coordinates assume nonzero values. The potential energy of a stable system is always positive. In addition, Eq. (7-5) shows that U is a homogeneous second-degree expression in n variables. For a stable system, U is said to be a *positive definite quadratic function* or a *positive definite quadratic form* of the coordinates. Since $k_{ij} = k_{ji}$, the stiffness matrix

$$K = [k_{ij}] \tag{7-6}$$

is symmetrical. U can be expressed in matrix notation as

$$U = \tfrac{1}{2}\lfloor q\rfloor K\{q\} \tag{7-7}$$

where $\lfloor q\rfloor$ is a row matrix defined as the transpose of the column matrix $\{q\}$.†

The kinetic energy function can be expressed as a homogeneous quadratic function of the generalized velocities.

$$T = \tfrac{1}{2}\sum_{i=1}^{n}\sum_{j=1}^{n}m^*_{ij}\dot{q}_i\dot{q}_j$$

In general, the coefficients m^*_{ij} are functions of the coordinates. Let us expand m^*_{ij} in power series in terms of the coordinates and define m_{ij} as the constant terms in these expansions; that is, for $q_1 = q_2 = \ldots, = q_n = 0$. The second-order terms in T are $m_{ij}\dot{q}_i\dot{q}_j$, and all other terms in T are at least of the third order. For small oscillations about equilibrium we neglect the third and higher order terms. Thus

$$T = \tfrac{1}{2}\sum_{i=1}^{n}\sum_{j=1}^{n}m_{ij}\dot{q}_i\dot{q}_j \tag{7-8}$$

† It can be shown that if U in Eq. (7-7) is a real positive definite quadratic form, then every leading principal minor of K is positive. For example,

$$K = \begin{bmatrix} 2 & 2 & -1 \\ 2 & 3 & -2 \\ -1 & -2 & 2 \end{bmatrix}, \quad |2| = 2, \quad \begin{vmatrix} 2 & 2 \\ 2 & 3 \end{vmatrix} = 2, \quad \begin{vmatrix} 2 & 2 & -1 \\ 2 & 3 & -2 \\ -1 & -2 & 2 \end{vmatrix} = 1$$

where m_{ij} are constants and $m_{ij} = m_{ji}$. Thus the inertia matrix

$$M = [m_{ij}] \tag{7-9}$$

is symmetrical, and T can be expressed in matrix notation as

$$T = \tfrac{1}{2}\lfloor \dot{q} \rfloor M\{\dot{q}\} \tag{7-10}$$

Since the kinetic energy is always positive, T is a positive definite quadratic function of the velocities $\dot{q}_i$.

From Eqs. (7-5) and (7-8) the Lagrangian function becomes

$$L = T - U = \tfrac{1}{2}\sum_{i=1}^{n}\sum_{j=1}^{n} m_{ij}\dot{q}_i\dot{q}_j - \tfrac{1}{2}\sum_{i=1}^{n}\sum_{j=1}^{n} k_{ij}q_iq_j \tag{7-11}$$

Substituting this equation in Eq. (7-4) gives

$$M\{\ddot{q}\} + K\{q\} = \{0\} \tag{7-12}$$

where the matrices M and K are defined in Eqs. (7-9) and (7-6). For example, if the system considered has three degrees of freedom, the equations of motion have the same form as shown in Eq. (7-1). It is observed that T and U are quadratic functions only if terms higher than the second order are neglected. The resultant equations of motion, Eq. (7-12), are second-order linear ordinary differential equations. In other words, a necessary condition for a system to be linear is that T is a quadratic function of the velocities and U is a quadratic function of the coordinates.

7-3. UNDAMPED FREE VIBRATION: PRINCIPAL MODES

It was stated in Chap. 3 that the geometric configuration of a vibratory system can be specified by more than one set of generalized coordinates $\{q\}$. It is possible to select a particular set of coordinates $\{p\}$ such that there is neither static nor dynamic coupling in the resultant equations of motion. Each of the equations thus obtained contains only one dependent variable, and it can be solved independent of the others. The solution of each equation is a harmonic function. The coordinates $\{p\}$ are called the *principal coordinates* and the modes of vibration described by the coordinates $\{p\}$ are called the *principal modes*. We shall use matrix notation to show that the general motions of the masses of a multidegree-of-freedom system can be expressed as superpositions of its principal modes of vibration. (See Sec. 3-2.) Some of the equations in this section are also written out explicitly for a three-degree-of-freedom system, for purposes of comparison.

Consider a system with the equations of motion

$$M\{\ddot{q}\} + K\{q\} = \{0\} \tag{7-13}$$

Let us define a dynamic matrix $H = M^{-1}K$. Premultiplying Eq. (7-13) by M^{-1} gives

$$\{\ddot{q}\} + H\{q\} = \{0\} \tag{7-14}$$

It is desired to find a set of coordinates $\{p\}$ such that the system described by Eq. (7-14) is also described by the equations of motion

$$\{\ddot{p}\} + \Lambda\{p\} = \{0\} \tag{7-15}$$

where Λ is a diagonal matrix with λ_j as its diagonal elements. Let us assume λ_j are distinct. For a three-degree-of-freedom system, Eq. (7-15), written explicitly, becomes

$$\begin{bmatrix} \ddot{p}_1 \\ \ddot{p}_2 \\ \ddot{p}_3 \end{bmatrix} + \begin{bmatrix} \lambda_1 & 0 & 0 \\ 0 & \lambda_2 & 0 \\ 0 & 0 & \lambda_3 \end{bmatrix} \begin{bmatrix} p_1 \\ p_2 \\ p_3 \end{bmatrix} = \begin{bmatrix} 0 \\ 0 \\ 0 \end{bmatrix} \tag{7-16}$$

or

$$\ddot{p}_j + \lambda_j p_j = 0 \tag{7-17}$$

It is evident that the solutions of Eqs. (7-15), (7-16), and (7-17) are harmonic functions. Hence $\{p\}$ are the principal coordinates.

Since the same physical system is described by the coordinates $\{p\}$ and $\{q\}$, these coordinates can be transformed into one another. Let us assume that there exists a linear transformation such that

$$\{q\} = [\mu]\{p\} \tag{7-18}$$

or

$$\begin{bmatrix} q_1 \\ q_2 \\ q_3 \end{bmatrix} = \begin{bmatrix} \mu_{11} & \mu_{12} & \mu_{13} \\ \mu_{21} & \mu_{22} & \mu_{23} \\ \mu_{31} & \mu_{32} & \mu_{33} \end{bmatrix} \begin{bmatrix} p_1 \\ p_2 \\ p_3 \end{bmatrix} \tag{7-19}$$

where $[\mu]$ is a square matrix of order n corresponding to the degree of freedom of the system, and its elements μ_{ij} are constants. Substituting Eq. (7-18) in Eq. (7-14) gives

$$[\mu]\{\ddot{p}\} + H[\mu]\{p\} = \{0\}$$

Premultiplying this equation by $[\mu]^{-1}$ yields

$$\{\ddot{p}\} + [\mu]^{-1}H[\mu]\{p\} = \{0\} \tag{7-20}$$

Comparing Eqs. (7-15) and (7-20), $\{p\}$ is a set of principal coordinates if

$$[\mu]^{-1}H[\mu] = \Lambda \tag{7-21}$$

We shall proceed to show that the transformation matrix $[\mu]$ exists and it is defined as a modal column matrix of H. (See Appendix C.)

Returning to Eq. (7-14), harmonic solutions are of the form $\{q\} = \{\mu\}e^{j\omega t}$, where $\{\mu\}$ is a column of constants and $j = \sqrt{-1}$. Substituting this equation in Eq. (7-14) and factoring out $e^{j\omega t}$ gives

$$(-\omega^2 I + H)\{\mu\} = \{0\} \tag{7-22}$$

or

$$\begin{aligned} (h_{11} - \omega^2)\mu_1 + h_{12}\mu_2 + h_{13}\mu_3 &= 0 \\ h_{21}\mu_1 + (h_{22} - \omega^2)\mu_2 + h_{23}\mu_3 &= 0 \\ h_{31}\mu_1 + h_{32}\mu_2 + (h_{33} - \omega^2)\mu_3 &= 0 \end{aligned} \tag{7-23}$$

Equation (7-22) is a set of algebraic equations. Consistency requires that the matrix $[\omega^2 I - H]$ be singular. (See Theorem 2, Appendix C.) In other words, this set of equations has a solution other than the trivial one, $\{\mu\} = \{0\}$, only if the determinant $|\omega^2 I - H| = 0$. Thus

$$\Delta(\omega) = |\omega^2 I - H| = 0 \tag{7-24}$$

or

$$\Delta(\omega) = \begin{vmatrix} h_{11} - \omega^2 & h_{12} & h_{13} \\ h_{21} & h_{22} - \omega^2 & h_{23} \\ h_{31} & h_{32} & h_{33} - \omega^2 \end{vmatrix} = 0 \tag{7-25}$$

This is called the characteristic or the frequency equation of the system, and it is of the nth order in ω^2. $(\omega^2 I - H)$ is called the *characteristic matrix* of H.

Let us define $\lambda = \omega^2$.† Rewriting Eq. (7-22) and (7-24), we have

$$(\lambda I - H)\{\mu\} = \{0\} \tag{7-26}$$

$$\Delta(\lambda) = |\lambda I - H| = 0 \tag{7-27}$$

Assume that the roots λ_j of Eq. (7-27) are distinct. We shall discuss the cases involving zero and repeated roots in the sections to follow. Since Eq. (7-26) is of the same form as Eq. (C-34) in Appendix C, its solution is a modal column $\{\mu_{is}\}$, $i = 1, 2, \ldots, n$, appropriate to λ_s.

† We assume that the configuration of the system at $\{q\} = \{0\}$ is one of stable equilibrium; that is, U is a positive definite quadratic function of the coordinates. It is evident that the roots of $|\lambda M - K| = 0$ and those of $|\lambda I - H| = 0$ are identical. It can be shown that if M and K are real, symmetric, and positive definite, then the roots λ are real and positive.

A modal column matrix $[\mu]$, formed from a collection of these modal columns, will diagonalize H to give the property indicated in Eq. (7-21). [See Eq. (C-44), Appendix C.] Hence the required transformation specified in Eq. (7-18) exists, and the transformation matrix $[\mu]$ is a modal column matrix of H.

Since the motions described by the principal coordinates $\{p\}$ are harmonic, Eq. (7-18) indicates that the general solutions of a multi-degree-of-freedom system can be expressed as superpositions of the principal modes of vibration. Specifically, for a three-degree-of-freedom system, the corresponding equation, Eq. (7-19), can be expressed as

$$\begin{aligned} q_1 &= \mu_{11}p_1 + \mu_{12}p_2 + \mu_{13}p_3 \\ q_2 &= \mu_{21}p_1 + \mu_{22}p_2 + \mu_{23}p_3 \\ q_3 &= \mu_{31}p_1 + \mu_{32}p_2 + \mu_{33}p_3 \end{aligned} \tag{7-28}$$

From Eq. (7-17), the motion described by a principal coordinate is of the form

$$p_j = c_j \sin(\omega_j t + \psi_j)$$

where c_j and ψ_j are arbitrary. Substituting this equation in Eq. (7-28) gives

$$\begin{aligned} q_1 &= c_1\mu_{11} \sin(\omega_1 t + \psi_1) + c_2\mu_{12} \sin(\omega_2 t + \psi_2) + c_3\mu_{13} \sin(\omega_3 t + \psi_3) \\ q_2 &= c_1\mu_{21} \sin(\omega_1 t + \psi_1) + c_2\mu_{22} \sin(\omega_2 t + \psi_2) + c_3\mu_{23} \sin(\omega_3 t + \psi_3) \\ q_3 &= c_1\mu_{31} \sin(\omega_1 t + \psi_1) + c_2\mu_{32} \sin(\omega_2 t + \psi_2) + c_3\mu_{33} \sin(\omega_3 t + \psi_3) \end{aligned} \tag{7-29}$$

Hence every oscillation represented by each modal column is a principal mode of vibration, and this equation expresses the general motions $\{q\}$ as superpositions of the principal modes. It is noted that this equation and Eq. (3-12), Chap. 3, are identical in form if the value of the first element of each modal column is unity. Since a modal column is indeterminate to the extent that it can always be multiplied by an arbitrary constant, the absolute values of the elements in a modal column are secondary. (See Appendix C.) In other words, the amplitudes of oscillation of a principal mode are arbitrary, depending on the initial conditions.

The coefficients $\{c\}$ and $\{\psi\}$ in Eq. (7-29) can be evaluated by the direct substitution of the initial conditions $\{q(0)\}$ and $\{\dot{q}(0)\}$ in the equation. It is more convenient, however, to use the identity

$$\begin{aligned} c_j \sin(\omega_j t + \psi_j) &= c_j \cos\psi_j \sin\omega_j t + c_j \sin\psi_j \cos\omega_j t \\ &= b_j \sin\omega_j t + a_j \cos\omega_j t \end{aligned}$$

to express each of the harmonic components as a sum of two harmonic functions of the same frequency. Thus, for a multidegree-of-freedom system, we have

$$\{q\} = [\mu]\{b \sin \omega t\} + [\mu]\{a \cos \omega t\} \tag{7-30}$$

where $\{b \sin \omega t\}$ and $\{a \cos \omega t\}$ are column matrices with elements $b_j \sin \omega_j t$ and $a_j \cos \omega_j t$, respectively. Differentiating Eq. (7-30) with respect to time gives

$$\{\dot{q}\} = [\mu]\{\omega b \cos \omega t\} - [\mu]\{\omega a \sin \omega t\} \tag{7-31}$$

Substituting $\{q(0)\}$ in Eq. (7-30) gives

$$\{q(0)\} = [\mu]\{a\} \quad \text{or} \quad \{a\} = [\mu]^{-1}\{q(0)\} \tag{7-32}$$

Substituting $\{\dot{q}(0)\}$ in Eq. (7-31) yields

$$\{\dot{q}(0)\} = [\mu]\{\omega b\} \quad \text{or} \quad \{\omega b\} = [\mu]^{-1}\{\dot{q}(0)\} \tag{7-33}$$

where $\{a\}$ and $\{\omega b\}$ are column matrices with elements a_j and $\omega_j b_j$, respectively. Thus the coefficients $\{a\}$ and $\{b\}$ can be evaluated from the given initial conditions.

Example 1. A torsional system with three degrees of freedom is as shown in Fig. 7-1. If $J_1 = J_2 = J_3 = J$ and $k_{t1} = k_{t2} = k_{t3} = k_t$, determine the motions $\theta_j(t)$. (See Example 6(b), Chap. 3.)

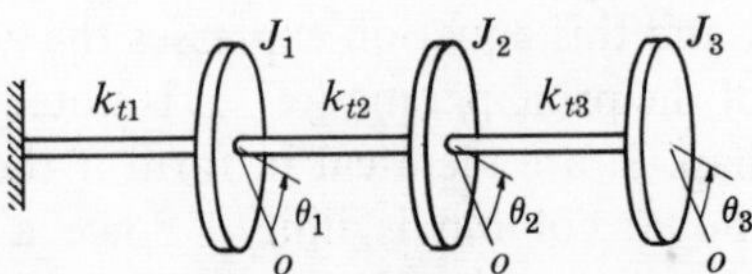

Fig. 7-1. *A three-disk torsional system*

Solution: The equations of motion of the system are

$$J\ddot{\theta}_1 + 2k_t\theta_1 - k_t\theta_2 = 0$$

$$J\ddot{\theta}_2 + 2k_t\theta_2 - k_t\theta_1 - k_t\theta_3 = 0$$

$$J\ddot{\theta}_3 + k_t\theta_3 - k_t\theta_2 = 0$$

or

$$M\{\ddot{\theta}\} + K\{\theta\} = \{0\}$$

where

$$M = \begin{bmatrix} J & 0 & 0 \\ 0 & J & 0 \\ 0 & 0 & J \end{bmatrix}, \quad M^{-1} = \frac{1}{J}\begin{bmatrix} 1 & 0 & 0 \\ 0 & 1 & 0 \\ 0 & 0 & 1 \end{bmatrix}, \quad K = k_t \begin{bmatrix} 2 & -1 & 0 \\ -1 & 2 & -1 \\ 0 & -1 & 1 \end{bmatrix}$$

For convenience, let us define a constant $h = k_t/J$. The dynamic matrix H is

$$H = M^{-1}K = h\begin{bmatrix} 2 & -1 & 0 \\ -1 & 2 & -1 \\ 0 & -1 & 1 \end{bmatrix}$$

Defining $\lambda = \omega^2$, from Eq. (7-22) we obtain

$$(\lambda I - H)\{\mu\} \equiv f(\lambda)\{\mu\} = \{0\}$$

From Eq. (7-24), the value of λ can be obtained from the characteristic equation $\Delta(\lambda) = |\lambda I - H| = |f(\lambda)| = 0$. From Example 6(*b*), Chap. 3, these values are $\lambda_1 = 0.198h$, $\lambda_2 = 1.555h$, and $\lambda_3 = 3.247h$.

Substituting the appropriate values of λ in the equation $f(\lambda)\{\mu\} = 0$, the modal columns $\{\mu\}$ can be determined by the methods described in Appendix C; that is, (1) by solving a set of algebraic equations, or (2) by finding the adjoint matrix $F(\lambda)$ or $f(\lambda)$. (See Examples 7 and 9 in Appendix C.) Using the adjoint matrix $F(\lambda)$ to find the modal columns, we have

$$f(\lambda) = \begin{bmatrix} \lambda - 2h & h & 0 \\ h & \lambda - 2h & h \\ 0 & h & \lambda - h \end{bmatrix},$$

$$F(\lambda) = \begin{bmatrix} (\lambda - h)(\lambda - 2h) - h^2 & -h(\lambda - h) & h^2 \\ -h(\lambda - h) & (\lambda - h)(\lambda - 2h) & -h(\lambda - 2h) \\ h^2 & -h(\lambda - 2h) & (\lambda - 2h)^2 - h^2 \end{bmatrix}$$

$$F(\lambda_1) = h^2 \begin{bmatrix} 0.445 & 0.802 & 1.0 \\ 0.802 & 1.445 & 1.802 \\ 1.0 & 1.802 & 2.25 \end{bmatrix}$$

$$= h^2 \begin{bmatrix} 1.0 \\ 1.802 \\ 2.25 \end{bmatrix} [0.445 \quad 0.802 \quad 1.0]$$

$$F(\lambda_2) = h^2 \begin{bmatrix} -1.247 & -0.555 & 1.0 \\ -0.555 & -0.247 & 0.445 \\ 1.0 & 0.445 & -0.802 \end{bmatrix}$$

$$= h^2 \begin{bmatrix} 1.0 \\ 0.445 \\ -0.802 \end{bmatrix} [-1.247 \quad -0.555 \quad 1.0]$$

$$F(\lambda_3) = h^2 \begin{bmatrix} 1.802 & -2.247 & 1.0 \\ -2.247 & 2.802 & -1.247 \\ 1.0 & -1.247 & 0.555 \end{bmatrix}$$

$$= h^2 \begin{bmatrix} 1.0 \\ -1.247 \\ 0.555 \end{bmatrix} [1.802 \quad -2.247 \quad 1.0]$$

The modal columns may be selected as $\{1.0 \quad 1.802 \quad 2.25\}$, $\{1.0 \quad 0.445 \quad -0.802\}$, and $\{1.0 \quad -1.247 \quad 0.555\}$.

From Eq. (7-29) the motions $\theta_j(t)$ are

$$\theta_1 = \theta_{11} \sin(\omega_1 t + \psi_1) + \theta_{12} \sin(\omega_2 t + \psi_2) + \theta_{13} \sin(\omega_3 t + \psi_3)$$

$$\theta_2 = 1.802\theta_{11} \sin(\omega_1 t + \psi_1) + 0.445\theta_{12} \sin(\omega_2 t + \psi_2) - 1.247\theta_{13} \sin(\omega_3 t + \psi_3)$$

$$\theta_3 = 2.25\theta_{11} \sin(\omega_1 t + \psi_1) - 0.802\theta_{12} \sin(\omega_2 t + \psi_2) + 0.555\theta_{13} \sin(\omega_3 t + \psi_3)$$

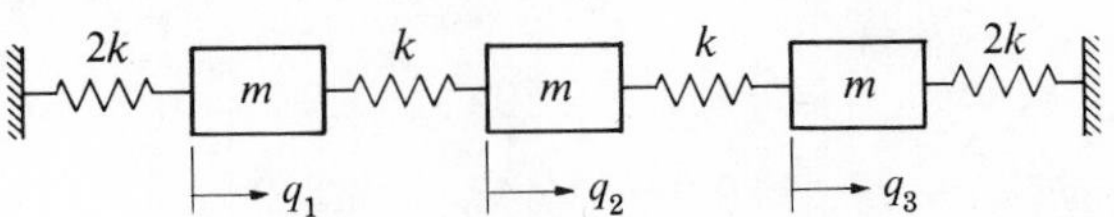

Fig. 7-2. *A three-degree-of-freedom system*

Example 2. A three-degree-of-freedom system is as shown in Fig. 7-2. Determine $\{q(t)\}$ for the initial conditions $\{q(0)\} = \{2 \quad 1 \quad -1\}$ and $\{\dot{q}(0)\} = \{1 \quad 0 \quad 2\}$.

Solution: The equations of motion of the system can be expressed as

$$M\{\ddot{q}\} + K\{q\} = \{0\}$$

where

$$M = m\begin{bmatrix} 1 & 0 & 0 \\ 0 & 1 & 0 \\ 0 & 0 & 1 \end{bmatrix}, \quad \text{and} \quad K = k\begin{bmatrix} 3 & -1 & 0 \\ -1 & 2 & -1 \\ 0 & -1 & 3 \end{bmatrix}$$

For convenience, let us define a constant $h = k/m$. The dynamic matrix and the characteristic matrix $(\lambda I - H)$ are

$$H = M^{-1}K = h\begin{bmatrix} 3 & -1 & 0 \\ -1 & 2 & -1 \\ 0 & -1 & 3 \end{bmatrix}$$

and

$$(\lambda I - H) = \begin{bmatrix} \lambda - 3h & h & 0 \\ h & \lambda - 2h & h \\ 0 & h & \lambda - 3h \end{bmatrix}$$

It can be shown that the latent roots of the characteristic equation $|\lambda I - H| = 0$ are $\lambda_1 = h$, $\lambda_2 = 3h$, and $\lambda_3 = 4h$. Hence the natural frequencies of the system are $\omega_1 = \sqrt{h}$, $\omega_2 = \sqrt{3h}$, and $\omega_3 = 2\sqrt{h}$.

Let us determine the modal columns from the adjoint $F(\lambda)$ of $(\lambda I - H)$.

$$F(\lambda) = \begin{bmatrix} (\lambda - 2h)(\lambda - 3h) - h^2 & -h(\lambda - 3h) & h^2 \\ -h(\lambda - 3h) & (\lambda - 3h)^2 & -h(\lambda - 3h) \\ h^2 & -h(\lambda - 3h) & (\lambda - 2h)(\lambda - 3h) - h^2 \end{bmatrix}$$

$$F(\lambda_1) = h^2\begin{bmatrix} 1 & 2 & 1 \\ 2 & 4 & 2 \\ 1 & 2 & 1 \end{bmatrix}, \quad F(\lambda_2) = h^2\begin{bmatrix} -1 & 0 & 1 \\ 0 & 0 & 0 \\ 1 & 0 & -1 \end{bmatrix},$$

$$F(\lambda_3) = h^2\begin{bmatrix} 1 & -1 & 1 \\ -1 & 1 & -1 \\ 1 & -1 & 1 \end{bmatrix}$$

Forming the modal column matrix $[\mu]$ from the nonzero columns of $F(\lambda_j)$ and finding its inverse, we have

$$[\mu] = \begin{bmatrix} 1 & 1 & 1 \\ 2 & 0 & -1 \\ 1 & -1 & 1 \end{bmatrix}, \quad \text{and} \quad [\mu]^{-1} = \tfrac{1}{6}\begin{bmatrix} 1 & 2 & 1 \\ 3 & 0 & -3 \\ 2 & -2 & 2 \end{bmatrix}$$

Substituting the given initial conditions in Eqs. (7-32) and (7-33) gives

$$\{a\} = [\mu]^{-1}\{q(0)\} = \tfrac{1}{6}\begin{bmatrix} 1 & 2 & 1 \\ 3 & 0 & -3 \\ 2 & -2 & 2 \end{bmatrix}\begin{bmatrix} 2 \\ 1 \\ -1 \end{bmatrix} = \begin{bmatrix} 1/2 \\ 3/2 \\ 0 \end{bmatrix}$$

$$\{\omega b\} = [\mu]^{-1}\{\dot{q}(0)\} = \tfrac{1}{6}\begin{bmatrix} 1 & 2 & 1 \\ 3 & 0 & -3 \\ 2 & -2 & 2 \end{bmatrix}\begin{bmatrix} 1 \\ 0 \\ 2 \end{bmatrix} = \begin{bmatrix} 1/2 \\ -1/2 \\ 1 \end{bmatrix}$$

Since $\{\omega b\} = \{\omega_1 b_1 \;\; \omega_2 b_2 \;\; \omega_3 b_3\}$, the constants $\{b\}$ can be determined as

$$\{b\} = \begin{bmatrix} b_1 \\ b_2 \\ b_3 \end{bmatrix} = \begin{bmatrix} 1/2\omega_1 \\ -1/2\omega_2 \\ 1/\omega_3 \end{bmatrix} = \begin{bmatrix} 1/2\sqrt{h} \\ -1/2\sqrt{3h} \\ 1/2\sqrt{h} \end{bmatrix}$$

The solution $\{q(t)\}$ is obtained by substituting the coefficients $\{a\}$ and $\{b\}$ in Eq. (7-30). Writing the solution explicitly, we have

$$q_1 = \frac{1}{2\sqrt{h}}\sin\sqrt{h}\,t - \frac{1}{2\sqrt{3h}}\sin\sqrt{3h}\,t + \frac{1}{2\sqrt{h}}\sin 2\sqrt{h}\,t + \frac{1}{2}\cos\sqrt{h}\,t + \frac{3}{2}\cos\sqrt{3h}\,t$$

$$q_2 = \frac{1}{\sqrt{h}}\sin\sqrt{h}\,t - \frac{1}{2\sqrt{h}}\sin 2\sqrt{h}\,t + \cos\sqrt{h}\,t$$

$$q_3 = \frac{1}{2\sqrt{h}}\sin\sqrt{h}\,t + \frac{1}{2\sqrt{3h}}\sin\sqrt{3h}\,t + \frac{1}{2\sqrt{h}}\sin 2\sqrt{h}\,t + \frac{1}{2}\cos\sqrt{h}\,t - \frac{3}{2}\cos\sqrt{3h}\,t$$

7-4. NORMAL COORDINATES

The principal coordinates discussed in the preceding section are often called the normal coordinates. Since a modal column can be multiplied by an arbitrary constant (see Appendix C), it is possible to normalize the principal coordinates by multiplying the modal columns by appropriate constants to obtain solutions in a convenient form. We shall call such coordinates the normal coordinates.

Applying the transformation $\{q\} = [\mu]\{p\}$ to the kinetic energy function T and the potential energy U in Eqs. (7-10) and (7-7), respectively, we obtain

$$T = \tfrac{1}{2}[\dot{p}][\mu]^T M[\mu]\{\dot{p}\} = \tfrac{1}{2}[\dot{p}][a_{ij}]\{\dot{p}\} \tag{7-34}$$

$$U = \tfrac{1}{2}[p][\mu]^T K[\mu]\{p\} = \tfrac{1}{2}[p][b_{ij}]\{p\} \tag{7-35}$$

where $[a_{ij}]$ is the inertia matrix and $[b_{ij}]$ the stiffness matrix corresponding to the principal coordinates $\{p\}$. Since the motions described by the principal coordinates are harmonic functions, the equations of motion will have neither dynamic nor elastic coupling terms. In other words, $[a_{ij}]$ and $[b_{ij}]$ are diagonal matrices. This statement can be verified as follows: The equations of motion in principal coordinates are

$$[a_{ij}]\{\ddot{p}\} + [b_{ij}]\{p\} = \{0\} \tag{7-36}$$

or

$$\{\ddot{p}\} + \Lambda\{p\} = \{0\} \tag{7-37}$$

where

$$\begin{aligned} \Lambda = [a_{ij}]^{-1}[b_{ij}] &= ([\mu]^T M[\mu])^{-1}([\mu]^T K[\mu]) \\ &= [\mu]^{-1}M^{-1}K[\mu] = [\mu]^{-1}H[\mu] \end{aligned}$$

Equations (7-34) and (7-35) show that $[a_{ij}]$ and $[b_{ij}]$ are symmetrical, since the transposes of these matrices are the matrices themselves. From Eqs. (7-36) and (7-37) we obtain

$$[a_{ij}]\Lambda = [b_{ij}] \tag{7-38}$$

or

$$\begin{bmatrix} a_{11} & a_{12} & a_{13} \\ a_{21} & a_{22} & a_{23} \\ a_{31} & a_{32} & a_{33} \end{bmatrix} \begin{bmatrix} \lambda_1 & 0 & 0 \\ 0 & \lambda_2 & 0 \\ 0 & 0 & \lambda_3 \end{bmatrix} = \begin{bmatrix} \lambda_1 a_{11} & \lambda_2 a_{12} & \lambda_3 a_{13} \\ \lambda_1 a_{21} & \lambda_2 a_{22} & \lambda_3 a_{23} \\ \lambda_1 a_{31} & \lambda_2 a_{32} & \lambda_3 a_{33} \end{bmatrix} = \begin{bmatrix} b_{11} & b_{12} & b_{13} \\ b_{21} & b_{22} & b_{23} \\ b_{31} & b_{32} & b_{33} \end{bmatrix} \tag{7-39}$$

Since $[b_{ij}]$ is symmetrical, Eqs. (7-38) and (7-39) require $\lambda_j a_{ij} = \lambda_i a_{ji}$. Since $[a_{ij}]$ is symmetrical and it is assumed that the latent roots of $\Delta(\lambda) = 0$ are distinct, the conditions are satisfied if $a_{ij} = a_{ji} = b_{ij} = b_{ji} = 0$ for $i \neq j$. Hence $[a_{ij}]$ and $[b_{ij}]$ are diagonal.

Let us define a particular set of principal coordinates by the transformation

$$\{q\} = [\nu]\{p\} \tag{7-40}$$

If the transformation matrix $[\nu]$ has the property

$$T = \tfrac{1}{2}\lfloor\dot{q}\rfloor M\{\dot{q}\} = \tfrac{1}{2}\lfloor\dot{p}\rfloor[\nu]^T M[\nu]\{\dot{p}\} = \tfrac{1}{2}\lfloor\dot{p}\rfloor\{\dot{p}\}$$
$$U = \tfrac{1}{2}\lfloor q\rfloor K\{q\} = \tfrac{1}{2}\lfloor p\rfloor[\nu]^T K[\nu]\{p\} = \tfrac{1}{2}\lfloor p\rfloor\Lambda\{p\}$$

or

$$[\nu]^T M[\nu] = I \tag{7-41}$$

$$[\nu]^T K[\nu] = \Lambda \tag{7-42}$$

then this particular set of principal coordinates is called the *normal coordinates.*

Comparing Eqs. (7-18) and (7-40), it is noted that the transformation matrix $[\nu]$ is a particular choice of the modal column matrix $[\mu]$. Since a modal column is indeterminate to the extent that it can always be multiplied by an arbitrary constant, a modal column matrix can always be postmultiplied by a nonsingular diagonal matrix of constants $[n]$; that is

$$[\nu] = [\mu][n] \tag{7-43}$$

where

$$[n] = \begin{bmatrix} n_1 & 0 & \dots & 0 & 0 \\ 0 & n_2 & \dots & 0 & 0 \\ \cdot & \cdot & \cdot\,\cdot\,\cdot & \cdot & \cdot \\ 0 & 0 & \dots & 0 & n_n \end{bmatrix}$$

Each of the constants in the diagonal matrix $[n]$ is called a normalizing factor n_s appropriate to λ_s. Substituting Eq. (7-43) in Eq. (7-41) gives

$$[n][\mu]^T M[\mu][n] = I \tag{7-44}$$

Since $[\mu]$ can be determined by the method of the last section, the normalizing factors are specified by Eq. (7-44). Recalling that $[\mu]^T M[\mu]$ is diagonal, it can be shown by simple algebraic operations that Eq. (7-44) requires

$$n_s^2 \sum_i \sum_j m_{ij}\mu_{is}\mu_{js} = 1$$

or

$$n_s = 1\Big/\sqrt{\sum_i \sum_j m_{ij}\mu_{is}\mu_{js}} \tag{7-45}$$

Example 3. A two-degree-of-freedom system is as shown in Fig. 7-3. (See Example 15, Chap. 3.) If $m_1 = m_2 = m$ and $k_1 = k_2 = k$, write the equations of motion of the system in terms of the principal and the normal coordinates.

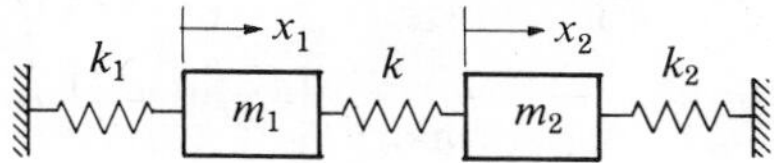

Fig. 7-3. *A two-degree-of-freedom system*

Solution: It can be shown that the equations of motion are

$$\begin{bmatrix} m & 0 \\ 0 & m \end{bmatrix}\begin{bmatrix} \ddot{x}_1 \\ \ddot{x}_2 \end{bmatrix} + \begin{bmatrix} 2k & -k \\ -k & 2k \end{bmatrix}\begin{bmatrix} x_1 \\ x_2 \end{bmatrix} = \begin{bmatrix} 0 \\ 0 \end{bmatrix}$$

Hence the dynamic matrix H and its characteristic matrix $f(\lambda) = (\lambda I - H)$ are

$$H = M^{-1}K = \begin{bmatrix} 2k/m & -k/m \\ -k/m & 2k/m \end{bmatrix}, \quad f(\lambda) = \begin{bmatrix} \lambda - 2k/m & k/m \\ k/m & \lambda - 2k/m \end{bmatrix}$$

It can be shown that the roots of the characteristic equation $\Delta(\lambda) = |f(\lambda)| = 0$ are $\lambda_1 = k/m$ and $\lambda_2 = 3k/m$.

Let us determine the modal columns from the adjoint $F(\lambda)$ of $f(\lambda)$.

$$F(\lambda) = \begin{bmatrix} \lambda - 2k/m & -k/m \\ -k/m & \lambda - 2k/m \end{bmatrix}$$

$$F(\lambda_1) = \frac{k}{m}\begin{bmatrix} -1 & -1 \\ -1 & -1 \end{bmatrix} = \frac{k}{m}\begin{bmatrix} 1 \\ 1 \end{bmatrix}[-1 \quad -1]$$

$$F(\lambda_2) = \frac{k}{m}\begin{bmatrix} 1 & -1 \\ -1 & 1 \end{bmatrix} = \frac{k}{m}\begin{bmatrix} 1 \\ -1 \end{bmatrix}[1 \quad -1]$$

Hence the modal columns may be selected as $\{1 \quad 1\}$ and $\{1 \quad -1\}$. The corresponding principal coordinates are

$$\{x\} = [\mu]\{p\} \quad \text{or} \quad \begin{bmatrix} x_1 \\ x_2 \end{bmatrix} = \begin{bmatrix} 1 & 1 \\ 1 & -1 \end{bmatrix} \begin{bmatrix} p_1 \\ p_2 \end{bmatrix}$$

Applying Eqs. (7-34) and (7-35) to obtain the inertia matrix $[a_{ij}]$ and the stiffness matrix $[b_{ij}]$ the equations of motion in principal coordinates are

$$\begin{bmatrix} 2m & 0 \\ 0 & 2m \end{bmatrix} \begin{bmatrix} \ddot{p}_1 \\ \ddot{p}_2 \end{bmatrix} + \begin{bmatrix} 2k & 0 \\ 0 & 6k \end{bmatrix} \begin{bmatrix} p_1 \\ p_2 \end{bmatrix} = \begin{bmatrix} 0 \\ 0 \end{bmatrix}$$

Applying Eq. (7-45), the normalizing factors are

$$n_1 = 1/\sqrt{2m} \quad \text{and} \quad n_2 = 1/\sqrt{2m}$$

Thus the transformation matrix for obtaining the normal coordinates is

$$[\nu] = [\mu][n] = \begin{bmatrix} 1 & 1 \\ 1 & -1 \end{bmatrix} \begin{bmatrix} 1/\sqrt{2m} & 0 \\ 0 & 1/\sqrt{2m} \end{bmatrix} = \frac{1}{\sqrt{2m}} \begin{bmatrix} 1 & 1 \\ 1 & -1 \end{bmatrix}$$

Obtaining the corresponding inertia and stiffness matrices, the equations in normal coordinates are

$$\begin{bmatrix} \ddot{p}_1 \\ \ddot{p}_2 \end{bmatrix} + \begin{bmatrix} k/m & 0 \\ 0 & 3k/m \end{bmatrix} \begin{bmatrix} p_1 \\ p_2 \end{bmatrix} = \begin{bmatrix} 0 \\ 0 \end{bmatrix}$$

7-5. ORTHOGONALITY OF THE PRINCIPAL MODES OF VIBRATION

Two column matrices X and Y are said to be orthogonal with respect to a symmetrical matrix A if they satisfy the equation

$$X^T A Y = Y^T A X = 0 \tag{7-46}$$

It was shown in Sec. 3 that every motion represented by each modal column is a principal mode of vibration. We shall show that these modal columns are orthogonal with respect to the inertia matrix M and the stiffness matrix K.

Consider a system with the equations of motion

$$M\{\ddot{q}\} + K\{q\} = \{0\} \tag{7-47}$$

In principal coordinates, the equations of motion, Eq. (7-36), become

$$[a_{ij}]\{\ddot{p}\} + [b_{ij}]\{p\} = \{0\} \tag{7-48}$$

where

$$[a_{ij}] = [\mu]^T M[\mu] \tag{7-49}$$

$$[b_{ij}] = [\mu]^T K[\mu] \tag{7-50}$$

It was shown in the last section that, since $\{p\}$ are principal coordinates, the matrices $[a_{ij}]$ and $[b_{ij}]$ are diagonal; that is,

$$a_{ij} = b_{ij} = 0, \quad \text{if} \quad i \neq j \tag{7-51}$$

Hence, for two modal columns $\{\mu_{ir}\}$ and $\{\mu_{is}\}$ appropriate to ω_r and ω_s, respectively, we have

$$\{\mu_{ir}\}^T M\{\mu_{is}\} = \{\mu_{ir}\}^T K\{\mu_{is}\} = 0, \quad \text{if} \quad r \neq s \tag{7-52}$$

The orthogonality of the principal modes of vibration was illustrated in Example 16, Chap. 3.

7-6. SEMIDEFINITE SYSTEMS

If a system has one or more of its natural frequencies equal to zero, it may move as a rigid body without disturbing the forces acting upon it. This class of systems is called *semidefinite systems*. (See Sec. 3-3, Chap. 3.) In other words, the configuration of the system at $\{q\} = \{0\}$ is such that its potential energy U remains unchanged when any or all of its generalized coordinates $\{q\}$ assume nonzero values; that is, $\{q\} = \{0\}$ corresponds to a condition of neutral equilibrium.†

Consider the equations of motion of a system

$$M\{\ddot{q}\} + K\{q\} = \{0\} \tag{7-53}$$

Using the transformation $\{q\} = [\mu]\{p\}$ to express Eq. (7-53) in terms of the principal coordinates $\{p\}$, Eq. (7-15), we have

$$\{\ddot{p}\} + \Lambda\{p\} = \{0\} \tag{7-54}$$

or

$$\ddot{p}_j + \lambda_j p_j = 0 \tag{7-55}$$

where Λ is a diagonal matrix and its diagonal elements λ_j are the roots of the characteristic equation $\Delta(\lambda) = |\lambda I - H| = 0$. Let us assume

† Since $U = \frac{1}{2}[q]K\{q\}$ and it is possible to have U vanish without $\{q\} = \{0\}$, the stiffness matrix K is necessarily singular. Hence, if K is singular, the system is semidefinite.

that the roots are distinct and consider the case of repeated roots in the next section.

It is evident from Eq. (7-55) that $\omega_j = \sqrt{\lambda_j}$ is a natural frequency of the system. If one of the roots $\lambda = \lambda_s = 0$, we have $\ddot{p}_s = 0$; that is,

$$p_s = a_s + b_s t \tag{7-56}$$

where a_s and b_s are constants. The motion described by p_s is aperiodic. Since the general motions of a multidegree-of-freedom system are the superpositions of the principal modes, p_s is a component of the motions described by every generalized coordinate q_j. Using the method outlined in Sec. 7-3, the general motions $\{q(t)\}$ can be expressed in terms of the principal modes $\{p(t)\}$ as $\{q(t)\} = [\mu]\{p(t)\}$, Eq. (7-18). Hence with the exception of the zero frequency term, or the *zero mode*, the solutions can be expressed in the form of Eq. (7-30).

For example, if a system has three degrees of freedom and the frequencies are $\omega = 0$, ω_2, and ω_3, Eq. (7-30) can be modified and written explicitly as

$$\begin{bmatrix} q_1 \\ q_2 \\ q_3 \end{bmatrix} = \begin{bmatrix} \mu_{11} & \mu_{12} & \mu_{13} \\ \mu_{21} & \mu_{22} & \mu_{23} \\ \mu_{31} & \mu_{32} & \mu_{33} \end{bmatrix} \left[\begin{bmatrix} t & 0 & 0 \\ 0 & \sin\omega_2 t & 0 \\ 0 & 0 & \sin\omega_3 t \end{bmatrix} \begin{bmatrix} b_1 \\ b_2 \\ b_3 \end{bmatrix} + \begin{bmatrix} 1 & 0 & 0 \\ 0 & \cos\omega_2 t & 0 \\ 0 & 0 & \cos\omega_3 t \end{bmatrix} \begin{bmatrix} a_1 \\ a_2 \\ a_3 \end{bmatrix} \right] \tag{7-57}$$

Since the system is semidefinite, the modal column $\{\mu_{i1}\}$ is $\{1 \quad 1 \quad 1\}$.

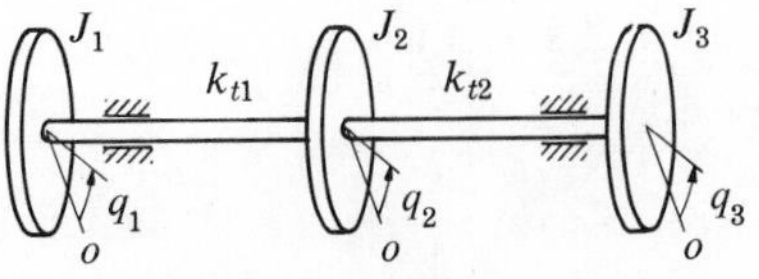

Example 4. A semidefinite system is as shown in Fig. 7-4. If $J_1 = J_2 = J_3 = J$ and $k_{t1} = k_{t2} = k_t$, determine $\{q(t)\}$ for the initial conditions $\{q(0)\} = \{5 \quad 4 \quad 2\}$ and $\{\dot{q}(0)\} = \{0\}$.

Fig. 7-4. *A semidefinite system*

Solution: The equations of motion can be expressed in the form

$$M\{\ddot{q}\} + K\{q\} = \{0\}$$

where

$$M = \begin{bmatrix} J & 0 & 0 \\ 0 & J & 0 \\ 0 & 0 & J \end{bmatrix}, \quad K = k_t \begin{bmatrix} 1 & -1 & 0 \\ -1 & 2 & -1 \\ 0 & -1 & 1 \end{bmatrix}$$

Defining $H = M^{-1}K$ and $h = k_t/J$, a constant, the characteristic matrix of H is

$$f(\lambda) = (\lambda I - H) = \begin{bmatrix} \lambda - h & h & 0 \\ h & \lambda - 2h & h \\ 0 & h & \lambda - h \end{bmatrix}$$

$$\Delta(\lambda) = |f(\lambda)| = \lambda(\lambda - h)(\lambda - 3h)$$

Hence, corresponding to λ_1, λ_2, and λ_3, the natural frequencies of the system are $\omega_1 = 0$, $\omega_2 = \sqrt{h}$, and $\omega_3 = \sqrt{3h}$.

Let us determine the modal columns from the adjoint $F(\lambda)$ of $f(\lambda)$.

$$F(\lambda) = \begin{bmatrix} (\lambda - h)(\lambda - 2h) - h^2 & -h(\lambda - h) & h^2 \\ -h(\lambda - h) & (\lambda - h)^2 & -h(\lambda - h) \\ h^2 & -h(\lambda - h) & (\lambda - h)(\lambda - 2h) - h^2 \end{bmatrix}$$

$$F(\lambda_1) = h^2 \begin{bmatrix} 1 & 1 & 1 \\ 1 & 1 & 1 \\ 1 & 1 & 1 \end{bmatrix}, \quad F(\lambda_2) = h^2 \begin{bmatrix} -1 & 0 & 1 \\ 0 & 0 & 0 \\ 1 & 0 & -1 \end{bmatrix},$$

$$F(\lambda_3) = h^2 \begin{bmatrix} 1 & -2 & 1 \\ -2 & 4 & -2 \\ 1 & -2 & 1 \end{bmatrix}$$

The modal column matrix and its inverse may be selected as

$$[\mu] = \begin{bmatrix} 1 & 1 & 1 \\ 1 & 0 & -2 \\ 1 & -1 & 1 \end{bmatrix} \quad \text{and} \quad [\mu]^{-1} = \tfrac{1}{6} \begin{bmatrix} 2 & 2 & 2 \\ 3 & 0 & -3 \\ 1 & -2 & 1 \end{bmatrix}$$

Applying Eq. (7-57) and the initial conditions $\{q(0)\} = \{5 \quad 4 \quad 2\}$, we have

$$\{q(0)\} = [\mu]I\{a\} \quad \text{or} \quad \{a\} = [\mu]^{-1}\{q(0)\}$$

$$\begin{bmatrix} a_1 \\ a_2 \\ a_3 \end{bmatrix} = \tfrac{1}{6} \begin{bmatrix} 2 & 2 & 2 \\ 3 & 0 & -3 \\ 1 & -2 & 1 \end{bmatrix} \begin{bmatrix} 5 \\ 4 \\ 2 \end{bmatrix} = \tfrac{1}{6} \begin{bmatrix} 22 \\ 9 \\ -1 \end{bmatrix}$$

Applying the initial conditions $\{\dot{q}(0)\} = \{0\}$, we obtain

$$\begin{bmatrix} 0 \\ 0 \\ 0 \end{bmatrix} = \begin{bmatrix} 1 & 1 & 1 \\ 1 & 0 & -2 \\ 1 & -1 & 1 \end{bmatrix} \begin{bmatrix} 1 & 0 & 0 \\ 0 & \omega_2 & 0 \\ 0 & 0 & \omega_3 \end{bmatrix} \begin{bmatrix} b_1 \\ b_2 \\ b_3 \end{bmatrix} \quad \text{or} \quad \{b\} = \{0\}$$

Hence the solutions are

$$\begin{aligned} q_1 &= (1/6)(22 + 9 \cos \sqrt{h}\, t - \cos \sqrt{3h}\, t) \\ q_2 &= (1/6)(22 + 2 \cos \sqrt{3h}\, t) \\ q_3 &= (1/6)(22 - 9 \cos \sqrt{h}\, t - \cos \sqrt{3h}\, t) \end{aligned}$$

7-7. SYSTEMS WITH EQUAL FREQUENCIES

Owing to symmetry or some other factors, the characteristic equation of a system may have repeated roots; that is, more than one of the natural frequencies of the system are of equal value. For example, Fig. 7-5 shows a single mass attached to four identical springs. If the springs are at right angles to one another and the mass is constrained to move in a horizontal plane, the system has two equal natural frequencies. A root occurring more than once is said to represent a degenerate mode of vibration. The more common case of distinct roots was discussed in the previous sections; in this section we shall briefly consider systems with repeated roots.

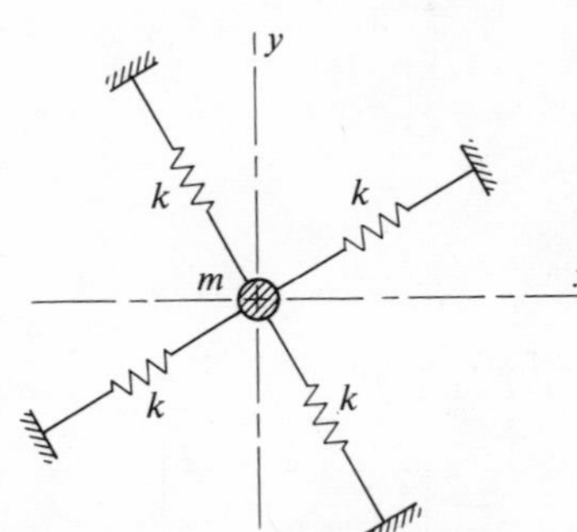

Fig. 7-5. *System with two equal natural frequencies*

Consider a system with the equations of motion

$$\{\ddot{q}\} + H\{q\} = \{0\} \tag{7-58}$$

where $\{q\}$ is a set of generalized coordinates and $H = M^{-1}K$. In the previous sections the solutions $\{q\}$ were obtained from the transformation

$$\{q\} = [\mu]\{p\} \tag{7-59}$$

where $[\mu]$ is a modal column matrix and $\{p\}$ represents the principal oscillations. Since the principal oscillations are harmonic, the problem is one of finding the appropriate modal column matrix $[\mu]$.

For simplicity, let us consider a system with two equal natural frequencies; that is, two latent roots of the characteristic equation $\Delta(\lambda) = |\lambda I - H| = 0$ equal λ_s. It was stated in Appendix C that the adjoint matrix $F(\lambda_s)$ of $(\lambda I - H)$ is null and the derived adjoint, defined as $(d/d\lambda)F(\lambda) \equiv F^{(1)}(\lambda)$, satisfies the equation [see Theorem 8 and Eq. (C-56), Appendix C]

$$(\lambda_s I - H)F^{(1)}(\lambda_s) = 0 \tag{7-60}$$

It was also stated that $F^{(1)}(\lambda_s)$ is of rank 2. Hence the required modal column matrix $[\mu]$ can be formed from the nonzero independent columns of $F^{(1)}(\lambda_s)$ corresponding to the repeated roots and the modal columns corresponding to the distinct roots.

Example 5. Two unit masses are attached to six springs inside a rigid frame, as shown in Fig. 7-6. Assume that the masses are at their static equilibrium positions at the origins of the generalized coordinates $\{q\}$. Determine $\{q(t)\}$.

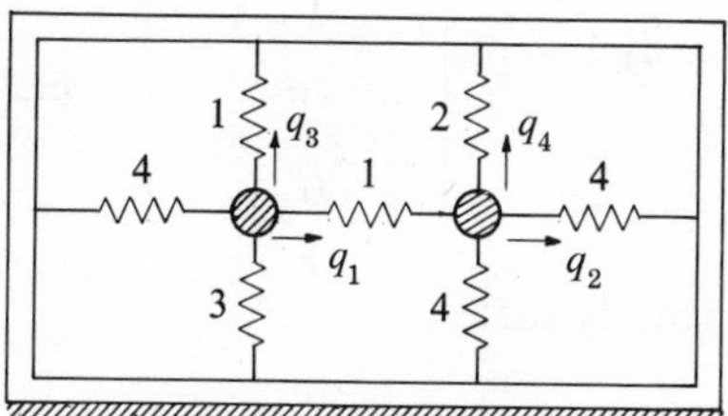

Fig. 7-6. *System with equal natural frequencies*

Solution: It can be shown that the equations of motion are

$$M\{\ddot{q}\} + K\{q\} = \{0\}$$

where

$$M = \begin{bmatrix} 1 & 0 & 0 & 0 \\ 0 & 1 & 0 & 0 \\ 0 & 0 & 1 & 0 \\ 0 & 0 & 0 & 1 \end{bmatrix} \qquad K = \begin{bmatrix} 5 & -1 & 0 & 0 \\ -1 & 5 & 0 & 0 \\ 0 & 0 & 4 & 0 \\ 0 & 0 & 0 & 6 \end{bmatrix}$$

From Fig. 7-6, it is evident that, for small oscillations, the coordinates q_1 and q_2 are elastically coupled but q_3 and q_4 are not coupled to any of the other coordinates. Hence the system can be analyzed easily by methods described in Chap. 3.

To illustrate the method outlined in this section, we define $H = M^{-1}K$ and write the characteristic matrix of H and the characteristic equation as

$$f(\lambda) = (\lambda I - H) = \begin{bmatrix} \lambda - 5 & 1 & 0 & 0 \\ 1 & \lambda - 5 & 0 & 0 \\ 0 & 0 & \lambda - 4 & 0 \\ 0 & 0 & 0 & \lambda - 6 \end{bmatrix}$$

$$\Delta(\lambda) = (\lambda - 4)^2(\lambda - 6)^2 = 0$$

Hence the natural frequencies are $\omega_1 = \omega_2 = 2$ and $\omega_3 = \omega_4 = \sqrt{6}$. Simple algebraic operations show that $f(\lambda_s)$ is of degeneracy 2. Hence $F(\lambda_s)$ is null (see Theorem 8, Appendix C), and the modal columns have to be determined from $F^{(1)}(\lambda_s)$.

$$F(\lambda) = (\lambda - 4)(\lambda - 6) \begin{bmatrix} \lambda - 5 & -1 & 0 & 0 \\ -1 & \lambda - 5 & 0 & 0 \\ 0 & 0 & \lambda - 6 & 0 \\ 0 & 0 & 0 & \lambda - 4 \end{bmatrix}$$

It can be verified readily that

$$F^{(1)}(\lambda_1) = \begin{bmatrix} 2 & 2 & 0 & 0 \\ 2 & 2 & 0 & 0 \\ 0 & 0 & 4 & 0 \\ 0 & 0 & 0 & 0 \end{bmatrix} \qquad F^{(1)}(\lambda_3) = \begin{bmatrix} 2 & -2 & 0 & 0 \\ -2 & 2 & 0 & 0 \\ 0 & 0 & 0 & 0 \\ 0 & 0 & 0 & 4 \end{bmatrix}$$

Forming the modal column matrix from the nonzero independent columns of $F^{(1)}(\lambda_s)$, the solution can be expressed as

$$\begin{bmatrix} q_1 \\ q_2 \\ q_3 \\ q_4 \end{bmatrix} = \begin{bmatrix} 1 & 0 & 1 & 0 \\ 1 & 0 & -1 & 0 \\ 0 & 1 & 0 & 0 \\ 0 & 0 & 0 & 1 \end{bmatrix} \begin{bmatrix} c_1 \sin(2t + \psi_1) \\ c_2 \sin(2t + \psi_2) \\ c_3 \sin(\sqrt{6}\, t + \psi_3) \\ c_4 \sin(\sqrt{6}\, t + \psi_4) \end{bmatrix}$$

or

$$
\begin{aligned}
q_1 &= c_1 \sin(2t + \psi_1) + c_3 \sin(\sqrt{6}\,t + \psi_3) \\
q_2 &= c_1 \sin(2t + \psi_1) - c_3 \sin(\sqrt{6}\,t + \psi_3) \\
q_3 &= c_2 \sin(2t + \psi_2) \\
q_4 &= c_4 \sin(\sqrt{6}\,t + \psi_4)
\end{aligned}
$$

7-8. INFLUENCE COEFFICIENTS

In the previous sections the stiffness matrix K was used to describe the elastic properties of a system. It was shown in Sec. 3-7, Chap. 3, that the elastic properties can also be described by influence coefficients. We shall show the relationship between these two descriptions of the same properties.

An influence coefficient δ_{ij} is defined as the static deflection at station i owing to a unit force applied at station j of a system. It was shown that $\delta_{ij} = \delta_{ji}$. By the principle of superposition, the deflection at any station of a system can be obtained by summing the product of the influence coefficients and the respective forces. For example, if the system considered has three degrees of freedom, we have

$$
\begin{aligned}
q_1 &= \delta_{11}Q_1 + \delta_{12}Q_2 + \delta_{13}Q_3 \\
q_2 &= \delta_{21}Q_1 + \delta_{22}Q_2 + \delta_{23}Q_3 \\
q_3 &= \delta_{31}Q_1 + \delta_{32}Q_2 + \delta_{33}Q_3
\end{aligned} \tag{7-61}
$$

or

$$\{q\} = [\delta_{ij}]\{Q\} \tag{7-62}$$

where $\{q\}$ are the generalized coordinates and $\{Q\}$ the generalized forces, respectively. $[\delta_{ij}]$ is symmetrical; it is called the *flexibility matrix*, and its elements are the flexibility influence coefficients.

Conversely, the forces corresponding to the elastic deflections of a system can be expressed as

$$\{Q\} = [k_{ij}]\{q\} \tag{7-63}$$

Comparing Eqs. (7-62) and (7-63), it is evident that

$$[\delta_{ij}] = [k_{ij}]^{-1} \tag{7-64}$$

It can be shown that the potential energy U of a system can be expressed as a quadratic form of the coordinates as well as the generalized forces; that is,

$$U = \tfrac{1}{2}\lfloor q \rfloor K\{q\} = \tfrac{1}{2}\lfloor Q \rfloor [\delta_{ij}]\{Q\} \tag{7-65}$$

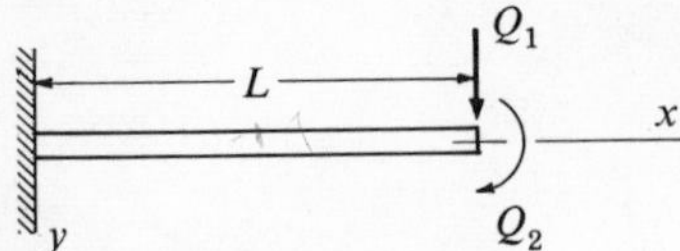

Fig. 7-7. *Cantilever with force and torque applied*

Example 6. A uniform cantilever has a force Q_1 and a moment Q_2 applied at its free end, as shown in Fig. 7-7. Determine the flexibility and the stiffness matrices and show that $[k_{ij}] = [\delta_{ij}]^{-1}$.

Solution: From Example 14, Chap. 3, the flexibility matrix and its inverse for the cantilever are

$$[\delta_{ij}] = \frac{L}{EI}\begin{bmatrix} L^2/3 & L/2 \\ L/2 & 1 \end{bmatrix}, \qquad [\delta_{ij}]^{-1} = \frac{EI}{L}\begin{bmatrix} 12/L^2 & -6/L \\ -6/L & 4 \end{bmatrix}$$

Let us determine the stiffness coefficients directly from Eq. (7-63).

$$Q_1 = k_{11}q_1 + k_{12}q_2$$
$$Q_2 = k_{21}q_1 + k_{22}q_2$$

If the system is constrained, as shown in Fig. 7-8(*a*), so that $q_1 = 1$ and $q_2 = 0$, then $Q_1 = k_{11}$ and $Q_2 = k_{21}$. From elementary beam theory, we have

$$EI\, y^{(2)}(x) = k_{11}(L - x) + k_{21}$$

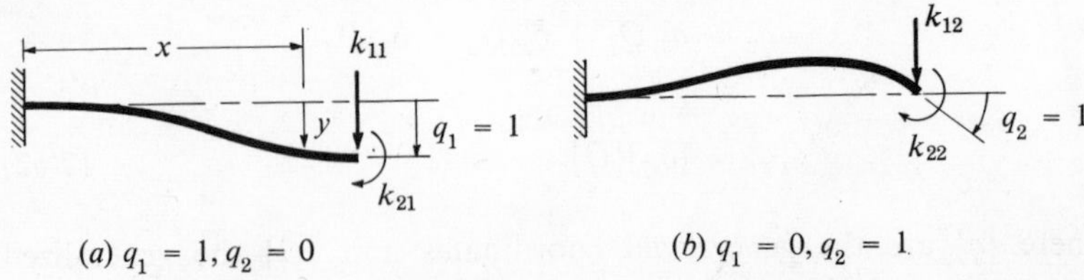

Fig. 7-8. *Determination of stiffness coefficients k_{ij}*

where $y(x)$ is the beam deflection at an intermediate point x and $y^{(2)}(x) \equiv d^2y/dx^2$. By successive integration and applying the boundary conditions $y^{(1)}(0) = y^{(1)}(L) = y(0) = 0$ and $y(L) = q_1 = 1$, we obtain

$$0 = k_{11}L^2/2 + k_{21}L$$
$$EI = k_{11}L^3/3 + k_{21}L^2/2$$

Hence $k_{11} = 12EI/L^3$ and $k_{21} = -6EI/L^2$.

Similarly, if the beam is constrained, as shown in Fig. 7-8(*b*), so that $q_1 = 0$ and $q_2 = 1$, the generalized forces applied to the system are k_{12} and k_{22}, respectively. Using the boundary conditions $y^{(1)}(0) = y(0) = y(L) = 0$ and $y^{(1)}(L) = q_2 = 1$, it can be shown that $k_{12} = -6EI/L^2$ and $k_{22} = 4EI/L$. Collecting the stiffness coefficients, we have

$$[k_{ij}] = \frac{EI}{L}\begin{bmatrix} 12/L^2 & -6/L \\ -6/L & 4 \end{bmatrix} = [\delta_{ij}]^{-1}$$

7-9. NATURAL FREQUENCIES AND PRINCIPAL MODES BY MATRIX ITERATION†

Matrix iteration is a convenient method for finding the natural frequencies and the principal modes of systems with a reasonably large number of degrees of freedom. Through the use of electronic digital computers, the solution of large scale problems becomes practical. We shall briefly discuss the method which first gives the fundamental frequency and then the higher harmonics. For convenience, a three-degree-of-freedom system is used to illustrate the method, although the equations can easily be rewritten for an *n*-degree-of-freedom system.

Consider a system with the equations of motion

$$M\{\ddot{q}\} + K\{q\} = \{0\} \tag{7-66}$$

The general motions $\{q\}$ are the superpositions of the principal modes. The modal columns, representing the principal modes, can be obtained from Eq. (7-22):

$$(-\omega^2 I + H)\{\mu\} = \{0\} \tag{7-67}$$

To determine the fundamental frequency, let us rewrite this equation in the form

$$(zI - G)\{\mu\} = \{0\} \tag{7-68}$$

where $z = 1/\omega^2$ and $G = H^{-1} = K^{-1}M$ is called the *dynamic matrix*, which is the inverse of the dynamic matrix defined in the previous sections. The fundamental frequency is obtained from the dominant root of the equation

$$\Delta(z) = |zI - G| = 0 \tag{7-69}$$

† W. J. Duncan and A. R. Collar, "A Method for the Solution of Oscillation Problems by Matrices," *Phil. Mag.*, Ser. 7, Vol. 17, 1934, p. 865.

Let $\{\mu_{is}\} = \{\mu_{1s}\ \mu_{2s}\ \mu_{3s}\}$ be a modal column appropriate to z_s. Eq. (7-68) can be written as

$$G\{\mu_{is}\} = z_s\{\mu_{is}\} \tag{7-70}$$

Premultiplying this equation by G gives

$$G^2\{\mu_{is}\} = z_s G\{\mu_{is}\} = z_s^2\{\mu_{is}\}$$

Repeated premultiplication of Eq. (7-70) n times by G yields

$$G^n\{\mu_{is}\} = z_s^n\{\mu_{is}\} \tag{7-71}$$

For a three-degree-of-freedom system, there are three equations of this type, one for each modal column $\{\mu_{is}\}$. Combining the three equations, we obtain

$$G^n \begin{bmatrix} \mu_{11} & \mu_{12} & \mu_{13} \\ \mu_{21} & \mu_{22} & \mu_{23} \\ \mu_{31} & \mu_{32} & \mu_{33} \end{bmatrix} = \begin{bmatrix} \mu_{11} & \mu_{12} & \mu_{13} \\ \mu_{21} & \mu_{22} & \mu_{23} \\ \mu_{31} & \mu_{32} & \mu_{33} \end{bmatrix} \begin{bmatrix} z_1^n & 0 & 0 \\ 0 & z_2^n & 0 \\ 0 & 0 & z_3^n \end{bmatrix} \tag{7-72}$$

where $[\mu] = [\mu_{ij}]$ is the modal column matrix. For convenience, we define

$$[v] = [\mu]^{-1} \tag{7-73}$$

Postmultiplying Eq. (7-72) by $[v]$ gives

$$G^n = [\mu] \begin{bmatrix} z_1^n & 0 & 0 \\ 0 & z_2^n & 0 \\ 0 & 0 & z_3^n \end{bmatrix} [v] \tag{7-74}$$

Let the roots z_j be distinct and $z_1 > z_2 > z_3$. If n is sufficiently large, we have $z_1^n \gg z_2^n \gg z_3^n$. Hence

$$G^n = \begin{bmatrix} \mu_{11} & \mu_{12} & \mu_{13} \\ \mu_{21} & \mu_{22} & \mu_{23} \\ \mu_{31} & \mu_{32} & \mu_{33} \end{bmatrix} \begin{bmatrix} z_1^n & 0 & 0 \\ 0 & 0 & 0 \\ 0 & 0 & 0 \end{bmatrix} \begin{bmatrix} v_{11} & v_{12} & v_{13} \\ v_{21} & v_{22} & v_{23} \\ v_{31} & v_{32} & v_{33} \end{bmatrix} = z_1^n \begin{bmatrix} \mu_{11} \\ \mu_{21} \\ \mu_{31} \end{bmatrix} [v_{11}\ \ v_{12}\ \ v_{13}] \tag{7-75}$$

We shall make use of this equation to obtain the dominant root z_1 and the corresponding modal column.

To start the iteration process, let us select an arbitrary 3×1

column matrix $\{\mu\}_o$. Repeated premultiplication of $\{\mu\}_o$ n times by G gives

$$G^n\{\mu\}_o = \{\mu\}_n \tag{7-76}$$

Comparing Eqs. (7-75) and (7-76), we obtain

$$\{\mu\}_n = G^n\{\mu\}_o = z_1^n \begin{bmatrix} \mu_{11} \\ \mu_{21} \\ \mu_{31} \end{bmatrix} [v_{11} \quad v_{12} \quad v_{13}]\{\mu\}_o = z_1^n c_1 \begin{bmatrix} \mu_{11} \\ \mu_{21} \\ \mu_{31} \end{bmatrix} \tag{7-77}$$

where $c_1 = [v_{11}\ v_{12}\ v_{13}]\{\mu\}_o$ is a scalar constant.

If n is sufficiently large, each further iteration results in the multiplication of the column matrix in Eq. (7-77) by a constant z_1. Since z_1 is the dominant root of the characteristic equation $|zI - G| = 0$, the fundamental frequency of the system is

$$\omega_1 = \sqrt{1/z_1} \tag{7-78}$$

Since a modal column is indeterminate to the extent that it can be multiplied by an arbitrary constant, the corresponding modal column is proportional to $\{\mu\}_n$.

Recapitulating, the iteration process involves (1) assuming an arbitrary first mode $\{\mu\}_o$, (2) substituting $\{\mu\}_o$ for $\{\mu_{is}\}$ on the left side of Eq. (7-70), (3) performing the multiplication, and (4) using the computed mode shape, right side of Eq. (7-70), for the iteration. If the mode assumed is not identical to one of the higher modes, the iteration converges to the first mode, that is, the mode with the lowest natural frequency.

If the first mode could be removed, however, convergence would be towards the second mode, that is, the mode with the second lowest natural frequency. This can be accomplished by constraining the coordinates so that the first mode is absent. The given generalized coordinates $\{q\}$ and a corresponding set of principal coordinates $\{p\}$ are related by the transformation

$$\{q\} = [\mu]\{p\} \tag{7-79}$$

Defining $[v] = [\mu]^{-1}$, and premultiplying Eq. (7-79) by $[v]$ gives

$$\{p\} = [v]\{q\} \tag{7-80}$$

From Eq. (7-80), the first mode is absent if we define

$$p_1 = v_{11}q_1 + v_{12}q_2 + v_{13}q_3 = 0 \tag{7-81}$$

Using this equation, the coordinate system with the required constraint can be expressed as

$$\begin{bmatrix} q_1 \\ q_2 \\ q_3 \end{bmatrix} = \begin{bmatrix} 0 & -\dfrac{v_{12}}{v_{11}} & -\dfrac{v_{13}}{v_{11}} \\ 0 & 1 & 0 \\ 0 & 0 & 1 \end{bmatrix} \begin{bmatrix} q_1 \\ q_2 \\ q_3 \end{bmatrix} \quad \text{or} \quad \{q\} = S\{q\} \tag{7-82}$$

where the square matrix S is as defined in the equation.

The row matrix $[v_{11}\ v_{12}\ v_{13}]$ in Eq. (7-81) can be obtained from the modal column $\{\mu_{11}\ \mu_{21}\ \mu_{31}\}$, which is already determined, and the relation shown in Eq. (7-49).

$$[\mu]^T M[\mu] = [a_{ij}]$$

Premultiplying this equation by $[a_{ij}]^{-1}$ and postmultiplying the resultant equation by $[\mu]^{-1}$ yields

$$[a_{ij}]^{-1}[\mu]^T M = [\mu]^{-1} = [v] \tag{7-83}$$

Recalling that $[a_{ij}]$ is diagonal, its inverse is another diagonal matrix. Thus, considering the first row of Eq. (7-83), we obtain

$$[v_{11} \quad v_{12} \quad v_{13}] = (\text{constant})\ [\mu_{11} \quad \mu_{21} \quad \mu_{31}]M \tag{7-84}$$

where M is a square matrix of order 3. Since the ratios of the elements in $[v_{11}\ v_{12}\ v_{13}]$ are required in Eq. (7-82), the constant in Eq. (7-84) need not be considered.

Substituting Eq. (7-82) in the general matrix equation (7-70), we obtain

$$GS\{\mu_{is}\} = z_s\{\mu_{is}\}$$

or

$$G_1\{\mu_{is}\} = z_s\{\mu_{is}\} \tag{7-85}$$

Since the first mode is absent in the coordinates specified, the new dynamic matrix G_1 can be used for the iteration process to obtain the second mode. Similarly, the higher harmonics and modes can be found.

Example 7. Use the matrix iteration method to verify the natural frequencies and modes of the system discussed in Example 2. (See Fig. 7-2.)

Solution: From Example 2 the natural frequencies and the modal column matrix are

$$\omega_1 : \omega_2 : \omega_3 = \sqrt{k/m} : \sqrt{3k/m} : 2\sqrt{k/m}$$

$$[\mu] = \begin{bmatrix} 1 & 1 & 1 \\ 2 & 0 & -1 \\ 1 & -1 & 1 \end{bmatrix}$$

To find the fundamental frequency and mode, we determine the dynamic matrix G and apply the iteration process. Let us define

$$G^* = K^{-1}M = \frac{m}{12k}\begin{bmatrix} 5 & 3 & 1 \\ 3 & 9 & 3 \\ 1 & 3 & 5 \end{bmatrix} = \frac{m}{12k} G$$

Applying Eq. (7-70), we obtain

$$G^*\{\mu_{is}\} = \frac{m}{12k} G\{\mu_{is}\} = z_s^*\{\mu_{is}\}$$

For convenience, we define $z = z^* 12k/m$; that is, $z = 12k/m\omega^2$. Thus we obtain

$$G\{\mu_{is}\} = z_s\{\mu_{is}\}$$

To begin the iteration process, we choose an arbitrary column

$$\{\mu\}_o = \{1 \quad 1 \quad 1\}$$

$$G\{\mu\}_o = \begin{bmatrix} 5 & 3 & 1 \\ 3 & 9 & 3 \\ 1 & 3 & 5 \end{bmatrix}\begin{bmatrix} 1 \\ 1 \\ 1 \end{bmatrix} = \begin{bmatrix} 9 \\ 15 \\ 9 \end{bmatrix} = 9\begin{bmatrix} 1 \\ 1.67 \\ 1 \end{bmatrix}$$

The computed column is normalized as shown and the common factor 9 is discarded. Continuing the iteration, we have

$$\begin{bmatrix} 5 & 3 & 1 \\ 3 & 9 & 3 \\ 1 & 3 & 5 \end{bmatrix}\begin{bmatrix} 1 \\ 1.67 \\ 1 \end{bmatrix} = \begin{bmatrix} 11.01 \\ 21.03 \\ 11.01 \end{bmatrix} = 11.01\begin{bmatrix} 1 \\ 1.91 \\ 1 \end{bmatrix}$$

We can discard the common factor 11.01. After 5 iterations we obtain

$$\begin{bmatrix} 5 & 3 & 1 \\ 3 & 9 & 3 \\ 1 & 3 & 5 \end{bmatrix}\begin{bmatrix} 1 \\ 1.998 \\ 1 \end{bmatrix} = \begin{bmatrix} 11.994 \\ 23.982 \\ 11.994 \end{bmatrix} = 11.994\begin{bmatrix} 1 \\ 2.0 \\ 1 \end{bmatrix}$$

We may choose $\{\mu_{11}\ \mu_{21}\ \mu_{31}\} = \{1\quad 2\quad 1\}$ and $z = 12$. Hence

$$\omega_1^2 = \frac{12k}{m}\frac{1}{z} = \frac{k}{m}$$

To determine the second mode, we construct the dynamic matrix G_1 from Eqs. (7-84), (7-82), and (7-85).

$$[v_{11}\quad v_{12}\quad v_{13}] = [1\quad 2\quad 1]\begin{bmatrix} m & 0 & 0 \\ 0 & m & 0 \\ 0 & 0 & m \end{bmatrix}, \qquad S = \begin{bmatrix} 0 & -2 & -1 \\ 0 & 1 & 0 \\ 0 & 0 & 1 \end{bmatrix}$$

$$G_1 = GS = \begin{bmatrix} 5 & 3 & 1 \\ 3 & 9 & 3 \\ 1 & 3 & 5 \end{bmatrix}\begin{bmatrix} 0 & -2 & -1 \\ 0 & 1 & 0 \\ 0 & 0 & 1 \end{bmatrix} = \begin{bmatrix} 0 & -7 & -4 \\ 0 & 3 & 0 \\ 0 & 1 & 4 \end{bmatrix}$$

Beginning the iteration with an arbitrary column $\{\mu\}_o = \{1\quad 1\quad 1\}$, we have

$$\begin{bmatrix} 0 & -7 & -4 \\ 0 & 3 & 0 \\ 0 & 1 & 4 \end{bmatrix}\begin{bmatrix} 1 \\ 1 \\ 1 \end{bmatrix} = \begin{bmatrix} -11 \\ 3 \\ 5 \end{bmatrix} = 5\begin{bmatrix} -2.2 \\ 0.6 \\ 1 \end{bmatrix};$$

$$\begin{bmatrix} 0 & -7 & -4 \\ 0 & 3 & 0 \\ 0 & 1 & 4 \end{bmatrix}\begin{bmatrix} -2.2 \\ 0.6 \\ 1 \end{bmatrix} = \begin{bmatrix} -8.2 \\ 1.8 \\ 4.6 \end{bmatrix} = 4.6\begin{bmatrix} -1.78 \\ 0.39 \\ 1 \end{bmatrix}$$

After 14 iterations the column repeats itself, $\{-1.0\quad 0.0\quad 1.0\}$, and $z = 4$. Hence the modal column may be selected as $\{1\quad 0\quad -1\}$ and

$$\omega_2^2 = \frac{12k}{m}\frac{1}{z} = \frac{3k}{m}$$

To determine the third mode we made use of the relations in Eq. (7-80) to suppress the first and the second. If $p_1 = p_2 = 0$, we have

$$[v_{11}\quad v_{12}\quad v_{13}]\{q\} = 0$$
$$[v_{21}\quad v_{22}\quad v_{23}]\{q\} = 0$$

The values of these row vectors can be found from Eq. (7-83). Selecting $[v_{11}\ v_{12}\ v_{13}] = [1\quad 2\quad 1]$ and $[v_{21}\ v_{22}\ v_{23}] = [1\quad 0\quad -1]$, we have

$$q_1 + 2q_2 + q_3 = 0$$
$$q_1 \qquad\quad - q_3 = 0$$

Eliminating q_1 from these equations, we obtain $q_2 = -q_3$. Writing these relations of the coordinates with matrix notation gives

$$\begin{bmatrix} q_1 \\ q_2 \\ q_3 \end{bmatrix} = \begin{bmatrix} 1 & 0 & 0 \\ 0 & 0 & -1 \\ 0 & 0 & 1 \end{bmatrix} \begin{bmatrix} q_1 \\ q_2 \\ q_3 \end{bmatrix} \quad \text{or} \quad \{q\} = S_1\{q\}$$

Let us construct a dynamic matrix G_2 that has the first and second modes absent.

$$G_2 = G_1S_1 = \begin{bmatrix} 0 & -7 & -4 \\ 0 & 3 & 0 \\ 0 & 1 & 4 \end{bmatrix} \begin{bmatrix} 1 & 0 & 0 \\ 0 & 0 & -1 \\ 0 & 0 & 1 \end{bmatrix} = \begin{bmatrix} 0 & 0 & 3 \\ 0 & 0 & -3 \\ 0 & 0 & 3 \end{bmatrix}$$

Selecting an arbitrary column $\{\mu\}_o = \{1 \quad 1 \quad 1\}$, we have

$$\begin{bmatrix} 0 & 0 & 3 \\ 0 & 0 & -3 \\ 0 & 0 & 3 \end{bmatrix} \begin{bmatrix} 1 \\ 1 \\ 1 \end{bmatrix} = \begin{bmatrix} 3 \\ -3 \\ 3 \end{bmatrix} = 3 \begin{bmatrix} 1 \\ -1 \\ 1 \end{bmatrix}$$

Hence the modal column is $\{1 \quad -1 \quad 1\}$, and $z = 3$.

$$\omega_3^2 = \frac{12k}{m}\frac{1}{z} = \frac{4k}{m}$$

7-10. DAMPED FREE VIBRATION

Conservative systems were discussed in the previous sections. If a system is nonconservative, a *dissipation function D* [Eq. (4-44), Chap. 4] can be defined to characterize the damping in the system.

$$D = \tfrac{1}{2}\sum_i\sum_j c_{ij}\dot{q}_i\dot{q}_j \quad \text{or} \quad D = \tfrac{1}{2}[\dot{q}]C\{\dot{q}\} \tag{7-86}$$

where C is called the viscous *damping matrix*. The function D represents half the rate of energy dissipation in the system. For free vibrations the equations of motion become

$$M\{\ddot{q}\} + C\{\dot{q}\} + K\{q\} = \{0\} \tag{7-87}$$

This equation can be solved by the general method discussed previously, but the algebraic manipulations can be quite involved. In many practical problems the damping force in the system is small, and a system with light damping can be treated as a conservative system in obtaining the natural frequencies and modes. We shall briefly discuss this type of system.

Using the differential operator $D \equiv d/dt$, Eq. (7-87) can be written as (D denotes the differential operator for the remainder of this chapter.)

$$(MD^2 + CD + K)\{q\} = \{0\} \tag{7-88}$$

or

$$f(D)\{q\} = \{0\} \tag{7-89}$$

where $f(D)$ is a linear differential operator, and this is called the *D matrix*. The standard solution of Eqs. (7-88) and (7-89) is of the form $\{q\} = \{\mu\}e^{\lambda t}$, where $\{\mu\}$ is a column of constants. (*Note:* $\lambda \neq \omega^2$, as defined in the previous sections.) Substituting this expression in Eq. (7-89) and factoring out $e^{\lambda t}$ gives

$$f(\lambda)\{\mu\} = \{0\} \tag{7-90}$$

This is a set of linear homogeneous algebraic equations. Consistency requires that

$$\Delta(\lambda) = |f(\lambda)| = 0 \tag{7-91}$$

If the dynamic system has n degrees of freedom, Eq. (7-91) is an algebraic equation of $2n$th degree in λ. Let us assume that the roots of $\Delta(\lambda) = 0$ are distinct. From Eq. (7-90), appropriate to each root λ_s, we have a modal column $\{\mu_{is}\}$, $i = 1, 2, \ldots, n$. Corresponding to the $2n$ roots, there are $2n$ such modal columns.

The modal columns can be obtained from the adjoint $F(\lambda)$ of $f(\lambda)$. Consider the equation

$$f(D)[e^{\lambda_s t}F(\lambda_s)] = e^{\lambda_s t}f(\lambda_s)F(\lambda_s) = [0] \tag{7-92}$$

$F(\lambda_s)$ is of unit rank. Comparing Eqs. (7-90) and (7-92), a modal column appropriate to λ_s may be selected proportional to any nonzero column of $F(\lambda_s)$.

If a system is stable and lightly damped, the roots λ_i of its characteristic equation are complex with negative real parts. Since complex roots occur in conjugate pairs, the modal columns associated with the complex roots must also form conjugate pairs in order that the motions $\{q\}$ be real. It is evident that, by combining these complex functions, the motions can be expressed as sine and cosine functions with amplitudes diminishing exponentially. (See Sec. 3-5.)

Alternatively, by combining the modal columns $\{\mu\}$ to obtain a modal column matrix $[\mu]$, the formal solution of Eq. (7-87) can be expressed as

$$\underset{n\times 1}{\{q\}} = \underset{n\times 2n}{[\mu]}\underset{2n\times 1}{\{c\ e^{\lambda t}\}} \tag{7-93}$$

where $\{c\,e^{\lambda t}\}$ is a column matrix with elements $c_i e^{\lambda_i t}$. In order to evaluate the constants c_i, it is convenient to express Eq. (7-93) as

$$\underset{n\times 1}{\{q\}} = \underset{n\times 2n}{[\mu]}\ \underset{2n\times 2n}{[e^{\lambda t}]}\ \underset{2n\times 1}{\{c\}} \tag{7-94}$$

where

$$[e^{\lambda t}] = \begin{bmatrix} e^{\lambda_1 t} & 0 & \cdots & 0 \\ 0 & e^{\lambda_2 t} & \cdots & 0 \\ \cdot & \cdot & \cdots & \cdot \\ 0 & 0 & \cdots & e^{\lambda_{2n} t} \end{bmatrix} \tag{7-95}$$

Differentiating Eq. (7-94) with respect to time gives

$$\{\dot{q}\} = [\mu]\Lambda[e^{\lambda t}]\{c\} \equiv [\mu\Lambda][e^{\lambda t}]\{c\} \tag{7-96}$$

where

$$[\mu\Lambda] = \begin{bmatrix} \mu_{11}\lambda_1 & \mu_{12}\lambda_2 & \cdots & \mu_{12n}\lambda_{2n} \\ \cdot & \cdot & \cdots & \cdot \\ \mu_{n1}\lambda_1 & \mu_{n2}\lambda_2 & \cdots & \mu_{n2n}\lambda_{2n} \end{bmatrix} \tag{7-97}$$

Combining Eqs. (7-94) and (7-96), we obtain

$$\underset{2n\times 1}{\begin{bmatrix} q \\ -- \\ \dot{q} \end{bmatrix}} = \underset{2n\times 2n}{\begin{bmatrix} \mu \\ -- \\ \mu\Lambda \end{bmatrix}} \underset{2n\times 2n}{[e^{\lambda t}]}\ \underset{2n\times 1}{\{c\}} \tag{7-98}$$

Given the initial conditions $\{q(0)\}$ and $\{\dot{q}(0)\}$, we have $[e^{\lambda t}] = I$ a unit matrix. Thus the constant $\{c\}$ can be evaluated from the equation

$$\{c\} = \begin{bmatrix} \mu \\ -- \\ \mu\Lambda \end{bmatrix}^{-1} \begin{bmatrix} q(0) \\ -- \\ \dot{q}(0) \end{bmatrix} \tag{7-99}$$

In general, the algebra involved in finding the complex roots and the modal columns is very tedious. We shall use an undamped system to illustrate the method discussed in this section.

Example 8. The equations of motion of a given dynamic system are

$$\begin{bmatrix} 2 & 0 \\ 0 & 2 \end{bmatrix} \begin{bmatrix} \ddot{q}_1 \\ \ddot{q}_2 \end{bmatrix} + \begin{bmatrix} 5 & -3 \\ -3 & 5 \end{bmatrix} \begin{bmatrix} q_1 \\ q_2 \end{bmatrix} = \begin{bmatrix} 0 \\ 0 \end{bmatrix}$$

Determine $\{q\}$ for the initial conditions $\{q(0)\} = \{0\}$ and $\{\dot{q}(0)\} = \{1 \;\; 0\}$.

Solution: Applying Eq. (7-91), we obtain

$$\Delta(\lambda) = |f(\lambda)| = \begin{vmatrix} 2\lambda^2 + 5 & -3 \\ -3 & 2\lambda^2 + 5 \end{vmatrix} = 4(\lambda^2 + 1)(\lambda^2 + 4)$$

The roots of $\Delta(\lambda) = 0$ are j, $-j$, $2j$, and $-2j$. The adjoint of $f(\lambda)$ is

$$F(\lambda) = \begin{bmatrix} 2\lambda^2 + 5 & 3 \\ 3 & 2\lambda^2 + 5 \end{bmatrix}$$

$$F(j) = F(-j) = \begin{bmatrix} 3 & 3 \\ 3 & 3 \end{bmatrix}, \quad F(2j) = F(-2j) = \begin{bmatrix} -3 & 3 \\ 3 & -3 \end{bmatrix}$$

Hence Eq. (7-98) becomes

$$\begin{bmatrix} q_1 \\ q_2 \\ \dot{q}_1 \\ \dot{q}_2 \end{bmatrix} = \begin{bmatrix} 1 & 1 & 1 & 1 \\ 1 & 1 & -1 & -1 \\ j & -j & 2j & -2j \\ j & -j & -2j & 2j \end{bmatrix} \begin{bmatrix} e^{jt} & 0 & 0 & 0 \\ 0 & e^{-jt} & 0 & 0 \\ 0 & 0 & e^{2jt} & 0 \\ 0 & 0 & 0 & e^{-2jt} \end{bmatrix} \begin{bmatrix} c_1 \\ c_2 \\ c_3 \\ c_4 \end{bmatrix}$$

From Eq. (7-99) the constants $\{c\}$ may be evaluated as

$$\begin{bmatrix} c_1 \\ c_2 \\ c_3 \\ c_4 \end{bmatrix} = \frac{1}{8} \begin{bmatrix} 2 & 2 & -2j & -2j \\ 2 & 2 & 2j & 2j \\ 2 & -2 & -j & j \\ 2 & -2 & j & -j \end{bmatrix} \begin{bmatrix} 0 \\ 0 \\ 1 \\ 0 \end{bmatrix} = \frac{1}{8} \begin{bmatrix} -2j \\ 2j \\ -j \\ j \end{bmatrix}$$

Using $\{c\}$ thus obtained and applying Eq. (7-93), we have

$$q_1 = (1/8)(-2je^{jt} + 2je^{-jt} - je^{2jt} + je^{-2jt})$$

$$q_2 = (1/8)(-2je^{jt} + 2je^{-jt} + je^{2jt} - je^{-2jt})$$

or

$$q_1 = (1/4)(2 \sin t + \sin 2t)$$

$$q_2 = (1/4)(2 \sin t - \sin 2t)$$

7-11. FORCED VIBRATIONS

Consider the equations of motion of a system

$$M\{\ddot{q}\} + C\{\dot{q}\} + K\{q\} = \{Q\} \tag{7-100}$$

where $\{Q\}$ are the generalized forces applied to the system corresponding to the generalized coordinates $\{q\}$. Using the differential operator $D \equiv d/dt$, this equation can be written as

$$(MD^2 + CD + K)\{q\} = \{Q\} \tag{7-101}$$

or

$$f(D)\{q\} = \{Q\} \tag{7-102}$$

where $f(D)$ is a linear differential operator. Since $f(D)F(D) = I\,\Delta(D)$ [Eq. (C-46), Appendix C], the particular integral $\{q\}$ of Eq. (7-102) is of the form

$$\{q\} = [f(D)]^{-1}\{Q\} = \frac{F(D)}{\Delta(D)}\{Q\} \tag{7-103}$$

It is necessary to specify $\{Q\}$ in order to solve this equation.

Case 1. Excitation Force $\{Q\}$ Exponential

Many functions commonly encountered can be expressed in the exponential form. Let us consider a single function Q of $\{Q\}$.

$$Q = e^{\theta t}a \tag{7-104}$$

where θ and a are constants but may assume real or complex values. This expression can be used to represent the following functions: (1) If $\theta = 0$, Q is a constant excitation force. (2) If θ is real, Q is an exponential force. (3) If θ is imaginary, Q represents the excitation functions $F_o \sin(\omega t + \gamma)$ and $F_o \cos(\omega t + \gamma)$. (4) If θ is complex, Q may be used to represent the functions $F_o e^{-\sigma t} \sin(\omega t + \gamma)$ and $F_o e^{-\sigma t} \cos(\omega t + \gamma)$. (5) Since a periodic function can be expanded into a Fourier series, Q may be regarded as one of the harmonic components of the series.

If Q is exponential, from Eq. (7-103) we have

$$\{q\} = e^{\theta t}\frac{F(\theta)}{\Delta(\theta)}\{a\} \tag{7-105}$$

This is the formal solution of Eq. (7-100) except when $\Delta(\theta) = 0$. Such a condition occurs when the excitation frequency coincides with one of the natural frequencies of an undamped system.

Let $\theta = \lambda_s$ be one of the roots of $\Delta(D) = 0$. The equations of motion, Eq. (7-102), become

$$f(D)\{q\} = e^{\lambda_s t}\{a\} \tag{7-106}$$

Assume the solution of this equation to be of the form

$$\{q\} = \frac{\partial}{\partial \lambda}[e^{\lambda t}F(\lambda)]\{b\}|_{\lambda=\lambda_s} = e^{\lambda_s t}[F^{(1)}(\lambda_s) + tF(\lambda_s)]\{b\} \tag{7-107}$$

where $\{b\}$ is a column of undetermined coefficients. Let us form the expression

$$\begin{aligned} f(D)\frac{\partial}{\partial \lambda}[e^{\lambda t}F(\lambda)] &= f\left(\frac{\partial}{\partial t}\right)\frac{\partial}{\partial \lambda}[e^{\lambda t}F(\lambda)] = \frac{\partial}{\partial \lambda}f\left(\frac{\partial}{\partial t}\right)[e^{\lambda t}F(\lambda)] \\ &= \frac{\partial}{\partial \lambda}[e^{\lambda t}f(\lambda)F(\lambda)] = \frac{\partial}{\partial \lambda}[e^{\lambda t}\Delta(\lambda)]I \\ &= e^{\lambda t}[\Delta^{(1)}(\lambda) + t\,\Delta(\lambda)]I \end{aligned}$$

If $\lambda = \lambda_s$, $\Delta(\lambda_s) = 0$. Using this expression and Eqs. (7-106) and (7-107), we obtain

$$\{b\} = \{a\}/\Delta^{(1)}(\lambda_s) \tag{7-108}$$

Case 2. Excitation Force $\{Q\}$ General

For a given general excitation force $\{Q\}$, Eq. (7-103) can be solved by expanding $1/\Delta(D)$ into a sum of simple partial fractions and then making use of the integral operator $1/(D - \lambda_s)$, where λ_s is a root of $\Delta(D) = 0$. Using the identities†

$$\frac{1}{\Delta(D)} = \sum_s \frac{1}{\Delta^{(1)}(\lambda_s)(D - \lambda_s)}$$

$$\frac{1}{D - \lambda_s}Q = e^{\lambda_s t}\int e^{-\lambda_s t}Q(t)\,dt$$

† Assume that $1/\Delta(D)$ can be expanded in the form

$$\frac{1}{\Delta(D)} = \frac{A_1}{D - \lambda_1} + \cdots + \frac{A_i}{D - \lambda_i} + \cdots + \frac{A_n}{D - \lambda_n}$$

where $A_{1,2,\ldots,n}$ are constants. To evaluate a typical coefficient A_i, we multiply both sides of this equation by $(D - \lambda_i)$ and take the limit as $D \to \lambda_i$. Since both $(D - \lambda_i)$ and $\Delta(\lambda_i)$ vanish at $D = \lambda_i$, the quantity $(D - \lambda_i)/\Delta(\lambda_i)$ is indeterminate of the form 0/0. The limit is obtained by applying L'Hospital's rule. Hence $A_i = 1/\Delta^{(1)}(\lambda_i)$.

Eq. (7-103) can be written as

$$\{q\} = F(D) \sum_s \frac{e^{\lambda_s t}}{\Delta^{(1)}(\lambda_s)} \int_0^t e^{-\lambda_s t}\{Q(t)\}\, dt \tag{7-109}$$

The arbitrary constant from the integration in this equation can be assumed to be zero, because it can be merged with the complementary function.

Example 9. The equations of motion of a given dynamic system are

$$\begin{aligned} m_1\ddot{x}_1 + (k_1 + k)x_1 - kx_2 &= F \sin \omega t \\ -kx_1 + m_2\ddot{x}_2 + (k_2 + k)x_2 &= 0 \end{aligned}$$

Determine the steady-state motions of x_1 and x_2. (See Sec. 3-4, Chap. 3.)

Solution: Since $F \sin \omega t = \text{Im}[Fe^{j\omega t}]$, the equations can be expressed as

$$\begin{bmatrix} m_1 & 0 \\ 0 & m_2 \end{bmatrix} \begin{bmatrix} \ddot{x}_1 \\ \ddot{x}_2 \end{bmatrix} + \begin{bmatrix} k_1 + k & -k \\ -k & k_2 + k \end{bmatrix} \begin{bmatrix} x_1 \\ x_2 \end{bmatrix} = e^{j\omega t} \begin{bmatrix} F \\ 0 \end{bmatrix}$$

or

$$\begin{bmatrix} m_1 D^2 + k_1 + k & -k \\ -k & m_2 D^2 + k_2 + k \end{bmatrix} \begin{bmatrix} x_1 \\ x_2 \end{bmatrix} \equiv f(D) \begin{bmatrix} x_1 \\ x_2 \end{bmatrix} = e^{j\omega t} \begin{bmatrix} F \\ 0 \end{bmatrix}$$

$$F(D) = \text{adj.} f(D) = \begin{bmatrix} m_2 D^2 + k_2 + k & k \\ k & m_1 D^2 + k_1 + k \end{bmatrix}$$

Applying Eq. (7-105) and taking the imaginary component of the solutions, we obtain

$$\begin{bmatrix} x_1 \\ x_2 \end{bmatrix} = \text{Im} \left\{ e^{j\omega t} \frac{\begin{bmatrix} k_2 + k - \omega^2 m_2 & k \\ k & k_1 + k - \omega^2 m_1 \end{bmatrix} \begin{bmatrix} F \\ 0 \end{bmatrix}}{\begin{vmatrix} k_1 + k - \omega^2 m_1 & -k \\ -k & k_2 + k - \omega^2 m_2 \end{vmatrix}} \right\}$$

The general solution of Eq. (7-100) consists of the complementary function, Eq. (7-94), and the particular integral, Eq. (7-103). Denoting these solutions by the symbols $\{q_c\}$ and $\{q_p\}$, respectively, the general solution is

$$\{q\} = \{q_c\} + \{q_p\} \tag{7-110}$$

or

$$\{q\} = \{\mu\}[e^{\lambda t}]\{c\} + \{q_p\} \tag{7-111}$$

where $\{c\}$ is a column of constants.

To evaluate the coefficients $\{c\}$, we make use of the technique shown in Eq. (7-96). Differentiating Eq. (7-111) with respect to time gives

$$\{\dot{q}\} = [\mu\Lambda][e^{\lambda t}]\{c\} + \{\dot{q}_p\} \tag{7-112}$$

Combining Eqs. (7-111) and (7-112), we obtain

$$\begin{bmatrix} q \\ \text{--} \\ \dot{q} \end{bmatrix} = \begin{bmatrix} \mu \\ \text{---} \\ \mu\Lambda \end{bmatrix} [e^{\lambda t}]\{c\} + \begin{bmatrix} q_p \\ \text{--} \\ \dot{q}_p \end{bmatrix}$$

Rearranging this equation and using the initial conditions, we have

$$\{c\} = \begin{bmatrix} \mu \\ \text{----} \\ \mu\Lambda \end{bmatrix}^{-1} \begin{bmatrix} q(0) - q_p(0) \\ \text{-------} \\ \dot{q}(0) - \dot{q}_p(0) \end{bmatrix} \tag{7-113}$$

SUGGESTED READING

Frazer, R. A., W. J. Duncan, and A. R. Collar, *Elementary Matrices* (London: Cambridge University Press, 1957), chaps. 5, 6, 9, and 10.

Karman, von, T., and M. A. Biot, *Mathematical Methods in Engineering* (New York: McGraw-Hill Book Co., Inc., 1940), chaps. 5 and 6.

Pipes, L. A., *Applied Mathematics for Engineers and Physicists* (New York: McGraw-Hill Book Co., Inc., 2nd ed., 1958), chap. 8.

Scanlan, R. H., and R. Rosenbaum, *Aircraft Vibration and Flutter* (New York: The Macmillan Co., 1951), chaps. 1, 2, 9, and 10.

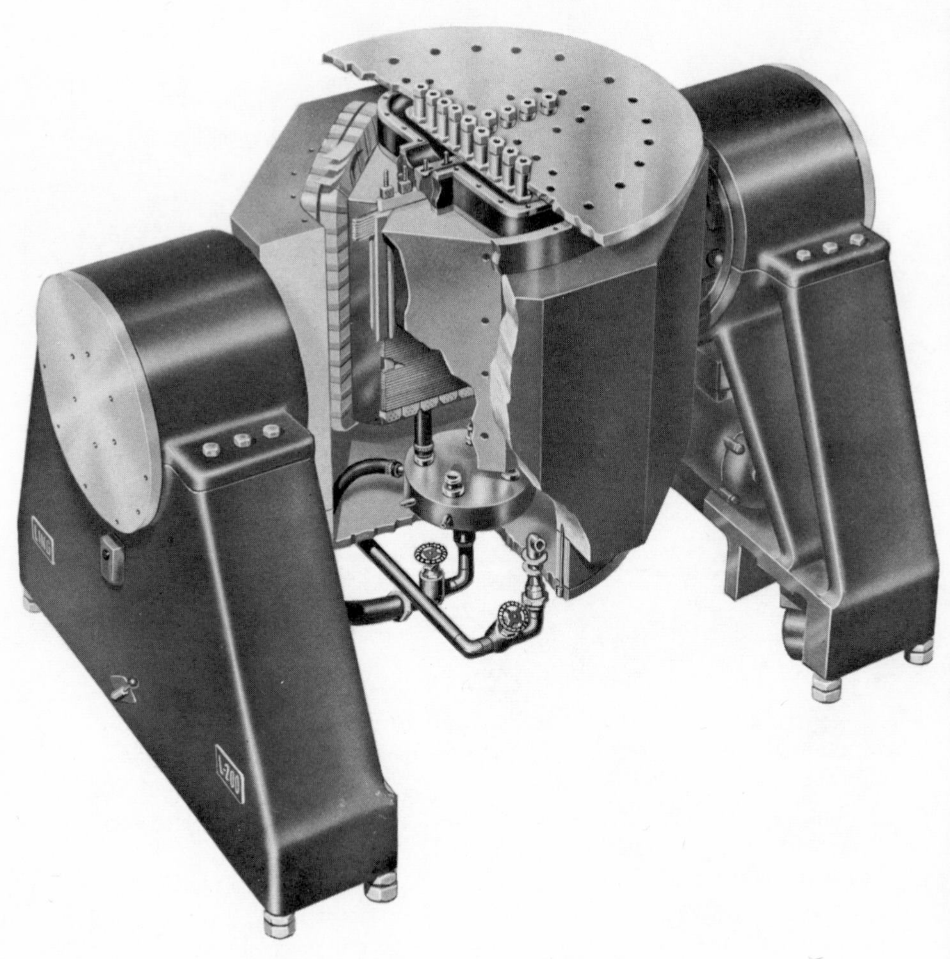

Cutaway drawing shows internal construction and cruciform armature of Model L-200 22,000 pound-force shaker. This exciter can take a 100 pound package to 100 "g." (courtesy Ling Electronics Division, Ling-Temco-Vought, Inc.)

8 TRANSIENTS

8-1. INTRODUCTION

When a periodic or aperiodic excitation is applied to a system, the resultant motion consists of two components, namely, the steady-state and the transient motions. The steady state is defined as a periodic motion or one that is invariant with time. If the steady-state motion is periodic, it is at the excitation frequency. A motion that is not at steady state is called a transient motion. It represents the behavior of the system as it relaxes from a state of constraint to its equilibrium state. Thus the transient motion oscillates at the natural frequencies of the system.

Dynamic systems are often subjected to the abrupt application of excitation. During normal operations, a punch press is subjected to periodic abrupt loading. Delicate equipment transported by vehicles may be damaged because of the irregularities of the roadbed. Military equipment may have to be designed to withstand certain shock tests.

In any case, the abrupt loading may be sufficient to set up strong transient vibrations in the system considered.

Transient motion is important even when the excitation is not abruptly applied. For example, not all equipment is operated at constant speeds. Conceivably, a piece of equipment may operate satisfactorily at constant speed, but may not be satisfactory under varying operating conditions. Furthermore, it is always necessary to bring a device from a state of rest to its operating speed.

Mechanical and electromechanical systems are discussed in this chapter. Except for the excitation functions, no new concepts are necessary for the discussion of the mechanical systems. The Laplace transform method is used to solve the equations of motion instead of the method of ordinary differential equations, or the so-called classical method. This is due to the fact that the classical method is not well suited to handle the types of excitation functions considered, and the evaluation of the constants of integration is not easy unless the number of constants is small. Network equations are developed for the electromechanical systems. Once the equations are derived, they can be solved by the method prescribed for the mechanical systems.

8-2. SHOCK AND IMPACT

The terms "shock" and "impact" are not precisely defined. Generally, "shock" denotes a rapid application of excitation to a system, such as a sharp blow at the mass or a sudden movement of the spring. This sudden change of the excitation force or displacement is characterized by an excitation function with first derivatives of high magnitude and usually aperiodic. The excitation may induce strong transient vibrations in the system. The air pressure pulse caused by gunfire, the dropping of a package on a hard floor, and the landing of an aircraft on a runway are examples of shock applied to dynamic systems.

Design data are not easy to obtain from shock test data. First of all, mechanical transients are difficult to measure, and their recordings are not easily interpreted for design applications. Often, the shock excitation is somewhat random in nature. The actual shock excitation function must be used if the analysis of a problem is to be realistic.

Shock machines are used extensively for the testing of equipment. The machine consists essentially of a shock table and a hammer. The equipment being tested is mounted on the shock table, and the hammer delivers the shock to the table. It is likely that the machine shock and the field shock are not similar. Although machine tests may correlate well with certain field tests, at best they are of a comparative nature.

"Impact" implies the collision of two objects, such as in an automobile accident, or a bullet striking an object. The colliding bodies may be perfectly elastic, inelastic, or intermediate between these extremes. If the impact velocity is high, permanent local deformation of the colliding objects may result. The theory of impact has been investigated by an imposing array of scientific minds. No general impact theory, however, has been developed to date.†

In view of the extended nature of shock and impact, we shall limit our study to the effect of well-defined excitation functions on linear dynamic systems.

8-3. THE LAPLACE TRANSFORMATION

The Laplace transform method converts linear differential equations into algebraic equations which can be manipulated algebraically until the desired form is obtained. Then the algebraic expressions are reconverted to complete the solution of the original differential equations. We are all acquainted with some form of transformation. In arithmetic the product of two numbers can be determined by adding the logarithm of the numbers and then converting the sum by the reverse process (antilogarithm) to obtain the original product. Here, the transformation involved is to convert the original numbers into powers of 10; the original numbers are transformed into some other numbers. The problem is simplified because, through the transformation, the multiplication operation is replaced by addition.

The Laplace transform converts a given function $f(t)$ of the real variable t into a function $F(s)$ of the complex variable s by the operation

$$F(s) = \int_0^\infty f(t)\, e^{-st}\, dt = \mathscr{L} f(t) \tag{8-1}$$

where $s = \sigma + j\omega$ is complex, σ is a constant, and the symbol $\mathscr{L}$ denotes the Laplace transform operation. The value of $f(t)$ to be used at $t = 0$ is $f(0+)$. We assume that the transformation exists. (See

† W. Goldsmith, *Impact—The Theory and Physical Behavior of Colliding Solids*, Edward Arnold & Co., London, 1960, Chap. 1.

Appendix D.) Equation (8-1) is called the (direct) Laplace transformation, and $F(s)$ is the Laplace transform of $f(t)$. Conversely, a given function $F(s)$ can be transformed to give a corresponding function $f(t)$ by the relation

$$f(t) = \frac{1}{2\pi j}\int_{\sigma - j\infty}^{\sigma + j\infty} F(s)\, e^{ts}\, ds = \mathscr{L}^{-1} F(s) \tag{8-2}$$

Equation (8-2) is called the inversion integral, and the symbol $\mathscr{L}^{-1}$ denotes the inverse transform of $F(s)$. For the types of functions considered, we assume that the inverse transformation is unique. Equations (8-1) and (8-2) form a Laplace transform pair.

It should be mentioned that it is often unnecessary to use the Laplace transform integral, Eq. (8-1), to determine the functions $F(s)$ for the given functions $f(t)$ in a problem. Generally, the problems fall into repeating patterns. If the transform pairs $f(t)$ and $F(s)$ are tabulated, as illustrated in Table 8-1, then the direct transformation of a given problem becomes a matter of routine. Since the transform pairs are tabulated, it is also unnecessary to use Eq. (8-2) for the evaluation of the inverse transform. It is rarely necessary, however, to refer to elaborate tables of Laplace transform pairs; a large number of functions can be expressed in terms of a few basic pairs. It may be well to remember some of the basic pairs and the properties of the Laplace transform rather than to rely heavily on elaborate tables.

Example 1. Find the Laplace transform $F(s)$ when the function $f(t)$ is given as (*a*) A, a constant; (*b*) e^{-at}; (*c*) $\sin \omega t$; (*d*) $\cosh at$; (*e*) t; and (*f*) $e^{-at} f(t)$.

Solution: (*a*) Substituting $f(t) = A$ in the Laplace transform integral gives

$$F(s) = \mathscr{L}\, A = \int_0^\infty A e^{-st}\, dt = -\frac{A}{s} e^{-st}\Big|_0^\infty = \frac{A}{s}$$

If the value of the constant A is unity, we have a *unit step function* $u(t)$ which is defined by the relation

$$u(t) = \begin{cases} 0 & \text{for} \quad t < 0 \\ 1 & \phantom{\text{for}} \quad t > 0 \end{cases} \tag{8-3}$$

(*b*) $$F(s) = \mathscr{L}\, e^{-at} = \int_0^\infty e^{-at} e^{-st}\, dt = \int_0^\infty e^{-(s+a)t}\, dt$$

$$= -\frac{1}{s+a} e^{-(s+a)t}\Big|_0^\infty = \frac{1}{s+a}$$

(*c*) The sine function can be expressed in its exponential form and substituted in the Laplace transform integral as

$$F(s) = \mathscr{L} \sin \omega t = \int_0^\infty \frac{1}{2j}(e^{j\omega t} - e^{-j\omega t})e^{-st}\,dt$$

$$= \frac{1}{2j}\left(\frac{1}{s - j\omega} - \frac{1}{s + j\omega}\right) = \frac{\omega}{s^2 + \omega^2}$$

(*d*) Expressing the hyperbolic cosine in its exponential form and substituting in the Laplace transform integral gives

$$F(s) = \mathscr{L} \cosh at = \int_0^\infty \tfrac{1}{2}(e^{at} + e^{-at})e^{-st}\,dt$$

$$= \frac{1}{2}\left(\frac{1}{s - a} + \frac{1}{s + a}\right) = \frac{s}{s^2 - a^2}$$

(*e*) $$F(s) = \mathscr{L}t = \int_0^\infty te^{-st}\,dt$$

Integrating by parts, we let

$$u = t \qquad du = dt$$

$$dv = e^{-st}\,dt \qquad v = -e^{-st}/s$$

Substituting these expressions in the identity

$$\int u\,dv = uv - \int v\,du$$

we obtain

$$F(s) = -\frac{te^{-st}}{s}\bigg|_0^\infty + \frac{1}{s}\int_0^\infty e^{-st}\,dt = 1/s^2$$

(*f*) $$F(s) = \mathscr{L}\, e^{-at} f(t) = \int_0^\infty e^{-at} f(t)\, e^{-st}\,dt = \int_0^\infty f(t)\, e^{-(s+a)t}\,dt$$

$$= F(s + a)$$

Some of the function transform pairs commonly encountered in engineering problems are as shown in Table 8-1. We shall need additional properties of the Laplace transform, such as the transformation of the time derivative, in order to solve differential or integral-differential equations.

Theorem 1. Linearity

(*a*) If $F(s)$ is the Laplace transform of a given function $f(t)$, then

$$\mathscr{L}\,[a\,f(t)] = a\,F(s) \qquad \textbf{(8-4)}$$

TABLE 8-1

LAPLACE TRANSFORM PAIRS

$F(s)$	$f(t)$
$\frac{1}{s}$	$u(t)$
$\frac{1}{s^2}$	t
$\frac{n!}{s^{n+1}}$	t^n
$\frac{1}{s+a}$	e^{-at}
$\frac{\omega}{s^2+\omega^2}$	$\sin \omega t$
$\frac{s}{s^2+\omega^2}$	$\cos \omega t$
$\frac{a}{s^2-a^2}$	$\sinh at$
$\frac{s}{s^2-a^2}$	$\cosh at$
$F(s+a)$	$e^{-at} f(t)$

Proof: $$\mathscr{L}\,[a\,f(t)] = \int_0^\infty [a\,f(t)]e^{-st}\,dt = a\int_0^\infty f(t)\,e^{-st}\,dt = a\,F(s)$$

(*b*) If $F_1(s)$ and $F_2(s)$ are the Laplace transforms of $f_1(t)$ and $f_2(t)$, respectively, then

$$\mathscr{L}\,[f_1(t) \pm f_2(t)] = F_1(s) \pm F_2(s) \tag{8-5}$$

Proof: $$\begin{aligned}\mathscr{L}\,[f_1(t) \pm f_2(t)] &= \int_0^\infty [f_1(t) \pm f_2(t)]e^{-st}\,dt \\ &= \int_0^\infty f_1(t)\,e^{-st}\,dt \pm \int_0^\infty f_2(t)\,e^{-st}\,dt \\ &= F_1(s) \pm F_2(s)\end{aligned}$$

Theorem 2. Real Differentiation

If $f(t)$ has the Laplace transform $F(s)$, then the Laplace transform of its derivative $\frac{df(t)}{dt} = f'(t)$ is

$$\mathscr{L}\,f'(t) = s\,F(s) - f(0+) \tag{8-6}$$

Proof: From the definition of the Laplace transform, we have

$$F(s) = \mathscr{L} f(t) = \int_0^\infty f(t)\, e^{-st}\, dt$$

Integrating by parts, we let

$$u = f(t) \qquad du = f'(t)\, dt$$

$$dv = e^{-st}\, dt \qquad v = -\epsilon^{-st}/s$$

Substituting these expressions in the identity

$$\int u\, dv = uv - \int v\, du$$

we obtain

$$F(s) = -\frac{1}{s} f(t)\, e^{-st}\Big|_0^\infty + \frac{1}{s}\int_0^\infty f'(t)\, e^{-st}\, dt$$

$$F(s) = \frac{1}{s}\,[f(0+) + \mathscr{L} f'(t)]$$

This equation can be rearranged to give Eq. (8-6).

It can be shown that the transformation of the nth derivative of a function $f(t)$ can be expressed as

$$\mathscr{L} f^{(n)}(t) = s^n F(s) - s^{n-1} f(0+) - \cdots - s f^{(n-2)}(0+) - f^{(n-1)}(0+) \tag{8-7}$$

Theorem 3. Real Integration

If $F(s)$ is the Laplace transform of $f(t)$ the integral of which is

$$f^{(-1)}(t) = \int f(t)\, dt = \int_0^t f(t)\, dt + f^{(-1)}(0+)$$

then the Laplace transform of the integral of $f(t)$ is

$$\mathscr{L} f^{(-1)}(t) = \frac{F(s)}{s} + \frac{f^{(-1)}(0+)}{s} \tag{8-8}$$

Proof: From the definition of the Laplace transform, we have

$$F(s) = \int_0^\infty f(t)\, e^{-st}\, dt$$

Integrating by parts, we let

$$u = e^{-st} \qquad du = -se^{-st}\, dt$$

$$v = \int f(t)\, dt \qquad dv = f(t)\, dt$$

and obtain

$$F(s) = e^{-st}\int f(t)\,dt \Big|_0^\infty + s\int_0^\infty \left[\int f(t)\,dt\right] e^{-st}\,dt$$
$$= -f^{(-1)}(0+) + s\,\mathscr{L} f^{(-1)}(t)$$

This equation can be rearranged to give Eq. (8-8).

Example 2. The equation of motion of a simple spring-mass system with an excitation $f(t)$ is

$$m\ddot{x} + c\dot{x} + kx = f(t) \tag{8-9}$$

If the initial conditions are $x(0+) = x_o$ and $\dot{x}(0+) = \dot{x}_o$, determine the Laplace transform of $x(t)$.

Solution: Using Theorems 1 and 2, the transform of the left side of Eq. (8-9) is

$$\mathscr{L}\,[m\ddot{x} + c\dot{x} + kx] = m\,\mathscr{L}\,\ddot{x}(t) + c\,\mathscr{L}\,\dot{x}(t) + k\,\mathscr{L}\,x(t)$$
$$= m[s^2\,X(s) - sx_o - \dot{x}_o] + c[s\,X(s) - x_o] + k\,X(s)$$
$$= [ms^2 + cs + k]\,X(s) - (ms + c)x_o - m\dot{x}_o$$

The transform of the applied excitation function $f(t)$ is $F(s)$. Thus rearranging the transformed equation gives

$$X(s) = \frac{1}{ms^2 + cs + k}\,[F(s) + (ms + c)x_o + m\dot{x}_o] \tag{8-10}$$

The function $x(t)$ can be obtained from the inverse transform of $X(s)$. Equation (8-10) is called the *subsidiary equation* of the differential equation, Eq. (8-9). $X(s)$ is the *response transform*, and $(ms^2 + cs + k)$ is the *characteristic function* of the system. Equation (8-10) is typical of all transform solutions.

Generally, the response transform is expressed as

(Response transform) = (system function)(total excitation transform) **(8-11)**

This equation can be written symbolically as

$$H(s) = G(s)\,F(s) \tag{8-12}$$

Comparing Eqs. (8-10) and (8-12), the system function $G(s)$ is the reciprocal of the characteristic function. If the system function is a fraction, the characteristic function is its denominator. Since $G(s)$ contains all the essential specifications of the system, it describes the behavior of the system in response to an excitation $F(s)$. Hence $G(s)$ is the ratio of the response transform (output) to the total excitation transform (input) of a system. The total excitation transform consists of the transform of the externally applied excitation and the contributions which are due to the initial conditions. It should be noted that $F(s)$ in Eq. (8-12) is used in the general sense but $F(s)$ in Eq. (8-10) pertains to a particular problem and it is the transform of an excitation force $f(t)$. $G(s)$ is also called the *transfer function* of the system.

A transfer function is defined as the ratio of the operational output of a dynamic system to the operational input causing that output. Thus, in Eq. (8-10), if the external excitation alone is considered as the operational input, the transfer function is $1/(ms^2 + cs + k)$ and all the initial values are set equal to zero. Similarly, if the initial displacement x_o alone is considered as input, the corresponding transfer function is $(ms + c)/(ms^2 + cs + k)$. It should be noted that a transfer function is a ratio of two quantities and it does not have to be nondimensional.

Example 2 illustrates the use of the Laplace transform method for solving an initial value problem of a one-degree-of-freedom system. The steps involved are:

1. State the equation describing the system.
2. Transform each of the terms of the differential equation using the appropriate initial conditions.
3. Solve for the transform of the desired unknown.
4. Perform the inverse transform to obtain the problem solution.

The steps enumerated apply equally well to solving problems with multidegrees of freedom. Step 1 is essential in any method of analysis. Step 2 converts differential equations into algebraic equations. The algebraic equations are manipulated in Step 3 to give the transform of the desired unknown. Step 4 completes the solution of the problem by the Laplace transform method. We shall illustrate the first three steps of this procedure with a two-degree-of-freedom system before discussing the inverse transformation.

Example 3. The equations of motion of a two-degree-of-freedom system [see Eq. (3-27), Chap. 3] are

$$\begin{aligned} m_1\ddot{x}_1 + (k_1 + k)x_1 - kx_2 &= f(t) \\ -kx_1 + m_2\ddot{x}_2 + (k_2 + k)x_2 &= 0 \end{aligned} \qquad \textbf{(8-13)}$$

If the initial conditions are $x_1(0+) = x_{10}$, $\dot{x}_1(0+) = \dot{x}_{10}$, $x_2(0+) = x_{20}$, $\dot{x}_2(0+) = \dot{x}_{20}$, determine the Laplace transform of $x_1(t)$ and $x_2(t)$.

Solution: Transforming the given set of equations and rearranging yields

$$(m_1s^2 + k_1 + k)\, X_1(s) - k\, X_2(s) = F(s) + m_1(x_{10}s + \dot{x}_{10})$$
$$-k\, X_1(s) + (m_2s^2 + k_2 + k)\, X_2(s) = m_2(x_{20}s + \dot{x}_{20})$$

Solving for the response transforms $X_1(s)$ and $X_2(s)$ from these algebraic equations, we obtain

$$X_1(s) = \frac{\begin{vmatrix} F(s) + m_1(x_{10}s + \dot{x}_{10}) & -k \\ m_2(x_{20}s + \dot{x}_{20}) & m_2s^2 + k_2 + k \end{vmatrix}}{\Delta(s)}$$

$$X_2(s) = \frac{\begin{vmatrix} m_1s^2 + k_1 + k & F(s) + m_1(x_{10}s + \dot{x}_{10}) \\ -k & m_2(x_{20}s + \dot{x}_{20}) \end{vmatrix}}{\Delta(s)} \qquad \textbf{(8-14)}$$

where

$$\Delta(s) = \begin{vmatrix} m_1s^2 + k_1 + k & -k \\ -k & m_2s^2 + k_2 + k \end{vmatrix}$$

8-4. PARTIAL FRACTIONS

It is necessary to convert the response transforms, as indicated in Eqs. (8-10) and (8-14), into functions of t to complete the solutions. The technique used is to express the response transforms as partial fractions in order that the transform pairs $F(s)$ and $f(t)$, such as those indicated in Table 8-1, can be identified and used for the inverse transformation.

For the types of problems considered, the response transforms can be expressed as a ratio of two polynomials in s,

$$F(s) = \frac{A(s)}{B(s)} = \frac{a_ms^m + a_{m-1}s^{m-1} + \ldots + a_1s + a_o}{s^n + b_{n-1}s^{n-1} + \ldots + b_1s + b_o} \qquad \textbf{(8-15)}$$

where m and n are positive integers. We shall assume that $F(s)$ is a proper fraction, that is, $n > m$. If $F(s)$ is an improper fraction, that is, $n < m$, we can divide the numerator by the denominator and obtain $F(s)$ as the sum of a polynomial in s and a proper fraction. We shall show two methods of resolving $F(s)$ into partial fractions—Heaviside's expansion theorem and the elementary rules of partial fractions.

HEAVISIDE'S EXPANSION THEOREM

The response transform in Eq. (8-15) can be expressed as

$$F(s) = \frac{A(s)}{B(s)} = \frac{A(s)}{(s - s_1)(s - s_2) \ldots (s - s_k) \ldots (s - s_n)} \tag{8-16}$$

where $s_{1,2,\ldots,n}$ are the roots of the algebraic equation $B(s) = 0$. If this equation is of degree n, it has not more than n roots.† The values of the roots may be zero, real or complex, distinct or repeating. Treating the different types of roots simply as numbers, we have only two cases: (1) the roots are all distinct, and (2) some of the roots are repeating, that is, when $B(s) = 0$ has multiple roots.

Case 1. Roots of $B(s) = 0$ All Distinct

Let us write Eq. (8-15) as a sum of simple partial fractions.

$$F(s) = \frac{A(s)}{B(s)} = \frac{A_1}{s - s_1} + \frac{A_2}{s - s_2} + \ldots + \frac{A_k}{s - s_k} + \ldots + \frac{A_n}{s - s_n} \tag{8-17}$$

where $A_{1,2,\ldots,n}$ are constants as yet to be determined. To determine a typical constant A_k, we multiply both sides of Eq. (8-17) by $(s - s_k)$ and take the limit as $s \to s_k$. This operation results in the expression

$$A_k = \lim_{s \to s_k} (s - s_k) \frac{A(s)}{B(s)} \tag{8-18}$$

Since both $(s - s_k)$ and $B(s)$ vanish at $s = s_k$, the quantity $(s - s_k)\, A(s)/B(s)$ is indeterminate of the form 0/0 as s is set equal s_k. The limit indicated in Eq. (8-18) can be obtained by applying L'Hospital's rule. We differentiate the numerator and the denominator separately with respect to s and then let s approach s_k.

$$A_k = \lim_{s \to s_k} \frac{(d/ds)[(s - s_k)\, A(s)]}{(d/ds)\, B(s)} = \left[\frac{A(s)}{B'(s)}\right]_{s = s_k} \tag{8-19}$$

where $B'(s)$ denotes a differentiation with respect to s.

† R. V. Churchill, *Introduction to Complex Variables and Applications*, McGraw-Hill Book Co., Inc., New York, 1948, p. 96.

Alternatively, if both sides of Eq. (8-17) are multiplied by $(s - s_k)$, the quantity $(s - s_k)$ can be canceled with the corresponding factor in $B(s)$. Now, if s is set equal to s_k, we have

$$A_k = \left[(s - s_k)\frac{A(s)}{B(s)}\right]_{s=s_k} \tag{8-20}$$

since all the terms on the right side would be zero with the exception of A_k. Equation (8-20) is not indeterminate because it is assumed that $(s - s_k)$ is already canceled out before setting $s = s_k$.

Case 2. $B(s) = 0$ Has Multiple Roots

Let one of the roots s_1 of $B(s) = 0$ be of multiplicity p and the other roots be distinct. The response transform $F(s)$ can be expressed as

$$F(s) = \frac{A(s)}{B(s)} = \frac{A(s)}{(s - s_1)^p(s - s_2)\ldots}$$

This equation can be expanded into partial fractions as

$$F(s) = \frac{A_{11}}{(s - s_1)^p} + \frac{A_{12}}{(s - s_1)^{p-1}} + \ldots + \frac{A_{1p}}{(s - s_1)}$$
$$+ \frac{A_2}{(s - s_2)} + \frac{A_3}{(s - s_3)} + \ldots \tag{8-21}$$

To evaluate the constants A_{11} to A_{1p}, we multiply both sides of Eq. (8-21) by $(s - s_1)^p$ and obtain

$$(s - s_1)^p \frac{A(s)}{B(s)} = A_{11} + A_{12}(s - s_1) + \ldots + A_{1p}(s - s_1)^{p-1}$$
$$+ (s - s_1)^p\left(\frac{A_2}{s - s_2} + \frac{A_3}{s - s_3} + \ldots\right) \tag{8-22}$$

On the left-hand side of Eq. (8-22) the quantity $(s - s_1)^p$ can be canceled with the corresponding factor in $B(s)$. Now setting $s = s_1$ gives

$$A_{11} = [(s - s_1)^p A(s)/B(s)]_{s=s_1} \tag{8-23}$$

A_{12} is obtained by differentiating Eq. (8-22) with respect to s and substituting $s = s_1$.

$$A_{12} = \left\{\frac{d}{ds}\left[(s - s_1)^p \frac{A(s)}{B(s)}\right]\right\}_{s=s_1} \tag{8-24}$$

The typical coefficient A_{1k} is determined by the relation

$$A_{1k} = \frac{1}{(k-1)!}\left\{\frac{d^{k-1}}{ds^{k-1}}\left[(s-s_1)^p \frac{A(s)}{B(s)}\right]\right\}_{s=s_1} \tag{8-25}$$

The constants corresponding to the distinct roots can be found by the method shown in Case 1.

Example 4. Roots of $B(s) = 0$ All Distinct

Determine the inverse transform (*a*) of $F(s) = \dfrac{s-5}{s^2+5s+4}$, and (*b*) of $F(s) = \dfrac{14s^2+16s+30}{s^3+2s^2+5s}$

Solution: (*a*) The given function $F(s)$ can be expressed as

$$F(s) = \frac{s-5}{(s+1)(s+4)} = \frac{A_1}{s+1} + \frac{A_2}{s+4}$$

From Eq. (8-20) we have

$$A_1 = [(s+1)\,F(s)]_{s=-1} = \left[\frac{s-5}{s+4}\right]_{s=-1} = \frac{-6}{3} = -2$$

$$A_2 = [(s+4)\,F(s)]_{s=-4} = \left[\frac{s-5}{s+1}\right]_{s=-4} = \frac{-9}{-3} = 3$$

From Table 8-1 the inverse transform is

$$f(t) = \mathscr{L}^{-1}\,F(s) = \mathscr{L}^{-1}\left[\frac{-2}{s+1} + \frac{3}{s+4}\right]$$

$$= -2e^{-t} + 3e^{-4t}$$

(*b*) The given function $F(s)$ can be expressed as

$$F(s) = \frac{14s^2+16s+30}{s(s^2+2s+5)} = \frac{A_1}{s} + \frac{A_2}{s+1+j2} + \frac{A_3}{s+1-j2}$$

From Eq. (8-20) the constants are evaluated as

$$A_1 = [s\,F(s)]_{s=0} = \left[\frac{14s^2+16s+30}{s^2+2s+5}\right]_{s=0} = 6$$

$$A_2 = [(s+1+j2)\,F(s)]_{s=-(1+j2)} = \left[\frac{14s^2+16s+30}{s(s+1-j2)}\right]_{s=-(1+j2)}$$

$$= \frac{-28+j24}{(1+j2)(j4)} = \frac{7-j6}{2-j} = \frac{7-j6}{2-j}\,\frac{2+j}{2+j}$$

$$= 4 - j$$

We know that complex roots occur in conjugate pairs and that A_3 must be the conjugate of A_2 in order for $f(t)$ to be real. Hence, once a constant has been evaluated for one of the complex roots, the constant for the conjugate root need not be evaluated. The conjugate of A_2 is

$$A_2^* = A_3 = 4 + j$$

From Table 8-1 the inverse transform is

$$\begin{aligned} f(t) = \mathscr{L}^{-1} F(s) &= \mathscr{L}^{-1}\left(\frac{6}{s} + \frac{4-j}{s+1+j2} + \frac{4+j}{s+1-j2}\right) \\ &= 6 + (4-j)\, e^{-(1+j2)t} + (4+j)\, e^{-(1-j2)t} \\ &= 6 + 2e^{-t}(4\cos 2t - \sin 2t) \end{aligned}$$

It should be noted that if a complex number $z = a + jb$ has a conjugate $z^* = a - jb$, the sum of the complex number and its conjugate is $2a$; that is, their sum is equal to two times the real part of either the complex number itself or its conjugate. Similarly, the sum of two complex conjugate functions is two times the real part of the complex function itself or its conjugate. Hence the inverse transform in this problem can be written as

$$\begin{aligned} f(t) &= \mathscr{L}^{-1}\frac{6}{s} + 2\,\text{Re}\,\mathscr{L}^{-1}\left(\frac{4-j}{s+1+j2}\right) \\ &= 6 + 2\,\text{Re}\,[(4-j)e^{-(1+2j)t}] \\ &= 6 + 2e^{-t}(4\cos 2t - \sin 2t) \end{aligned}$$

where the symbol Re denotes the real part of a function.

Example 5. $B(s) = 0$ Has Multiple Roots

Find the inverse transform of $F(s) = \dfrac{s+6}{s^3(s+1)^2}$

Solution: The given function $F(s)$ can be expressed as

$$F(s) = \frac{A_{11}}{s^3} + \frac{A_{12}}{s^2} + \frac{A_{13}}{s} + \frac{A_{21}}{(s+1)^2} + \frac{A_{22}}{s+1}$$

Using Eq. (8-25), the constants are evaluated as

$$A_{11} = [s^3 F(s)]_{s=0} = \left[\frac{s+6}{(s+1)^2}\right]_{s=0} = 6$$

$$A_{12} = \left\{\frac{d}{ds}[s^3 F(s)]\right\}_{s=0} = \left[\frac{d}{ds}\frac{s+6}{(s+1)^2}\right]_{s=0} = -11$$

$$A_{13} = \frac{1}{2!}\left\{\frac{d^2}{ds^2}[s^3 F(s)]\right\}_{s=0} = \frac{1}{2!}\left[\frac{d^2}{ds^2}\frac{s+6}{(s+1)^2}\right]_{s=0} = 16$$

$$A_{21} = [(s+1)^2 F(s)]_{s=-1} = \left[\frac{s+6}{s^3}\right]_{s=-1} = -5$$

$$A_{22} = \left\{\frac{d}{ds}[(s+1)^2 F(s)]\right\}_{s=-1} = \left[\frac{d}{ds}\frac{s+6}{s^3}\right]_{s=-1} = -16$$

The inverse transform of $F(s)$ is

$$f(t) = \mathscr{L}^{-1}\left[\frac{A_{11}}{s^3} + \frac{A_{12}}{s^2} + \frac{A_{13}}{s} + \frac{A_{21}}{(s+1)^2} + \frac{A_{22}}{s+1}\right]$$

$$= 3t^2 - 11t + 16 - (5t + 16)e^{-t}$$

ELEMENTARY RULES OF PARTIAL FRACTIONS

The elementary rules for expressing a proper fraction as the sum of partial fractions may be easier to use than the Heaviside method when the number of constants to be determined is small. The method becomes increasingly laborious, however, if the number of constants is large. We shall illustrate these rules with the problems in Examples 4 and 5.

Case 1. The Roots of $B(s) = 0$ Are Real and Distinct

$$F(s) = \frac{s-5}{(s+1)(s+4)} = \frac{A_1}{s+1} + \frac{A_2}{s+4}$$

Multiplying the equation by $(s+1)(s+4)$ gives

$$s - 5 = A_1(s+4) + A_2(s+1)$$

Equating the coefficients of like powers of s in this equation yields

$$A_1 + A_2 = 1, \qquad 4A_1 + A_2 = -5.$$

Solving these equations simultaneously, we obtain

$$A_1 = -2, \qquad A_2 = 3.$$

Instead of solving simultaneously as indicated, it is observed in the equation

$$s - 5 = A_1(s + 4) + A_2(s + 1)$$

that A_1 can be evaluated by setting $s = -1$, and that A_2 can be evaluated by setting $s = -4$.

Setting $s = -1$ gives

$$-1 - 5 = A_1(-1 + 4) \quad \text{or} \quad A_1 = -2$$

Setting $s = -4$ gives

$$-4 - 5 = A_2(-4 + 1) \quad \text{or} \quad A_2 = 3$$

Case 2. The Roots of $B(s) = 0$ Are Real and Repeating

$$F(s) = \frac{s + 6}{s^3(s + 1)^2} = \frac{A_{11}}{s^3} + \frac{A_{12}}{s^2} + \frac{A_{13}}{s} + \frac{A_{21}}{(s + 1)^2} + \frac{A_{22}}{s + 1}$$

Multiplying the equation by $s^3(s + 1)^2$ gives

$$s + 6 = A_{11}(s + 1)^2 + A_{12}s(s + 1)^2 + A_{13}s^2(s + 1)^2 + A_{21}s^3 + A_{22}s^3(s + 1)$$

Setting $s = 0$ and $s = -1$ gives $A_{11} = 6$ and $A_{21} = -5$, respectively. Substituting these values in the above equation and simplifying, we have

$$5s^3 - 6s^2 - 11s = A_{12}s(s + 1)^2 + A_{13}s^2(s + 1)^2 + A_{22}s^3(s + 1)$$

Expanding this equation and equating the coefficients of like powers of s, we obtain

$$A_{13} + A_{22} = 0, \qquad 2A_{12} + A_{13} = -6,$$

$$A_{12} + 2A_{13} + A_{22} = 5, \qquad A_{12} = -11.$$

Solving simultaneously gives

$$A_{12} = -11, \quad A_{13} = 16, \quad A_{22} = -16$$

Case 3. Roots of $B(s) = 0$ Are Complex and Distinct

$$F(s) = \frac{14s^2 + 16s + 30}{s^3 + 2s^2 + 5s} = \frac{A_1}{s} + \frac{A_2 s + A_3}{s^2 + 2s + 5}$$

The quadratic factor $(s^2 + 2s + 5)$ which gives the complex roots cannot be separated into linear factors as illustrated in Case 1. We assume that the numerator of the partial fraction is of degree one less than the degree of the denominator. Multiplying this equation by $s(s^2 + 2s + 5)$ gives

$$14s^2 + 16s + 30 = A_1(s^2 + 2s + 5) + (A_2 s + A_3)s$$

Setting $s = 0$ gives $A_1 = 6$. Substituting this value in the above equation and simplifying, we have $A_2 = 8$ and $A_3 = 4$. Hence

$$F(s) = \frac{6}{s} + \frac{8s + 4}{s^2 + 2s + 5}$$

By simple algebraic manipulation, this equation can be expressed as

$$F(s) = \frac{6}{s} + 2\left[\frac{4(s + 1)}{(s + 1)^2 + 2^2} - \frac{2}{(s + 1)^2 + 2^2}\right]$$

It is noted in Table 8-1 that

$$\mathscr{L}^{-1}\frac{s}{s^2 + \omega^2} = \cos \omega t, \quad \mathscr{L}^{-1}\frac{\omega}{s^2 + \omega^2} = \sin \omega t,$$

and
$$\mathscr{L}^{-1} F(s + a) = e^{-at} f(t)$$

Using these transform pairs, the inverse transform of the given function $F(s)$ is

$$f(t) = \mathscr{L}^{-1} F(s) = 6 + 2e^{-t}(4 \cos 2t - \sin 2t)$$

Case 4. Roots of $B(s) = 0$ Are Complex and Repeating

Consider the example

$$F(s) = \frac{s^4 + s^3 - 3s^2 + 5s - 4}{(s + 1)(s^2 + 2s + 5)^2}$$

Assuming that $F(s)$ can be expressed as

$$F(s) = \frac{A_1}{s + 1} + \frac{A_2 s + A_3}{s^2 + 2s + 5} + \frac{A_4 s + A_5}{(s^2 + 2s + 5)^2}$$

the values of the constants can be evaluated by the methods outlined in the previous cases.

8-5. APPLICATIONS

The applications of the properties of Laplace transformation discussed in Sec. 8-3 are illustrated in this section. Further properties of the Laplace transform and their applications will be shown in the sections to follow.

Example 6. Determination of Steady-State and Transient Motions

The equation of motion of a one-degree-of-freedom system with sinusoidal excitation is

$$m\ddot{x} + c\dot{x} + kx = F_o \sin \omega t \tag{8-26}$$

If the initial conditions are $x(0+) = x_o$ and $\dot{x}(0+) = \dot{x}_o$, determine the steady-state and the transient responses of the system.

Solution: From Eq. (8-10), the subsidiary equation of Eq. (8-26) is

$$X(s) = \frac{1}{ms^2 + cs + k}\left[F_o \frac{\omega}{s^2 + \omega^2} + (ms + c)x_o + m\dot{x}_o\right]$$

Defining $c/m = 2\zeta\omega_n$ and $k/m = \omega_n^2$ and assuming $\zeta < 1$, we have

$$\begin{aligned} X(s) &= \frac{F_o\omega/m + (s^2 + \omega^2)[(s + 2\zeta\omega_n)x_o + \dot{x}_o]}{(s^2 + \omega^2)(s^2 + 2\zeta\omega_n + \omega_n^2)} \\ &= \frac{F_o\omega/m + (s^2 + \omega^2)[(s + 2\zeta\omega_n)x_o + \dot{x}_o]}{(s + j\omega)(s - j\omega)(s + \zeta\omega_n + j\omega_d)(s + \zeta\omega_n - j\omega_d)} \\ &= \frac{A_1}{s + j\omega} + \frac{A_2}{s - j\omega} + \frac{A_3}{s + \zeta\omega_n + j\omega_d} + \frac{A_4}{s + \zeta\omega_n - j\omega_d} \end{aligned} \tag{8-27}$$

where $\omega_d^2 = (1 - \zeta^2)\omega_n^2$. The constants A_1 to A_4 can be evaluated by Eq. (8-20). The formal solution is

$$\begin{aligned} x(t) &= \mathcal{L}^{-1} X(s) = 2 \operatorname{Re}(A_1 e^{-j\omega t}) + 2 \operatorname{Re}[A_3 e^{-(\zeta\omega_n + j\omega_d)t}] \\ &= 2 \operatorname{Re}(A_1 e^{-j\omega t}) + 2e^{-\zeta\omega_n t} \operatorname{Re}(A_3 e^{-j\omega_d t}) \end{aligned} \tag{8-28}$$

It is evident that the first term on the right of Eq. (8-28) describes a harmonic motion of frequency ω, and it is the steady-state solution. The second term with an exponential decay gives the transient motion. Referring to Eq. (8-27), it is obvious that A_1 can be evaluated without evaluating A_3. Hence the steady-state solution can be determined without considering the transient motion. Similarly, by determining A_3, the transient motion of the system can be evaluated without considering the steady-state motion.

To generalize the results of this example, it was shown in Sec. 8-3 [see Eq. (8-12)] that the response transform of a system can be expressed as

$$H(s) = G(s)\,F(s)$$

Expressing $H(s)$ in partial fractions gives

$$H(s) = \frac{A_1}{s - s_1} + \frac{A_2}{s - s_2} + \cdots \qquad \textbf{(8-29)}$$

It is possible to identify the steady-state and the transient terms in Eq. (8-29) from a knowledge of the Laplace transform pairs. Grouping these terms together, $H(s)$ can be expressed as

$$H(s) = H_s(s) + H_t(s) \qquad \textbf{(8-30)}$$

where $H_s(s)$ is the steady-state response transform and $H_t(s)$ is the transient. Thus the time response of the system is

$$h(t) = h_s(t) + h_t(t) \qquad \textbf{(8-31)}$$

where $h_s(t)$ and $h_t(t)$ are the inverse transforms of $H_s(s)$ and $H_t(s)$, respectively.

If the applied excitation is periodic of which $F_o \sin \omega t$ is one of the harmonic components, the Laplace transform of $F_o \sin \omega t$ is $F_o\omega/(s^2 + \omega^2)$. $H_s(s)$, corresponding to $F_o \sin \omega t$, must be of the form

$$H_s(s) = \frac{A_1}{s + j\omega} + \frac{A_2}{s - j\omega}$$

The steady-state response is

$$h_s(t) = 2\,\mathrm{Re}(A_1 e^{-j\omega t})$$

which is the same quantity given in Eq. (8-28). The transient response is obtained from the remaining partial fractions. Hence the steady-state or the transient response can be obtained without considering the other.

If the applied excitation is aperiodic and the steady-state response exists, the steady-state and the transient responses can be identified from $H(s)$ and a knowledge of the Laplace transform pairs. In fact, the time response of a system can generally be interpreted from its response transform and a knowledge of the Laplace transform pairs without finding the inverse transform.

Example 7. A spring-mass system is as shown in Fig. 8-1. If the system is initially relaxed and a step-function excitation is applied to the mass, find the motion of the mass.

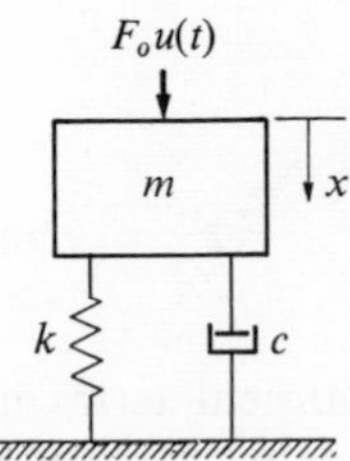

Fig. 8-1. *Mechanical system with step-function excitation*

Solution: From Newton's law of motion, the equation of motion of the system is

$$m\ddot{x} + c\dot{x} + kx = F_o u(t)$$

If the system is initially relaxed, the Laplace transform of this equation is

$$(ms^2 + cs + k)\, X(s) = F_o/s$$

Dividing this equation by m, defining $c/m = 2\zeta\omega_n$, $k/m = \omega_n^2$, and $\omega_d = \sqrt{1 - \zeta^2}\,\omega_n$, assuming $\zeta < 1$ and solving for the response transform $X(s)$, we have

$$X(s) = \frac{F_o}{m}\,\frac{1}{s(s^2 + 2\zeta\omega_n s + \omega_n^2)} = \frac{F_o}{m}\left(\frac{A_1}{s} + \frac{A_2 s + A_3}{s^2 + 2\zeta\omega_n s + \omega_n^2}\right)$$

By elementary rules of partial fractions, $A_1 = 1/\omega_n^2$, $A_2 = -1/\omega_n^2$, and $A_3 = -2\zeta\omega_n/\omega_n^2$. Substituting these values in the response transform and rearranging gives

$$X(s) = \frac{F_o}{m\omega_n^2}\left[\frac{1}{s} - \frac{(s + \zeta\omega_n)}{(s + \zeta\omega_n)^2 + \omega_d^2} - \frac{\zeta\omega_n}{(s + \zeta\omega_n)^2 + \omega_d^2}\right]$$

The inverse transform of these partial fractions can be obtained from the pairs shown in Table 8-1. The time response $x(t)$ is

$$x(t) = \frac{F_o}{k}\left[1 - e^{-\zeta\omega_n t}(\cos\omega_d t + \frac{\zeta}{\sqrt{1 - \zeta^2}}\sin\omega_d t)\right]$$

Physically, the solution indicates that the new equilibrium position is at a distance F_o/k below the original equilibrium position and the mass oscillates with diminishing amplitude about the new equilibrium position. Hence the steady-state solution is due to the first term in the response transform, and the transient solution is due to the second and third terms.

Example 8. An apparatus is rigidly bolted to a trailer which is traveling at a fair speed. If the trailer hits a 3-in. curb, estimate the peak acceleration to which the apparatus is subjected.

Data:

Weight of apparatus = 772 lb; trailer spring $k = 1{,}944$ lb-in.$^{-1}$; weight of trailer body = 1,544 lb; pneumatic tire $k_{eq} = 4{,}900$ lb-in.$^{-1}$; and weight of wheel and axle of trailer = 386 lb.

Solution: (*a*) As a first approximation, let us neglect the effect of the tire and assume that a shock displacement of 3 in. is applied to the trailer

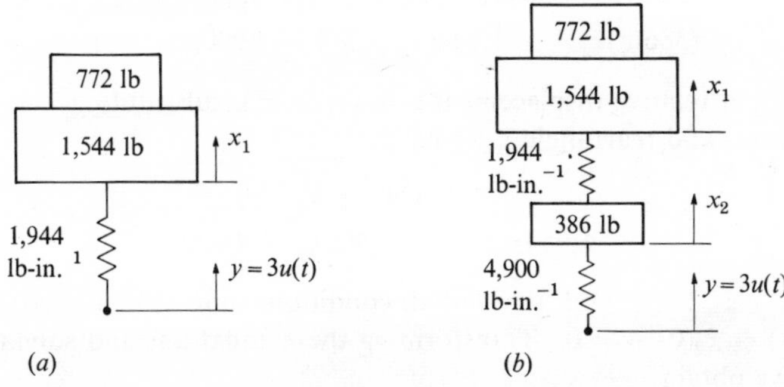

Fig. 8-2. *Schematic representation of a trailer hitting a 3-in. curb: (a) first approximation, neglecting effect of tires; (b) second approximation, including effect of tires*

spring. The equivalent system is as shown in Fig. 8-2(*a*). The equation of motion of the system is

$$(772 + 1{,}544)(1/g)\ddot{x}_1 = -1{,}944(x_1 - y)$$

or

$$6\ddot{x}_1 + 1{,}944(x_1 - y) = 0$$

Defining a relative displacement $z = (x_1 - y)$, substituting z in this equation, and rearranging, we have

$$\ddot{z} + 324z = -\ddot{y}$$

where $y = 3\,u(t)$ and the initial conditions are $z(0+) = -y_0$ and $\dot{z}(0+) = 0$. Transforming this equation and solving for the response transform $Z(s)$, we obtain

$$Z(s) = -3\,\frac{s}{s^2 + 324}$$

The corresponding time response is

$$z(t) = -3 \cos 18t$$

Alternatively, $x_1(t)$ can be evaluated from $\ddot{x}_1 + 324x_1 = 324y$ and $z(t)$ obtained from the relation $z = (x_1 - y)$.

The maximum force transmitted to the apparatus is equal to $kz_{\max}$. Equating (mass)(maximum acceleration) $= kz_{\max}$ gives

$$\text{Maximum acceleration} = (1{,}944)(3)g/(772 + 1{,}544) = 2.52 \text{ g's}$$

(*b*) Considering the effect of the tire, the equivalent system is as shown in Fig. 8-2(*b*). The equations of motion are

$$(772 + 1{,}544)(1/g)\ddot{x}_1 = -1{,}944(x_1 - x_2)$$
$$(386/g)\ddot{x}_2 = -1{,}944(x_2 - x_1) - 4{,}900(x_2 - y)$$

Defining a relative displacement $z = (x_1 - x_2)$, substituting z in these equations, and rearranging, we have

$$\ddot{z} + 324z + \ddot{x}_2 = 0$$
$$-1{,}944z + \ddot{x}_2 + 4{,}900x_2 = 4{,}900y$$

where $y = 3\,u(t)$ and the initial conditions are $z(0+) = \dot{z}(0+) = x_2(0+) = \dot{x}_2(0+) = 0$. Transforming these equations and solving for $Z(s)$, we obtain

$$Z(s) = \frac{\begin{vmatrix} 0 & s^2 \\ 14{,}700/s & s^2 + 4{,}900 \end{vmatrix}}{\begin{vmatrix} s^2 + 324 & s^2 \\ -1{,}944 & s^2 + 4{,}900 \end{vmatrix}} = -\frac{14{,}700s}{(s^2 + 229)(s^2 + 6{,}930)} = -2.19\left(\frac{s}{s^2 + 229} - \frac{s}{s^2 + 6{,}930}\right)$$

The corresponding time response is

$$z(t) = -2.19\,(\cos 15.1t - \cos 83.2t)$$

Hence the maximum possible displacement is the sum of these amplitudes, which is equal to 2(2.19), and the maximum possible acceleration is $(1{,}944)(2)(2.19)g/(722 + 1{,}544) = 3.68$ g's.

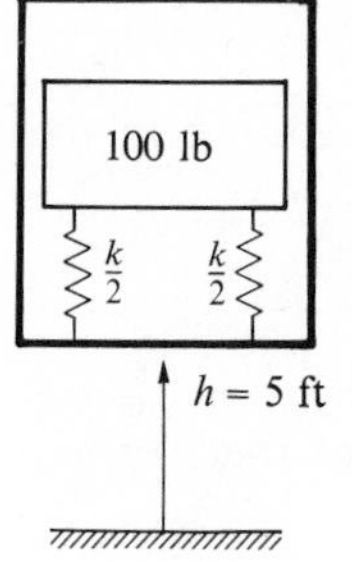

Fig. 8-3. *Dropping of package on hard floor*

Example 9. An apparatus weighing 100 lb is shipped in a container as shown in Fig. 8-3. It is anticipated that, in the process of unloading, the container will be dropped from a height of 5 ft to a hard floor. If the maximum acceleration to which the apparatus may be subjected without damage is 12 g's, specify the springs supporting the apparatus and estimate the maximum relative displacement of the apparatus in the container.

Solution: For the duration that the container is in contact with the floor, the equation of motion of the mass m is

$$m\ddot{x} + kx = 0$$

where x is measured from the static equilibrium position of the mass with the container in contact with the floor. Assume that the spring is unstressed at the instant $t = 0$ when the container comes in contact with the floor. The initial displacement $x(0+) = -$(static deflection) $= -W/k = -g/\omega_n^2$. The mass m has fallen through a distance $h' = h -$ (static deflection) $= h - g/\omega_n^2$, and the initial velocity of the mass m is $\dot{x}(0+) = \sqrt{2gh'}$. Transforming the equation of motion, the corresponding subsidiary equation is

$$X(s) = \frac{\sqrt{2gh'}}{s^2 + \omega_n^2} - \frac{g}{\omega_n^2}\frac{s}{s^2 + \omega_n^2}$$

The time response of the mass is

$$\begin{aligned} x(t) &= \frac{\sqrt{2gh'}}{\omega_n}\sin \omega_n t - \frac{g}{\omega_n^2}\cos \omega_n t \\ &= \sqrt{\frac{2gh'}{\omega_n^2} + \left(\frac{g}{\omega_n^2}\right)^2}\sin (\omega_n t - \phi) \\ &= \sqrt{\frac{2gh}{\omega_n^2} - \left(\frac{g}{\omega_n^2}\right)^2}\sin (\omega_n t - \phi) \end{aligned}$$

where

$$\phi = \tan^{-1}\frac{g}{\omega_n\sqrt{2gh'}}$$

The maximum acceleration in g's is

$$\frac{\ddot{x}_{\max}}{g} = 12 = \frac{\omega_n^2}{g}\sqrt{\frac{2gh}{\omega_n^2} - \left(\frac{g}{\omega_n^2}\right)^2} = \sqrt{\frac{2h\omega_n^2}{g} - 1}$$

Solving for ω_n^2 and k, we have

$$\begin{aligned} \omega_n^2 &= 145g/120 = 466 \\ k &= \omega_n^2 m = (466)(100/g) = 121 \text{ lb-in.}^{-1} \end{aligned}$$

It may be noted that $g/\omega_n^2 = W/k = 100/121$ in. is the static deflection and $\sqrt{2gh'/\omega_n^2} = 9.9$ in. is the dynamic deflection due to h'.

Frequently, the static deflection is small compared with the dynamic deflection. The maximum deflection from the static equilibrium position is $x_{max} = 9.9$ in. If the container does not leave the floor, the maximum relative displacement between the apparatus and the container is $2x_{max} = 19.8$ in. The container will leave the floor if the inertia force $m\ddot{x}$ is equal to the gravitational force on the mass m and the container. In any case, the container must be designed to accommodate the high relative displacement which may not be practical. By neglecting the static deflection, the deflection x_{max} is proportional to $1/\omega_n$, and the acceleration $\ddot{x}_{max} = \omega_n^2 x_{max}$. Hence a decrease in deflection by increasing ω_n can be accomplished only at the expense of increasing the acceleration.

8-6. ADDITIONAL PROPERTIES OF THE LAPLACE TRANSFORMATION

The properties of the Laplace transformation treated in Sec. 8-3 are sufficient to solve a number of physical problems. Additional properties will be discussed in this section to cover certain types of excitation functions as well as any arbitrary excitation. Techniques for extending the usefulness of the table of Laplace transform pairs will be discussed at the end of this section. We shall need a few concepts before discussing these properties.

GATE FUNCTION AND UNIT IMPULSE

The unit step function, defined by

$$u(t) = \begin{cases} 0 & \text{for } t < 0 \\ 1 & t > 0 \end{cases} \tag{8-32}$$

is shown in Fig. 8-4(a). It was shown in Sec. 8-3 that $\mathscr{L}\, u(t) = 1/s$. Let this function be translated to the right along the axis of the independent variable t, as shown in Fig. 8-4(b). The translated unit step function is defined as

$$u(t-a) = \begin{cases} 0 & \text{for } t < a \\ 1 & t > a \end{cases} \tag{8-33}$$

The Laplace transform of $u(t-a)$ is

$$\mathscr{L}\, u(t-a) = \int_o^\infty u(t-a)e^{-st}\, dt = \int_a^\infty e^{-st}\, dt = -\frac{1}{s}e^{-st}\Big|_a^\infty = \frac{e^{-as}}{s}$$

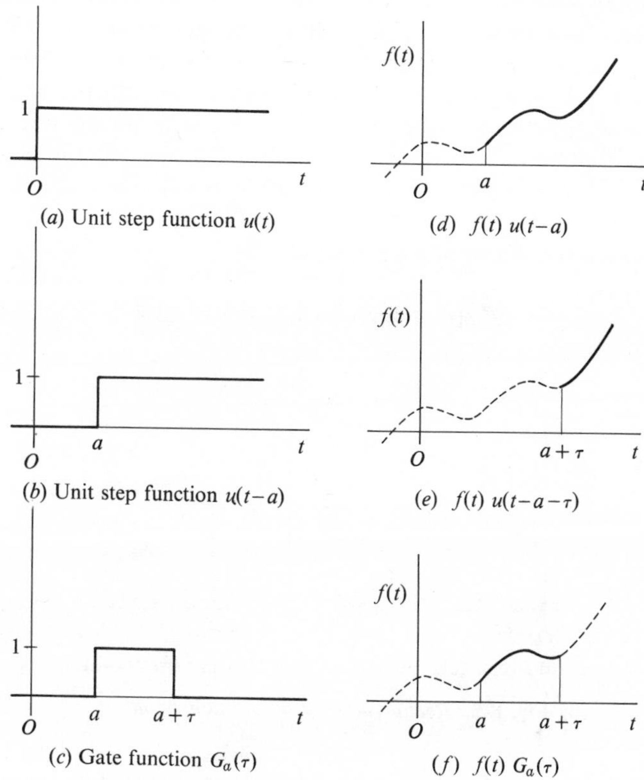

(*a*) Unit step function $u(t)$

(*d*) $f(t)\, u(t-a)$

(*b*) Unit step function $u(t-a)$

(*e*) $f(t)\, u(t-a-\tau)$

(*c*) Gate function $G_a(\tau)$

(*f*) $f(t)\, G_a(\tau)$

Fig. 8-4. *Gate function and its use*

A gate function $G_a(\tau)$, as shown in Fig. 8-4(*c*), can be obtained by subtracting one unit step function from another. The gate function is defined as

$$G_a(\tau) = u(t - a) - u(t - a - \tau) \tag{8-34}$$

It is illustrated in Fig. 8-4(*d*) that if a function $f(t)$ is multiplied by $u(t - a)$ the resultant function is zero for $t < a$ but has the same value as $f(t)$ for $t > a$. Figure 8-4(*e*), in which $f(t)$ is multiplied by $u(t - a - \tau)$, can be explained in like manner. Hence, multiplying $f(t)$ by a gate function, as indicated in Fig. 8-4(*f*), is equivalent to discarding the unwanted portions of $f(t)$ and considering only the portions between $a < t < (a + \tau)$. It will be shown that the gate function is useful in the construction of wave forms.

A rectangular pulse of pulse height F_o and duration t_o, such that the impulse of the force equals $F_o t_o$, is shown in Fig. 8-5(*a*). The rectangular pulse can be constructed from the difference of two step functions. As a special case, consider a rectangular pulse of unit area; that is, the area $F_o t_o = 1$ or $F_o = 1/t_o$. If the pulse duration t_o is decreased, the pulse height F_o is increased to maintain the unit area. In the limit, as t_o approaches zero, we have a *unit impulse* defined by

$$\delta(t) = \lim_{t_o \to 0} \frac{1}{t_o} [u(t) - u(t - t_o)] \tag{8-35}$$

The Laplace transform of a unit impulse is

$$\mathscr{L}\, \delta(t) = \lim_{t_o \to 0} \frac{1}{t_o s} (1 - e^{-t_o s}) = 1 \tag{8-36}$$

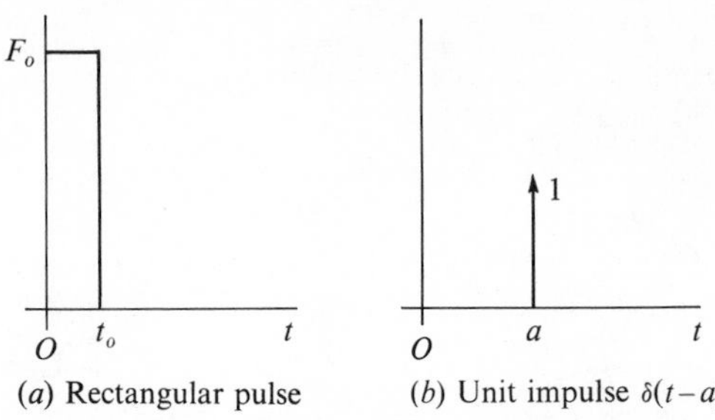

(*a*) Rectangular pulse (*b*) Unit impulse $\delta(t-a)$

Fig. 8-5. *Rectangular pulse and unit impulse*

A unit impulse translated by an amount a is represented by the symbol $\delta(t - a)$ as shown in Fig. 8-5(*b*).

The unit impulse represents a physical quantity such as an impact. If a unit impulse is multiplied by a constant, this constant is considered as the strength of the impulse. When the duration of the impact is very short compared with the natural frequency of the system, the response of the system to the impact can be obtained by considering it the response to an impulse of the appropriate strength.

The unit impulse function, as defined in Eq. (8-35), is often considered as the derivative of a unit step function, even though strictly speaking this limit does not exist. Avoiding the mathematical controversy, we shall define the unit impulse as a function having the properties

$$\delta(t - a) = 0 \quad \text{for } t \neq a$$

$$\int_0^\infty \delta(t - a)\, dt = 1 \tag{8-37}$$

Any function having these properties is called the Dirac delta function.

The product of a continuous function $f(t)$ and the unit impulse $\delta(t - a)$ is zero everywhere except at $t = a$. Hence we have

$$\int_0^\infty f(t)\, \delta(t - a)\, dt = f(a) \tag{8-38}$$

In particular, if $f(t) = e^{-st}$, we obtain

$$\int_0^\infty \delta(t - a)\, e^{-st}\, dt = \mathscr{L}\, \delta(t - a) = e^{-as} \tag{8-39}$$

Theorem 4. Real Translation (Shifting Theorem)
If $f(t)$ has the Laplace transform $F(s)$, then

$$\mathscr{L}\,[f(t - a)u(t - a)] = e^{-as}\, F(s) \tag{8-40}$$

Proof: From the definition of Laplace transform, Eq. (8-1), we have

$$\int_0^\infty f(\tau)u(\tau)e^{-s\tau}\, d\tau = F(s)$$

Since $u(\tau)$ is a unit step function, its addition will not affect the equation. Substituting $(t - a)$ for τ, where a is a positive constant, gives

$$\int_0^\infty f(t - a)u(t - a)e^{-s(t-a)}\, dt = e^{as} \int_0^\infty f(t - a)u(t - a)e^{-st}\, dt = F(s)$$

Since $f(t - a)u(t - a) = 0$ for $t < a$, the lower limit of the integral can be changed from a to zero. Rearranging, we obtain

$$\mathscr{L}\,[f(t - a)u(t - a)] = \int_0^\infty f(t - a)u(t - a)\, e^{-st}\, dt = e^{-as}\, F(s)$$

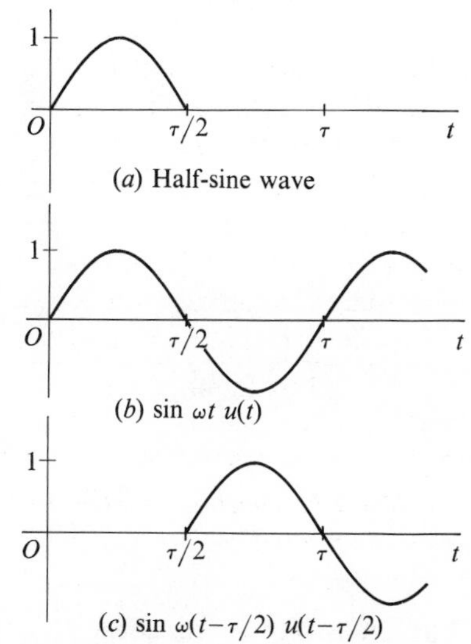

Fig. 8-6. *Construction of the half-sine wave as shown in* (a)

Example 10. Determine the Laplace transform of the single half-sine wave as shown in Fig. 8-6(a).

Solution: The angular frequency ω equals $2\pi/\tau$, where τ is the period of the sine wave. Referring to Figs. 8-6(b) and (c), the given function can be expressed as

$$f(t) = \sin \omega t \, u(t) + \sin \omega(t - \tau/2) \, u(t - \tau/2)$$

The Laplace transform of the second term on the right can be obtained from the real translation theorem.

$$\mathscr{L} f(t) = F(s) = \frac{\omega}{s^2 + \omega^2} + \frac{\omega}{s^2 + \omega^2} e^{-\tau s/2}$$

or

$$F(s) = \frac{\omega}{s^2 + \omega^2}(1 + e^{-\pi s/\omega})$$

In relation to the real translation theorem, it should be noted that the following functions are not identical:

(a) $f(t - a)$ (b) $f(t)u(t - a)$

(c) $f(t - a)u(t)$ (d) $f(t - a)u(t - a)$

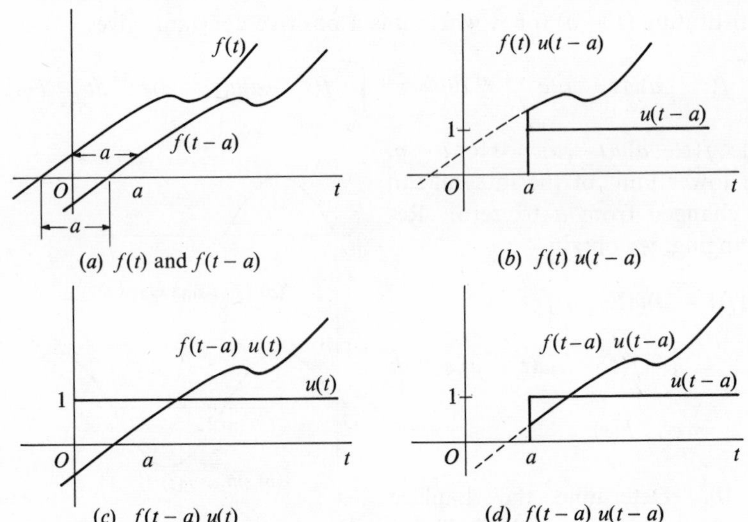

Fig. 8-7. *Functions relating to real translation; the real translation theorem applies only to* (*d*)

For an arbitrary function $f(t)$, the four functions enumerated are illustrated in Fig. 8-7. The real translation theorem applies only to functions described by (*d*).

Complementary to the real translation theorem, the equation $\mathcal{L}\, e^{-at} f(t) = F(s + a)$, developed in Example 1(*f*), may be called the complex translation theorem. In this equation $F(s)$ is translated, and s is complex.

Theorem 5. Periodic Functions

If the Laplace transform of the first period of a periodic function is $F_1(s)$, then the Laplace transform $F(s)$ of the periodic function with period τ is

$$F(s) = \frac{F_1(s)}{1 - e^{-\tau s}} \tag{8-41}$$

Proof: Let the periodic function $f(t)$ with period τ be expressed as

$$f(t) = f_1(t) + f_2(t) + f_3(t) + \ldots$$

where $f_1(t), f_2(t) \ldots$ are the functions describing the first, the second, and the subsequent cycles. Since $f(t)$ is periodic, the second cycle can be obtained by the translation of the first cycle by a period τ. All subsequent cycles can be obtained by translating the first cycle by the appropriate number of periods. Hence $f(t)$ can be expressed as

$$f(t) = f_1(t) + f_1(t - \tau)u(t - \tau) + f_1(t - 2\tau)u(t - 2\tau) + \ldots$$

By the real translation theorem, the Laplace transform of $f(t)$ is

$$F(s) = (1 + e^{-\tau s} + e^{-2\tau s} + \ldots)\, F_1(s)$$

$$= \frac{1}{1 - e^{-\tau s}}\, F_1(s)$$

Example 11. Determine the Laplace transform of the sawtooth periodic function as shown in Fig. 8-8(*a*).

Solution: The first cycle of the periodic function, as shown in Fig. 8-8(*b*), is defined by

$$f_1(t) = \frac{H}{\tau} t \quad \text{for} \quad 0 < t < \tau$$

It may be considered to consist of three components and be expressed as

$$f_1(t) = f_a(t) - f_b(t) - f_c(t)$$

where $f_a(t), f_b(t)$, and $f_c(t)$ are illustrated in Figs. 8-8(*c*) and (*d*). Hence

$$f_1(t) = \frac{H}{\tau}\, t\, u(t) - \frac{H}{\tau}(t - \tau)u(t - \tau) - Hu(t - \tau)$$

$$F_1(s) = \frac{H}{\tau s^2}\,[1 - (1 + \tau s)e^{-\tau s}]$$

The Laplace transform of the sawtooth period function is

$$F(s) = \frac{1}{1 - e^{-\tau s}} F_1(s) = \frac{H}{\tau s^2(1 - e^{-\tau s})} [1 - (1 + \tau s)e^{-\tau s}]$$

Let us derive $f_1(t)$ by the use of the gate function. The first cycle of the periodic function is

$$f_1(t) = f(t)\, G_o(\tau) = \frac{H}{\tau} t\, [u(t) - u(t - \tau)]$$

$$= \frac{H}{\tau} [tu(t) - (t - \tau)u(t - \tau) - \tau u(t - \tau)]$$

It is evident that the two descriptions of $f_1(t)$ are identical.

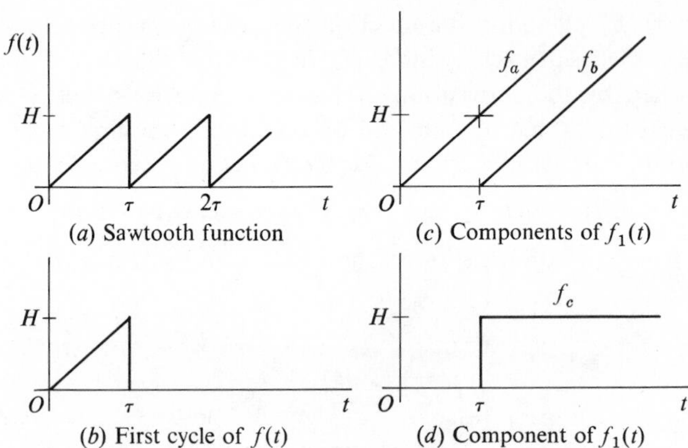

Fig. 8-8. *Components of a sawtooth function*

Theorem 6. Real Convolution

If $f_1(t)$ and $f_2(t)$ have, respectively, the Laplace transforms $F_1(s)$ and $F_2(s)$, then

$$\mathcal{L}\left[\int_0^t f_1(\tau)f_2(t - \tau)\, d\tau\right] = \mathcal{L}\left[\int_0^t f_1(t - \tau)f_2(\tau)\, d\tau\right] = F_1(s)\, F_2(s) \tag{8-42}$$

Proof: For convenience, let us write $f(t) = \int_0^t f_1(\tau)f_2(t - \tau)\, d\tau$ and $F(s) = \mathcal{L} f(t)$. From the definition of the Laplace transform, we have

$$F(s) = \int_0^\infty \left[\int_0^t f_1(\tau)f_2(t - \tau)\, d\tau\right] e^{-st}\, dt$$

To express $F(s)$ as the product of $F_1(s)$ and $F_2(s)$, we must extend the limits of both of the integrations from 0 to ∞. Let us multiply the integrand of the second integral by a unit step function $u(t-\tau)$,† which equals unity for $\tau < t$ and is zero for $\tau > t$. The value of the integral is unaffected by $u(t-\tau)$. The upper limit of the second integral can now be extended to ∞ after the insertion of $u(t-\tau)$ in the integrand.

$$F(s) = \int_0^\infty \left[\int_0^\infty f_1(\tau) f_2(t-\tau) u(t-\tau)\, d\tau \right] e^{-st}\, dt$$

Now let us interchange the order of the integrations with respect to t and τ.

$$F(s) = \int_0^\infty f_1(\tau) \left[\int_0^\infty f_2(t-\tau) u(t-\tau) e^{-st}\, dt \right] d\tau$$

From the real translation theorem the second integral equals $e^{-\tau s} F_2(s)$. Hence

$$\begin{aligned} F(s) &= \int_0^\infty f_1(\tau)[e^{-\tau s} F_2(s)]\, d\tau = F_2(s) \int_0^\infty f_1(\tau) e^{-s\tau}\, d\tau \\ &= F_2(s) F_1(s) = F_1(s) F_2(s) \end{aligned}$$

The convolution integral of two functions $f_1(t)$ and $f_2(t)$ is commonly written as

$$\int_0^t f_1(t-\tau) f_2(\tau)\, d\tau = \int_0^t f_1(\tau) f_2(t-\tau)\, d\tau = f_1(t) * f_2(t)$$

The convolution integral provides a method to evaluate the response of a system to any excitation function. By Eq. (8-12), the response transform $H(s)$ of a system to an excitation $F(s)$ is

$$H(s) = G(s)\, F(s) \qquad \textbf{(8-43)}$$

where $G(s)$ is the system function. Let us assume that the system is initially relaxed, that is, with initial conditions equal to zero. Now if a unit impulse $\delta(t)$ is applied to the system, the response transform is

$$H_\delta(s) = G(s)\, \mathscr{L}\, \delta(t) = G(s) \qquad \textbf{(8-44)}$$

and the time response of the system becomes

$$h_\delta(t) = \mathscr{L}^{-1}\, G(s) = g(t) \qquad \textbf{(8-45)}$$

Equation (8-44) states that the response of an initially relaxed system to

† Note that $u(t-\tau)$ and $u(\tau-t)$ are two different unit step functions. See Fig. 8-13.

a unit impulse is the inverse Laplace transform of the transfer function. $g(t)$ is called the *impulse response* of the system. The inverse transform of $H(s)$ from Eq. (8-43) is

$$h(t) = \mathscr{L}^{-1}[G(s)\,F(s)] = g(t)_* f(t)$$
$$= \int_0^t g(\tau) f(t-\tau)\,d\tau = \int_0^t g(t-\tau) f(\tau)\,d\tau \qquad \textbf{(8-46)}$$

Equation (8-46) indicates that the response of a linear system to an excitation function $f(t)$ is the convolution of its impulse response and the excitation function. This statement is called Borel's theorem. If $f(t)$ is an arbitrary excitation the convolution integral can be evaluated graphically or numerically.

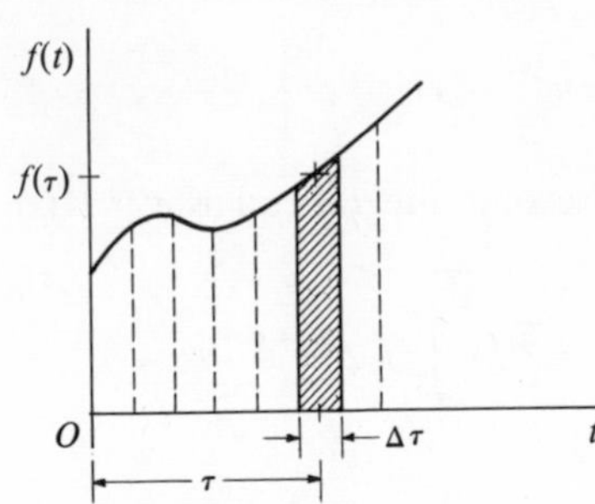

Fig. 8-9. *Approximation of $f(t)$ by a series of rectangular pulses*

The convolution integral can be formulated from physical reasoning, using the principle of superposition. In Fig. 8-9, consider an arbitrary excitation function $f(t)$ to be composed of a large number of rectangular pulses each of duration $\Delta\tau$. The strength of a typical pulse equals $f(\tau)\,\Delta\tau$. If $\Delta\tau$ is sufficiently small, the pulse may be considered as an impulse. The response of the system to the typical impulse $f(\tau)\,\Delta\tau\,\delta(t-\tau)$ is

$$f(\tau)\,\Delta\tau\,g(t-\tau)$$

Using the principle of superposition, the total response to the excitation $f(t)$ up to time t is

$$h(t) = \lim_{\Delta\tau\to 0} \sum_{t=0}^{t} f(\tau)\,\Delta\tau\,g(t-\tau) = \int_0^t g(t-\tau) f(\tau)\,d\tau$$

We have assumed initially relaxed conditions in our discussion of the impulse response. If the system is not initially relaxed, the response due to the externally applied excitation and that due to the initial conditions must be determined separately and the results superposed to find the resultant response.

The response of a system to an arbitrary excitation can also be examined by using the unit step input instead of the unit impulse. Let us define the *step response* $c(t)$ as the response of an initially relaxed linear system to a step input. $c(t)$ is commonly called the *indicial response.* By the principle of superposition, it can be shown that the response $h(t)$ to an excitation $f(t)$ can be expressed as

$$h(t) = f(0)c(t) + \int_o^t f'(\tau)c(t - \tau)\, d\tau, \quad \text{for } t \geq 0 \tag{8-47}$$

The proof of this equation is left as an exercise.

Theorem 7. Initial Value

If $f(t)$ and $f'(t)$ are Laplace transformable and the $\lim_{s\to\infty} sF(s)$ exists, then

$$\lim_{s\to\infty} s\,F(s) = \lim_{t\to 0+} f(t) \tag{8-48}$$

Proof: By the real differentiation theorem we have

$$\int_o^\infty f'(t)e^{-st}\, dt = s\,F(s) - f(0+)$$

Consider the integral on the left side of this equation. Since s is a parameter and independent of t, we can let s approach infinity before integrating. If s approaches infinity, the integral is zero. Hence

$$0 = \lim_{s\to\infty} [s\,F(s) - f(0+)]$$

or

$$\lim_{s\to\infty} s\,F(s) = f(0+)$$

Example 12. Consider the subsidiary equation [see Eq. (8-10)]

$$X(s) = \frac{1}{ms^2 + cs + k}[F(s) + (ms + c)x_o + m\dot{x}_o]$$

If $F(s)$ equals $\mathscr{L}\, F_o \sin \omega t$, determine the initial value of $x(t)$.

Solution: Substituting $F(s) = F_o\omega/(s^2 + \omega^2)$ and rearranging gives

$$X(s) = \frac{F_o\omega + (s^2 + \omega^2)[(ms + c)x_o + m\dot{x}_o]}{(s^2 + \omega^2)(ms^2 + cs + k)}$$

Applying Eq. (8-48), we have

$$\lim_{s\to\infty} s\,X(s) = \lim_{s\to\infty} s\,\frac{F_o\omega + (s^2 + \omega^2)[(ms + c)x_o + m\dot{x}_o]}{(s^2 + \omega^2)(ms^2 + cs + k)}$$

Dividing the numerator and the denominator by s^4 and then setting $s \to \infty$ gives

$$\lim_{s\to\infty} s\,X(s) = x_o = \lim_{t\to 0+} x(t)$$

It should be noted that the initial value theorem does not enable us to find the initial values of a problem. If the subsidiary equation is given, however, the theorem enables us to determine the initial conditions without having to evaluate the inverse transform.

Theorem 8. Final Value

If $f(t)$ and $f'(t)$ are Laplace transformable and the $\lim_{s\to 0} s\,F(s)$ exists,

$$\lim_{s\to 0} s\,F(s) = \lim_{t\to\infty} f(t) \tag{8-49}$$

Proof: By the real differentiation theorem we have

$$\int_0^\infty f'(t)e^{-st}\,dt = s\,F(s) - f(0+)$$

Consider the integral on the left side of this equation. Again, let s approach 0 before we integrate. Thus

$$\int_0^\infty f'(t)\,dt = \lim_{t\to\infty}\int_0^t f'(t)\,dt = \lim_{t\to\infty}[f(t) - f(0+)]$$

Equating the right side of the last two equations and taking limits gives

$$\lim_{t\to\infty}[f(t) - f(0+)] = \lim_{s\to 0}[s\,F(s) - f(0+)]$$

Since $f(0+)$ is independent of t or s, it can be canceled out from this equation. The resultant equation is Eq. (8-49).

Example 13. Consider the subsidiary equation [see Eq. (8-10)]

$$X(s) = \frac{1}{ms^2 + cs + k}[F(s) + (ms + c)x_o + m\dot{x}_o]$$

If $F(s)$ equals $\mathscr{L}\,F_o u(t)$, determine the final value of $x(t)$.

Solution: Substituting $F(s) = F_o/s$ and rearranging gives

$$X(s) = \frac{F_o + s[(ms + c)x_o + m\dot{x}_o]}{s(ms^2 + cs + k)}$$

Applying Eq. (8-49), we have

$$\lim_{s\to 0} s\,X(s) = \lim_{s\to 0}\frac{F_o + s[(ms + c)x_o + m\dot{x}_o]}{ms^2 + cs + k} = \frac{F_o}{k} = \lim_{t\to\infty} x(t)$$

It should be noted that the final value theorem is not applicable to periodic functions, because such functions do not reach a definite value as t approaches infinity.

Theorem 9. Complex Differentiation

If $f(t)$ has the Laplace transform $F(s)$, then

$$\mathscr{L}\,[t\,f(t)] = -\frac{d}{ds}\,F(s) \tag{8-50}$$

Proof: By definition,

$$\int_0^\infty f(t)\,e^{-st}\,dt = F(s)$$

Let us differentiate both sides of the equation with respect to s. Thus

$$\frac{d}{ds}\int_0^\infty f(t)\,e^{-st}\,dt = \int_0^\infty f(t)\,\frac{d}{ds}\,e^{-st}\,dt = -\int_0^\infty t\,f(t)\,e^{-st}\,dt = \frac{d}{ds}\,F(s)$$

This theorem states that the Laplace transform of the product of t and $f(t)$ equals the derivative of $F(s)$ with respect to s with a sign change. It may be used to extend the usefulness of the table of Laplace transform pairs.

Example 14. Determine the Laplace transform of (t sin ωt).

Solution: The Laplace transform of sin ωt is $\omega/(s^2 + \omega^2)$. Applying Eq. (8-50), the Laplace transform of (t sin ωt) is

$$\mathscr{L}\,(t \sin \omega t) = -\frac{d}{ds}\frac{\omega}{s^2 + \omega^2} = \frac{2\omega s}{(s^2 + \omega^2)^2}$$

Theorem 10. Complex Integration

If $f(t)$ has the Laplace transform $F(s)$, and if $\int_s^\infty F(s)\,ds$ exists, then

$$\mathscr{L}\left[\frac{f(t)}{t}\right] = \int_s^\infty F(s)\,ds \tag{8-51}$$

Proof: By definition,

$$\int_0^\infty f(t)\,e^{-st}\,dt = F(s)$$

Integrating both sides of this equation with respect to s between the limits of s and infinity gives

$$\int_s^\infty \int_0^\infty f(t)\,e^{-st}\,dt\,ds = \int_s^\infty F(s)\,ds$$

Changing the order of integration for the left side of this equation, we have

$$\int_s^\infty \int_0^\infty f(t)\, e^{-st}\, dt\, ds = \int_0^\infty f(t) \int_s^\infty e^{-st}\, ds\, dt = \int_0^\infty \frac{f(t)}{t}\, e^{-st}\, dt$$

The theorem is proved by equating the last integral to the right side of the previous equation.

Example 15. Determine the Laplace transform of $(\sin \omega t)/t$.

Solution: The Laplace transform of $\sin \omega t$ is $\omega/(s^2 + \omega^2)$. By the direct application of Eq. (8-51), the Laplace transform of $(\sin \omega t)/t$ is

$$\int_s^\infty \frac{\omega}{s^2 + \omega^2}\, ds = \tan^{-1} \frac{s}{\omega}\Big|_0^\infty = \frac{\pi}{2} - \tan^{-1} \frac{s}{\omega} = \tan^{-1} \frac{\omega}{s}$$

8-7. APPLICATIONS

The application of some of the properties of the Laplace transformation developed in the preceding section will be illustrated in this section.

Example 16. A spring-mass system with a movable support is shown in Fig. 8-10(*a*). (*a*) If the system is initially at rest and the support is given a step displacement $y_o u(t)$, determine the motion of the mass. (*b*) While the mass is in motion owing to the disturbance in part *a*, if the support is given a step displacement $-y_o u(t - t_o)$, determine the motion of the mass. (*c*) Determine the motion of the mass for $t > t_o$ if $t_o = 2\pi/\omega_n$ and $t_o = \pi/\omega_n$, where $\omega_n = \sqrt{k/m}$ is the natural frequency of the system.

Solution: From Newton's law of motion, the equation of motion is

$$m\ddot{x} = -k(x - y)$$

(*a*) Substituting $y(t) = y_o u(t)$ in this equation gives

$$m\ddot{x} + kx = ky_o u(t)$$

The subsidiary equation and the solution are

$$X(s) = y_o \frac{\omega_n^2}{s(s^2 + \omega_n^2)} = y_o \left(\frac{1}{s} - \frac{s}{s^2 + \omega_n^2}\right)$$

$$x(t) = y_o(1 - \cos \omega_n t)$$

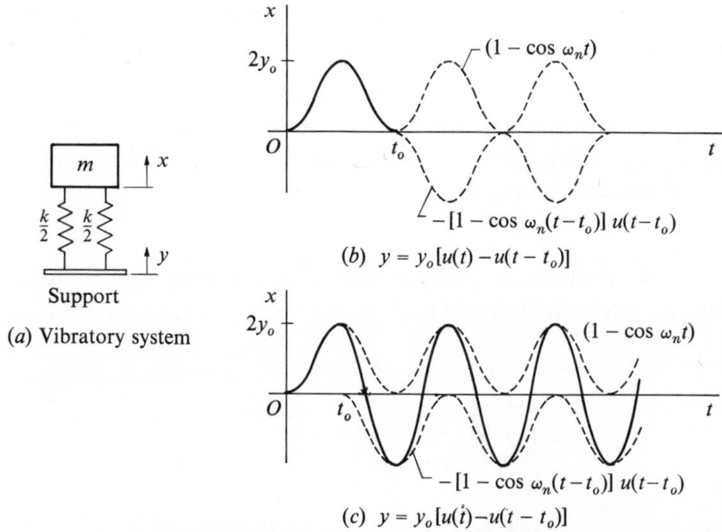

Fig. 8-10. *Response of system to step inputs*

(*b*) Substituting $y(t) = y_o[u(t) - u(t - t_o)]$ in the equation of motion gives

$$m\ddot{x} + kx = ky_o[u(t) - u(t - t_o)]$$

The subsidiary equation and the solution of this equation are

$$X(s) = y_o \frac{\omega_n^2(1 - e^{-t_o s})}{s(s^2 + \omega_n^2)} = y_o \left[\left(\frac{1}{s} - \frac{s}{s^2 + \omega_n^2}\right) - \left(\frac{1}{s} - \frac{s}{s^2 + \omega_n^2}\right)e^{-t_o s}\right]$$

$$x(t) = y_o\{(1 - \cos \omega_n t) - [1 - \cos \omega_n(t - t_o)]u(t - t_o)\}$$

(*c*) If $t_o = 2\pi/\omega_n$, the term $[1 - \cos \omega_n(t - t_o)]u(t - t_o)$ represents the quantity $(1 - \cos \omega_n t)$ translated by one period to the right of the origin. Thus, except for the translation, it is identical to the original function. For $t > t_o$, the algebraic sum of the right side of the last equation is zero and $x(t) = 0$. The physical interpretation is that, just before the support is displaced by an amount $-y_o$ at $t = t_o$, the mass is at its original static equilibrium position with $\dot{x}(t) = 0$. The spring force on the mass is ky_o. Now the support is displaced by an amount $-y_o$, leaving the spring unstressed. Thus the mass is back in its original static equilibrium position with zero velocity and the spring unstressed. Hence the mass becomes stationary.

If $t_o = \pi/\omega_n$, the function $[1 - \cos \omega_n(t - t_o)]u(t - t_o)$ represents

$(1 - \cos \omega_n t)$ translated by half a period to the right of the origin. For $t > t_o$, the motion of the mass is

$$x(t) = 2y_o \cos \omega_n(t - t_o)$$

Physically, this means that at $t = t_o+$ the mass has zero velocity and is displaced by an amount $2y_o$ from its original static equilibrium position.

The motions corresponding to the two values of t_o are illustrated in Fig. 8-10(*b*) and (*c*).

Example 17. A dynamic system consists of two masses and a coupling spring as shown in Fig. 8-11. If the system is initially at rest on a

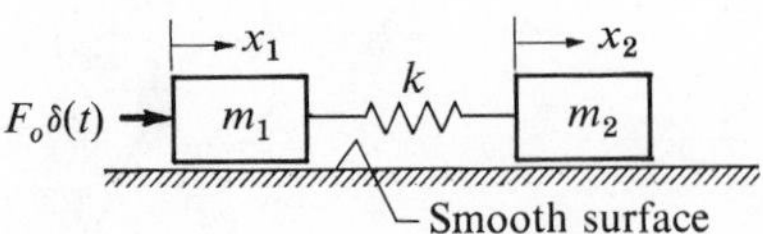

Fig. 8-11. *Impact applied to semi-definite system*

smooth horizontal surface and an impact $F_o\,\delta(t)$ is applied to the mass m_1, determine the motions of the masses.

Solution: From Newton's law of motion, the equations of motion are

$$m_1\ddot{x}_1 + kx_1 - kx_2 = F_o\,\delta(t)$$
$$-kx_1 + m_2\ddot{x}_2 + kx_2 = 0$$

Since the system is initially at rest on a horizontal surface, both the initial displacements and velocities are zero. Transforming these equations, we obtain

$$(m_1s^2 + k)\,X_1(s) - k\,X_2(s) = F_o$$
$$-k\,X_1(s) + (m_2s^2 + k)\,X_2(s) = 0$$

$X_1(s)$ and $X_2(s)$ can be solved from these equations by Cramer's rule.

$$X_1(s) = \frac{\begin{vmatrix} F_o & -k \\ 0 & m_2s^2 + k \end{vmatrix}}{\Delta(s)} = \frac{F_o(m_2s^2 + k)}{\Delta(s)}$$

$$X_2(s) = \frac{\begin{vmatrix} m_1s^2 + k & F_o \\ -k & 0 \end{vmatrix}}{\Delta(s)} = \frac{kF_o}{\Delta(s)}$$

where $$\Delta(s) = [m_1m_2s^2 + k(m_1 + m_2)]s^2$$

Defining $\omega^2 = k(m_1 + m_2)/m_1m_2$, substituting $\Delta(s)$ in the response transforms, and simplifying, we obtain

$$X_1(s) = \frac{F_o}{m_1 + m_2}\left(\frac{1}{s^2} + \frac{m_2}{\omega m_1}\frac{\omega}{s^2 + \omega^2}\right)$$

$$X_2(s) = \frac{F_o}{m_1 + m_2}\left(\frac{1}{s^2} - \frac{1}{\omega}\frac{\omega}{s^2 + \omega^2}\right)$$

Hence the time responses of the masses are

$$x_1(t) = \frac{F_o}{m_1 + m_2}\left(t + \frac{m_2}{\omega m_1}\sin \omega t\right)$$

$$x_2(t) = \frac{F_o}{m_1 + m_2}\left(t - \frac{1}{\omega}\sin \omega t\right)$$

The solutions indicate that (1) the masses will move together with the mean velocity of $F_o/(m_1 + m_2)$; (2) the masses will oscillate with the same frequency, which is the natural frequency of the system, but the motions are 180 deg out of phase; and (3) the amplitude ratio of the oscillations is inversely proportional to the masses. It should be noted that the transform of the second-order differential equation with an impulse applied is identical to the transform of the same equation with an equivalent initial velocity. Thus the same solutions can be obtained if an equivalent initial velocity is used.

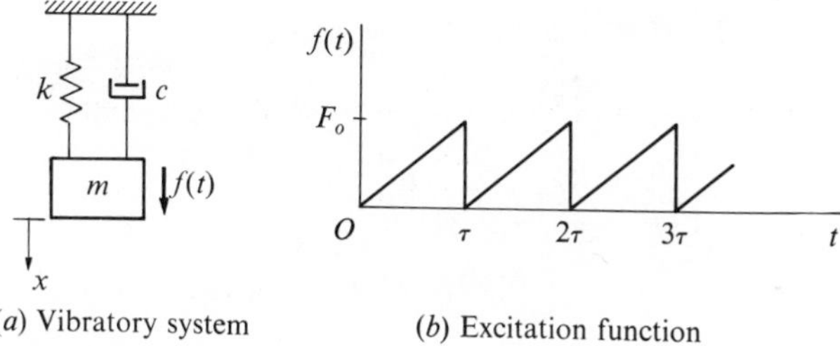

(*a*) Vibratory system (*b*) Excitation function

Fig. 8-12. *Sawtooth excitation applied to mechanical system*

Example 18. A sawtooth excitation force is applied to a spring-mass system as shown in Fig. 8-12. If the system is initially relaxed, determine the transient and the steady-state motions of the mass.

Solution: From Newton's law of motion, the equation of motion is

$$m\ddot{x} + c\dot{x} + kx = f(t)$$

Transforming this equation, we obtain

$$(ms^2 + cs + k)\,X(s) = \frac{F_o}{\tau}\left[\frac{1}{s^2} - \frac{\tau e^{-\tau s}}{s(1 - e^{-\tau s})}\right] \tag{8-52}$$

The right side of this equation is the Laplace transform of the sawtooth function. (See Example 11.) Dividing through by m, defining $c/m = 2\zeta\omega_n$ and $k/m = \omega_n^2$, and assuming $\zeta < 1$, we obtain the subsidiary equation

$$X(s) = \frac{F_o}{m\tau}\left[\frac{1}{s^2(s^2 + 2\zeta\omega_n s + \omega_n^2)} - \frac{\tau e^{-\tau s}}{s(s^2 + 2\zeta\omega_n s + \omega_n^2)(1 - e^{-\tau s})}\right] \tag{8-53}$$

It can be shown that the right side of Eq. (8-53) can be expressed in partial fractions as

$$\begin{aligned}\frac{1}{s^2(s^2 + 2\zeta\omega_n s + \omega_n^2)} &= \frac{1}{\omega_n^4}\left[\frac{\omega_n^2}{s^2} - \frac{2\zeta\omega_n}{s} + \frac{2\zeta\omega_n s + (4\zeta^2 - 1)\omega_n^2}{s^2 + 2\zeta\omega_n s + \omega_n^2}\right]\\ &= \frac{1}{\omega_n^4}\left[\frac{\omega_n^2}{s^2} - \frac{2\zeta\omega_n}{s} + \frac{2\zeta\omega_n(s + \zeta\omega_n) + (2\zeta^2 - 1)\omega_n^2}{s^2 + 2\zeta\omega_n s + \omega_n^2}\right]\\ &= \frac{1}{\omega_n^4}\left[\frac{\omega_n^2}{s^2} - \frac{2\zeta\omega_n}{s} + \frac{2\zeta\omega_n(s + \zeta\omega_n) + (2\zeta^2 - 1)\omega_n^2\,\omega_d/\omega_d}{(s + \zeta\omega_n)^2 + \omega_d^2}\right]\end{aligned}$$

where $\omega_d = \sqrt{1 - \zeta^2}\,\omega_n$, and

$$\begin{aligned}\frac{\tau e^{-\tau s}}{s(s^2 + 2\zeta\omega_n s + \omega_n^2)(1 - e^{-\tau s})} &= \frac{\tau}{\omega_n^2}\left[\frac{1}{s} - \frac{s + 2\zeta\omega_n}{s^2 + 2\zeta\omega_n s + \omega_n^2}\right]\frac{e^{-\tau s}}{1 - e^{-\tau s}}\\ &= \frac{\tau}{\omega_n^2}\left[\frac{1}{s} - \frac{(s + \zeta\omega_n) + \zeta\omega_n\omega_d/\omega_d}{(s + \zeta\omega_n)^2 + \omega_d^2}\right]\frac{e^{-\tau s}}{1 - e^{-\tau s}}\end{aligned}$$

Substituting these partial fractions in Eq. (8-53) and simplifying, the subsidiary equation becomes

$$\begin{aligned}X(s) = {}&\frac{F_o}{k\tau}\left[\frac{2\zeta}{\omega_n}\frac{s + \zeta\omega_n}{(s + \zeta\omega_n)^2 + \omega_d^2} + \frac{2\zeta^2 - 1}{\omega_d}\frac{\omega_d}{(s + \zeta\omega_n)^2 + \omega_d^2}\right]\\ &+ \frac{F_o}{k\tau}\left\{\frac{1}{s^2} - \frac{2\zeta}{\omega_n}\frac{1}{s} - \left[\frac{1}{s} - \frac{s + \zeta\omega_n}{(s + \zeta\omega_n)^2 + \omega_d^2}\right.\right.\\ &\left.\left.- \frac{\zeta\omega_n}{\omega_d}\frac{\omega_d}{(s + \zeta\omega_n)^2 + \omega_d^2}\right]\frac{\tau e^{-\tau s}}{1 - e^{-\tau s}}\right\}\end{aligned} \tag{8-54}$$

or

$$X(s) = X_t(s) + X_s(s)$$

where the subscripts t and s denote the transient and the steady-state responses, respectively.

From Eq. (8-54) it is easily recognized that the transient response is

$$x_t(t) = \frac{F_o}{k\tau}\left[\frac{2\zeta}{\omega_n}\cos\omega_d t + \frac{2\zeta^2 - 1}{\omega_d}\sin\omega_d t\right]e^{-\zeta\omega_n t} \qquad \textbf{(8-55)}$$

Let us rewrite $X_s(s)$ in a more convenient form before finding its inverse transform. Noting that

$$\frac{e^{-\tau s}}{1 - e^{-\tau s}} = e^{-\tau s} + e^{-2\tau s} + e^{-3\tau s} + \ldots$$

the steady-state term in Eq. (8-54) can be written as

$$\begin{aligned} X_s(s) = \frac{F_o}{k\tau}\Bigg[&\frac{1}{s^2} - \frac{2\zeta}{\omega_n}\frac{1}{s} - \frac{\tau}{s}(e^{-\tau s} + e^{-2\tau s} + e^{-3\tau s} + \ldots) \\ &+ \frac{\tau(s + \zeta\omega_n)}{(s + \zeta\omega_n)^2 + \omega_d^2}(e^{-\tau s} + e^{-2\tau s} + e^{-3\tau s} + \ldots) \\ &+ \frac{\tau\zeta\omega_n}{\omega_d}\frac{\omega_d}{(s + \zeta\omega_n)^2 + \omega_d^2}(e^{-\tau s} + e^{-2\tau s} + e^{-3\tau s} + \ldots)\Bigg] \end{aligned}$$

The inverse transform of $X_s(s)$ is

$$\begin{aligned} x_s(t) = \frac{F_o}{k\tau}\Bigg\{&t - \frac{2\zeta}{\omega_n} - \tau[u(t - \tau) + u(t - 2\tau) + u(t - 3\tau) + \ldots] \\ &+ \tau e^{-\zeta\omega_n(t-\tau)}\left[\cos\omega_d(t - \tau) + \frac{\zeta\omega_n}{\omega_d}\sin\omega_d(t - \tau)\right]u(t - \tau) \\ &+ \tau e^{-\zeta\omega_n(t-2\tau)}\left[\cos\omega_d(t - 2\tau) + \frac{\zeta\omega_n}{\omega_d}\sin\omega_d(t - 2\tau)\right]u(t - 2\tau) \\ &+ \ldots\Bigg\} \end{aligned} \qquad \textbf{(8-56)}$$

Equation (8-56) indicates that the response $x_s(t)$ for $0 < t < \tau$ is $(F_o/k\tau)(t - 2\zeta/\omega_n)$; the response for $\tau < t < 2\tau$ is $(F_o/k\tau)\{t - \tau - 2\zeta/\omega_n + \tau e^{-\zeta\omega_n(t-\tau)}[\cos\omega_d(t - \tau) + (\zeta\omega_n/\omega_d)\sin\omega_d(t - \tau)]\}$, etc. The response for $(n - 1)\tau < t < n\tau$ is

$$\begin{aligned} x_s(t) = \frac{F_o}{k\tau}\Bigg\{&t - (n - 1)\tau - \frac{2\zeta}{\omega_n} + \sum_{p=1}^{n-1}\tau e^{-\zeta\omega_n(t-p\tau)} \\ &\times\left[\cos\omega_d(t - p\tau) + \frac{\zeta\omega_n}{\omega_d}\sin\omega_d(t - p\tau)\right]u(t - p\tau)\Bigg\} \end{aligned} \qquad \textbf{(8-57)}$$

As t becomes large, the number n is large, therefore $0 < [t - (n-1)\tau] < \tau$. Depending on the rate of the exponential decay, the quantity under the summation sign may not be greatly affected by the terms associated with $t < (n-1)\tau$. In fact, if $e^{-\zeta\omega_n\tau} = e^{-3} = 1/20$, the exponential decay term would have decreased to 1/20th of its initial value before the subsequent sawtooth excitation is applied to the system. Hence, if $\zeta\omega_n\tau > 3$, from Eq. (8-57), $x_s(t)$ for a typical period can be approximated as

$$x_s(t) = \frac{F_o}{k\tau}\left\{t - \frac{2\zeta}{\omega_n} + \tau e^{-\zeta\omega_n t}\left[\cos\omega_d t + \frac{\zeta\omega_n}{\omega_d}\sin\omega_d t\right]\right\} \quad \textbf{(8-58)}$$

where $0 < t < \tau$. In Eq. (8-58) the number n is discarded, because retaining n in the equation does not reveal additional information. If $\zeta\omega_n\tau < 3$, an additional exponential term will have to be used in this equation.

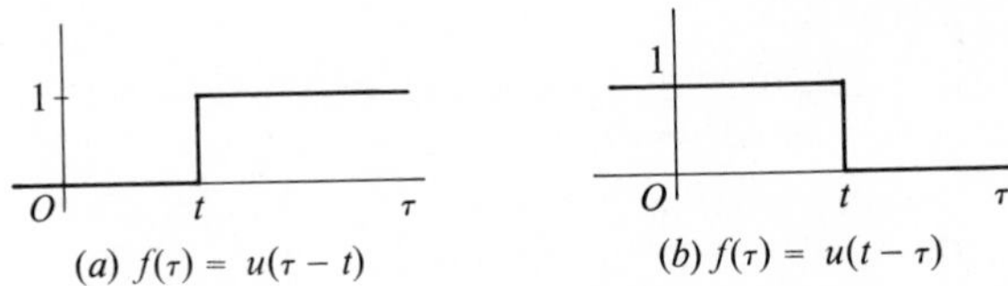

Fig. 8-13. *Two different unit step functions*

Example 19. The equation of motion of a system with a unit step excitation is

$$m\ddot{x} + c\dot{x} + kx = F_o u(t)$$

If the system is initially relaxed, determine the response $x(t)$ by the method of convolution.

Solution: The subsidiary equation of the given differential equation is

$$X(s) = \frac{1}{m[(s+\zeta\omega_n)^2 + \omega_d^2]}\frac{F_o}{s} = G(s)\,F(s)$$

From Eq. (8-45) the impulse response of the given system is

$$x_\delta(t) = \mathscr{L}^{-1}\,G(s) = \mathscr{L}^{-1}\frac{1}{m[(s+\zeta\omega_n)^2 + \omega_d^2]} = \frac{e^{-\zeta\omega_n t}}{m\omega_d}\sin\omega_d t$$

From Eq. (8-46) the response of the system to $f(t)$ is

$$x(t) = \int_o^t g(\tau)f(t-\tau)\,d\tau = \int_o^t\left(\frac{e^{-\zeta\omega_n\tau}}{m\omega_d}\sin\omega_d\tau\right)F_o u(t-\tau)\,d\tau$$

The functions $u(\tau - t)$ and $u(t - \tau)$, where t is a parameter, are as illustrated in Figs. 8-13(a) and (b), respectively. The function $u(t - \tau)$ may be interpreted as $u[-(\tau - t)]$, and it is the mirror image of $u(\tau - t)$ about $t = \tau$. Since $u(t - \tau) = 1$ for $\tau < t$, the response $x(t)$ can be written as

$$
\begin{aligned}
x(t) &= \frac{F_o}{m\omega_d}\int_o^t e^{-\zeta\omega_n\tau}\sin\omega_d\tau\,d\tau \\
&= \frac{F_o}{m\omega_d}\left[\frac{e^{-\zeta\omega_n\tau}}{\omega_n^2}(-\zeta\omega_n\sin\omega_d\tau - \omega_d\cos\omega_d\tau)\right]_o^t \\
&= \frac{F_o}{k}\left[1 - e^{-\zeta\omega_n t}\left(\frac{\zeta}{\sqrt{1-\zeta^2}}\sin\omega_d t + \cos\omega_d t\right)\right] \\
&= \frac{F_o}{k}\left[1 - \frac{e^{-\zeta\omega_n t}}{\sqrt{1-\zeta^2}}\sin(\omega_d t + \phi)\right]
\end{aligned}
$$

where $\phi = \tan^{-1}\sqrt{1-\zeta^2}/\zeta$.

We have shown the evaluation of the convolution of two analytical functions in this example. When the functions are arbitrary, the convolution can be obtained graphically. We shall illustrate this graphical method with a simple example.

Example 20. Evaluate $f_1(t)_* f_2(t) = t_* e^{-at}$ graphically.

Solution: By definition,

$$t_* e^{-at} = \int_o^t [(t - \tau)u(t - \tau)]e^{-a\tau}\,d\tau$$

Let t_1 be a particular value of t. The convolution integral can be written as

$$\int_o^{t_1} (t_1 - \tau)u(t_1 - \tau)e^{-a\tau}\,d\tau$$

Starting from the function $\tau u(\tau)$, as shown in Fig. 8-14(a), the function $(\tau - t_1)u(\tau - t_1)$ is the translation of $\tau u(\tau)$ as shown in Fig. 8-14(b). Following the explanation in the preceding example, the function $(t_1 - \tau)u(t_1 - \tau)$ is the mirror image of $(\tau - t_1)u(\tau - t_1)$ about t_1 as shown in Fig. 8-14(c). The function $e^{-a\tau}$ is shown in Fig. 8-14(d). To obtain the convolution integral, the functions in Fig. 8-14(c) and (d) are multiplied together and then integrated for $0 < \tau < t_1$. The product of the two functions is shown in Fig. 8-14(e), and the value of the integral is indicated in Fig. 8-14(f). For example, if $a = 0.5$ and $t_1 = 1$, the value of the integral is 0.42. This value is obtained by

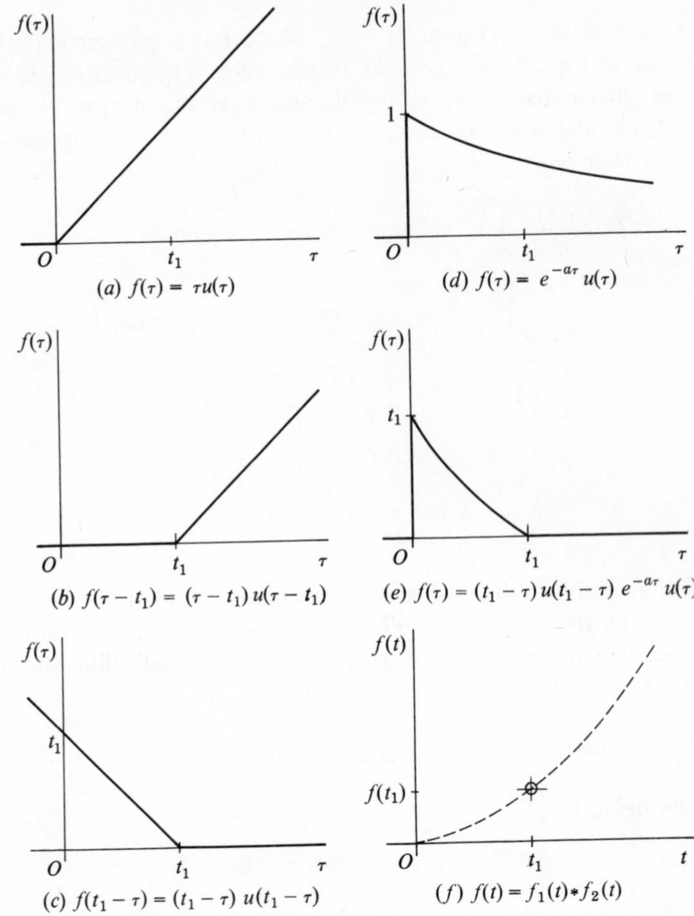

Fig. 8-14. *Graphical evaluation of convolution integral*

summing the area under the curve in Fig. 8-14(*e*). Similarly, the value of the convolution integral can be evaluated for other chosen values of t. Repeating this process, the function $f(t) = f_1(t) * f_2(t)$ is obtained as shown in the dashed line in Fig. 8-14(*f*).

8-8. ELECTRICAL NETWORKS

Elementary electrical networks are discussed in this section to provide a basis for the discussion of electromechanical analogy and electromechanical systems in the sections to follow. We shall first discuss the network elements and then use Kirchhoff's laws to derive the network equations of lumped-parameter systems.

Electrical networks are composed of active and passive elements which are symbolically represented as shown in Table 8-2. The active elements are idealized as voltage sources and current sources. The passive elements are idealized as resistors, capacitors, and inductors. We shall refer to Table 8-2 in our discussion of the circuit elements.

The active elements are the sources of electrical energy. A *voltage source* is an element that maintains the voltage $v(t)$ across its terminals independent of the current flowing in the element. The $(\pm)$ sign

TABLE 8-2

ELECTRICAL NETWORK ELEMENTS AND VOLTAGE CURRENT RELATIONS

ELEMENT	SYMBOL	VOLTAGE	CURRENT
Voltage source	− v (t) +	$v(t)$ independent of current flow	Current depending on network
Current source	i (t)	Voltage depending on network	$i(t)$ independent of voltage across terminals
Resistance	+ i R − v	$v = iR$	$i = v/R$
Capacitance	+ i C − v	$v = \frac{1}{C}\int_o^t i\,dt + v(0)$	$i = C\frac{dv}{dt}$
Inductance	+ i L − v	$v = L\frac{di}{dt}$	$i = \frac{1}{L}\int_o^t v\,dt + i(0)$

indicates the polarity of the source at the instant the voltage $v(t)$ has a positive value. A *current source* is an element that maintains the current $i(t)$ independent of the voltage across its terminals. The arrow indicates the direction of the current flow at the instant $i(t)$ is positive.

The passive elements do not generate electrical energy. The *resistor* is an energy-dissipating element. The voltage drop across a resistor in the direction of a positive current flow is†

$$v = iR \quad \text{or} \quad i = v/R \tag{8-59}$$

The energy dissipated in a resistor is

$$W = \int vi\,dt = \int i^2R\,dt$$

Hence the rate of energy dissipation in a resistor is

$$\frac{dW}{dt} = i^2R \tag{8-60}$$

The *capacitor* is the passive element in which electric energy is stored. The voltage drop across a capacitor in the direction of positive current flow is

$$v = \frac{1}{C}\int i\,dt = \frac{1}{C}\left[\int_o^t i\,dt + q(0)\right] = \frac{1}{C}\int_o^t i\,dt + v(0) \tag{8-61}$$

or

$$i = C\frac{dv}{dt} \tag{8-62}$$

where $q(0)$ is the initial accumulation of electric charge on the capacitor and $v(0)$ is the corresponding initial voltage across the capacitor. (It may be recalled that capacitance C is defined as $C = q/v$ and that current i is defined as $i = dq/dt$.) The polarity of $v(0)$ is the same as that of v; that is, if $v(0)$ is positive, it is a voltage drop in the direction of the positive current flow. The energy stored in a capacitor is

$$W = \int vi\,dt = \int \frac{q}{C}\frac{dq}{dt}\,dt = \frac{1}{C}\int q\,dq = \frac{q^2}{2C} = \frac{C}{2}v^2 \tag{8-63}$$

The *inductor* is the passive element in which electromagnetic energy is stored. From Faraday's law, the voltage drop across an inductor in the direction of a positive current flow is

$$v = L\frac{di}{dt} \tag{8-64}$$

† For convenience, the symbols v, i, etc., are used to represent the instantaneous values in the equations. The voltage source and the current source are denoted as $v(t)$ and $i(t)$, respectively.

or

$$i = \frac{1}{L}\int v\,dt = \frac{1}{L}\int_0^t v\,dt + i(0) \tag{8-65}$$

where L is the self-inductance of the element and $i(0)$ is the initial current flowing across the element. The current $i(0)$ is considered positive if it is in the direction of the assumed current flow. The energy stored in an inductor is

$$W = \int vi\,dt = \int L\frac{di}{dt}\,i\,dt = \frac{L}{2}\,i^2 \tag{8-66}$$

Since the charge on a capacitor and consequently the voltage across it cannot be changed instantaneously, the initial conditions of a capacitor can be evaluated as $q(0+) = q(0-)$ and $v(0+) = v(0-)$. Similarly, the magnetic energy in an inductor cannot be changed instantaneously, and we deduce the initial conditions $i(0+) = i(0-)$.

In the foregoing discussions, if the unit of the independent variable t is time in seconds and the units of v, i, and q are in volts, amperes, and coulombs, respectively, the units of R, C, and L are ohms, farads, and henrys. We shall assume that the values of R, C, and L are constants for the elements considered.

When two coils, or inductors, are adjacent to one another, the magnetic flux of one coil may link with the other. The coils are said to be magnetically coupled together. From Faraday's law, the induced voltage in a coil is proportional to the time rate of change of magnetic flux linkage. Hence the change of magnetic flux due to the change of current in one coil will induce a voltage in the other. The mutual inductance M between two inductors is symbolically represented in Fig. 8-15. If i_1 in coil 1 is varying, the open-circuit voltage across the terminals of coil 2 is

$$v_2 = M\frac{di_1}{dt} \tag{8-67}$$

The dots on the terminals indicate the sense of the winding of the coils. Hence the sign of M can be positive or negative depending on the winding sense. The dots are placed on the terminals of the coils whose potentials due to the mutual inductance rise and fall together. Thus if i_1 enters the terminal of coil 1 marked with a dot, as shown in Fig. 8-15, the dot indicates the high potential side of coil 2. We shall illustrate this sign convention in Example 21. Similarly, if i_2 in coil 2 is varying, the open-circuit voltage in coil 1 is

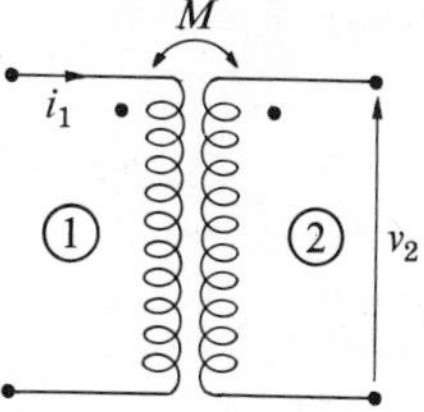

Fig. 8-15. *Symbol representing mutual inductance*

$$v_1 = M\frac{di_2}{dt}$$

Although the symbol M is used for mutual inductance and it may appear as a parameter in network equations, mutual inductance is not considered a network element.

KIRCHHOFF'S LAWS

When the circuit elements described in the previous paragraphs are connected together to form an electrical network, the network equations can be formulated by using Kirchhoff's voltage law or Kirchhoff's current law. The former gives the so-called loop equations and the latter the node equations.

Kirchhoff's voltage law may be stated as follows: Around a closed loop (path, or mesh) in an electrical network, the algebraic sum of the instantaneous voltage drops is zero; that is,

$$\sum_k v_k(t) \text{ around a closed loop} = 0 \tag{8-68}$$

Consider the one-loop network shown in Fig. 8-16 in which the switch S is closed at $t = 0$. The flow of current is assumed positive in the clockwise direction. Summing the voltage drops across the elements at $t = 0+$ gives

$$v_L + v_R + v_C - v(t) = 0$$

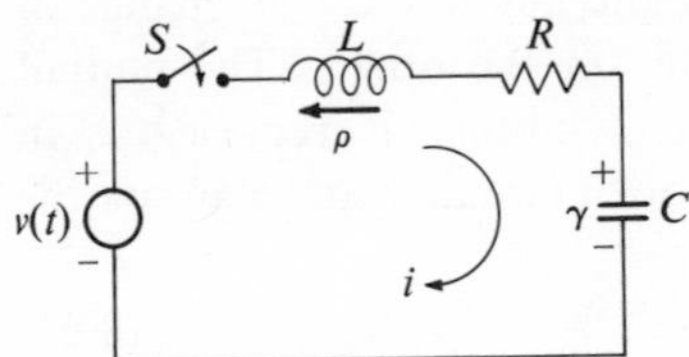

Fig. 8-16. *One-loop network*

Substituting the voltage drops for the respective elements gives

$$L\frac{di}{dt} + Ri + \frac{1}{C}\int i\,dt = v(t) \tag{8-69}$$

Let the initial current ρ in the inductance be in the direction indicated in the figure and the initial voltage γ across the capacitor have the polarity shown. The Laplace transform of Eq. (8-69) is

$$L[s\,I(s) - i(0)] + R\,I(s) + \frac{1}{C}\left[\frac{I(s)}{s} + \frac{i^{(-1)}(0)}{s}\right] = V(s)$$

Since $i(0) = -\rho$ and $i^{(-1)}(0)/C = q(0)/C = \gamma$, the subsidiary equation is

$$I(s) = \frac{V(s) - L\rho - \gamma/s}{Ls + R + 1/Cs} \tag{8-70}$$

For convenience, Eq. (8-69) can be written as

$$L\frac{di}{dt} + Ri + \frac{1}{C}\int_o^t i\,dt = v(t) - \gamma \tag{8-71}$$

It can be shown that the subsidiary equations to Eqs. (8-69) and (8-71) are identical. When the capacitor is common to two loops, Eq. (8-71) is more convenient to use.

Example 21. Determine the loop equations for the two-loop network shown in Fig. 8-17.

Solution: Let the currents be positive in the clockwise direction. Applying Kirchhoff's voltage law to loops 1 and 2 and rearranging, we obtain

$$(L_1 + L_3)\frac{di_1}{dt} + R_1 i_1 + \left(\frac{1}{C_1} + \frac{1}{C_2}\right)\int_o^t i_1\,dt - (L_3 + M)\frac{di_2}{dt} - \frac{1}{C_1}\int_o^t i_2\,dt = v(t) - \gamma$$

$$-(L_3 + M)\frac{di_1}{dt} - \frac{1}{C_1}\int_o^t i_1\,dt + (L_2 + L_3)\frac{di_2}{dt} + R_2 i_2 + \frac{1}{C_1}\int_o^t i_2\,dt = \gamma$$

It is noted in loop 1 that the voltage drop across C_1 is $\frac{1}{C_1}\int (i_1 - i_2)\,dt$ $= \frac{1}{C_1}\int_o^t (i_1 - i_2)\,dt + \gamma = \frac{1}{C_1}\int_o^t i_1\,dt - \frac{1}{C_1}\int_o^t i_2\,dt + \gamma.$ Using the definite integral in the equation avoids bringing in the initial conditions $i_1^{(-1)}(0)$ and $i_2^{(-1)}(0)$. The value of M is negative in this example. The sign convention is that if both currents enter or both currents leave the terminal marked with the dot, the sign of M is positive; otherwise, it is negative. In this illustration, i_1 is leaving the dotted terminal, and i_2 is entering. Thus the sign of M is negative.

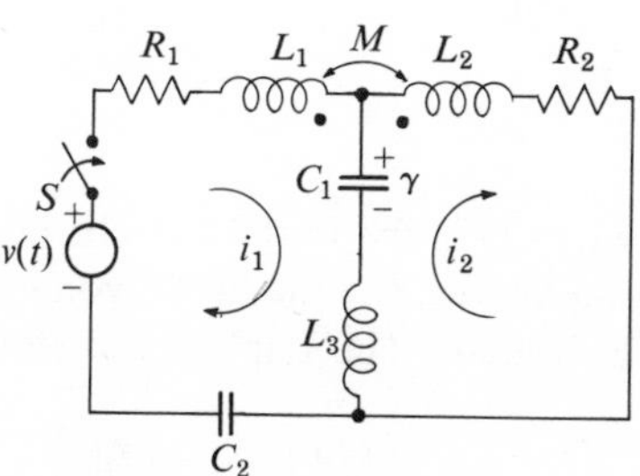

Fig. 8-17. *A two-loop network*

As shown in the last two examples, the loop currents are the dependent variables in the loop equations. If these currents are known, the voltage drop across every circuit element can be determined accordingly. The steps involved in the loop analysis may be summarized as follows:

1. Determine the number of independent loops in the network. For simple networks, this can usually be done by inspection.†

2. Assign clockwise directions to the loop currents for the selected loops.

3. Use Kirchhoff's voltage law to write the loop equations.

4. Solve the loop equations simultaneously for the loop currents.

Kirchhoff's current law may be stated as follows: At a common junction, or node, in an electrical network, the algebraic sum of the instantaneous currents flowing into a node is zero; that is,

$$\sum_k i_k(t) \text{ at a common node} = 0 \tag{8-72}$$

Consider the one-node-pair network of Fig. 8-18 in which the switch S is closed at $t = 0$. Let us mark the reference node negative (−) and the other node positive (+); that is, the voltage v is assumed positive with respect to the reference node. Summing the currents flowing into the (+) node gives

$$i(t) - i_C - i_R - i_L = 0$$

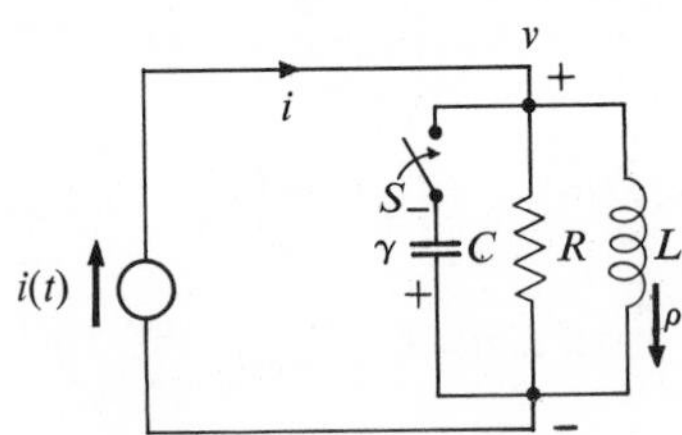

Fig. 8-18. *A one-node-pair network*

Substituting the expressions for the individual currents gives

$$C\frac{dv}{dt} + \frac{1}{R}v + \frac{1}{L}\int v\,dt = i(t) \tag{8-73}$$

or

$$C\frac{dv}{dt} + \frac{1}{R}v + \frac{1}{L}\int_0^t v\,dt = i(t) - \rho \tag{8-74}$$

where ρ is the initial current in the inductor in the direction shown in the figure. Let the initial voltage across the capacitor have the polarity indicated. The Laplace transform of Eq. (8-74) is

$$C[s\,V(s) - v(0)] + \frac{1}{R}V(s) + \frac{1}{Ls}V(s) = I(s) - \frac{\rho}{s}$$

† The number of independent loops in a network can be determined by the theory of linear graphs. If N, Ne, Nn, and Ns are the number of independent loops, the number of elements, the number of nodes (junctions of the circuit elements), and the number of separate parts, respectively, then the number of independent loops in a network is

$$N = Ne - Nn + Ns$$

For example, in the one-loop network, $Ne = 4$, $Nn = 4$, and $Ns = 1$. Thus there is only one independent loop. In the two-loop network, $Ne = 8$, $Nn = 7$, and $Ns = 1$, and there are two independent loops in the network.

Since $v(0) = -\gamma$, the subsidiary equation is

$$V(s) = \frac{I(s) - \gamma C - \rho/s}{Cs + 1/R + 1/Ls} \tag{8-75}$$

Example 22. Write the node equations for the two-node-pair network shown in Fig. 8-19.

Solution: Let us assign the reference node $-$ and the other nodes $+v_1$ and $+v_2$. Applying Kirchhoff's current law and rearranging, we obtain

$$C_1 \frac{dv_1}{dt} + \frac{1}{L_1} \int_o^t (v_1 - v_2)\, dt = i(t)$$

$$\frac{1}{R} v_2 + \frac{1}{L_1} \int_o^t (v_2 - v_1)\, dt + \frac{1}{L_2} \int_o^t v_2\, dt = 0$$

It is noted in the last two examples that the node voltages are the dependent variables in the node equations. If the node voltages are known, the current through every circuit element can be determined accordingly. The steps involved in the node analysis may be summarized as follows:

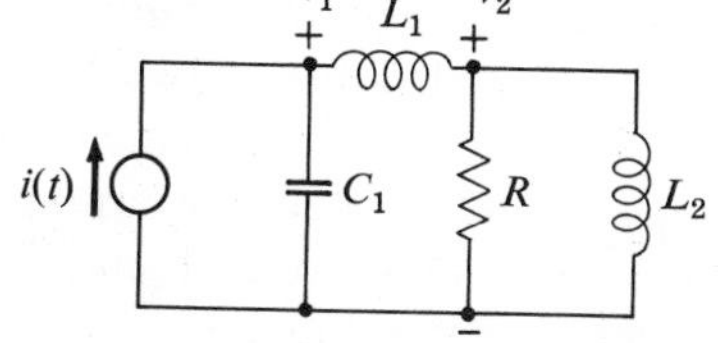

Fig. 8-19. *A two-node-pair network*

1. Determine the number of independent node pairs in the network. For simple networks, this can usually be done by inspection.†
2. Assign the reference node $-$ and all the other nodes $+$.
3. Use Kirchhoff's current law to write the node equations.
4. Solve the node equations simultaneously for the node voltages.

DUAL NETWORKS

Two electrical networks that are governed by sets of equations of the same form are called dual networks. Consider the one-loop network

† The number of independent node pairs in a network can be determined by the theory of linear graphs. If N, Nn, and Ns are the number of independent node pairs, the number of nodes (junctions), and the number of separate parts, respectively, then the number of independent node pairs in a network is

$$N = Nn - Ns$$

In the one-node-pair network, $Nn = 2$ and $Ns = 1$; therefore, $N = 1$. In the two-node-pair network, $Nn = 3$ and $Ns = 1$; therefore, $N = 2$.

shown in Fig. 8-16. The loop equation, Eq. (8-69), is

$$L\frac{di}{dt} + Ri + \frac{1}{C}\int i\,dt = v(t) \tag{8-76}$$

For the one-node-pair network of Fig. 8-18, the node equation, Eq. (8-73), is

$$C\frac{dv}{dt} + \frac{1}{R}v + \frac{1}{L}\int v\,dt = i(t) \tag{8-77}$$

It is evident that these two equations are of the same form, and that the networks shown in Figs. 8-16 and 8-18 are dual networks. Comparing these equations, we note that the loop current in Eq. (8-76) is the dual of the node-pair voltage drop in Eq. (8-77). Other quantities can be compared in like manner. The conversions for the dual quantities are tabulated in Table 8-3.

If a network is planar,† its dual network can be constructed graphically. Consider the one-loop network, as shown in Fig. 8-20(*a*), of which the one-node-pair network of Fig. 8-20(*b*) is its dual. Let us place dot 1 inside the loop and dot 2 outside it. These dots are to be the nodes of the dual network. Connect the two dots with dashed lines crossing every circuit element of the loop network. If a dashed line crosses a voltage source, we substitute a current source in the node network joining nodes 1 and 2. Other circuit elements are similarly substituted with the use of Table 8-3. The resultant circuit is as shown in Fig. 8-20(*b*).

TABLE 8-3

CONVERSIONS FOR DUAL ELECTRICAL NETWORKS

LOOP ANALYSIS	NODE ANALYSIS
Kirchhoff's voltage law	Kirchhoff's current law
Loop current	Node-pair voltage drop
Voltage source	Current source
Inductance, L	Capacitance, C
Resistance, R	Conductance, $1/R$
Capacitance, C	Inductance, L

† Networks which can be laid flat on a plane without branches crossing one another are called planar networks. Only planar networks have duals.

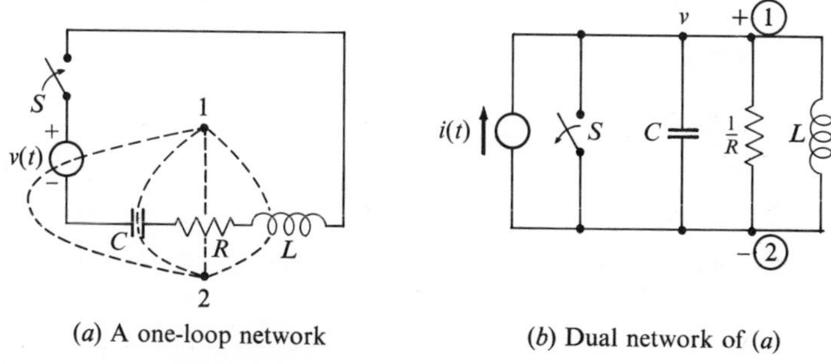

(a) A one-loop network

(b) Dual network of (a)

Fig. 8-20. *Construction of dual network*

Example 23. Construct the dual network for the two-loop network shown in Fig. 8-21(*a*).

Solution: Let us place dots 1 and 2 inside loops 1 and 2, respectively, and dot 3 outside the loops. These dots are to be the nodes in the dual network. Connect dots 1, 2, and 3 with dashed lines crossing every circuit element. Using node 3 as the reference node and the conversions in Table 8-3 to obtain the dual-circuit elements, the resultant circuit is as shown in Fig. 8-21(*b*). It can be shown that the network equations of these dual networks are of the same form.

Example 24. Construct the dual network for the two-loop network shown in Fig. 8-22(*a*).

Solution: Previously it was mentioned that mutual inductance is not considered a circuit element, and no provision is made for its conversion in dual networks in Table 8-3. The given circuit, however, can be modified as shown in Fig. 8-22(*b*); an inductance L_M corresponding to the

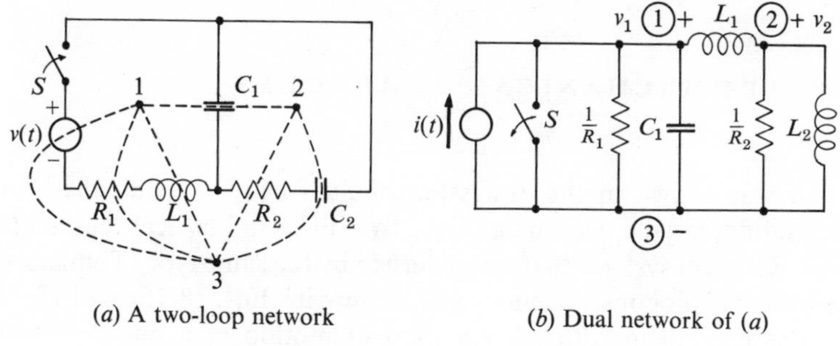

(a) A two-loop network

(b) Dual network of (a)

Fig. 8-21. *Construction of dual network*

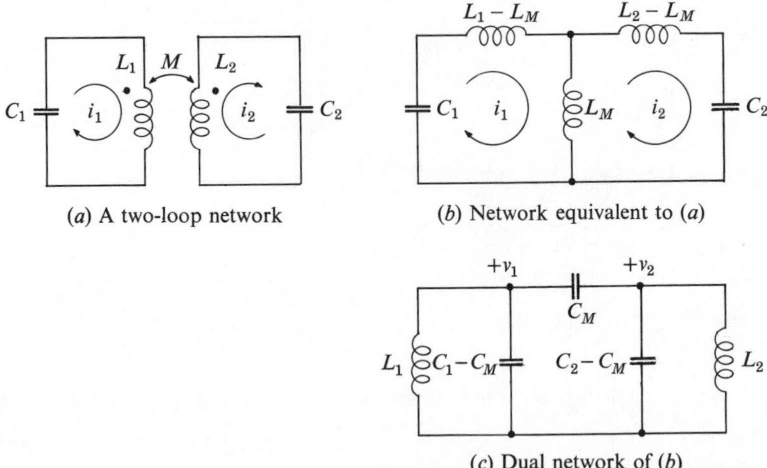

(*a*) A two-loop network

(*b*) Network equivalent to (*a*)

(*c*) Dual network of (*b*)

Fig. 8-22. *Dual network containing mutual inductance*

mutual inductance is introduced in this circuit. The loop equations are

$$(L_1 - L_M + L_M)\frac{di_1}{dt} + \frac{1}{C_1}\int i_1\,dt - L_M\frac{di_2}{dt} = 0$$

$$-L_M\frac{di_1}{dt} + (L_2 - L_M + L_M)\frac{di_2}{dt} + \frac{1}{C_2}\int i_2\,dt = 0$$

It is evident that these loop equations are identical to those for the network of Fig. 8-22(*a*). The steps for obtaining the dual network of Fig. 8-22(*b*) are the same as those outlined in the preceding example. Figure 8-22(*c*) shows its dual network.

8-9. ELECTROMECHANICAL ANALOGUES

It was shown in the discussion of dual networks that the same integrodifferential equation can have two different physical representations. Physical systems that are governed by the same type of equations are called analogous systems. Let us rewrite Eqs. (8-76) and (8-77) and compare them with the equation of motion of a one-degree-of-freedom mechanical system.

$$L\frac{di}{dt} + Ri + \frac{1}{C}\int i\,dt = v(t) \tag{8-78}$$

$$C\frac{dv}{dt} + \frac{1}{R}v + \frac{1}{L}\int v\,dt = i(t) \tag{8-79}$$

$$m\frac{d\dot{x}}{dt} + c\dot{x} + k\int \dot{x}\,dt = f(t) \tag{8-80}$$

Equation (8-80) describes the mechanical system in which $\dot{x} = dx/dt$ or $x = \int \dot{x}\,dt$, and $f(t)$ is the excitation force. It is apparent that the three systems are analogous and that the mechanical system has at least two electrical analogues.

Comparing Eqs. (8-78) and (8-80), if the excitation voltage $v(t)$ is analogous to the excitation force $f(t)$, the loop current is analogous to the velocity of the mass. The other analogous quantities in these equations can be compared in like manner. This is called the voltage-force analogy. Similarly, comparing Eqs. (8-79) and (8-80), we have the current-force analogy. The analogous quantities for the two electrical analogues of the mechanical system are tabulated in Table 8-4. It is obvious that, with a change of units, the same analogies will apply to rotational mechanical systems.

Although the electrical potential, or voltage, versus mechanical-force analogy is more intuitive than the current-force analogy, the latter gives a network identical to the arrangement of the mechanical system,

TABLE 8-4

ELECTROMECHANICAL ANALOGOUS QUANTITIES

Mechanical Quantity		Electrical Quantity			
		Voltage-Force Analogy		*Current-Force Analogy*	
Force (lb)	f	Voltage (volt)	v	Current (ampere)	i
Velocity (in.-sec^{-1})	$\dot{x}$	Current (ampere)	i	Voltage (volt)	v
Displacement (in.)	$x = \int_o^t \dot{x}\,dt$	Charge (coulomb)	$q = \int_o^t i\,dt$	Flux linkage (weber)	$\phi = \int_o^t v\,dt$
Mass (lb-sec^2-in.)	m	Inductance (henry)	L	Capacitance (farad)	C
Viscous damping (lb-sec-in.$^{-1}$)	c	Resistance (ohm)	R	Conductance (mho)	$1/R$
Compliance (in.-lb^{-1})	$1/k$	Capacitance (farad)	C	Inductance (henry)	L

or the mechanical network. Consider the one-degree-of-freedom system shown in Fig. 8-23(*a*). Since the excitation $f(t)$ and the inertia force $m\ddot{x}$ are measured with respect to the earth, the mechanical network is constructed as shown in Fig. 8-23(*b*). Using the current-force analogy, the electrical analogue of the mechanical system is as shown in Fig. 8-23(*c*). Comparing (*b*) and (*c*) of Fig. 8-23, it is evident that a mechanical connection is analogous to a node in the electrical circuit and that the two diagrams are identical in form. Mechanical networks

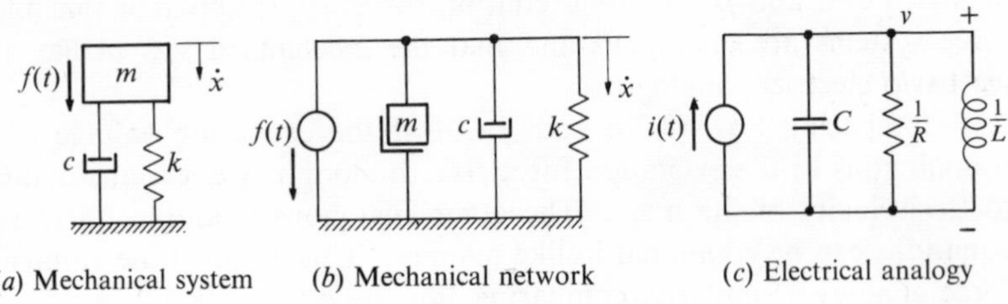

(*a*) Mechanical system (*b*) Mechanical network (*c*) Electrical analogy

Fig. 8-23. *Comparison of mechanical network and current-force analogy*

are used in the mobility method which is frequently employed for steady-state analysis.

We have shown two electrical analogues of a mechanical system. It is possible to write Eqs. (8-78) to (8-80) in different forms to obtain other analogies. For example, differentiating Eq. (8-79) and writing Eq. (8-80) in a different form gives

$$C\frac{d^2v}{dt^2} + \frac{1}{R}\frac{dv}{dt} + \frac{1}{L}v = \frac{di(t)}{dt} \tag{8-81}$$

$$m\frac{d^2x}{dt^2} + c\frac{dx}{dt} + kx = f(t) \tag{8-82}$$

Thus we derive $di(t)/dt$ analogous to force and voltage v analogous to displacement x. The circuit diagram corresponding to Eq. (8-81) is the same as that shown in Fig. 8-23(*c*) except that the time derivative of the current source is prescribed instead of the current.

Example 25. A mechanical system is shown in Fig. 8-24(*a*). (*a*) Write the equations of motion of the system, (*b*) use the voltage-force analogy to derive the electrical-analogue circuit, and (*c*) use the current-force analogy to derive the electrical-analogue circuit.

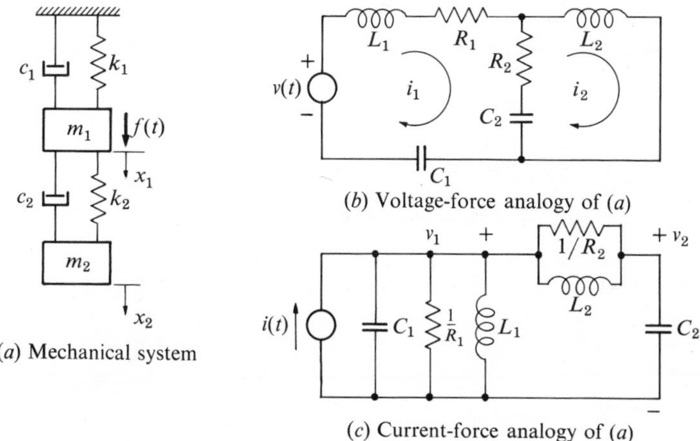

Fig. 8-24. *Electrical analogue of mechanical system*

Solution: (*a*) Using Newton's law of motion, the equations of motion are

$$m_1\ddot{x}_1 + (c_1 + c_2)\dot{x}_1 + (k_1 + k_2)x_1 - c_2\dot{x}_2 - k_2x_2 = f(t)$$

$$-c_2\dot{x}_1 - k_2x_1 + m_2\ddot{x}_2 + c_2\dot{x}_2 + k_2x_2 = 0$$

(*b*) Using the voltage-force analogy in Table 8-4, the analogue equations are

$$L_1\frac{di_1}{dt} + (R_1 + R_2)i_1 + \left(\frac{1}{C_1} + \frac{1}{C_2}\right)\int i_1\,dt - R_2i_2 - \frac{1}{C_2}\int i_2\,dt = v(t)$$

$$-R_2i_1 - \frac{1}{C_2}\int i_1\,dt + L_2\frac{di_2}{dt} + R_2i_2 + \frac{1}{C_2}\int i_2\,dt = 0$$

These are the loop equations of the circuit shown in Fig. 8-24(*b*).

(*c*) Using the current-force analogy in Table 8-4, the analogue equations are

$$C_1\frac{dv_1}{dt} + \left(\frac{1}{R_1} + \frac{1}{R_2}\right)v_1 + \left(\frac{1}{L_1} + \frac{1}{L_2}\right)\int v_1\,dt - \frac{1}{R_2}v_2 - \frac{1}{L_2}\int v_2\,dt = i(t)$$

$$-\frac{1}{R_2}v_1 - \frac{1}{L_2}\int v_1\,dt + C_2\frac{dv_1}{dt} + \frac{1}{R_2}v_2 + \frac{1}{L_2}v_2\,dt = 0$$

These are the node equations of the circuit shown in Fig. 8-24(*c*). It can be shown that the networks shown in Fig. 8-24(*b*) and (*c*) are dual networks.

Example 26. A geared system is as shown in Fig. 8-25. Using the voltage-torque analogy, obtain the electrical-analogue network of the system.

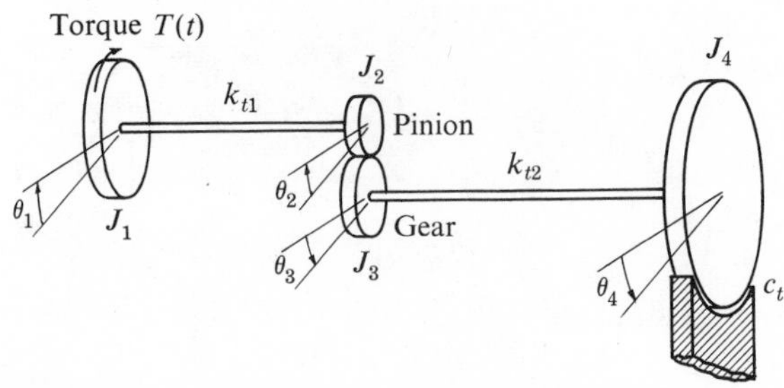

Fig. 8-25. *A geared system*

Solution: The voltage-force analogy was discussed in the previous examples. For the rotational system, we use voltage analogous to torque, loop current analogous to angular velocity $\dot{\theta}$, etc. The speed-reduction system is analogous to an ideal transformer. Let N_2 and N_3 be the number of teeth of the pinion and the gear, respectively, and let the corresponding torques be T_2 and T_3. For the mechanical system we have the relations

$$\frac{T_2}{T_3} = \frac{N_2}{N_3} = \frac{\dot{\theta}_3}{\dot{\theta}_2} \tag{8-83}$$

If n_2 and n_3 are the number of turns of the primary and the secondary of the ideal transformer, we have the relations

$$\frac{v_2}{v_3} = \frac{n_2}{n_3} = \frac{i_3}{i_2} \tag{8-84}$$

The analogous relations indicated in Eqs. (8-83) and (8-84) are shown in Fig. 8-26(*a*) representing an ideal transformer.

The equations of motion of the mechanical system are

$$J_1\ddot{\theta}_1 + k_{t1}\theta_1 - k_{t1}\theta_2 = T(t)$$
$$J_2\ddot{\theta}_2 + k_{t1}\theta_2 - k_{t1}\theta_1 = -T_2$$
$$J_3\ddot{\theta}_3 + k_{t2}\theta_3 - k_{t2}\theta_4 = T_3$$
$$J_4\ddot{\theta}_4 + c_t\dot{\theta}_4 + k_{t2}\theta_4 - k_{t2}\theta_3 = 0$$

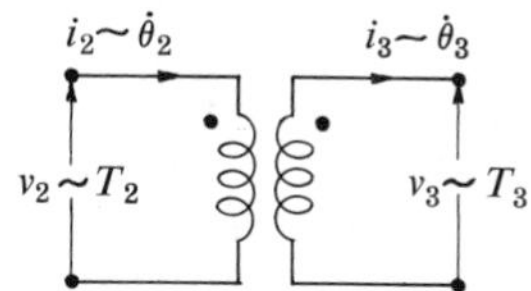

(*a*) Ideal transformer representing a gear system

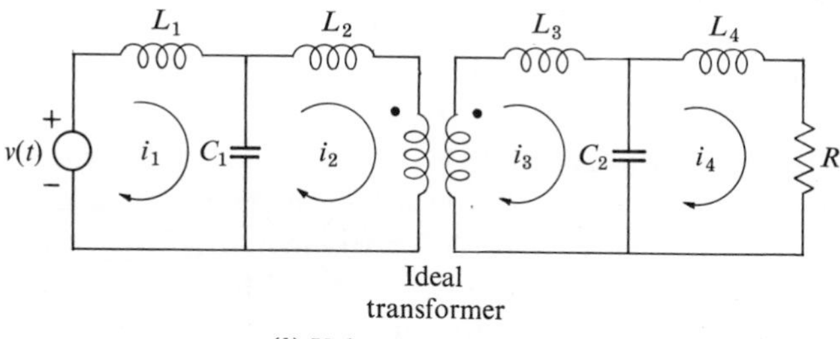

(*b*) Voltage-torque analogue

Fig. 8-26. *Voltage-torque analogue of mechanical system shown in Fig.* 8-25

The torque at the pinion is a torque output and T_2 is negative. The voltage-torque-analogue network is as shown in Fig. 8-26(*b*). Using the symbol v to denote a voltage rise, the network equations are

$$L_1 \frac{di_1}{dt} + \frac{1}{C_1} \int i_1 \, dt - \frac{1}{C_1} \int i_2 \, dt = v(t)$$

$$L_2 \frac{di_2}{dt} + \frac{1}{C_1} \int i_2 \, dt - \frac{1}{C_1} \int i_1 \, dt = -v_2$$

$$L_3 \frac{di_3}{dt} + \frac{1}{C_2} \int i_3 \, dt - \frac{1}{C_2} \int i_4 \, dt = v_3$$

$$L_4 \frac{di_4}{dt} + Ri_4 + \frac{1}{C_2} \int i_4 \, dt - \frac{1}{C_2} \int i_3 \, dt = 0$$

8-10. ELECTROMECHANICAL SYSTEMS

Systems in which electrical and mechanical elements are combined as a working unit are called electromechanical systems. Examples of such systems are the electrical motor and generator, loudspeakers, and vibration tables, to mention just a few. These systems employ some means of energy conversion. The most common types of electromechanical converting systems are the electromagnetic, the electrostatic, the piezoelectric, and the magnetostrictive systems. In this section we shall limit our discussion to the electromagnetic converting system and obtain an all-electrical analogous network for the electromechanical system.

In the electromagnetic system the force acting on a conductor of length l carrying a current i in a magnetic field with flux intensity B is

$$f = Bli \tag{8-85}$$

In the meter-kilogram-second (mks) system of units, the force f is in newtons, the flux intensity B is in weber-meter^{-2}, the length l is in meters, and the current i is in amperes. One newton equals 10^5 dynes or 2.248 lb. The factor Bl, which is the ratio of the force to the applied current, is called the force factor. It is the direct relation between the mechanical and the electrical circuits. The current, the magnetic flux, and the force are mutually perpendicular to each other, as indicated in Fig. 8-27(a).

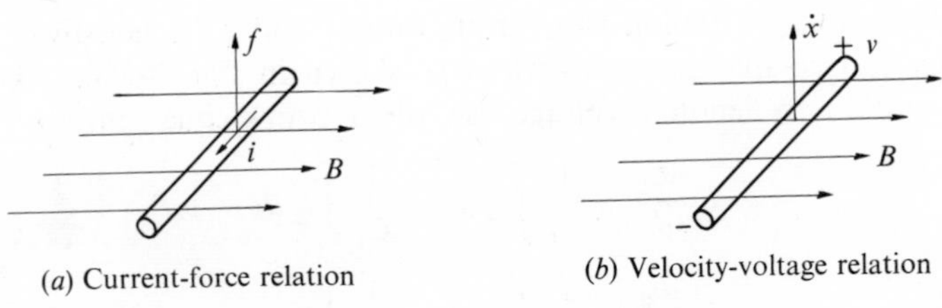

(a) Current-force relation (b) Velocity-voltage relation

Fig. 8-27. *Conductor in a uniform magnetic field*

The inverse relation is that an open-circuit voltage (electromotive force) is induced in a conductor moving in a magnetic field. This voltage is

$$v = Bl\dot{x} \tag{8-86}$$

where v is in volts and $\dot{x}$ in meter-sec^{-1}. The directions of the motion and the magnetic flux and the polarity of the induced voltage are indicated in Fig. 8-27(b).

Example 27. A galvanometer used in an oscillograph is as shown in Fig. 8-28(a). The galvanometer consists of a single-turn phosphor-bronze ribbon placed between the poles of a powerful magnet. A small mirror is cemented to the loop, and a definite tension is applied to the loop

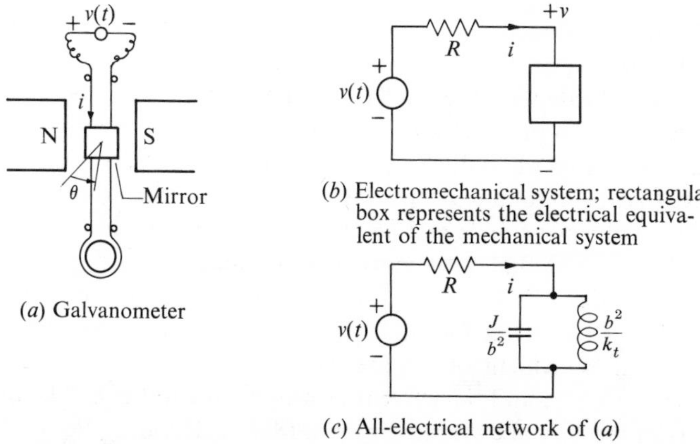

Fig. 8-28. *Electromechanical system*

by means of a spring. When a current is passed through the loop, the two sides of the loop move in opposite directions, thus turning the mirror about a vertical axis. Neglecting the self-inductance and the capacitance of the coil and the viscous damping of the moving parts, determine the equation of motion of the system if a voltage $v(t)$ is applied. Obtain an electrical analogue network of the system.

Solution: From Eq. (8-85) the torque developed by the galvanometer loop is

$$T(t) = (Blr)i = bi$$

where B is the magnetic flux intensity, l the effective length of the loop, and r the effective radius of the loop. The quantity (Blr) may be regarded as a torque factor. For convenience, let us define $(Blr) = b$, a constant. From Eq. (8-85) and Newton's law of motion, we have

$$J\ddot{\theta} + k_t\theta = bi \tag{8-87}$$

Since the velocity of the conductor normal to the magnetic flux is $\dot{x} = r\dot{\theta}$, from Eq. (8-86) the back emf induced in the loop owing to its rotation is

$$v = (Blr)\dot{\theta} = b\dot{\theta}$$

Let R be the total resistance of the loop. From Kirchhoff's voltage law we obtain

$$Ri + b\dot{\theta} = v(t) \tag{8-88}$$

Eliminating the current i from Eqs. (8-87) and (8-88) and simplifying gives

$$J\ddot{\theta} + \frac{b^2}{R}\dot{\theta} + k_t\theta = \frac{b}{R}v(t) \tag{8-89}$$

We have assumed that the torque bi and the voltage $b\dot{\theta}$ have the proper signs in Eqs. (8-87) and (8-88). Let i and θ be positive in the directions indicated in Fig. 8-28(a). The current i in the direction indicated tends to produce a torque to increase θ. A positive $\dot{\theta}$ in the direction indicated will induce a voltage in the loop opposing the applied voltage, that is, a voltage drop in the direction of the positive current flow. Thus the signs of bi and $b\dot{\theta}$ are correctly assigned.

The electrical-analogue network is obtained by substituting the induced voltage $v = b\dot{\theta}$ in Eq. (8-88), and the corresponding loop network is as shown in Fig. 8-28(b). The rectangular box represents the electrical equivalent of the mechanical system. The voltage drop across the box is v, and the current flowing through it is i. The branch currents in the box are specified by Eq. (8-87). Dividing Eq. (8-87) by b and substituting $v/b = \dot{\theta}$ gives

$$\frac{J}{b^2}\frac{dv}{dt} + \frac{k_t}{b^2}\int v\,dt = i \tag{8-90}$$

The all-electrical network based on Eqs. (8-88) and (8-90) is as shown in Fig. 8-28(c).

Example 28. The electromechanical system shown in Fig. 8-29 consists of a spring-supported mass, a damper, and a moving coil in a uniform magnetic field of flux intensity B. The inductance and the resistance of the coil are L and R, respectively. If a voltage $v(t)$ is applied across the coil, obtain an all-electric analogue network equation of the system.

Solution: From Eq. (8-85) the force developed by the moving coil is

$$f = Bli = bi$$

where l is the effective length of the coil and b, a constant, is defined as a force factor. The dot in the conductor indicates the direction of the coil current. The force developed tends to pull the coil downward in the direction of a positive displacement. Thus the equation of motion of the mass m is

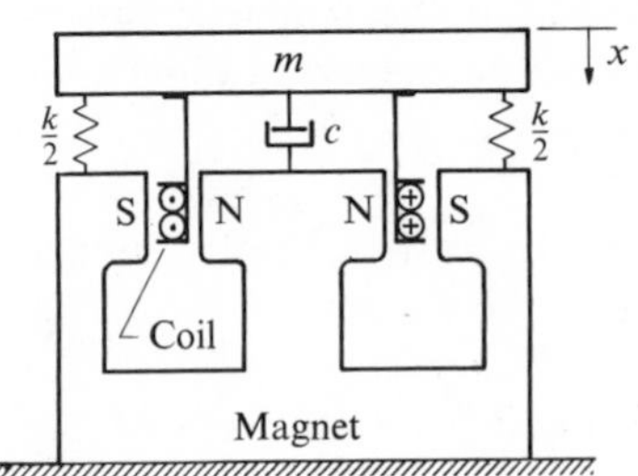

Fig. 8-29. *Electromechanical system*

$$m\ddot{x} + c\dot{x} + kx = bi \tag{8-91}$$

From Eq. (8-86) the voltage induced in the moving coil is

$$v = Bl\dot{x} = b\dot{x}$$

If $\dot{x}$ is positive, the induced voltage is a voltage drop in the direction of the positive current flow. From Kirchhoff's voltage law, the loop equation is

$$L\frac{di}{dt} + Ri + v = v(t) \tag{8-92}$$

Substituting $v/b = \dot{x}$ in Eq. (8-91) and dividing by b gives

$$\frac{m}{b^2}\frac{dv}{dt} + \frac{c}{b^2}v + \frac{k}{b^2}\int v\,dt = i \tag{8-93}$$

The electrical-analogue network based on Eqs. (8-92) and (8-93) is similar to that shown in Fig. 8-29(*c*), except that we have both L and R in the loop, corresponding to Eq. (8-92), and three branch currents in the electrical equivalent of the mechanical system, corresponding to Eq. (8-93).

SUGGESTED READING

Carslaw, H. S., and J. C. Jaeger, *Operational Methods in Applied Mathematics* (New York: Oxford University Press, Inc., 1941).

Cheng, D. K., *Analysis of Linear Systems* (Massachusetts: Addison-Wesley Publishing Company, Inc., 1959).

Churchill, R. V., *Operational Mathematics* (New York: McGraw-Hill Book Co., Inc., 1958).

Gardner, M. F., and J. L. Barns, *Transients in Linear Systems* (New York: John Wiley & Sons, Inc., 1942), vol. 1.

Mason, W. P., *Electromechanical Transducers and Wave Filters* (New York: D. Van Nostrand Co., Inc., 2nd ed., 1948).

Thomson, W. T., *Laplace Transformation* (New Jersey: Prentice-Hall, Inc., 2nd ed., 1960).

A thirty-two amplifier desk-top electronic analogue computer (courtesy Applied Dynamics, Inc.)

9 ELECTRONIC ANALOGUE COMPUTER

9-1. INTRODUCTION

The first analogue computer was probably the slide rule. When two numbers are multiplied together by a slide rule, the product is another number. If the original numbers are assigned the units of force in pounds and distance in feet, the unit of work in their product is in foot-pounds. If the original numbers are in volts and amperes, the answer is in watts. Hence these multiplications are analogous, and they differ only by a scale factor. The term "analogy" is defined to mean similarity of relation without identity. The relation indicated here is that the answer is the product of two quantities.

When a physical problem is expressed in terms of an equation or a set of equations, we have, in effect, transformed the problem into a mathematical model. The relation of the dependent and independent variables is specified by the equation. If two problems are expressible by the same type of equations, the dependent and independent variables of the problems have similar relationships, and the problems possess the same mathematical model. The solution of one problem may very well be the solution of the analogous problem if the proper scaling is applied. Hence the study of analogues is a mathematical study. A physical problem may be examined by means of its analogue in some other field. A problem in mechanical vibrations can be studied by means of an analogue computer, the setup of which has the same type of equation that describes the vibratory system.

The purpose of this chapter is to introduce the application of the electronic analogue computer for the study of mechanical vibrations. It is not intended to cover all phases of analogue-computer operation. It may be used for the laboratory section of a course in mechanical vibrations. The material is organized to permit the laboratory work to accompany the classroom work. The ordinary differential equations discussed are classified as linear and nonlinear. Nonlinear differential equations are not part of the text. The ease with which they can be handled by the analogue computer, however, may well justify their inclusion in the chapter. The first part of this chapter is devoted to the solution of linear ordinary differential equations with constant coefficients, the second part to some nonlinear applications.

Laboratory work on a computer is an essential condition of the subject matter studied. A set of experiments is included in Appendix A. All the experiments suggested can be performed on a relatively simple, inexpensive, commercially available computer.

9-2. APPLICATIONS

A computer is characterized by its ability to perform certain mathematical operations. Computers may be broadly classified as digital or analogue. The digital computer operates with discrete numbers. The desk calculator is a digital computer in its simple form. A variety of analogue devices may be classified as computers. For example, the slide rule is an analogue device and the electrical-network analyzer is a special-purpose analogue computer which may be designed to study problems in electrical power distribution or fluid flow. In contrast to the digital computer, an analogue computer is essentially a continuous device. The solution of a problem is usually displayed in the form of a continuous chart. The first general-purpose analogue computer was a mechanical device known as a differential analyzer. With the advent of electronic computers, both digital and analogue computers have enjoyed accelerated growth and widespread acceptance in engineering and scientific applications.

An analogue computer may be used (1) to solve mathematical equations, (2) to simulate a physical system, or (3) to control a physical process. In the first application the mathematical equations describing a physical system are formulated, and the computer is used to reduce the tedious task of obtaining a solution to the problem. The object of the second application is to use the computer as a physical model of a system under study. The behavior of the model is directly related to the original system. The behavior of the model itself, or in conjunction with other systems, is explored. Simulation is a valuable tool in investigations, especially where the system under study is so complex and nonlinear that an analytical study is extremely difficult. The so-called simulators used for training of technical personnel, such as aircraft pilots or nuclear reactor operators, may be regarded as special-purpose analogue computers. In the control of a physical process, the computer is required to perform as part of a physical system. Our present interest is in the solving of equations.

The electronic analogue computer is best suited for solving systems of ordinary differential equations, with constant or variable coefficients, linear or nonlinear. The problem set up is slightly more complex for equations with variable coefficients and for nonlinear equations. Analogue computers have been used successfully for solving partial differential equations, simultaneous algebraic equations, and eigenvalue problems.

Many problems can be solved on either the analogue or the digital computer. In general, a large-scale digital computer can handle any problem that can be solved on an analogue computer. Many problems can be handled adequately and more easily by the analogue computer, however, and the programming effort required may be substantially less than that for a digital computer. In view of the progress being made in these machines, it is difficult to make a flat statement in regard to their respective applications in engineering. In some large problems, analogue computers have been combined with digital computers, both operating simultaneously. Since 1950, machines possessing both digital and analogue features, called digital differential analyzers, have been commercially available.

It is often said that the analogue computer is less accurate than the digital. Within the limits of the machine, the solution of a problem on a digital computer can be displayed to as many significant digits as desired. The solution on an analogue computer is in the form of a continuous chart, which can be read to only three or four significant digits. The number of significant digits, or the precision of the answer, however, should not be confused with the accuracy of the solution of a problem. Accuracy should be viewed from the standpoint of the overall accuracy of the problem; namely, the available information on the variables and the parameters involved, the statement of the problem, the mathematical model selected, the conformity of the mathematical model to the physical model, and the accuracy of the computing equipment. Furthermore, accuracy should also be considered from the requirement of the problem and the cost and labor involved in obtaining the solution.

In using a computer, it should be remembered that the computer is a tool. It can only perform the mathematical operations in accordance with its instructions. The computer can only alleviate the laborious task of finding the solutions of a mathematical model. The analysis of the problem and the selection of the proper model still cannot be entrusted to the machine.

9-3. LINEAR COMPUTER ELEMENTS

The linear elements used in an electronic analogue computer are the *resistor*, the *capacitor*, and the high-gain *d-c amplifier*, which is also called the *operational amplifier*. It will be shown in the next section that the operations required for the solution of linear ordinary differential equations with constant coefficients can be obtained by the various combinations of these elements. The symbols representing the elements enumerated and the operational equations of these elements are tabulated in Table 9-1. We shall refer to the sketches in this table in the discussion of these elements. The voltages indicated are measured with respect to the ground potential.

The *resistor* is a passive element. The flow of electric current i through a resistor R is proportional to the difference in voltage e applied across the resistor. This relation is expressed as

$$e = Ri \tag{9-1}$$

where e is in volts, i is in amperes, and R is in ohms. In computer work, it is customary to express the value of the resistor in megohms. One megohm is 10^6 ohms.

TABLE 9-1

LINEAR COMPUTER ELEMENTS

ELEMENT	SYMBOL	OPERATIONAL EQUATION
Resistor R	e_i, i, e_o, $e = e_i - e_o$	$e = Ri$
Potentiometer a	e_i Top, R, Arm, e_o, aR, Bottom; or	$e_o = ae_i;\ 0 \leq a \leq 1$
Potentiometer (with load resistor)	e_i, a, e_o	$e_o = \dfrac{a}{1 + a(1 - a)R/R_L} e_i$
	e_i, R, aR, e_o, R_L	$0 \leq a \leq 1$
Capacitor C	e_i, i, e_o, $e = e_i - e_o$	$e = \dfrac{1}{Cs} i$ or
		$e = \dfrac{1}{C}\displaystyle\int_o^t i\,dt + E_o$
Impedance Z	e_i, i, Z, e_o, $e = e_i - e_o$	$e = Zi$
Amplifier	e_g, $-A$, e_o	$e_o = -Ae_g$

The *potentiometer* makes use of a resistor as a voltage divider. The potentiometer has three terminals; the input voltage e_i is applied to the "top" terminal, the "bottom" terminal is usually connected to ground, and the output voltage e_o is obtained from the "arm" terminal which is connected to a movable contact, or wiper, intermediate between the top and the ground terminals. If R is the total resistance of the potentiometer and the resistance between the arm and ground is aR, where $0 \leq a \leq 1$, then the output voltage e_o is

$$e_o = ae_i \quad \text{or} \quad \frac{e_o}{e_i} = a \tag{9-2}$$

If a *load resistor* R_L is connected to the arm of the potentiometer as shown in the sketch in Table 9-1, the output voltage of the potentiometer will be lower than that indicated in Eq. (9-2). Since R_L and the resistance aR are in parallel, it can be shown that the input-output voltage relation can be expressed as

$$\frac{e_o}{e_i} = \frac{a}{1 + a(1 - a)R/R_L} \tag{9-3}$$

The importance of this error can be seen by inserting typical values into Eq. (9-3). For a potentiometer of 100,000 ohms set at $a = 0.5$ and loaded with a 100,000-ohm resistor, from Eq. (9-3) the output voltage is $0.4e_i$. Comparing with $e_o/e_i = 0.5$ from Eq. (9-2), the error in the output voltage is 20 percent. The loading effect can be compensated for by setting the loaded potentiometer against a high-precision potentiometer with a null indicator. Potentiometer setting will be discussed in Experiment 1.

The *capacitor* is a passive element. The voltage drop across a capacitor is proportional to the accumulation of electric charge in the capacitor. Since electric current i is the rate of flow of electric charge, the voltage drop across a capacitor C is

$$e = \frac{1}{C}\int_o^t i\,dt + E_o \tag{9-4}$$

where C is in farads and the voltage E_o is due to the initial charge accumulated in the capacitor at time $t = 0$. It is customary to express the value of a capacitor C in microfarads (μf) which is 10^{-6} farad. For the purpose of manipulation, it is convenient to use the differential operator s to indicate the time derivative of a function, and the integrator operator $1/s$ to denote integration with respect to time.† Hence Eq. (9-4) can be written as

$$e = \frac{1}{Cs} i \tag{9-5}$$

To generalize the discussion on resistors and capacitors, let us introduce a quantity Z, called the impedance of the passive element. The relation between the instantaneous voltage drop across an element and the instantaneous current flowing through it is expressed as

$$e = Zi \tag{9-6}$$

Comparing this equation with Eqs. (9-1) and (9-5), the impedance of a resistor is R and that of a capacitor is $1/Cs$. It should be noted that Eq. (9-6) is equally applicable to a network or a combination of passive elements. The impedance of a network is the ratio of the voltage drop across the network to the total current flowing through the network.

The high-gain *d-c amplifier* is the most important element of the computer. It is characterized by the relation

$$e_o = -Ae_g \tag{9-7}$$

where e_o and e_g are the output and input voltages, respectively, and A is the gain of the amplifier. The voltages e_o and e_g are measured with respect to a common-ground potential. Hence the schematic representation shows only the ungrounded terminals. The magnitude of A ranges from 10^4 to 10^8. The output of the amplifier is usually limited to plus and minus 100 volts.

9-4. LINEAR OPERATIONS

The basic linear operations of an analogue computer are *sign inversion, multiplication by a constant, integration with respect to time,* and *summation.* The operations are generally performed on voltages by the use of the operational amplifier and the circuit as shown in Fig. 9-1. The voltages e_i, e_o, and e_g are the input, output, and grid

† Here the operator s may be used in the same manner as the differential operator $D = d/dt$. $1/s$ may be interpreted by comparing Eqs. (9-4) and (9-5). The use of the operator s and the Laplace transform complex variable s in Chap. 8 should be noted.

voltage of the first tube of the amplifier, respectively. Let us develop the general equation relating the output and input voltages before discussing these operations individually.

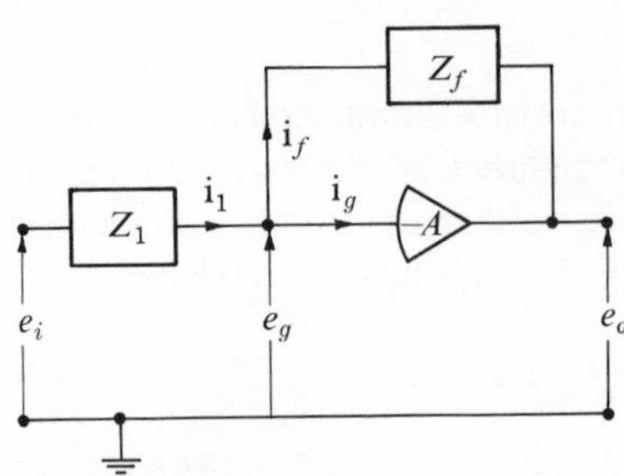

Fig. 9-1. *Basic amplifier circuit for linear operations*

Referring to Fig. 9-1, the voltage drop across the input impedance Z_i is $(e_i - e_g)$, and that across the feedback impedance Z_f is $(e_g - e_o)$. Assuming that the flow of currents is in the directions indicated and applying the Kirchhoff current law, we have

$$i_1 = i_f + i_g \tag{9-8}$$

Since i_g is the grid current of the first tube of the amplifier, it is negligible as compared with i_1 and i_f. The grid current is on the order of 10^{-9} amp, while the other currents are on the order of 10^{-3} amp. Neglecting the grid current gives

$$i_1 = i_f \tag{9-9}$$

Applying Eq. (9-6) and substituting the appropriate values of Z and e, we obtain

$$\frac{e_i - e_g}{Z_i} = \frac{e_g - e_o}{Z_f} \tag{9-10}$$

Substituting Eq. (9-7) in this equation and simplifying, we have

$$\frac{e_o}{e_i} = -\frac{Z_f}{Z_i} \frac{1}{1 + \frac{1}{A}\left(1 + \frac{Z_f}{Z_i}\right)} \tag{9-11}$$

If A is sufficiently large and the maximum value of Z_f/Z_i is approximately 50, this equation can be approximated by the relation

$$\frac{e_o}{e_i} = -\frac{Z_f}{Z_i} \tag{9-12}$$

This is the basic operational equation from which the linear operations enumerated and more complex operations can be derived. It should be noted that there is always a sign inversion associated with each operation with the amplifier.

The derivation of Eq. (9-12) can be simplified if it is assumed that

(1) i_g is negligible as compared with i_1 and i_f, and (2) e_g is negligible as compared with e_i and e_o. Since $e_o = -Ae_g$, if A is on the order of 10^4 or greater and e_o maximum is 100 volts, the magnitude of e_g is less than 0.01 volt. Hence e_g is essentially at ground potential. Neglecting e_g in Eq. (9-10) gives

$$\frac{e_i}{Z_i} = -\frac{e_o}{Z_f} \qquad \textbf{(9-13)}$$

Equation (9-12) can be obtained by rearranging this equation.

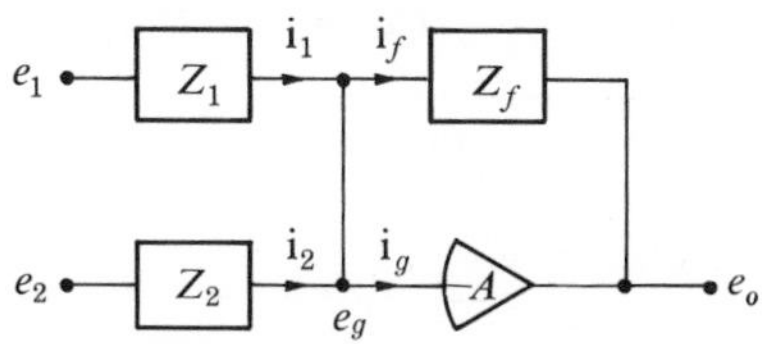

Fig. 9-2. *Amplifier circuit for summation*

If the amplifier circuit has more than one voltage input, as shown in Fig. 9-2, the operational equation can be derived easily by assuming that the grid current i_g is negligible and that the grid voltage e_g is essentially at ground potential. Applying the Kirchhoff current law, we have

$$i_1 + i_2 = i_f \qquad \textbf{(9-14)}$$

or

$$\frac{e_1}{Z_1} + \frac{e_2}{Z_2} = -\frac{e_o}{Z_f} \qquad \textbf{(9-15)}$$

Rearranging this equation gives

$$e_o = -\left(\frac{Z_f}{Z_1} e_1 + \frac{Z_f}{Z_2} e_2\right) \qquad \textbf{(9-16)}$$

which can be generalized as

$$e_o = -\sum_{i=1}^{n} \frac{Z_f}{Z_i} e_i \qquad \textbf{(9-17)}$$

The basic linear operations may be considered as special cases of Eq. (9-17). Their operational equations and the corresponding circuits are tabulated in Table 9-2.

Case 1. Sign Inversion

The sign-inversion operation is obtained by substituting R_f and R_i of equal value for Z_f and Z_i in Eq. 9-12.

$$e_o = -\frac{R_f}{R_i} e_i = -e_i \qquad \textbf{(9-18)}$$

TABLE 9-2

Basic Linear Computer Operations and Circuits

Operation	Computer Circuit	Operational Equation
Sign inversion	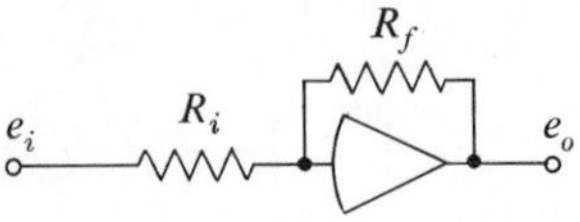	$e_o = -e_i;\ R_f = R_i$
Multiplication by a constant (with sign inversion)	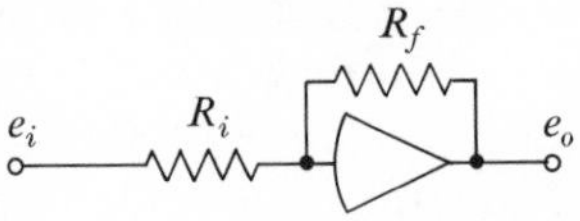	$e_o = -\dfrac{R_f}{R_i}\, e_i$
	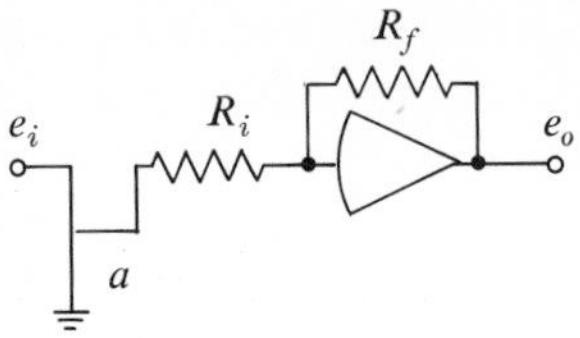	$e_o = -\dfrac{R_f}{R_i}\, ae_i;\ 0 \leqq a \leqq 1$
Integration (with sign inversion and multiplication by a constant)	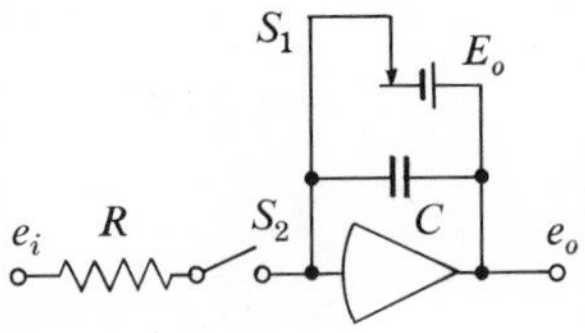	$e_o = -\dfrac{1}{RCs}\, e_i$ or $e_o = -\dfrac{1}{RC}\displaystyle\int_o^t e_i\, dt + E_o$
Summation (with sign inversion and multiplication by a constant)	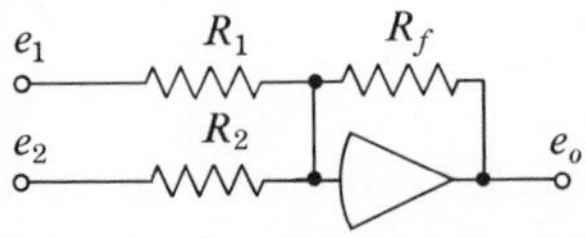	$e_o = -\left(\dfrac{R_f}{R_1}\, e_1 + \dfrac{R_f}{R_2}\, e_2\right)$
Summation-integration (with sign inversion and multiplication by a constant)	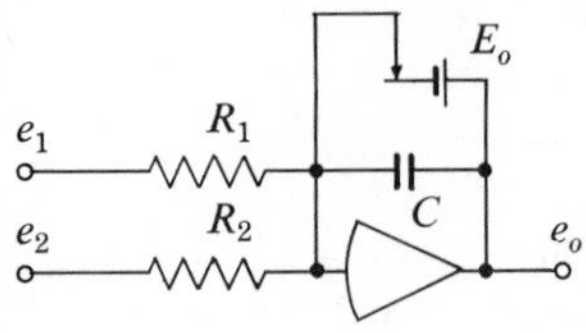	$e_o = -\dfrac{1}{s}\left(\dfrac{1}{R_1C}\, e_1 + \dfrac{1}{R_2C}\, e_2\right)$ or $e_o = -\displaystyle\int_o^t \left(\dfrac{e_1}{R_1C} + \dfrac{e_2}{R_2C}\right) dt + E_o$

It is apparent that the sign-inversion operation is independent of the value of these resistors as long as the ratio R_f/R_i is unity.

Case 2. Multiplication by a Constant (with Sign Inversion)

If R_f and R_i substituted for Z_f and Z_i in Eq. 9-12 are unequal, we obtain

$$e_o = -\frac{R_f}{R_i} e_i \tag{9-19}$$

Thus the output voltage e_o is equal to the input voltage e_i multiplied by the constant R_f/R_i. The ratio R_f/R_i can be greater than, equal to, or less than unity. The negative sign is associated with each operation with the amplifier.

The magnitude of the constant can be modified by the use of a potentiometer. For example, if $e_o = -1.49e_i$ is desired, the potentiometer may be set at $a = 1.49/2 = 0.745$, and the ratio R_f/R_i may be selected equal to 2/1. Thus the output voltage of the potentiometer is $0.745e_i$, and the output of the amplifier is $-(2)(0.745e_i)$, which is the desired value. In setting the potentiometer, it should be remembered that the grid voltage of the amplifier is essentially at ground potential, and that the input resistor R_i becomes a load resistor to the potentiometer. (See Sect. 9-3 and Experiment 1.)

Case 3. Integration with Respect to Time (with Sign Inversion and Multiplication by a Constant)

Let the passive elements Z_i and Z_f of Fig. 9-1 be R and $1/Cs$ respectively. From Eq. (9-12) the operational equation of the amplifier circuit becomes

$$e_o = -\frac{1}{RCs} e_i \quad \text{or} \quad e_o = -\frac{1}{RC}\int_0^t e_i \, dt + E_o \tag{9-20}$$

where $1/s$ is an integral operator, and E_o is the initial value of e_o at time $t = 0$. This equation shows that the input voltage e_i is integrated with respect to time. The unit of the time constant RC is in seconds if R is in megohms and C is in microfarads. Thus the unit of the output is in volts. The constant $1/RC$ can be greater than, equal to, or less than unity. Practical values of R and C are $0.1 \leq R \leq 10$ megohm and $0.01 \leq C \leq 1$ μf, respectively. The sign inversion is associated with an operation with the amplifier.

The initial condition E_o is due to the initial charge stored in the capacitor. A method to obtain E_o is shown schematically in the sketch in Table 9-2. A d-c voltage of appropriate value and polarity is connected across the capacitor. If the initial condition of e_o is positive, the positive terminal of the d-c voltage supply is connected to the output terminal of the amplifier. The switch S_1 is open and S_2 is closed at time $t = 0$ when the system is ready to compute. The switches in a computing circuit may be connected by an elaborate system of relays.

Case 4. Summation (with Sign Inversion and Multiplication by a Constant)

The basic circuit of a summer is shown in Fig. 9-2. With the appropriate passive elements substituted for the impedances Z_1, Z_2, and Z_f, this circuit can be used as a summer or a summer-integrator of several input voltages.

ALGEBRAIC SUMMATION: Let R_1, R_2, and R_f be substituted for Z_1, Z_2, and Z_f in the circuit of Fig. 9-2. If e_1 and e_2 are the input voltages, from Eq. (9-16) the output voltage is

$$e_o = -\left(\frac{R_f}{R_1} e_1 + \frac{R_f}{R_2} e_2\right) \tag{9-21}$$

The input voltages e_1 and e_2 are multiplied by the constants R_f/R_1 and R_f/R_2, respectively, and then added algebraically. The values of the constants can be greater than, equal to, or less than one. The sign inversion is due to the operation with an amplifier.

SUMMATION OF INTEGRALS: Let R_1, R_2, and $1/Cs$ be substituted for Z_1, Z_2, and Z_f in Eq. 9-16, corresponding to the input voltages e_1 and e_2, the output voltage is

$$e_o = -\left(\frac{1}{R_1Cs} e_1 + \frac{1}{R_2Cs} e_2\right) \tag{9-22}$$

or

$$e_o = -\int_o^t \left(\frac{1}{R_1C} e_1 + \frac{1}{R_2C} e_2 \right) dt + E_o \tag{9-23}$$

This equation shows that each of the input voltages is multiplied by a constant and then integrated with respect to time. The operation also involves a sign inversion. Although Eqs. (9-21) and (9-23) appear different in form, they both stem from the summation operation of Eq. (9-16).

Case 5. Differentiation

Theoretically, it is possible to perform the differentiation operation with the circuit shown in Fig. 9-1. Substituting $1/Cs$ and R for the impedances Z_i and Z_f in Eq. (9-12) the operational equation becomes

$$e_o = -RCse_i \quad \text{or} \quad e_o = -RC\frac{de_i}{dt} \tag{9-24}$$

Since differentiation is a noise-amplifying process, it is usually avoided in computer work. The equations in a problem can usually be rearranged so that the integration operation, rather than differentiation, is used. If differentiation cannot be avoided, an approximate differentiation circuit may be used in order to keep the noise at a reasonable level.

9-5. COMPUTER DIAGRAM NOTATIONS

The symbols used to represent the computer diagrams differ somewhat for different computers because of design differences. Two types of notations are in common use.

In the computer diagrams presented, it is assumed that the resistors and capacitors are connected externally to the amplifier. Hence each of the computer elements and its connection to the amplifier are shown in the diagram. On the other hand, if a computer has the resistors and capacitors connected internally, it is not necessary to indicate the internal connections, and the constant with which the input voltage is multiplied is indicated in the diagram. Hence a different type of notation is used. Examples of these two types of notations are illustrated and compared in Table 9-3.

TABLE 9-3

Comparison of Computer Diagram Notations (Resistors Are in Megohms and Capacitors in Microfarads)

Sign inverter

$$e_o = -e_i$$

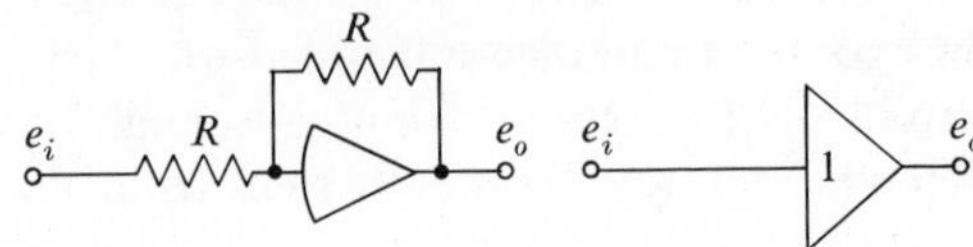

Integrator

$$e_o = -0.6\int_o^t e_i\,dt + E_o$$

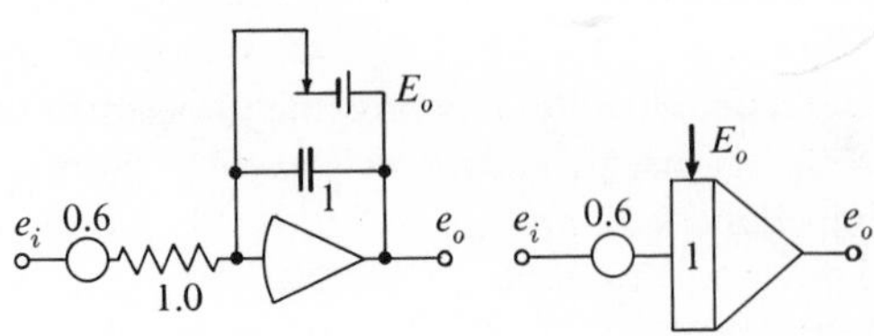

Summer

$$e_o = -(2e_1 + e_2)$$

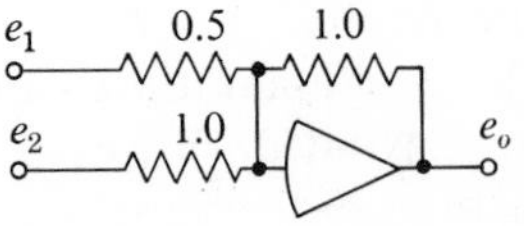

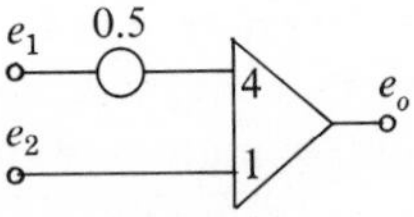

Summer-integrator

$$e_o = -\int_o^t (5e_1 + 2e_2)\,dt + E_o$$

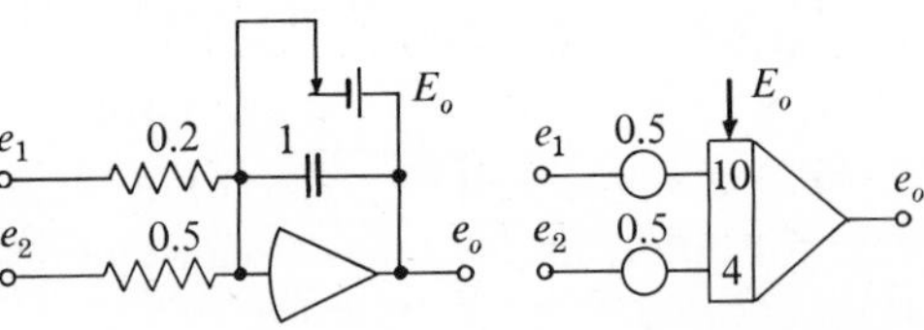

9-6. LINEAR COMPUTER CIRCUIT: ONE-DEGREE-OF-FREEDOM SYSTEM

The steps involved in using the electronic analogue computer for solving problems in vibrations are (1) the setting up of the differential equations of motion, (2) the rearranging of the equations suitable for computer operations, and (3) the selection of the proper scaling to convert the computer units to the units of the physical problem. The setting up of the equations of motion was discussed in previous chapters. In this section we shall discuss the rearranging of the equations in order to set up the computer circuit. The subject of scaling will be treated in the next four sections.

Consider a one-degree-of-freedom system, shown in Fig. 9-3, in which k is the spring constant of the spring, c is the damping coefficient of the damper, and m is the mass of the system. The excitation force $F \sin \omega t$ is applied to the mass. The equation of motion of the system is

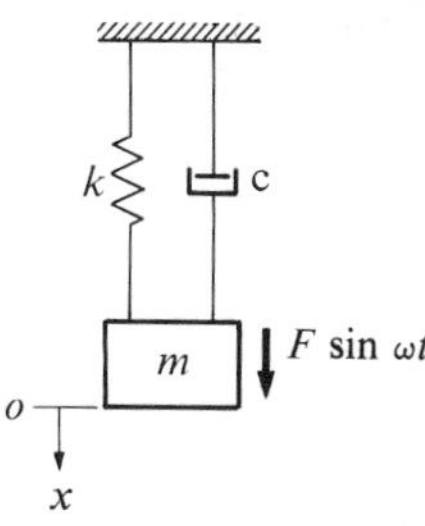

Fig. 9-3. *A one-degree-of-freedom vibratory system*

$$m\ddot{x} + c\dot{x} + kx = F \sin \omega t \tag{9-25}$$

To write this equation in a more convenient form, we divide the equation by m. Defining $c/m = 2\zeta\omega_n = a_1$, $k/m = \omega_n^2 = a_2$, and $(F/m) \sin \omega t = f(t)$, we obtain

$$\ddot{x} + 2\zeta\omega_n\dot{x} + \omega_n^2 x = f(t) \tag{9-26}$$

or

$$\ddot{x} + a_1\dot{x} + a_2 x = f(t) \tag{9-27}$$

Equation (9-27) can be rearranged for computer solution by writing the equation so that the highest-order derivative of the dependent variable appears alone on the left side of the equation.

$$\ddot{x} = -[a_1\dot{x} + a_2 x - f(t)] \tag{9-28}$$

This arrangement indicates that the acceleration $\ddot{x}(t)$ is obtained from the summation of the three quantities on the right side of Eq. (9-28). It is assumed in this method of solution that (1) a suitable arrangement of the computer elements ensures the existence of the correct mathematical relations between the voltages in the computer, (2) the voltages corresponding to the velocity and displacement can be obtained from the acceleration voltage by successive integrations, and (3) the quantities in the square bracket are available as inputs to a summer in order to obtain the acceleration. The equation is solved when the dependent variable $x(t)$ is explicit in terms of the independent variable t.

The computer circuit shown in Fig. 9-4 is arranged corresponding to Eq. (9-28). It should be noted that there is always a sign inversion associated with an operation with the amplifier. For simplicity, let the values of the R and C elements in the circuit be 1 megohm and 1 μf, respectively, and let the coefficients a_1 and a_2 be obtained from the potentiometer settings. It is easy to verify that the relation indicated in Eq. (9-28) is satisfied by the summing amplifier 1. The integration of $\ddot{x}(t)$ with amplifier 2 gives $-\dot{x}(t)$, and the integration of $-\dot{x}(t)$ with amplifier 3 yields $x(t)$. Amplifier 4 is used as a sign changer to obtain $+\dot{x}(t)$ from $-\dot{x}(t)$. The proper inputs to the summing amplifier 1 are obtained by multiplying $\dot{x}(t)$ by the constant a_1, multiplying $x(t)$ by a_2, and obtaining the excitation function $-f(t)$ from an external source. The acceleration, velocity, and displacement voltages can be measured

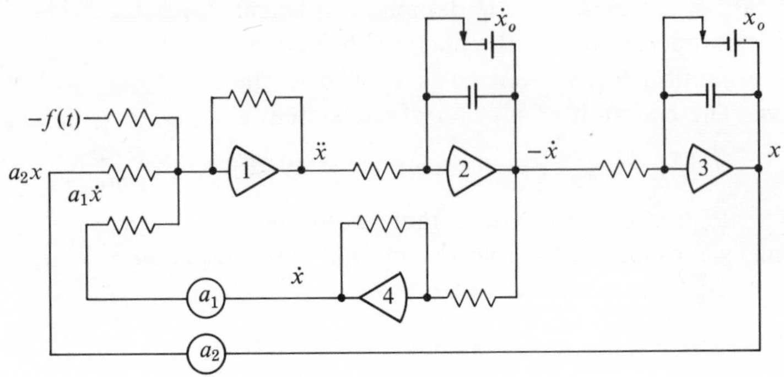

Fig. 9-4. *Computer circuit for* $\ddot{x} + 2\zeta\omega_n\dot{x} + \omega_n{}^2x = f(t)$

from the outputs of amplifiers 1, 4, and 3 and recorded by means of a suitable recorder.

It is noted in Fig. 9-4 that the voltages $a_1\dot{x}(t)$ and $a_2x(t)$, which must appear at the summing-amplifier input terminals, are not independent of $\ddot{x}(t)$. In fact, they must be obtained from $\ddot{x}(t)$ through successive integrations and multiplications by constants. The voltage $\ddot{x}(t)$ is treated as a known quantity, and we close the feedback paths to compel agreement between the right- and left-hand sides of Eq. (9-28). In other words, the output of amplifier 1 is needed in order to have available its inputs. The question is "whether the egg or the chick comes first." The answer to the dilemma is that a computer circuit deals with instantaneous values; given the instantaneous value of one quantity, we have the instantaneous values of the other quantities in the circuit. When thinking of a sequence of events, there is a time interval between the happening of one event and the subsequent one. With electronic circuits, the time interval between the input to an amplifier and its corresponding output is negligibly small. Hence the circuit works with instantaneous values. For example, it is not necessary to have the complete solution for velocity before it can be integrated to give the displacement, and there is no time interval between the instantaneous values of these quantities in the circuit.

Alternatively, Eq. (9-27) can be rearranged for computer solution as

$$\dot{x} = -\int_o^t [a_1\dot{x} + a_2x - f(t)]\,dt + \dot{x}(o) \tag{9-29}$$

This arrangement indicates that the equation of motion can be solved by making the three quantities in the bracket of Eq. (9-29) available as inputs to a summer integrator, as shown in Fig. 9-5. The output of the summer integrator amplifier 1 is $\dot{x}(t)$. The integration of $\dot{x}(t)$ gives

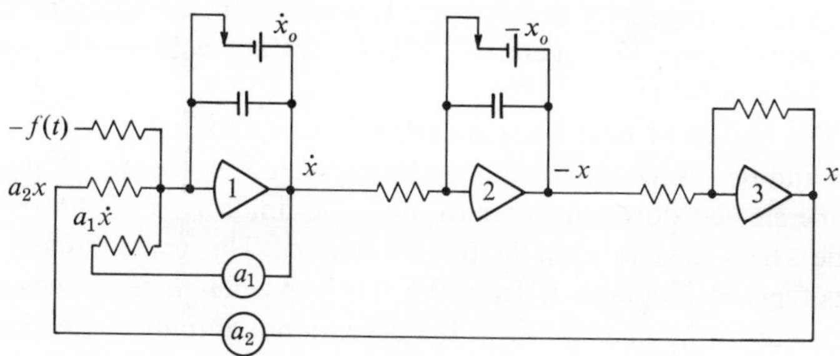

Fig. 9-5. *Computer circuit for* $\ddot{x} + 2\zeta\omega_n\dot{x} + \omega_n^2x = f(t)$

$-x(t)$. Amplifier 3 is used as a sign changer to convert $-x(t)$ to $+x(t)$. The appropriate inputs to the summer integrator are made available by multiplying $\dot{x}(t)$ and $x(t)$ with constants a_1 and a_2, respectively, and obtaining $-f(t)$ from an external voltage source.

Comparing Figs. 9-4 and 9-5, it is observed that we have two analogue-computer circuits for solving the same differential equation. Both circuits are practical, and an analogue-computer circuit is not unique. Since $\ddot{x}(t)$ does not appear in the circuit shown in Fig. 9-5, this circuit can be used if the acceleration is not observed.

9-7. TIME AND MAGNITUDE SCALING

The dependent and independent variables of the analogue computer are volts and time in seconds. Scaling is necessary to relate the variables of a physical problem with the computer variables. The general procedure is to time scale the independent problem variable before an attempt is made to magnitude scale the dependent variables. It is desirable to scale the given equations to conform to the capabilities of the computer and its associated equipment.

Time-scale change is necessary if the independent variable of the physical problem is other than time, such as displacement or angle. When time is the independent variable of a physical problem, the problem time may be in microseconds or hours. The computer time is always in seconds. The generally acceptable computer solution time varies from a few seconds to 15 min. Hence the problem may be slowed down or speeded up on a computer, and a time-scale change may be necessary even when time is the independent variable of the problem. Time-scale change may be performed either on the equation or on the computer. These two methods of time scaling will be discussed in Secs. 9-8 and 9-11.

The choice of time scale is dictated by the limitation of the computer and its associated equipment at high and low speeds. A problem may be slowed down in order to meet the limitations of the recorder or the servo devices used in the computer. The usual range of the stylus-type oscillograph is from 0.1 to 10 cps. High frequencies may contribute to the phase shift in the operational amplifiers. A problem may be speeded up to overcome the limitations of the equipment and to make maximum use of the computer time. For example, errors occur because of the very small but finite grid current at the input of the amplifier and the drift of zero of the reference level in the various parts of the computer. The effect of these errors is decreased by keeping the computing time small. Even if the integrators were perfect, the percentage error due to the imperfect setting of voltages will increase with a larger computing time. For example, if it is desired to integrate a constant voltage E and if ε is a small but constant error associated with the voltage setting, then the output of the integrator is

$$e_o = -\int_o^t (E + \varepsilon)\,dt = -(Et + \varepsilon t) \tag{9-30}$$

This equation shows that if e_o maximum is ± 100 volts, the voltage E can be large if the integration time is short. Hence the percentage error is less for a shorter computing time.

The scaling of the dependent variable is called magnitude or amplitude scaling. The selection of the scale factor is dictated by the maximum and minimum voltage limitations of the computer. The normal output of an amplifier is ± 100 volts. Operation beyond this range will cause the amplifiers to saturate, and the amplifier output may not be linear with the input. On the other hand, the magnitude of the operating voltage should be above a reasonable minimum in order that the random-noise voltages and the error voltages be minimized.

9-8. CHANGING THE TIME SCALE OF THE EQUATION

To change the time scale of an equation, it is only necessary to make a substitution for the independent variable. Let τ be the computer time and t be the problem time, which is often called the "real" time. The term "real time" has no particular significance, because the computer time, as measured by the clock, is just as real. The time scale α is defined by the relation

$$\tau = \alpha t \tag{9-31}$$

If α is greater than unity, the solution is slowed down by a factor α. If α is less than unity, the solution is speeded up by a factor α.

In changing the time scale of a differential equation, the time derivative becomes

$$\frac{d}{dt} = \frac{d}{d\tau/\alpha} = \alpha \frac{d}{d\tau} \tag{9-32}$$

In general, we have

$$\frac{d^n}{dt^n} = \alpha^n \frac{d^n}{d\tau^n} \tag{9-33}$$

Conversely, in the computer solution, the recorded derivatives are derivatives with respect to τ rather than t. All the results from the computer solution are recorded with respect to τ. Hence the relations in Eqs. (9-31) and (9-33) should be used in the interpretation of the computer solution.

Example 1. Assuming that $\tau = 5t$, change the time scale of the second-order differential equation

$$\ddot{x} + 2.4\dot{x} + 100x = 40 \sin 7t \tag{9-34}$$

with the initial conditions

$$x(o) = 1 \quad \text{and} \quad \dot{x}(o) = 2 \tag{9-35}$$

Solution: To change the time scale, we substitute $t = \tau/5$, $\dfrac{d}{dt} = 5\dfrac{d}{d\tau}$, and $\dfrac{d^2}{dt^2} = 5^2\dfrac{d^2}{d\tau^2}$ in Eq. (9-34) and obtain

$$5^2\ddot{x}(\tau) + (2.4)(5)\dot{x}(\tau) + 100x(\tau) = 40 \sin 7(\tau/5) \quad \textbf{(9-36)}$$

or

$$\ddot{x}(\tau) + 0.48\dot{x}(\tau) + 4x(\tau) = 1.6 \sin 1.4\tau \quad \textbf{(9-37)}$$

The corresponding initial conditions become

$$x(o) = 1 \quad \text{and} \quad \dot{x}(o) = 2/5 = 0.4 \quad \textbf{(9-38)}$$

It should be observed that the derivatives in the initial conditions are also changed by the relation shown in Eq. (9-33). The dependent variable $x(\tau)$ and its derivatives are now functions of τ.

The natural frequency from Eq. (9-34) is $\omega_n = \sqrt{100} = 10$ rad-sec^{-1} and that from Eq. (9-37) is $\omega_n = \sqrt{4} = 2$ rad-sec^{-1}. Hence the original equation is slowed down by a factor of 5 as specified by $\tau = 5t$. The damping factor from Eq. (9-34) is $\zeta = 2.4/2\omega_n = 2.4/(2)(10) = 0.12$, and that from Eq. (9-37) is $\zeta = 0.48/2\omega_n = 0.48/(2)(2) = 0.12$. Hence the damping characteristics of the systems described by these equations are identical. It should be emphasized that time scaling does not change the characteristics of a problem other than time; otherwise, we would be faced with an entirely new problem.

9-9. MAGNITUDE-SCALE FACTOR

Magnitude scaling consists of replacing the dependent variable of the problem by that of the computer. For example, if a problem variable is $x(t)$ in inches and the corresponding computer variable is $e_x(t)$ in volts, these variables can be related as

$$e_x = K_x x \quad \textbf{(9-39)}$$

where K_x is the magnitude-scale factor of the variable $x(t)$ and it has the units of volts-in.$^{-1}$ Similarly, the magnitude scaling of velocity, or other quantities, can be expressed as

$$e_{\dot{x}} = K_{\dot{x}}\dot{x}$$

where $K_{\dot{x}}$ has the units of volts-in.$^{-1}$-sec.

The normal maximum output of an amplifier is ± 100 volts. This range should not be exceeded in order to avoid the saturation, or the "overloading," of the amplifiers. Most computers have some signaling device to indicate this overloading. For accuracy of computing and recording, the peak voltages should not be near zero. It is recommended that the output voltages of all amplifiers range between ± 10 volts and ± 100 volts. Since errors may occur in estimating the range of the problem variables, the scale factor should be selected to allow an appropriate margin for this error.

One method of magnitude scaling is to associate a scaling factor K with each physical unit. The scale factor K_x for a problem variable $x(t)$ is determined by the relation

$$K_x \leq \frac{e_{\max}}{x_{\max}}, \text{ volts/physical unit} \tag{9-40}$$

For example, if $e_{\max}$ is 100 volts and $x_{\max}$ is estimated to be 2.5 in., then K_x should be equal to or less than $100/2.5 = 40$ volts-in.$^{-1}$ It is preferable to choose K_x less than 40 to allow for possible errors in estimating $x_{\max}$. If the K_x selected is 40 volts-in.$^{-1}$, the variable is labeled as $40x$ on the computer diagram, and the scale factor with dimensional units is noted in a corner of the diagram. Similarly, other scale factors are selected for the rest of the variables in the problem. This method of scaling is illustrated in Example 2.

An alternative method is to define the scale factor K for every problem variable by the relation

$$K = 1, \text{ volt/physical unit} \tag{9-41}$$

For example, the scale factor for displacement is 1 volt-in.$^{-1}$ and that for velocity is 1 volt-in.$^{-1}$-sec. It will be shown in Example 3 that if the output of an amplifier is labeled as $80x$ and 80 volts is measured, the corresponding displacement is 1 in. Having selected this *unity scale factor*, the voltage level of the problem on the computer is adjusted by multiplying the equation through by an appropriate constant. The constant is selected so that the output of the amplifier satisfying the differential equation does not exceed the desired limit of ± 100 volts. [See Eqs. (9-28) and (9-29).] This method of scaling is illustrated in Example 3.

Although the two schemes of magnitude scaling are similar in some respects, the second method has the advantage of focusing attention on the problem variables instead of the computer variables which are in volts. If the magnitude-scale factor is unity, the voltage output of an amplifier can be considered simply as a number without dimension. When using a slide rule for multiplication, we do not necessarily assign a dimension to the numbers on the slide rule. If the computer variable is considered as a number, we are more apt to think in terms of the problem variables than volts.

Since an estimation of the maximum values, or the magnitude, of the problem variables is required for magnitude scaling, we shall discuss this subject before presenting the examples on scaling.

9-10. ESTIMATION OF MAXIMUM VALUES

Great accuracy in estimating the magnitudes of the problem variables is not usually required. Since the operating voltages may be adjusted up or down, it is necessary only to change the magnitude scale to remedy the errors in the estimation. The magnitude of the problem variables and their derivatives may be estimated from the knowledge of the physical problem or the equations of the problem statement. It may only be necessary to make reasonable guesses. A few quick runs on the computer will reveal whether the guesses are adequate: it is hardly justifiable to go through a large number of numerical calculations.

Consider the equation

$$\ddot{x} + a_1\dot{x} + a_2 x = f(t) \tag{9-42}$$

Assume that f_m is the magnitude of $f(t)$ and that it is known from a knowledge of the physical problem. As a first approximation, let us neglect the damping effect, that is, $a_1 = 0$, and assume that $f(t) = f_m$. Assuming zero initial conditions, it can be shown that the displacement $x(t)$ is

$$x = \frac{f_m}{a_2}(1 - \cos \omega_n t) \tag{9-43}$$

where $\omega_n = \sqrt{a_2}$. Hence the magnitude of $x(t)$ is estimated to be

$$x_m = 2f_m/a_2 \tag{9-44}$$

From Eq. (9-43) it is also deduced that the magnitudes of the velocity and acceleration are

$$\dot{x}_m = \omega_n f_m/a_2 \tag{9-45}$$

and

$$\ddot{x}_m = \omega_n^2 f_m/a_2 \tag{9-46}$$

It should be restated that one method of estimating the maximum values of the problem variables and their derivatives is as good as another. If it is uncertain whether the estimation is adequate, magnitude-scale factors of lower values should be used. If an attempt is made to keep all peak voltages in the neighborhood of ± 50 volts, satisfactory operation will usually be achieved even when considerable error is made in the estimation.

Example 2. A vibratory system is described by the equation

$$\ddot{x} + 2.4\dot{x} + 100x = 0$$

with the initial conditions $x(0) = 1$ and $\dot{x}(0) = 2$. (*a*) Rewrite the equation to slow down the problem by a factor of 5, (*b*) select the proper magnitude-scale factor for the computer solution of the problem, and (*c*) sketch the computer diagram.

Solution: (*a*) The slowed-down equation and the corresponding initial conditions are (see Example 1)

$$\begin{aligned} \ddot{x}(\tau) + 0.48\dot{x}(\tau) + 4x(\tau) &= 0 \\ x(0) = 1 \quad \text{and} \quad \dot{x}(0) &= 0.4 \end{aligned} \tag{9-47}$$

(*b*) It is necessary to determine the magnitude of $x(\tau)$ and its derivatives before the magnitude-scale factor can be assigned. As an approximation, assume that $x_m = x(0) = 1$. The undamped natural frequency of the system is $\omega_n = \sqrt{4} = 2\,\text{rad-sec}^{-1}$. Neglecting damping, the magnitudes of the velocity and acceleration are estimated to be $\dot{x}_m = \omega_n x_m = 2$ and $\ddot{x}_m = \omega_n^2 x_m = 4$. Hence the magnitude-scale factors can be selected by the relations

$$K_x \leq 100/1 = 100 \text{ volts-in.}^{-1}$$
$$K_{\dot{x}} \leq 100/2 = 50 \text{ volts-in.}^{-1}\text{-sec}$$
$$K_{\ddot{x}} \leq 100/4 = 25 \text{ volts-in.}^{-1}\text{-sec}^2$$

Let $K_x = 80$, $K_{\dot{x}} = 40$, and $K_{\ddot{x}} = 20$ be selected for the problem.

(*c*) Equation (9-47) is rearranged for computer operation as

$$\ddot{x} = -(0.48\dot{x} + 4x)$$

Multiplying the equation by $K_{\ddot{x}} = 20$ gives

$$20\ddot{x} = -(9.6\dot{x} + 80x)$$

This equation is further modified by $K_x = 40$ and $K_{\dot{x}} = 80$ to yield

$$20\ddot{x} = -[(0.24)(40\dot{x}) + (1)(80x)]$$

The scaled computer diagram corresponding to this equation is shown in Fig. 9-6. The outputs of amplifiers 1, 2, and 3 are labeled as $20x$, $-40x$, and $80x$, respectively. The appropriate initial conditions are assigned to the integrators. Since the output of amplifier 3 is labeled as $80x$ and $x(0)$ is equal to 1, the capacitor of amplifier 3 is charged to +80 volts. The output of amplifier 2 is labeled as $-40x$ and $\dot{x}(0)$ is equal to 0.4; hence the capacitor of amplifier 2 is charged to −16 volts. The coefficient 0.24 is obtained from a potentiometer at the output of amplifier 4.

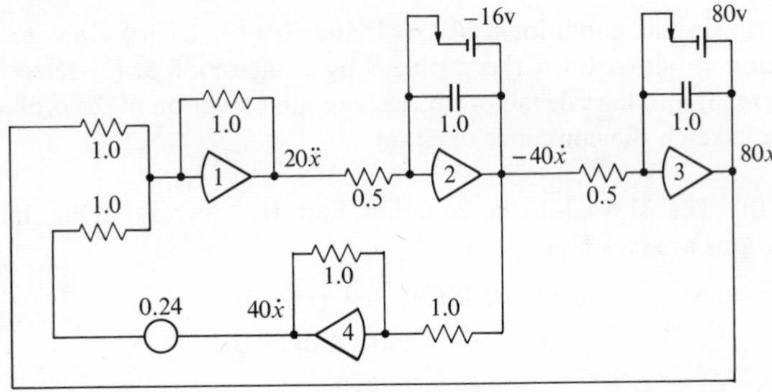

Fig. 9-6. *Computer circuit for* $\ddot{x} + 0.48\dot{x} + 4x = 0$ *with initial conditions* $x(0) = 1$ *and* $\dot{x}(0) = 0.4$, $K_x = 80$, $K_{\dot{x}} = 40$, *and* $K_{\ddot{x}} = 20$

Example 3. Employing the data of Example 2, sketch the computer diagram with the use of the unity magnitude-scale factor.

Solution: With the data from Example 2, we assume that $x_m = 1$, $\dot{x}_m = 2$, and $\ddot{x}_m = 4$, and we rearrange Eq. (9-47) to give

$$\ddot{x} = -(0.48\dot{x} + 4x)$$

To raise the voltage level of the circuit and to keep the output of the summing amplifiers within ±100 volts, this equation is multiplied by a factor of 20.

$$20\ddot{x} = -(9.6\dot{x} + 80x)$$

The computer diagram corresponding to this equation is shown in Fig. 9-7. The output of amplifier 1 is $20\ddot{x}$. It is multiplied by a constant of (9.6/20) through a potentiometer to obtain $(9.6/20)(20\ddot{x}) = 9.6\ddot{x}$. This quantity is integrated to give $-9.6\dot{x}$. The rest of the circuit diagram is self-evident.

The last two examples indicate that the differential equation $\ddot{x} = -(0.48\dot{x} + 4x)$ can be satisfied by either of the computer circuits. In both cases the relation is satisfied by the summer amplifier 1. It is evident that the circuit elements and the voltage level of a computer circuit can be adjusted at will as long as the original equation is satisfied and the output voltage of each amplifier is at a reasonable level.

Consider the circuit shown in Fig. 9-7. The output of amplifier 2 can be changed from $-9.6\dot{x}$ to $-40\dot{x}$ by a change of its circuit elements.

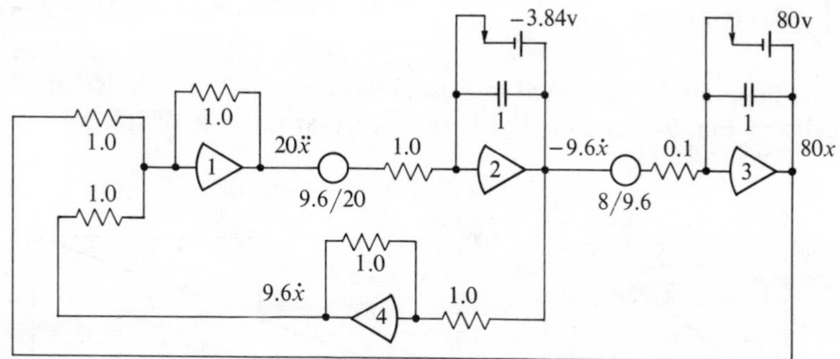

Fig. 9-7. *Computer circuit for* $\ddot{x} + 0.48\dot{x} + 4x = 0$ *with initial conditions* $x(0) = 1$ *and* $\dot{x}(0) = 0.4$ *and unity magnitude-scale factor*

Correspondingly, the circuit elements of amplifier 3 can be changed to obtain $80\dot{x}$ at its output. The resultant circuit can be made identical to that shown in Fig. 9-6. Since the circuits shown in Figs. 9-6 and 9-7 are derived from two methods of magnitude scaling and the circuits can be changed to become identical, it may be concluded that the two approaches to magnitude scaling are essentially two ways of looking at the same problem. It is convenient to use the unity magnitude-scale factor and to adjust the output voltage of an amplifier when it is necessary or expedient to do so.

9-11. CHANGING THE TIME SCALE OF THE COMPUTER

Time scaling, as performed on the equation of the physical system, was illustrated in Sec. 9-8. Alternatively, time scaling can be performed on the computer by changing its rate of integration. Consider the integration

$$\int_o^t 20\,dt = \int_o^5 20\,dt = (20)(t)\Big|_o^5 = 100 \tag{9-48}$$

Let the independent variable t be changed to τ by the relation $\tau = 2t$. Since the range of t is from 0 to 5, the range of τ is from 0 to 10. Substituting $t = \tau/2$ in Eq. (9-48) gives

$$\int_o^{\tau/2} 20\,d(\tau/2) = \frac{1}{2}\int_o^{\tau} 20\,d\tau = \frac{1}{2}\int_0^{10} 20\,d\tau = \frac{1}{2}(20)\,(\tau)\Big|_0^{10} = 100 \tag{9-49}$$

The integrations expressed in Eqs. (9-48) and (9-49) are shown graphically in Fig. 9-8(a) and (b). It is noted that the rate of integration is

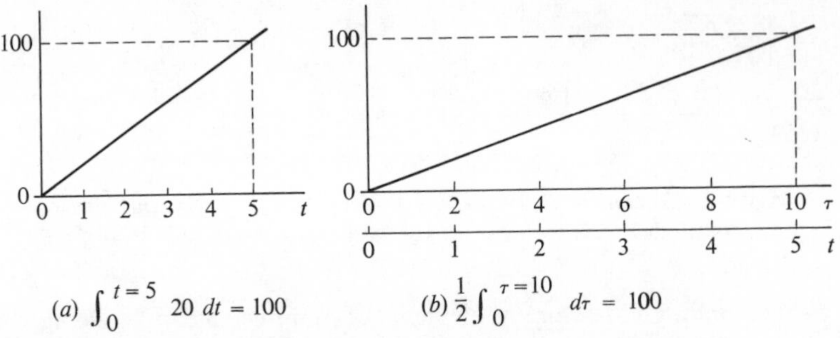

Fig. 9-8. *Change of time scale of integration* $\tau = 2t$

changed by the substitution of τ for t, and that the two curves are identical if the τ scale in Fig. 9-8(b) is relabeled in terms of t. Comparing Eqs. (9-48) and (9-49), we observe that the two integrals are identical in form, with the exception that one of them is multiplied by a constant of 1/2.

An integrator in the computer performs the operation

$$e_o(t) = -\frac{1}{RC}\int_o^t e_i(t)\,dt + E_o \tag{9-50}$$

Let τ be the new independent variable specified by the relation $\tau = \alpha t$. Substituting $t = \tau/\alpha$ in Eq. (9-50) gives

$$e_o(\tau/\alpha) = -\frac{1}{RC}\int_o^{\tau/\alpha} e_i(\tau/\alpha)\,d(\tau/\alpha) + E_o \tag{9-51}$$

or

$$e_o(\tau/\alpha) = -\frac{1}{\alpha RC}\int_0^{\tau} e_i(\tau/\alpha)\,d\tau + E_o \tag{9-52}$$

Hence the time scale of the integrator is changed by the substitution of the time constant αRC for RC.

Since a differential equation is solved on an analogue computer by means of successive integrations, the time scale of a computer circuit can be changed by the substitution of a new time constant αRC for all the integrators in the circuit. If $\alpha > 1$, the computer solution is slowed down by a factor of α. If $\alpha < 1$, the computer solution is speeded up by a factor of α.

The machine time τ is the independent variable in Eq. (9-52). If $e_o(\tau/\alpha)$ is recorded, the time axis of the recording is in τ. If the time axis is relabeled in terms of $\tau/\alpha = t$, however, the recorded curve becomes a plot of the function $e_o(t)$ versus t. Hence, by relabeling the τ axis, the solution obtained is the same as if the time scale had not been changed. The functions $e_i(\tau/\alpha)$ and $e_o(\tau/\alpha)$ in Eq. (9-52) may represent velocity and displacement, or acceleration and velocity. Since their values at a given time, t or τ, are not influenced by the time scaling, it is not necessary to change the derivatives by a factor of α^n as indicated by the method shown in Sec. 9-8. Furthermore, the initial condition E_o remains unchanged.

In summary, the steps involved in solving an equation by this method of time scaling may be briefly outlined as follows: (1) estimating the magnitudes of the problem variables and their derivatives, (2) magnitude scaling the equations of the physical problem, (3) setting up the computer circuit, (4) changing the RC time constant of all integrators to time-scale change the computer, and (5) relabeling the time axis of the recordings by the relation $t = \tau/\alpha$. We shall illustrate these steps with an example.

Example 4. A vibratory system is described by the equation

$$\ddot{x} + 2.4\dot{x} + 100x = 0$$

with the initial conditions $x(0) = 1$ and $\dot{x}(0) = 2$. Slow down the problem by a factor of 5 by time scaling on the computer.

Solution: In estimating the magnitude of the problem variables, assume that

$$x_m = x(0) = 1$$

$$\dot{x}_m = \omega_n x_m = (10)(1) = 10$$

$$\ddot{x}_m = \omega_n^2 x_m = (10)^2(1) = 100$$

The given equation can be expressed as $\ddot{x} = -(2.4\dot{x} + 100x)$. Using the unity magnitude-scale factor and attempting to limit the voltage corresponding to the acceleration within ± 100 volts, we multiply this equation by 0.8 and obtain $0.8\ddot{x} = -(1.92\dot{x} + 80x)$.

The computer circuit of this equation is shown in Fig. 9-9. The output of amplifier 1 is $0.8\ddot{x}$, as specified by this equation. It is expedient, however, to adjust the circuit elements of amplifier 2 to give an output of $-8\dot{x}$ instead of $-1.92\dot{x}$ as specified. The quantity

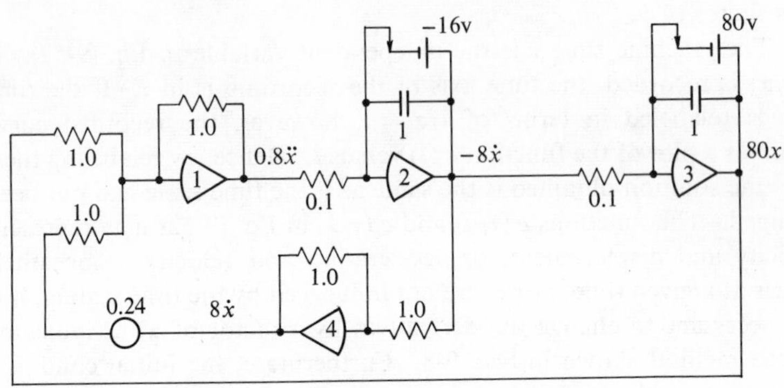

Fig. 9-9. *Computer circuit for* $\ddot{x} + 2.4\dot{x} + 100x = 0$ *with initial conditions* $x(0) = 1$ *and* $\dot{x}(0) = 2$

$1.92\dot{x}$ is made available to the summer amplifier 1 by the sign-changing amplifier 4 and a potentiometer set at 0.24. Since $8\dot{x}_m$ and $80x_m$ are both estimated equal to 80 volts, the outputs of amplifiers 2 and 3 are well within the ± 100-volt range. The initial conditions of the integrators are set according to the values of $x(0)$ and $\dot{x}(0)$, as given in the problem.

To slow down the computer operation by a factor of 5, replace the 0.1-megohm input resistor of the integrating amplifier 2 with a 0.5-megohm resistor, and do likewise for amplifier 3.

The displacement and velocity curves of the problem solution are as shown in Fig. 9-10. Figures 9-10(*a*) and (*b*) show the curves with $\tau = t$. Figures 9-10(*c*) and (*d*) show the same curves with $\tau = 5t$. The computer time τ, which is also the clock time, is used for the abscissa scale for the second set of curves. If the τ-axis is relabeled by the relation $t = \tau/5$, the two sets of curves in Fig. 9-10 will be identical.

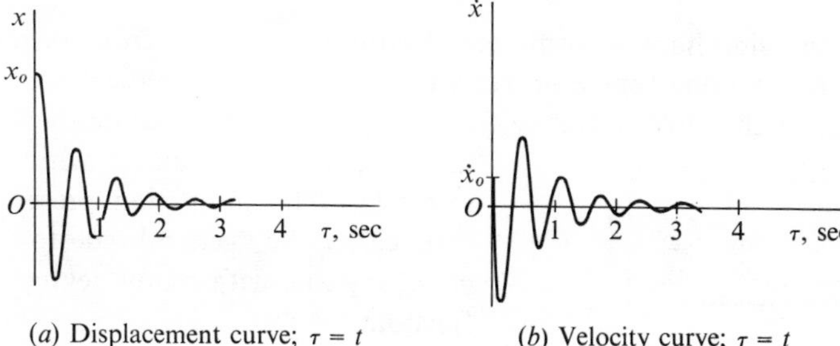

(*a*) Displacement curve; $\tau = t$ (*b*) Velocity curve; $\tau = t$

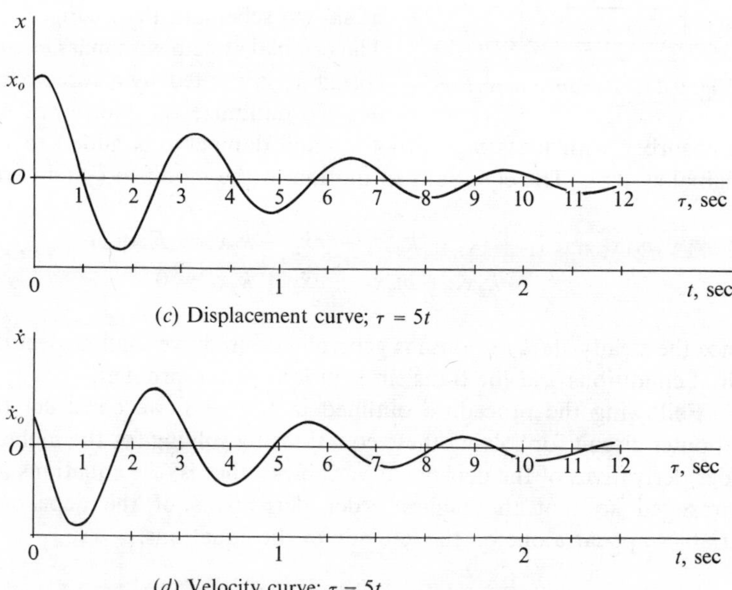

(*c*) Displacement curve; $\tau = 5t$

(*d*) Velocity curve; $\tau = 5t$

Fig. 9-10. *Computer solution of* $\ddot{x} + 2.4\dot{x} + 100x = 0$ *with initial conditions* $x(0) = 1$ *and* $\dot{x}(0) = 2$; *time-scale change on the computer*

9-12. LINEAR COMPUTER CIRCUIT: MORE-THAN-ONE-DEGREE-OF-FREEDOM SYSTEM

The computer circuit for a one-degree-of-freedom system was discussed in Sec. 9-6. The method presented is equally applicable for solving the simultaneous differential equations arising from systems with more than one degree of freedom. We shall discuss the computer circuit through a two-degree-of-freedom system and illustrate the time and magnitude scaling with an example. The discussion presented can easily be generalized to include systems with many degrees of freedom.

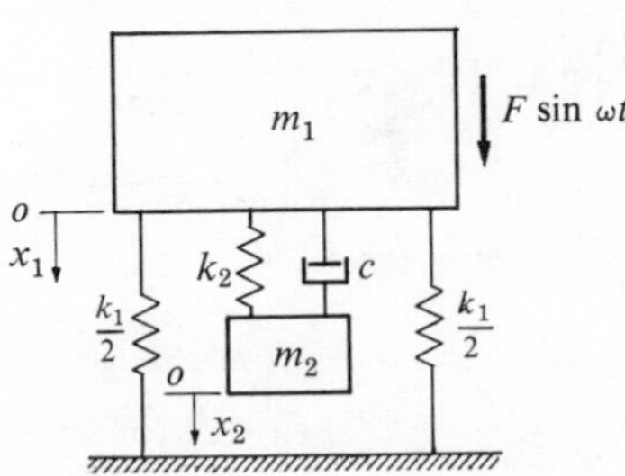

Fig. 9-11. *Dynamic absorber*

Consider the dynamic absorber with damping (see Sec. 3-6, Chap. 3) as shown schematically in Fig. 9-11. The original system with mass m_1 and spring k_1 is excited by a force $F \sin \omega t$. To minimize the motion of m_1, an absorber with mass m_2, spring k_2, and damper c is added to the original system. The equations of motion of the modified system are

$$\begin{aligned} m_1\ddot{x}_1 + c\dot{x}_1 + (k_1 + k_2)x_1 - c\dot{x}_2 - k_2x_2 &= F \sin \omega t \\ -c\dot{x}_1 - k_2x_1 + m_2\ddot{x}_2 + c\dot{x}_2 + k_2x_2 &= 0 \end{aligned} \tag{9-53}$$

Since the steady-state response is generally desired, we shall neglect the initial conditions and the transient solutions of the problem.

Following the procedure outlined in Sec. 9-6, we construct the computer circuit for solving these equations by solving for the highest-order derivatives of the dependent variables; that is, the equations are rearranged so that the highest-order derivatives of the dependent variables appear alone on the left side of the equations.

$$\begin{aligned} \ddot{x}_1 &= -\left[\frac{c}{m_1}\dot{x}_1 + \frac{(k_1 + k_2)}{m_1}x_1 - \frac{c}{m_1}\dot{x}_2 - \frac{k_2}{m_1}x_2 - \frac{F}{m_1}\sin \omega t\right] \\ \ddot{x}_2 &= -\left(-\frac{c}{m_2}\dot{x}_1 - \frac{k_2}{m_2}x_1 + \frac{c}{m_2}\dot{x}_2 + \frac{k_2}{m_2}x_2\right) \end{aligned} \tag{9-54}$$

For convenience, these equations are rewritten as

$$\begin{aligned} \ddot{x}_1 &= -[a_1\dot{x}_1 + a_2x_1 - a_3\dot{x}_2 - a_4x_2 - f(t)] \\ \ddot{x}_2 &= -(-a_5\dot{x}_1 - a_6x_1 + a_7\dot{x}_2 + a_8x_2) \end{aligned} \tag{9-55}$$

This arrangement of the equations indicates that the acceleration $\ddot{x}_1$ can be obtained by summing the five quantities in the bracket on the right side of the first equation. Similarly, $\ddot{x}_2$ can be obtained. If it is not necessary to measure the accelerations $\ddot{x}_1$ and $\ddot{x}_2$, the equations may be alternatively rearranged as

$$\begin{aligned} \dot{x}_1 &= -\int [a_1\dot{x}_1 + a_2x_1 - a_3\dot{x}_2 - a_4x_2 - f(t)]\,dt \\ \dot{x}_2 &= -\int (-a_5\dot{x}_1 - a_6x_1 + a_7\dot{x}_2 + a_8x_2)\,dt \end{aligned} \tag{9-56}$$

With either of the arrangements, the correct mathematical relations are assured if the quantities on the right side of the equations are made available as inputs to the amplifier circuits to obtain the quantities on the left side of the equations.

The computer circuit shown in Fig. 9-12 is arranged corresponding to Eq. (9-56). For simplicity, assume that the values of the R and C

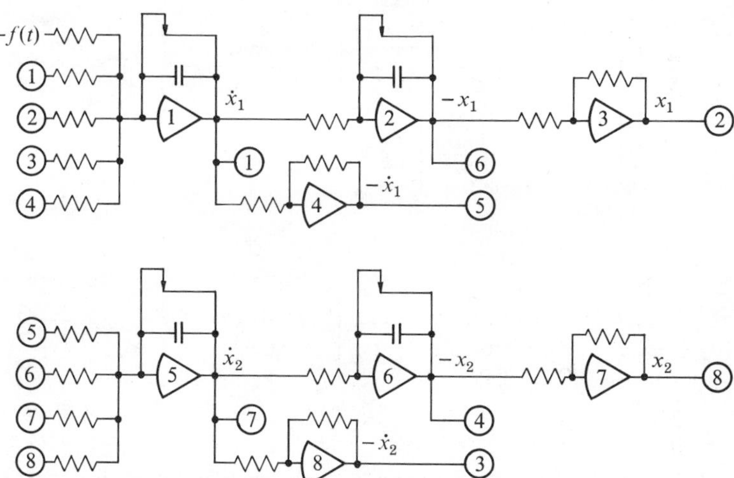

Fig. 9-12. *Computer circuit for solving the equations of motion of the dynamic absorber of Fig.* 9-11

elements in the circuit are 1 megohm and 1 μf, and that the coefficients a_1 to a_8 can be obtained from potentiometer settings. To avoid the use of long interconnection lines, the output and the input terminals of the amplifier circuits are labeled as terminals 1 to 8. For example, it is understood that the output terminal 1 is to connect to the input terminal 1. The input terminal, with a circle around a number, can be regarded as a notation for a potentiometer. For example, the potentiometer setting a_1 can be noted directly adjacent to the input terminal 1. As indicated in Fig. 9-12, the output of amplifier 1 is $\dot{x}_1$. Its inputs are obtained by multiplying $\dot{x}_1$ with a_1, x_1 with a_2, $-\dot{x}_2$ with a_3, $-x_2$ with a_4, and securing $-f(t)$ from an external voltage supply. Hence the mathematical relation shown in the first equation of Eq. (9-56) is satisfied by the summer integrator circuit of amplifier 1. The integration of $\dot{x}_1$ with amplifier 2 gives $-x_1$. Amplifiers 3 and 4 are used as sign changers. The arrangement of amplifiers 5 to 8 can be explained in like manner. It is essentially a duplication of the arrangement of amplifiers 1 to 4.

Comparing Figs. 9-5 and 9-12, it is noted that the arrangement of the amplifiers in Fig. 9-5 is identical to that of amplifiers 1 to 3 in

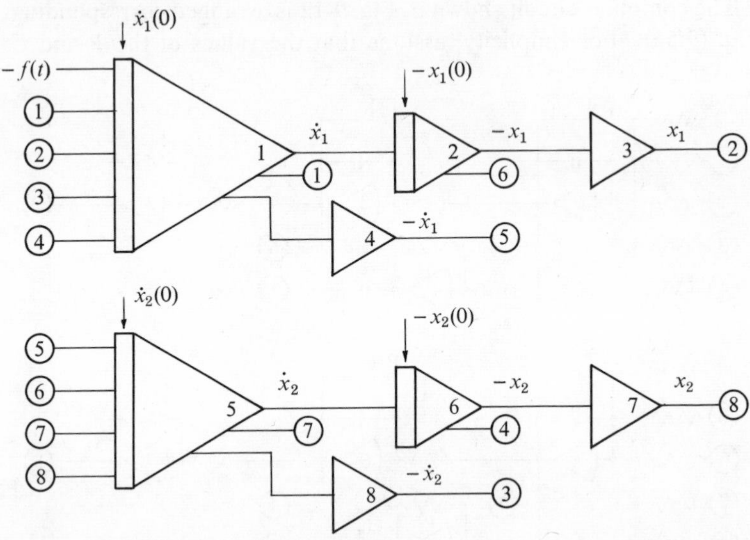

Fig. 9-13. *Computer circuit for solving the equations of motion of the dynamic absorber of Fig.* 9-11

Fig. 9-12. It requires two integrators and the necessary sign changers to solve a second-order linear differential equation with constant coefficients. Hence the circuit shown in Fig. 9-12 can be regarded as possessing two duplicating circuits, each having two integrators for the solution of a second-order differential equation. Since we have two simultaneous second-order differential equations in a two-degree-of-freedom system, the proper interconnections are made between the two parts of the circuit to solve the equations simultaneously. It is evident that this discussion can be extended to include systems with more than two degrees of freedom.

The computer circuit diagram of this problem for a computer with internally connected circuit elements is as shown in Fig. 9-13.

Example 5. A mass-spring system with $m_1 = 386$ lb and $k_1 = 8{,}100$ lb-in.$^{-1}$ and an excitation force of 200 sin $30\pi t$ lb is shown in Fig. 9-11. Since the system is being excited near resonance, a dynamic absorber with the parameters m_2, k_2, and c is added to the system. The mass of m_2 is selected to be one fifth of m_1. For a correctly "tuned" absorber, the spring constant k_2 can be determined by the relation

$$\frac{k_2}{k_1} = \frac{m_1 m_2}{(m_1 + m_2)^2} \tag{9-57}$$

The value of the damping coefficient c is to be determined by an analogue-computer solution of the problem. Set up an analogue-computer circuit to solve this problem.

Solution: The values of the parameters of the dynamic absorber are

$$m_1 = 386/g = 1 \text{ lb-sec}^2\text{-in.}; \quad k_1 = 8{,}100 \text{ lb-in.}^{-1}$$

$$m_2 = m_1/5 = 1/5 \text{ lb-sec}^2\text{-in.}; \quad k_2 = (5/36)k_1 = 1{,}125 \text{ lb-in.}^{-1}$$

Substituting these values in Eq. (9-53) gives

$$\begin{aligned} \ddot{x}_1 + c\dot{x}_1 + (8{,}100 + 1{,}125)x_1 - c\dot{x}_2 - 1{,}125x_2 &= 200 \sin 30\pi t \\ -c\dot{x}_1 - 1{,}125x_1 + (1/5)\ddot{x}_2 + c\dot{x}_2 + 1{,}125x_2 &= 0 \end{aligned} \tag{9-58}$$

It is necessary to estimate the magnitudes of the variables involved and to time and magnitude scale the equations before the circuit elements can be specified.

Since the excitation frequency is almost equal to the resonance frequency of the original system, it is difficult to estimate the magnitude of x_1 without some knowledge of the physical problem. The static deflection of k_1 due to a force of 200 lb is $x_{1(st)} = \dfrac{200}{8{,}100} = 0.0247$ in. It may be assumed that

$$x_{1(max)} = 4x_{1(st)} = 4(0.0247) \doteq 0.1 \text{ in.}$$

This assumption is justified from a knowledge of the problem. Alternatively, it may be argued that if the dynamic absorber is properly tuned, the amplitude of m_1 should not be excessive. Perhaps four times the static deflection is the right order of magnitude. Should this estimation prove to be wrong, the output voltages of the amplifiers can always be adjusted by changing the amplitude of the excitation voltage to the circuit. The magnitude of x_2 can be estimated from the second equation in Eq. (9-58). Neglecting all the terms with derivatives gives

$$x_{2(\max)} \doteq x_{1(\max)} = 0.1 \text{ in.}$$

Let us assume $x_{1(\max)} = x_{2(\max)} = 0.1$, $\dot{x}_{1(\max)} = \dot{x}_{2(\max)} = (30\pi)x_{1(\max)} \doteq 10$ in.-sec^{-1}, and that the accelerations are not required in the solution.

To time scale the equations, let us assume that the problem is slowed down by a factor of 10. Substituting $\tau = 10t$, $\dfrac{d}{dt} = 10\dfrac{d}{d\tau}$, and $\dfrac{d^2}{dt^2} = 10^2\dfrac{d^2}{d\tau^2}$ in Eq. (9-58) and rearranging, we have

$$\begin{aligned} \dot{x}_1 &= -\textstyle\int(0.1c\dot{x}_1 + 92.25x_1 - 0.1c\dot{x}_2 - 11.25x_2 - 2\sin 3\pi\tau)\,d\tau \\ \dot{x}_2 &= -\textstyle\int(-0.5c\dot{x}_1 - 56.25x_1 + 0.5c\dot{x}_2 + 56.25x_2)\,d\tau \end{aligned} \tag{9-59}$$

with the initial conditions

$$x_{1(\max)} = x_{2(\max)} = 0.1, \quad \dot{x}_{1(\max)} = \dot{x}_{2(\max)} = 1.0$$

It should be noted that slowing down a problem also has the effect of reducing the gain requirement of the amplifiers. For example, in Eq. (9-58), if $\ddot{x}(t)$ is given and integrated twice to obtain $(8{,}100 + 1{,}125)x_1$, the gain requirement is 9,225. If the problem is slowed down by a factor of 10, as shown in Eq. (9-59), however, the corresponding gain requirement is only $(1/10)^2(9{,}225) = 92.25$.

To magnitude scale the equations, let us assume that the circuit shown in Fig. 9-12 is used for the problem solution. Since the estimated maximum velocities are $\dot{x}_{1(\max)} = \dot{x}_{2(\max)} = 1.0$, let each of the equations in Eq. (9-59) be multiplied by 80.

$$\begin{aligned} 80\dot{x}_1 &= -\textstyle\int(8c\dot{x}_1 + 7{,}380x_1 - 8c\dot{x}_2 - 900x_2 - 160\sin 3\pi\tau)\,d\tau \\ 80\dot{x}_2 &= -\textstyle\int(-40c\dot{x}_1 - 4{,}500x_1 + 40c\dot{x}_2 + 4{,}500x_2)\,d\tau \end{aligned} \tag{9-60}$$

The computer circuit for solving these equations is as shown in Fig. 9-14. The circuit elements are selected so that the output voltages of the amplifiers do not exceed the ± 100-volts range. It should be noted that, for this particular problem, we multiplied each of the equations in Eq. (9-59) by 80. For other problems, each of the equations may be multiplied by an appropriate constant, but different constants may be used for different equations.

In the circuit shown in Fig. 9-14, it is assumed that the value of the damping coefficient c can be changed by varying the potentiometers 1, 3, 5, and 7 and adjusting the corresponding input resistors to amplifiers 1 and 5. This is a workable scheme if the potentiometers are mechanically ganged together.

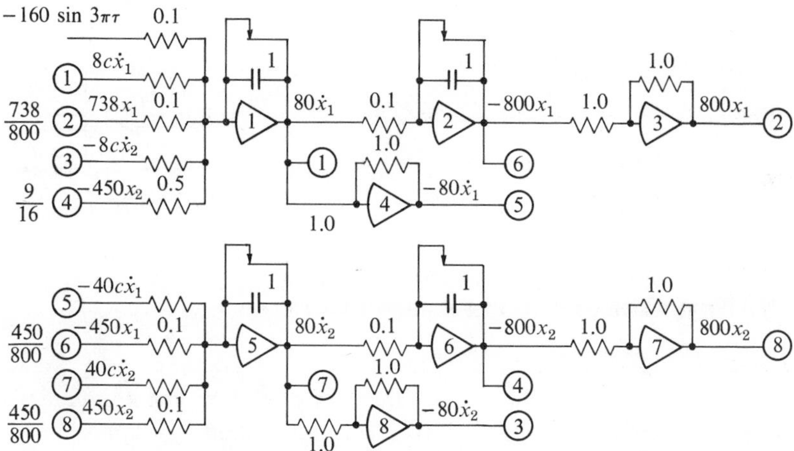

Fig. 9-14. *Computer circuit; Example* 5

Let us rearrange Eq. (9-59) in a more convenient form.

$$\begin{aligned} \dot{x}_1 &= -\int [0.1c(\dot{x}_1 - \dot{x}_2) + 92.25x_1 - 11.25x_2 - 2 \sin 3\pi\tau]\, d\tau \\ -\dot{x}_2 &= -\int [0.5c(\dot{x}_1 - \dot{x}_2) + 56.25x_1 - 56.25x_2]\, d\tau \end{aligned} \quad \textbf{(9-61)}$$

These equations indicate that if $c(\dot{x}_1 - \dot{x}_2)$ can be obtained as a single variable, the value of c can be varied by adjusting a single potentiometer. The corresponding computer circuit is shown in Fig. 9-15. The voltages in the circuit can be adjusted by changing the amplitude of the excitation voltage. The value of c, from 0 to 10, can be obtained by adjusting the potentiometer between amplifiers 4 and 5. For example, if the potentiometer setting is unity, the input to amplifier 1 through lead number 1 is $(\dot{x}_1 - \dot{x}_2)$. Since the first equation in Eq. (9-61) indicates that the corresponding quantity is $0.1c(\dot{x}_1 - \dot{x}_2)$, the value of c is 10. It is evident that the value of c can be varied over a wider range by changing the circuit elements of amplifier 5.

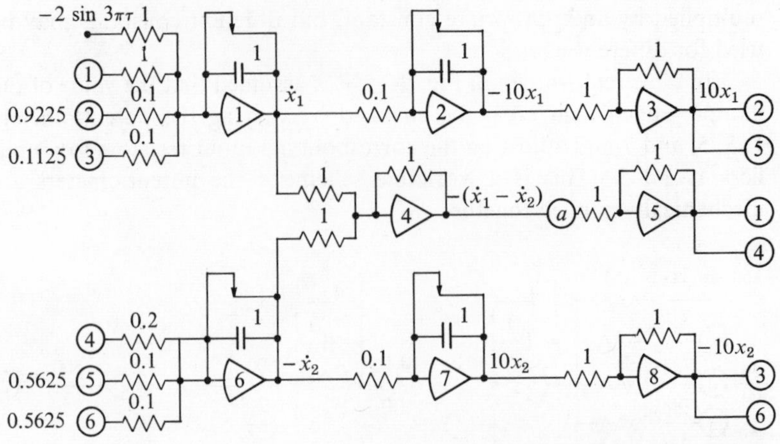

Fig. 9-15. *Computer circuit; Example* 5 [*see Eq.* (9-61)]

9-13. APPROXIMATE DIFFERENTIATION

It was mentioned in Sec. 9-4 that differentiation is a noise-amplifying process and is usually avoided. A method for approximate differentiation is discussed here for completeness and for introducing the so-called *implicit technique* which has general applications in the use of an analogue computer.

Consider the circuit of a sign inverter as shown in Fig. 9-16(*a*). Redrawing the circuit as shown in Fig. 9-16(*b*) and recalling that the grid current is negligible as compared with i_1 and i_2, we have

$$i_1 + i_2 = 0 \tag{9-62}$$

Since the voltage at the summing junction is essentially at ground potential, these currents can be expressed as

$$\frac{e_i}{R_i} + \frac{e_o}{R_f} = 0 \tag{9-63}$$

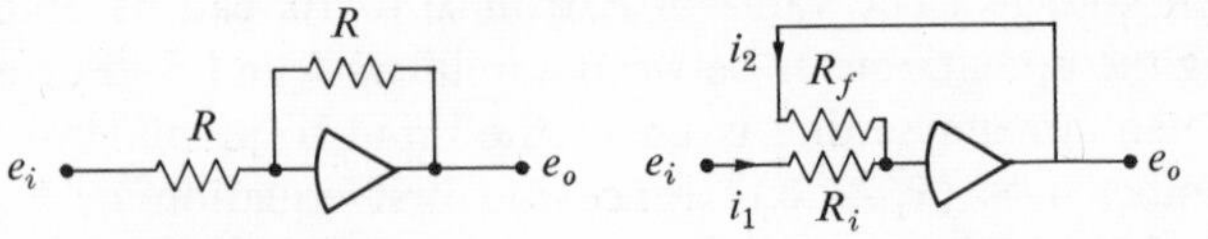

(*a*) Sign inverter $e_o = -e_i$ (*b*) Implicit equation $e_o + e_i = 0$

Fig. 9-16. *Circuit illustrating the implicit technique*

If $R_i = R_f$, Eq. (9-63) becomes the implicit equation

$$e_i + e_o = 0 \tag{9-64}$$

An approximate differentiation circuit is shown in Fig. 9-17. From Eq. (9-63), summing the current at the input of amplifier 1 gives

$$z + x + \int z\,dt - az = 0 \tag{9-65}$$

Rearranging this equation, we obtain

$$\int z\,dt + (1 - a)z = -x \tag{9-66}$$

If the potentiometer setting a is adjusted to as near unity as the noise level permits, we obtain

$$\lim_{a\to 1} z = -\frac{dx}{dt} \tag{9-67}$$

Hence the approximate differentiation of the function x can be obtained from the output of amplifier 1.

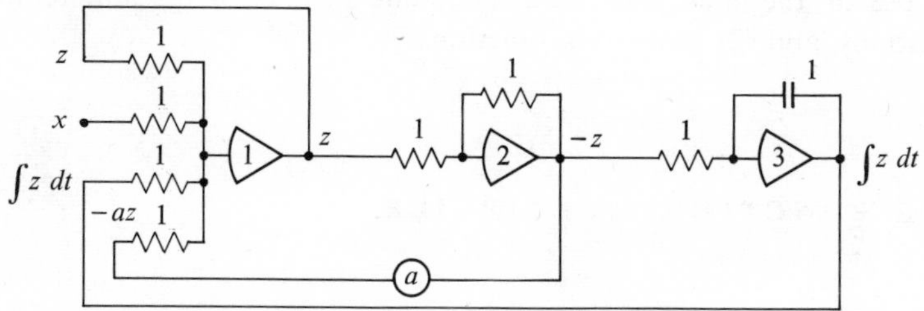

Fig. 9-17. *Approximate differentiation circuit*

9-14. NONLINEAR OPERATIONS

The solution of linear ordinary differential equations with constant coefficients was discussed in the previous sections. With the addition of the function multiplier and the function generator, the analogue computer can be used to solve linear ordinary differential equations with variable coefficients as well as nonlinear ordinary differential equations. The operations with these two computer elements are called *nonlinear operations*.

Consider the linear ordinary differential equation

$$\frac{d^2x}{dt^2} + a_1(t)\frac{dx}{dt} + a_2(t)x = f(t) \qquad \textbf{(9-68)}$$

in which $a_1(t)$ and $a_2(t)$ are functions of the independent variable t. This equation can be solved on the computer by generating the functions $a_1(t)$ and $a_2(t)$ and finding the products $a_1(t)\, dx/dt$ and $a_2(t)x$. Van der Pol's equation

$$\frac{d^2x}{dt^2} + \varepsilon(x^2 - 1)\frac{dx}{dt} + x = 0 \qquad \textbf{(9-69)}$$

is an example of a nonlinear ordinary differential equation. This equation can be solved if x^2 and the product $x^2\, dx/dt$ can be generated. It will be shown that these equations can be solved readily on the computer.

A large number of devices have been developed to perform the multiplication and function-generation operations. Descriptions of their design and operation can be found in most books on analogue computers. In the remainder of this chapter, we shall briefly discuss the use of the multiplier and the diode in (1) solving differential equations, and (2) generating functions.

9-15. THE FUNCTION MULTIPLIER

A function multiplier accepts two voltages e_1 and e_2 and generates an output voltage e_o with a scale factor, commonly equal to 0.01, and a sign inversion.

$$e_o = -0.01e_1e_2 \qquad \textbf{(9-70)}$$

The multiplier is represented by the symbol as shown in Fig. 9-18. Although multiplication can be performed by a number of devices, the generally used ones are the all-electronic and servo-driven multipliers. The former is capable of high speed and high accuracy, and the latter is inherently a slow-speed device. The servo multiplier is still commonly used, however, and we shall discuss it in some detail.

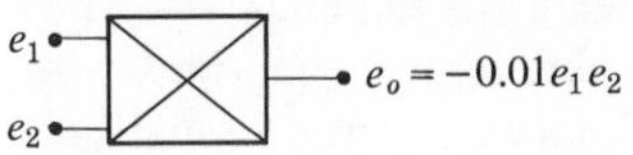

Fig. 9-18. *Function-multiplier symbol*

A servo multiplier is shown schematically in Fig. 9-19. The electrical connections are shown as solid lines, and the mechanical ones are dashed. The multiplier consists of a servo-driven follow-up potentiometer and a multiplying potentiometer. In actual practice there may be a number of multiplying potentiometers in a multiplier. Let F and X be the position of the wipers. Since the wipers are mechanically connected together, they will assume the same geometric positions with respect to the individual potentiometers. The wipers are moved by a servomotor driven by a differential amplifier.

The differential amplifier has the characteristic that if $e_1 > e_F$, one polarity of the output voltage is produced, causing the servomotor to move the wiper F toward the $+100$-volt terminal. If $e_1 = e_F$, the output voltage of the amplifier is zero, and the wiper F is stationary.

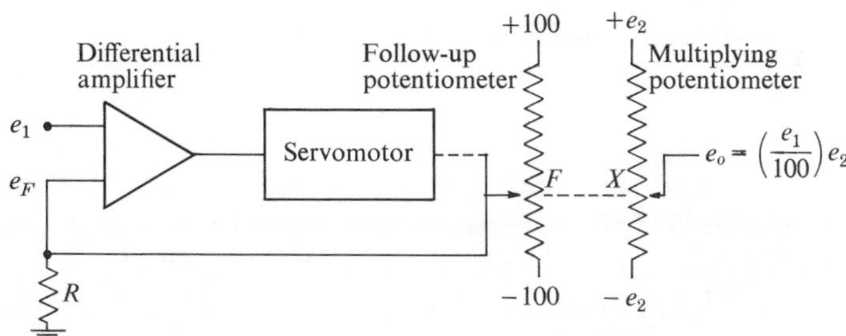

Fig. 9-19. *Servo-driven function multiplier*

If $e_1 < e_F$, the polarity of the amplifier output voltage is reversed, causing the servomotor to move F toward the -100-volt terminal. Thus, if e_1 is applied to the amplifier, the wiper F will move until e_F is equal to e_1. Since ± 100 volts are applied to the ends of the follow-up potentiometer, the voltage at its midpoint is zero. Hence the geometric position of F, above or below the midpoint, gives the corresponding fraction of ± 100 volts. For example, if F is halfway between the midpoint and $+100$ volts, e_F is $+50$ volts. Let $\pm e_2$ be applied to the ends of the multiplying potentiometer. Similarly, the geometric position of X, above or below the midpoint, gives the corresponding fraction of $\pm e_2$ volts. Now let a positive voltage e_1 be applied to the multiplier. The positions of F and X are $e_1/100$ above their midpoints. Hence the output voltage e_o of the multiplier is $(e_1/100)e_2$; that is, we obtain the product of e_1 and e_2 with a scale factor. The same reasoning applies if e_1 is negative. A sign inversion is obtained if the polarity of the multiplying potentiometer is reversed.

Since potentiometers are used for multiplication, the loading of potentiometers must be considered. The loading effect is compensated for if the follow-up potentiometer is loaded with a resistor R identical to the load on the multiplying potentiometer. Then the geometric distances along the potentiometers are not the same as the electrical distances. The electrical distances, however, are identical with identical loads, and the multiplier will operate correctly.

9-16. EXAMPLES: USE OF MULTIPLIER FOR SOLVING DIFFERENTIAL EQUATIONS

Example 6. Consider Mathieu's equation

$$\ddot{x} + (a - b \cos \omega t)x = 0 \tag{9-71}$$

in which a, b, and ω are constants. This is a linear ordinary differential equation with variable coefficients. The computer circuit for solving this equation can be derived by the method outlined in Sec. 9-6. Rearranging Eq. (9-71) gives

$$\ddot{x} = -(a - b \cos \omega t)x \tag{9-72}$$

Thus the mathematical relation is satisfied if the quantities ax and $(-b \cos \omega t)x$ are made available as inputs to a summer.

The function $\cos \omega t$ can be obtained from a low-frequency sine generator. Difficulties would be encountered in synchronizing the sine generator with the computer, however, since at the start of the problem at $t = 0$, the cosine function has to be equal to unity. It is expedient to generate the cosine function by the computer circuit for the undamped free vibration of a one-degree-of-freedom system. The circuit gives a sine and a cosine function. The resultant motion is a cosine function if a mass is displaced with an initial displacement and released with zero initial velocity.

The unscaled computer circuit for the solution of Eq. (9-72) is as shown in Fig. 9-20. Amplifiers 1 to 3 are arranged to solve the differential equation $\ddot{y} + \omega^2 y = 0$. Thus $y = b \cos \omega t$. As shown, the value of ω^2 is obtained from a potentiometer setting, and the value of ω is less than unity. The values of ω greater than 1 can be obtained by increasing the gain of the amplifier circuits. The product $(-b \cos \omega t)x$ is obtained from the output of the multiplier. The output of amplifier 4 is $-(a - b \cos \omega t)x$ which is $\ddot{x}$.

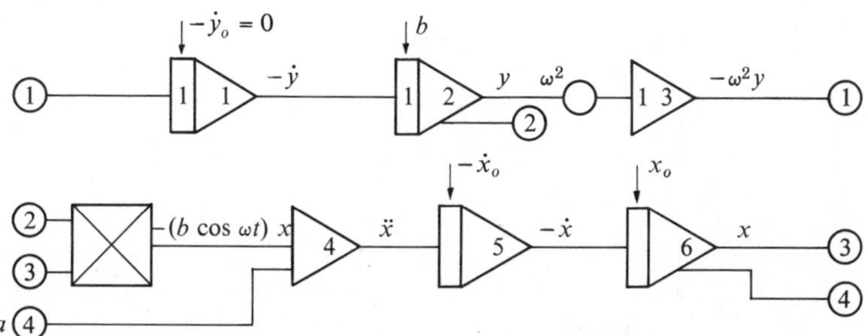

Fig. 9-20. *Unscaled circuit diagram for Mathieu's equation*

Example 7. To illustrate the solution of a nonlinear differential equation, consider Van der Pol's equation

$$\ddot{x} + \varepsilon(x^2 - 1)\dot{x} + x = 0 \tag{9-73}$$

where ε is a constant. Assume that $0 \leqslant \varepsilon \leqslant 1$. This equation resembles the equation of motion for the free vibration of a one-degree-of-freedom system, but with a nonlinear damping term. If $x < 1$, the quantity $\varepsilon(x^2 - 1)$ is negative, which would correspond to negative damping in the system. The amplitude of x will increase with time. If $x > 1$, the damping term is positive, and the amplitude of x will decrease with time. Thus, if the system is given a small initial displacement or velocity, the motion will build up and will eventually become periodic with constant amplitude. Conversely, if a large initial displacement or velocity is imparted to the system, its motion will diminish until the same periodic motion with constant amplitude is attained. Let us derive the computer circuit of this equation before discussing the method for presenting the nonlinear data.

The computer circuit can be obtained by rearranging Eq. (9-73).

$$\dot{x} = -\int [\varepsilon(x^2 - 1)\dot{x} + x]\, dt \tag{9-74}$$

Since the undamped natural frequency is 1 rad-sec^{-1}, it is unnecessary to time scale the equation. Through a trial-and-error process, it may be assumed that x_m is approximately 3. It is necessary to consider the scale factor of 0.01 introduced by the multiplier in the computer circuit. Multiplying Eq. (9-74) by a factor of 10, the scaled computer circuit obtained is as shown in Fig. 9-21.

Data from the study of second-order nonlinear differential equations without excitation are commonly presented in a *phase plane* in which the displacement x and the velocity $\dot{x}$ are used as Cartesian coordinates. The differential equation specifies a definite curve in the

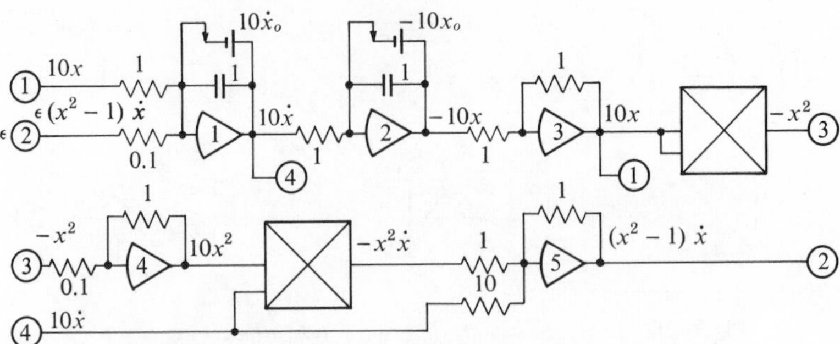

Fig. 9-21. *Computer circuit for Van der Pol's equation*

phase plane, and the curve is called the *phase trajectory* of the differential equation.

Consider a second-order nonlinear ordinary differential equation of the form

$$\ddot{x} + f(x,\dot{x}) + x = 0 \tag{9-75}$$

where $f(x, \dot{x})$ is nonlinear. To obtain the phase trajectory, the equation is written as two first-order equations. Letting

$$y = \frac{dx}{dt} \tag{9-76}$$

Eq. (9-75) can be written as

$$\frac{dy}{dt} = -f(x,y) - x \tag{9-77}$$

The set of equations shown in Eqs. (9-76) and (9-77) is a special case of the more general system.

$$\frac{dy}{dt} = Q(x,y) \tag{9-78}$$

$$\frac{dx}{dt} = P(x,y) \tag{9-79}$$

where $Q(x, y)$ and $P(x, y)$ are functions of x and y. Dividing Eq. (9-78) by Eq. (9-79) gives

$$\frac{dy}{dx} = \frac{Q(x,y)}{P(x,y)} \tag{9-80}$$

This equation specifies the phase trajectory of the differential equation in the phase plane.

As a special case of a nonlinear system, consider the linear differential equation

$$m\ddot{x} + kx = 0 \tag{9-81}$$

Following the procedure outlined in the previous paragraph, we have

$$\frac{dy}{dt} = -\frac{k}{m}x; \quad \frac{dx}{dt} = y \tag{9-82}$$

and

$$\frac{dy}{dx} = -\frac{k}{m}\frac{x}{y} \tag{9-83}$$

Separating the variables in Eq. (9-83) and integrating, we obtain

$$\frac{my^2}{2} + \frac{kx^2}{2} = E \tag{9-84}$$

where E is an arbitrary constant determined by the initial conditions x_o and y_o. Substituting the initial conditions yields

$$\frac{my_o^2}{2} + \frac{kx_o^2}{2} = E \tag{9-85}$$

Since the system considered is conservative, the quantity E is also the total energy of the system. Dividing through by E and defining $2E/k = a^2$ and $2E/m = b^2$, Eq. (9-84) becomes

$$\frac{x^2}{a^2} + \frac{y^2}{b^2} = 1 \tag{9-86}$$

Thus the phase trajectory of Eq. (9-81) is an ellipse.

Returning now to Van der Pol's equation, Eq. (9-73), if $\varepsilon = 0$, the equation has the same form as Eq. (9-81). Hence the phase trajectory is an ellipse as shown in Fig. 9-22(*a*). If $\varepsilon > 0$, it was explained before that the motion tends to build up for small oscillations and to decrease for large oscillations. Hence, after the initial transient, the motion becomes periodic, represented by a closed trajectory. This closed trajectory is called a *limit cycle*. The phase trajectories of Van der Pol's equation for various values of ε are as shown in Figs. 9-22(*b*) to (*d*). In each of these figures, the limit cycle is indicated by a heavy line. For a given ε, the same limit cycle is obtained whether the system is set into motion with the initial conditions inside or outside the limit cycle.

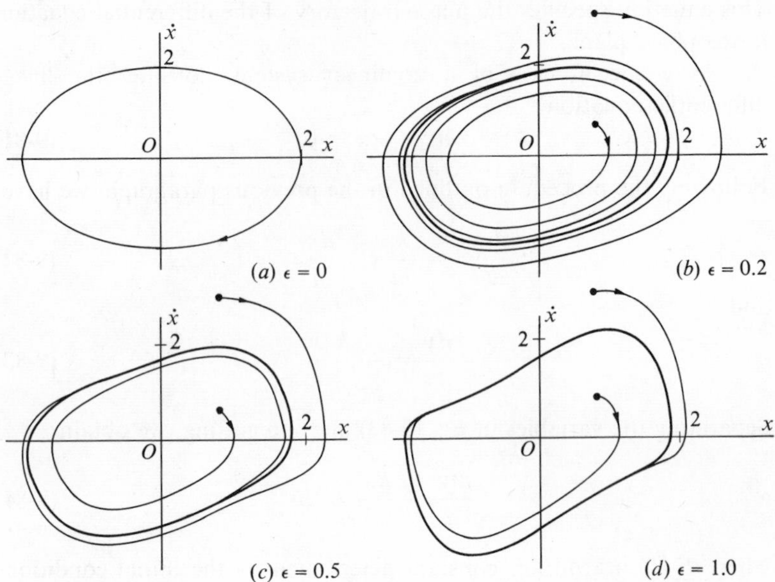

Fig. 9-22. *Phase trajectories of Van der Pol's equation for various values of* ε

9-17. DIODE FUNCTION GENERATOR

A function generator accepts a function x and generates a function $f(x)$. Usually, $f(x)$ is a single-valued function. The function generator is represented by the symbol shown in Fig. 9-23. Some function generators accept two variables x and y and generate a function $f(x,y)$. A large number of devices are used for function generation, and the methods that can be used for this purpose are limited only by the imagination of the designer. The object of this section is to discuss the diode as an element in a computer circuit for solving differential equations.

A diode is essentially a switching device. It offers a very low resistance to the flow of current in one direction and a very high resistance to the current flow in the reversed direction. A vacuum-tube diode is often referred to as a thermionic or hot diode, and the solid-state type is called a cold diode. The symbols used

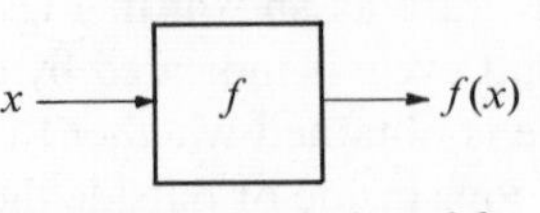

Fig. 9-23. *Symbol used for function generator*

for diodes are shown in Fig. 9-24. We shall use the symbol for cold diodes to represent diodes in all the figures. The static conduction characteristics of the two types of diodes are illustrated in Fig. 9-25. The figures indicate that if e is positive, that is, the plate P is more positive than the cathode K, the current i flows from P to K easily in this *forward* direction. Hence a diode has relatively low forward resistance. Over a reasonable voltage range, if K is more positive than P, the current flow in this reversed direction is very low. Hence the resistance of a diode to a reverse current flow is very high. The characteristics of the hot and cold diodes are illustrated in Figs. 9-25(b) and (c), respectively. Generally, a hot diode has lower forward resistance and higher resistance for reverse current flow than has a cold diode.

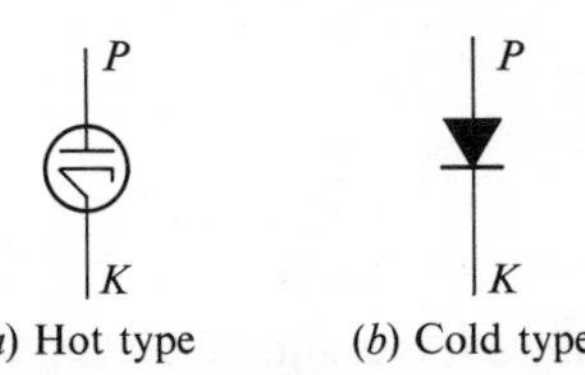

(a) Hot type (b) Cold type

Fig. 9-24. *Symbols used for diodes*

A diode function generator is based on the concept of straight-line approximation to a function. Consider the circuit of Fig. 9-26(a).

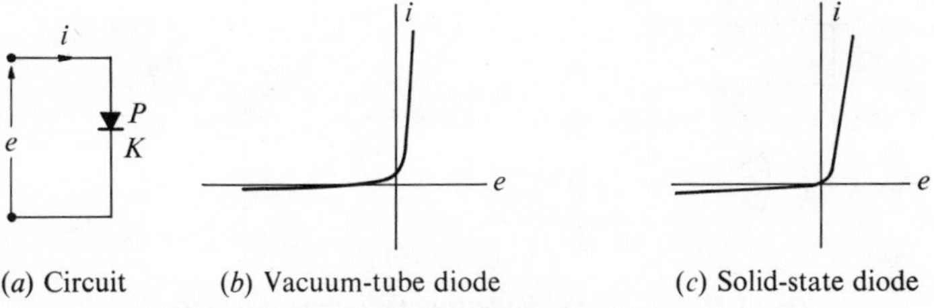

(a) Circuit (b) Vacuum-tube diode (c) Solid-state diode

Fig. 9-25. *Static conduction characteristics of diodes*

Recalling that a diode will conduct only if the plate is more positive than the cathode, and that e_g is essentially at ground potential, diode V_1 will conduct only if e_o is positive and greater than E_1. While e_o is positive, diode V_2 is not conducting and is virtually an open circuit. Similarly, diode V_2 will conduct only if e_o is negative and less than $-E_2$. With e_o being negative, V_1 is not conducting. Now, when $E_1 > e_o > -E_2$, none of the diodes are conducting and we have

$$e_o = -\frac{R_1}{R} e_1 \tag{9-87}$$

where e_1 can be positive or negative. When $e_o > E_1$ and V_1 is conducting, the current relation at the grid of the amplifier is

$$\frac{e_1}{R} = -\frac{e_o}{R_1} - \frac{e_o - E_1}{R_2}$$

or

$$e_o = -\frac{R_1 R_2}{R(R_1 + R_2)} e_1 + \frac{R_1 E_1}{R_1 + R_2} \tag{9-88}$$

Since it is assumed that $e_o > E_1$, e_1 has to be negative. When $e_o < -E_2$ and V_2 is conducting, the output voltage is

$$e_o = -\frac{R_1 R_3}{R(R_1 + R_3)} e_1 - \frac{R_1 E_2}{R_1 + R_3} \tag{9-89}$$

where e_1 is positive. Thus the function as shown in Fig. 9-26(*b*) and prescribed by Eqs. (9-87) to (9-89) is generated by the given circuit.

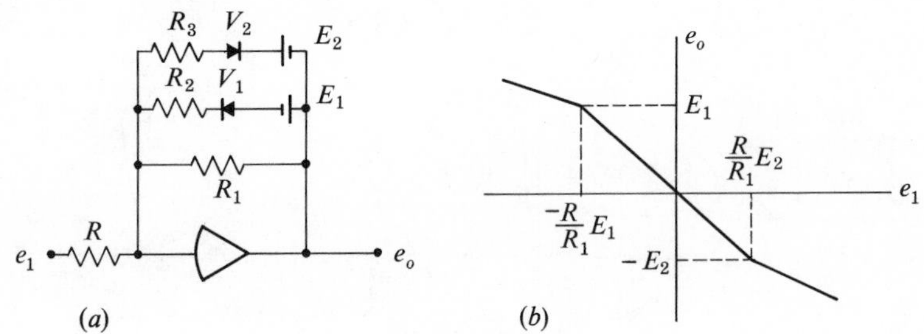

Fig. 9-26. *A simple diode function-generator circuit*

9-18. EXAMPLES: USE OF DIODES FOR SOLVING DIFFERENTIAL EQUATIONS

Since the diode is a switching device, it can be adapted to solve problems that are discontinuous in nature. We shall use three examples to illustrate the use of diodes.

Example 8. Consider the one-degree-of-freedom system of Fig. 9-27(*a*) in which the mass m is placed between the two springs k_1 and k_2. Assume that the mass, as shown, is in its equilibrium position, and that the springs are unstressed. The equations of motion of the system are

$$\begin{aligned} m\ddot{x} + k_1 x &= 0 \quad \text{for} \quad x > 0 \\ m\ddot{x} + k_2 x &= 0 \quad \text{for} \quad x < 0 \end{aligned} \tag{9-90}$$

These equations are programmed on the computer as shown in Fig. 9-27(b). Amplifiers 1 and 2 are used as integrators. When $x > 0$, diode V_1 does not conduct, and V_2 is conducting and thereby holding

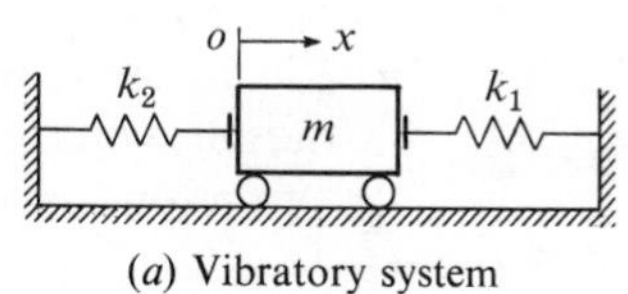

(a) Vibratory system

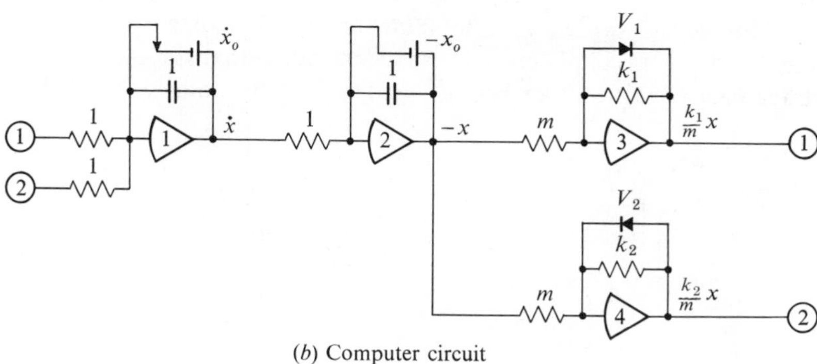

(b) Computer circuit

Fig. 9-27. *Computer circuit of Eq.* (9-90)

the output of amplifier 4 at the ground potential. Hence only amplifiers 1 to 3 need to be considered when $x > 0$. Similarly, when $x < 0$, only amplifiers 1, 2, and 4 need to be considered.

Example 9. The equations of motion of the one-degree-of-freedom system with Coulomb friction, as shown in Fig. 9-28, can be expressed as

$$m\ddot{x} + kx = \pm F \tag{9-91}$$

where F is the magnitude of the Coulomb frictional force. If the mass m is moving from left to right, its velocity is positive and the direction of F is from right to left; that is, the frictional force is equal to $-F$.

The converse is true if m is moving from right to left. To solve Eq. (9-91) on the computer, we rearrange the equation to obtain

$$\dot{x} = -\int \left(\frac{k}{m} x \mp \frac{F}{m}\right) dt \qquad \textbf{(9-92)}$$

Coulomb friction can be simulated by the circuit of Fig. 9-29(a). Assume that the magnitude of the frictional force is E, and that $\dot{x}$ is the input voltage to the circuit. If $\dot{x} > 0$, the output voltage of the circuit is negative, causing the diode V_1 to conduct. Since e_g is essentially at ground potential, the output voltage is $-E$. Similarly, if $\dot{x} < 0$, the output voltage of the circuit is $+E$. Since the input resistor R does not enter into this analysis, a resistor of 1 megohm or higher can be used. The voltage relation of the circuit is illustrated in Fig. 9-29(b).

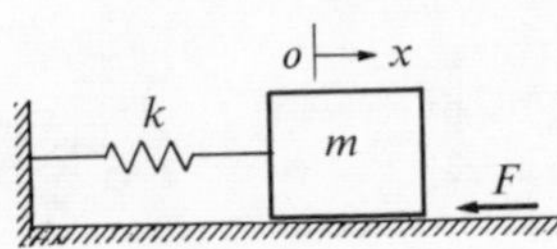

Fig. 9-28. *System with Coulomb friction*

Figure 9-29(c) shows a computer circuit for solving Eq. (9-91). The circuit with amplifiers 1 to 3 is identical to that for the study of undamped free vibration of a one-degree-of-freedom system. The

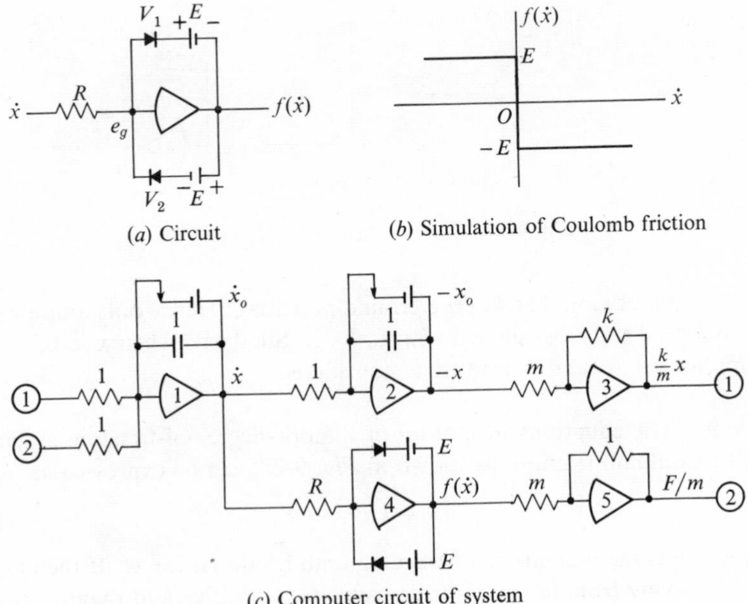

(a) Circuit

(b) Simulation of Coulomb friction

(c) Computer circuit of system

Fig. 9-29. *Computer simulation of system with Coulomb friction*

circuit with amplifier 4 was discussed in the previous paragraph. Amplifier 5 is used to change the sign of F and to multiply F by the constant $1/m$.

Example 10. A commonly encountered nonlinearity in physical systems is that of dead space. Consider the vibratory system of Fig. 9-30, in which the mass m is placed on a frictionless surface and is free to shuttle back and forth between two elastic stoppers. Let k_1 and k_2 be the elastic constants of the stoppers. The equation of motion of the system is

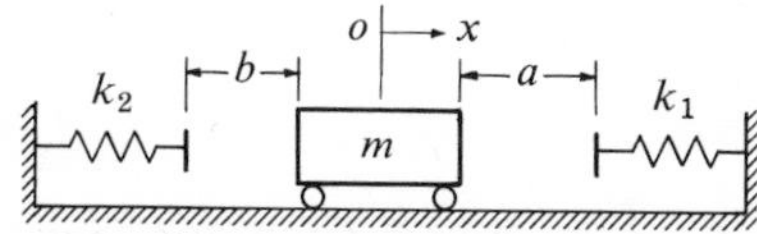

Fig. 9-30. *Vibratory system with dead space*

$$m\ddot{x} = -f(x) \tag{9-93}$$

where

$$f(x) = \begin{cases} 0 & \text{for} \quad -b < x < a \\ k_1(x - a) & x \geqslant a \\ k_2(x + b) & x \leqslant -b \end{cases} \tag{9-94}$$

The function $-f(x)$ can be generated by the circuit as shown in Fig. 9-31(a). If $-b < x < a$, none of the diodes will conduct. With zero voltage input, the amplifier output voltage is nil. If $x > a$, the

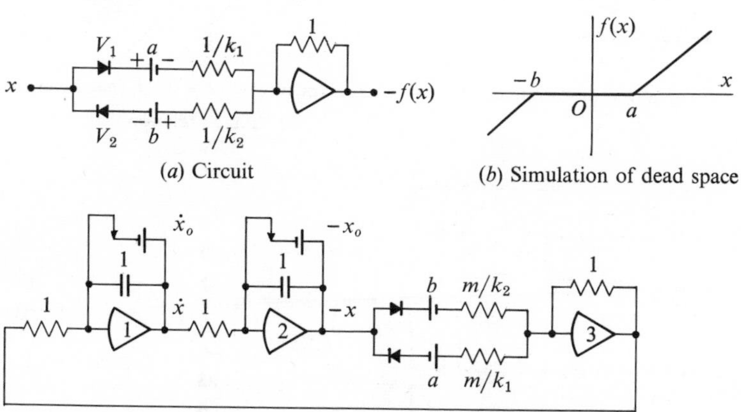

Fig. 9-31. *Computer simulation of system with dead space*

diode V_1 conducts and V_2 is off. The net input voltage to the amplifier circuit is $(x - a)$, and the output voltage is $-k_1(x - a)$. If $x < -b$, the corresponding input voltage is $(x + b)$, and the output voltage is $-k_2(x + b)$. It should be noted that, since $(x + b)$ is negative, the quantity $-k_2(x + b)$ is positive. The voltage relation of this circuit is shown in Fig. 9-31(*b*).

Equation (9-93) can be rearranged as

$$\dot{x} = -\int\left[\frac{1}{m}f(x)\right]dx \tag{9-95}$$

for the computer solution. Since $f(x)$ is required as input to the integrating amplifier 1 instead of $-f(x)$, and $-x$ is the input to the dead space circuit, the d-c voltages and the resistors in Fig. 9-31(*c*) are adjusted accordingly to accommodate the sign changes.

Since the diode is not perfect as a switching device, more precise switching can be obtained from the "idealized" diode circuit shown in Fig. 9-32(*a*). If the bias voltage E is zero and the input voltage x is positive, the amplifier output voltage is negative. The diode V_1 will conduct, and V_2 will not. With the values of the circuit elements specified, the output voltage is $-x$. If E is zero and x negative, the amplifier

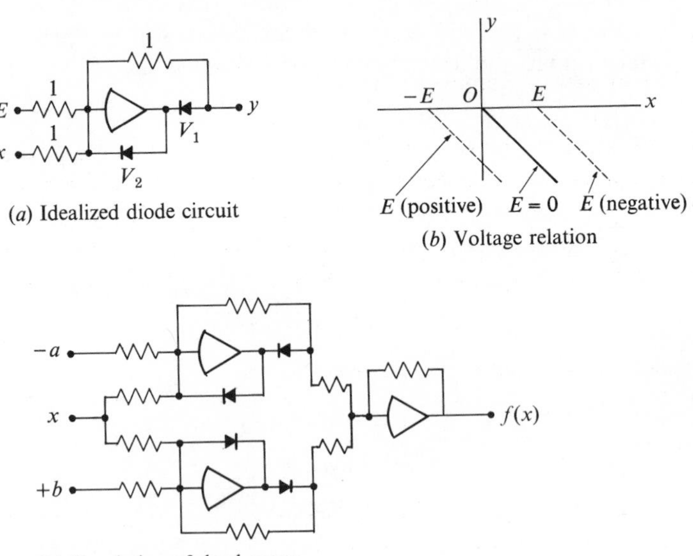

Fig. 9-32. *Idealized diode circuit for simulation of dead space*

output voltage is positive. The diode V_2 will conduct, and V_1 will not. V_2 holds the amplifier output voltage at essentially ground potential, and the amplifier output voltage is zero. This voltage relation is illustrated in Fig. 9-32(*b*). If a positive bias voltage E is introduced and the quantity $(x + E)$ is positive, the amplifier output voltage is $-(x + E)$. The output voltage is zero when $(x + E)$ is equal to zero or negative. This voltage relation is shown as dashed lines in Fig. 9-32(*b*). Similarly, the effect of a negative bias voltage can be deduced.

Using a combination of two idealized diode circuits, a circuit for dead-space simulation, as shown in Fig. 9-32(*c*), is obtained. If the bias voltages are $-a$ and $+b$, the output voltage $f(x)$ of the circuit is as shown in Fig. 9-31(*b*).

9-19. FUNCTION GENERATION

The use of nonlinear computer elements for solving differential equations was illustrated in the previous sections. Often, it may be necessary to perform the division operation and to generate functions, analytical or arbitrary, continuous or discontinuous. Some of the methods for function generation are discussed in this section.

Case 1. Use of Linear Computer Elements

The linear computer elements, as defined in Sec. 9-3, are the resistor, the capacitor, and the operational amplifier. Examples of some simple functions that can be generated are shown in Fig. 9-33. Figure 9-33(*a*) and (*b*) show the generation of functions by integration. Part (*c*) shows that a polynomial $f(t)$ can be generated. If the polynomial is

$$f(t) = a_1t^2 + a_2t + a_3 \tag{9-96}$$

by successive differentiation we obtain

$$\begin{aligned} f'(t) &= 2a_1t + a_2 \\ f''(t) &= 2a_1 \end{aligned} \tag{9-97}$$

Hence the constant $2a_1$ can be used as input to the first integrator with $-a_2$ as the initial condition. By successive integration with the appropriate initial conditions, the original polynomial, Eq. (9-96), is obtained. This method can be used to generate polynomials of higher order. It is

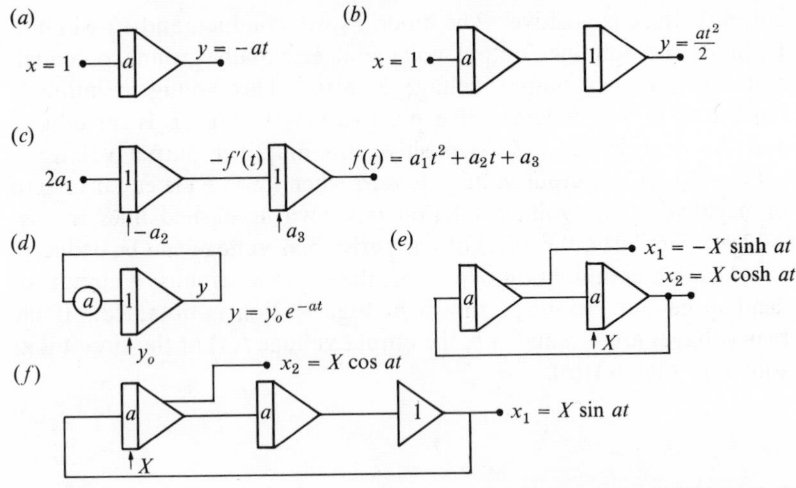

Fig. 9-33. *Generation of simple functions with linear computer elements*

interesting to note that if $f(t) = 0$, we obtain the roots of an algebraic equation. If the curve $f(t)$ is plotted versus time t on a recorder, the time at which $f(t) = 0$ gives the positive roots of the equation. Part (*d*) generates an exponential function from the equation

$$y = -\int_o^t ay\,dt + y_o \tag{9-98}$$

or

$$\dot{y} + ay = 0$$

The solution of this first-order differential equation is

$$y = y_o e^{-at} \tag{9-99}$$

Parts (*e*) and (*f*) can be analyzed quite easily.

The linear computer elements, described in Sec. 9-4, can be combined in an amplifier network to generate a large number of transfer functions. Although this class of function generation is less often used in solving differential equations, it is exceedingly useful in simulation and control studies. Referring to Fig. 9-34(*a*), the transfer function is

$$\frac{e_o}{e_i} = -\frac{Z_f}{Z_i} \tag{9-100}$$

Since the impedances in this equation are not specified, it is feasible to substitute networks of resistors and capacitors for the input and feedback impedances to obtain complex relations between the output

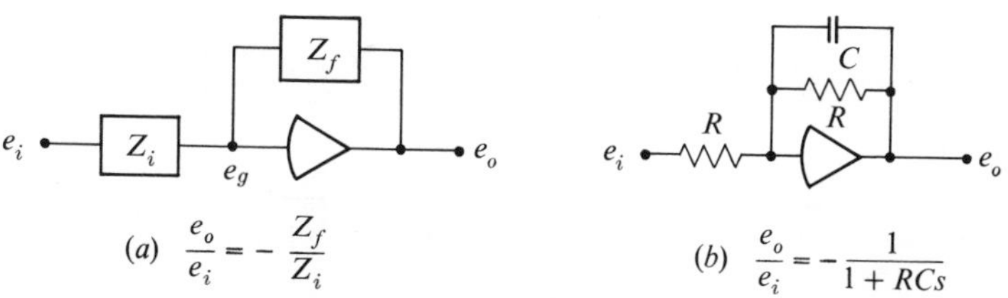

(a) $\frac{e_o}{e_i} = -\frac{Z_f}{Z_i}$ (b) $\frac{e_o}{e_i} = -\frac{1}{1+RCs}$

Fig. 9-34. *Networks illustrating the transfer function e_o/e_i*

and input voltages. For example, in Fig. 9-34(*b*) the input impedance is R and the feedback impedance, with R and C in parallel, is $R/(1 + RCs)$. The corresponding transfer function is

$$\frac{e_o}{e_i} = -\frac{R}{R(1 + RCs)} = -\frac{1}{1 + RCs} \qquad \textbf{(9-101)}$$

Now if $e_i = -E$, a constant, we obtain the differential equation

$$(RCs + 1)e_o = E \qquad \textbf{(9-102)}$$

where s is a differential operator. The solution of this equation is

$$e_o = E(1 - e^{-t/RC}) \qquad \textbf{(9-103)}$$

Case 2. Use of Multiplier

The use of the multiplier as a circuit element for function generation is illustrated by the examples shown in Fig. 9-35. For simplicity, the scale factor, associated with a function multiplier, is assumed equal to -1.

Figure 9-35(*a*) shows that the multiplication of x by itself gives $-x^2$. Part (*b*) shows that x^3 can be generated. Part (*c*) illustrates the implicit technique in function generation. The circuit has two inputs x and y, and an output z. The multiplication of y and z gives $-yz$ at the output of the multiplier. From Eq. (9-63) the inputs to the amplifier are related by the implicit equation.

$$\frac{x}{R_2} - \frac{yz}{R_1} = 0 \qquad \textbf{(9-104)}$$

Since each term in this equation represents a current, for satisfactory operation of the implicit-function circuit, one of the terms must be positive and the other negative. This circuit is the basis for the function

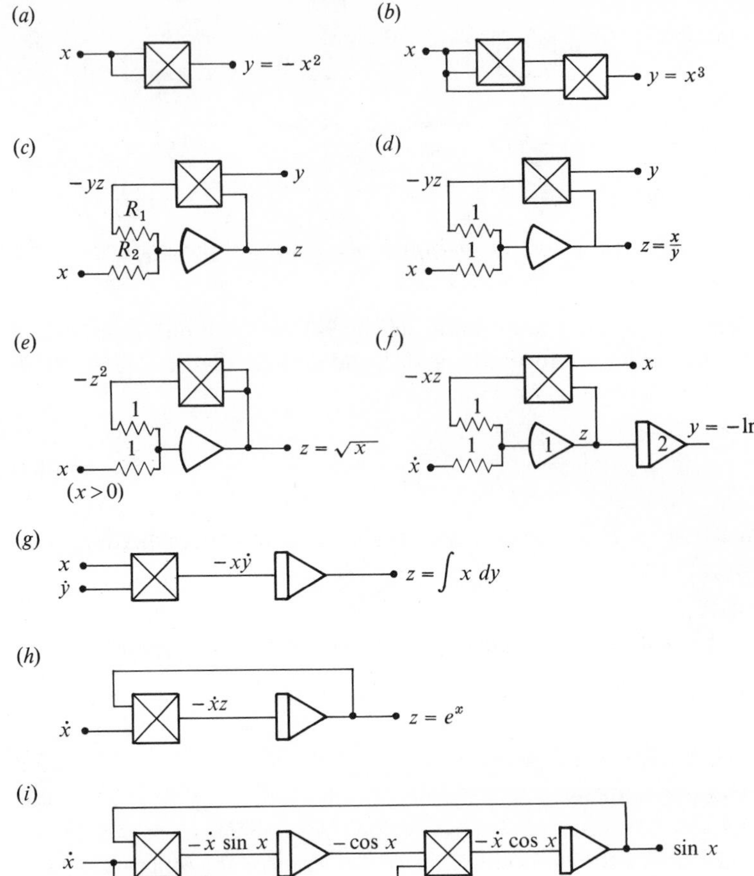

Fig. 9-35. *Use of multiplier for function generation*

generators of Fig. 9-35(d), (e), and (f). From Eq. (9-104), if $R_1 = R_2$, the implicit equation gives $z = x/y$, which is a division operation. If $R_1 = R_2$ and $y = z$, we obtain $z = \sqrt{x}$, which is a square-root operation. In (f), let the output of amplifier 1 be equal to z. From the implicit circuit, we have

$$\dot{x} - xz = 0 \tag{9-105}$$

or

$$z = \frac{\dot{x}}{x} \tag{9-106}$$

The integration of this equation with amplifier 2 gives

$$y = -\int z\,dt = -\int \frac{1}{x}\frac{dx}{dt}\,dt = -\int \frac{dx}{x} = -\ln x \qquad \textbf{(9-107)}$$

Part (g) indicates that a function x can be integrated with respect to the function y. Since time t is the only independent variable in a computer, the operation may be regarded as an integration with respect to a dependent variable. The function generator of (h) is based on the equation

$$z = -\int\left(-z\frac{dx}{dt}\right)dt = \int z\,dx \qquad \textbf{(9-108)}$$

or

$$\frac{dz}{z} = dx$$

Integrating the last equation gives $\ln z = x$ or $z = e^x$. The circuit for the generation of a sine and cosine function of a dependent variable x is given in Fig. 9-35(i).

Case 3. Use of Diodes

Function generation with the use of diodes can be accomplished by the various combinations of four diode limiting circuits. Diodes can be used as limiters either in series or in shunt with the input or the feedback path of the operational amplifier. The voltage at which a diode is made to conduct can be adjusted by the bias voltage setting. The four basic arrangements and output-input voltage relations of the circuits are shown in Fig. 9-36.

Consider the input shunt limiting circuit of Fig. 9-36(a). If neither of the diodes is conducting, the output voltage is

$$e_o = -\frac{R_3}{R_1 + R_2}e_i \qquad \textbf{(9-109)}$$

This relation holds if R_4 is infinite. Let us consider when R_4 is zero and when it is finite separately.

First, let R_4 equal zero. If the diodes are not conducting, the output voltage e_o is given by Eq. (9-109); and the voltage e_1 is

$$e_1 = \frac{R_2}{R_1 + R_2}e_i \qquad \textbf{(9-110)}$$

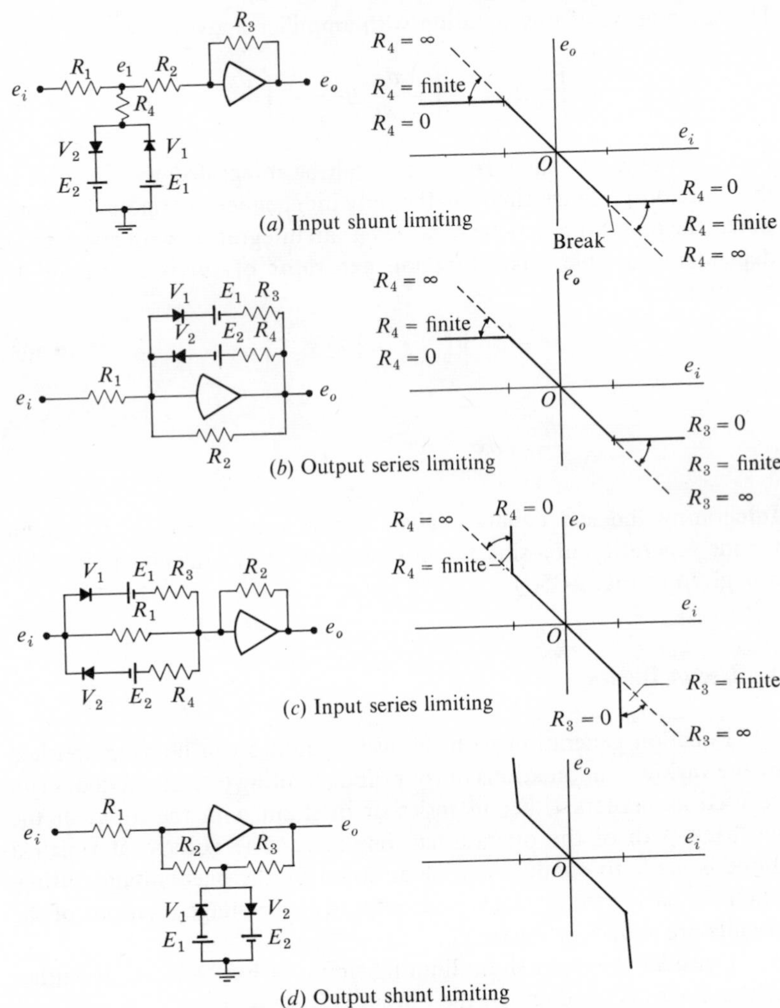

(*a*) Input shunt limiting

(*b*) Output series limiting

(*c*) Input series limiting

(*d*) Output shunt limiting

Fig. 9-36. *Basic diode-limiting circuits*

If e_i is positive and increasing, the diode V_2 will begin to conduct when e_1 is equal to E_2. Hence the breakpoint in the output voltage e_o occurs when $e_1 = E_2$; that is, when $e_i = E_2(R_1 + R_2)/R_2$. The corresponding value of e_o is $-E_2R_3/R_2$. Further increase in e_i will cause additional current to flow through R_1. This additional current is diverted through the conducting diode V_2 instead of flowing through R_2. Hence the output voltage remains constant. Similarly, the circuit characteristics can be explained when e_i is negative and decreasing.

Second, let R_4 be finite and e_i be positive and increasing. If the diodes are not conducting, the output voltage is given by Eq. (9-109). The breakpoint occurs when e_o is $-E_2R_3/R_2$. With further increase in e_i, V_2 will conduct, and summing the current at the node gives

$$\frac{e_i - e_1}{R_1} = \frac{e_1}{R_2} + \frac{e_1 - E_2}{R_4} \tag{9-111}$$

Solving for e_1, we obtain

$$e_1 = \frac{R_2R_4e_i + R_1R_2E_2}{R_1R_2 + R_2R_4 + R_4R_1} \tag{9-112}$$

The output voltage beyond the breakpoint is

$$e_o = -\frac{R_3}{R_2}e_1 = -\frac{R_3R_4e_i + R_1R_3E_2}{R_1R_2 + R_2R_4 + R_4R_1} \tag{9-113}$$

Since E_2 is a constant, the output e_o is linear with the input e_i with a negative slope equal to $-(R_3R_4)/(R_1R_2 + R_2R_4 + R_4R_1)$. This slope is steeper than when $R_4 = 0$ and less than when $R_4 = \infty$. Similarly, the circuit characteristics can be explained when e_i is negative and decreasing.

The analysis of the rest of the circuits in Fig. 9-36 is left as an exercise for the reader. It is interesting to note that the input shunt and the output series circuits have similar characteristics and that the input series and output shunt limiting circuits are similar.

Characteristics other than those represented in Fig. 9-36 can be obtained by the modification of the basic diode limiting circuits. For example, if R_1 is omitted in the input series limiting circuit of Fig. 9-36(*c*), we obtain a dead-space simulating circuit as shown in Fig. 9-31(*a*). The Coulomb friction simulating circuit of Fig. 9-29(*a*) can be obtained from the modification of the output series limiting circuit of Fig. 9-36(*b*).

Generally, the functions generated with diode limiting circuits are single-valued functions. The simulation of gear backlash or hysteresis, as shown in Fig. 9-37(*a*), requires a multivalued function. A circuit for the generation of this function is shown in Fig. 9-37(*b*). Let e_i be initially zero and the capacitor C be uncharged. As e_i increases, the output of amplifier 1 is negative. Diode V_1 will not conduct until the amplifier 1 output is less than $-E$. This portion of the curve is shown dashed in Fig. 9-37(*a*). With further increase in e_i, diode V_1 conducts, and the input voltage to the integrating amplifier 2 is $[-(e_i - e_o) + E]$.

If the RC time constant of the integrator is small (for example, $R = 50K$ and $C = 1\ \mu f$), the integrator output builds up rapidly to give

$$e_o = e_i - E \qquad \textbf{(9-114)}$$

If e_i reaches a maximum value and reverses its direction, the polarity of V_1 is reversed, and it does not conduct. Thus the integrator acts as a storage device since there is no path to discharge the capacitor. When

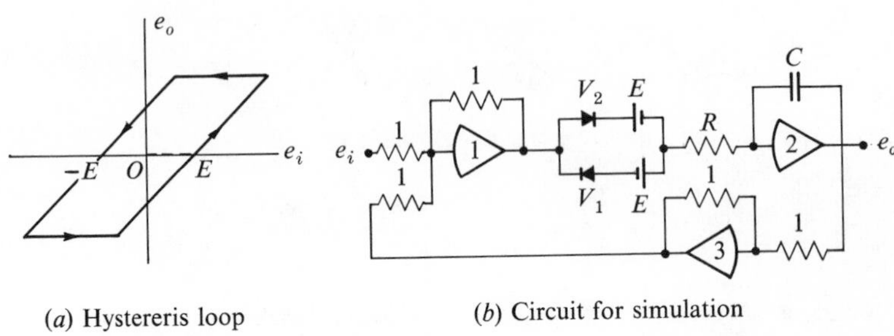

(a) Hystereris loop (b) Circuit for simulation

Fig. 9-37. *Computer diagram for simulation of hysteresis*

e_i is decreased by an amount $2E$, diode V_2 will conduct. Hence a horizontal portion of the hysteresis loop is obtained when neither of the diodes is conducting. The output voltage e_o, corresponding to decreasing e_i with V_2 conducting is

$$e_o = e_i + E \qquad \textbf{(9-115)}$$

Case 4. Use of Differential Relays

The diode is essentially a switching device. Hence a relay-operated switch can be constructed to perform the functions discussed in Case 3. A differential relay DR, as shown symbolically in Fig. 9-38(a), has two signal input terminals G and G' and two contacts K and K' with the relay arm A. If G is more positive than G', the relay closes the contact between A and K. If G' is more positive than G, the arm A is closed with K'.

Four examples of differential relay circuits are shown in Fig. 9-38. The circuit for simulating Coulomb friction is shown in Fig. 9-38(a). Let e_i represent the velocity of a mass and e_o the Coulomb frictional force. If $e_i > 0$, the relay closes the contact between A and K. Thus $e_o = -E$ for all positive values of e_i. The converse is true if e_i is

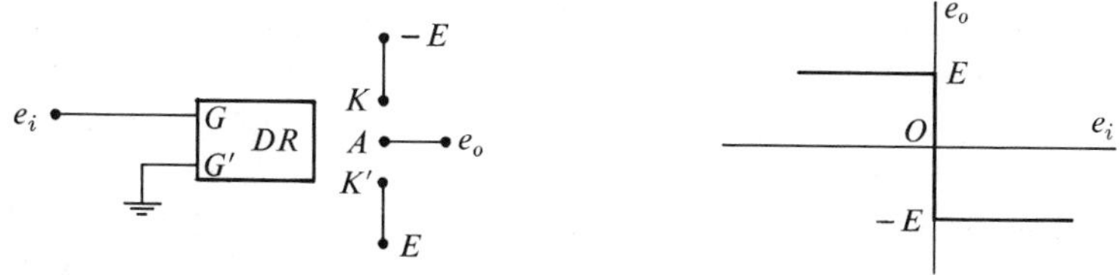

(*a*) Simulation of Coulomb friction

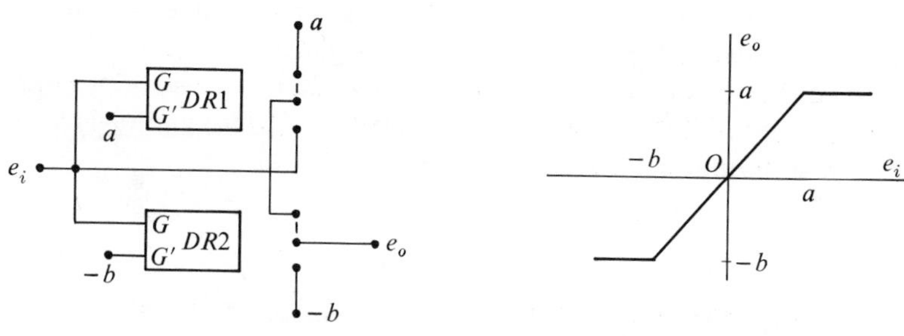

(*b*) Limiting circuit

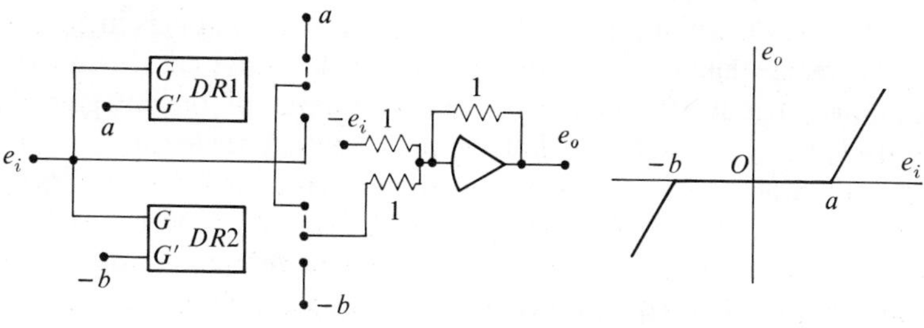

(*c*) Simulation of dead space

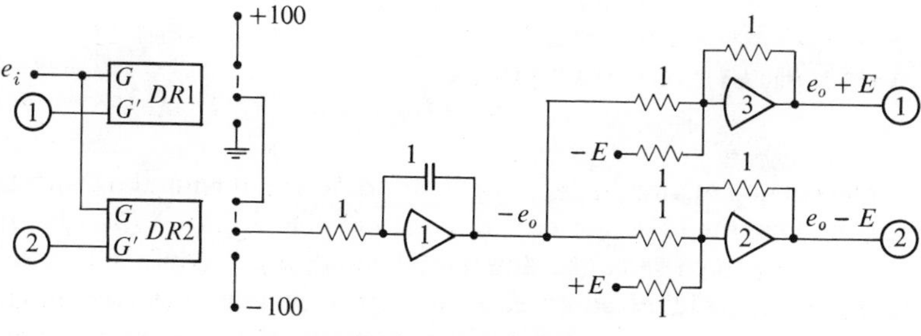

(*d*) Simulation of hysteresis

Fig. 9-38. ***Differential relay circuits***

negative. The limiting circuit, as shown in Fig. 9-38(*b*), consists of two differential relays. If $e_i > a$, e_o is connected to the voltage supply a and the value of the output is $e_o = a$. If $-b < e_i < a$, e_o is connected to e_i. If $e_i < -b$, e_o is connected to the voltage supply $-b$ and the value of the output is $e_o = -b$.

The dead-space simulating circuit of Fig. 9-38(*c*) uses two differential relays and a summing amplifier circuit. If $e_i > a$, the input voltage to the summer is $(a - e_i)$. Thus the output voltage e_o is equal to $(e_i - a)$. If $-b < e_i < a$, the input voltage to the summer is $(e_i - e_i)$ and e_o is zero. If $e_i < -b$, the input voltage to the summer is $(-b - e_i)$ and e_o is equal to $(b + e_i)$, where e_i is a negative voltage.

The hysteresis loop, as shown in Fig. 9-37(*a*), can also be generated with the use of differential relays. From Eqs. (9-114) and (9-115), the hysteresis loop is described by the equations

$$\begin{aligned} e_i &= e_o + E \quad && \text{if } e_i \text{ is increasing} \\ e_i &= e_o - E \quad && \text{if } e_i \text{ is decreasing} \\ e_o &= \text{constant} \quad && \text{if } e_{i(\max)} > e_i > e_{i(\max)} - 2E \text{ and} \\ & && \quad e_{i(\min)} < e_i < e_{i(\min)} + 2E \end{aligned} \tag{9-116}$$

The computer circuit to generate the function described by these equations is shown in Fig. 9-38(*d*). First, consider that e_i is increasing. If e_i is greater than $(e_o + E)$, the contacts in the relays are closed to the positions as indicated in the figure. The input of the integrator is connected to the positive voltage supply, thereby increasing e_o. If e_i is less than $(e_o + E)$, the integrator input is grounded, and e_o remains unchanged until e_i is further increased. Second, after e_i reaches a maximum and starts to decrease, the input of the integrator remains grounded until e_i is equal to $(e_o - E)$. The operation of the circuit for the remaining paths of the loop can be explained in like manner.

9-20. ALGEBRAIC EQUATIONS

The solving of simultaneous ordinary differential equations with the analogue computer was discussed in the previous sections. In using the matrix method to solve the differential equations, it is often necessary to manipulate with algebraic equations. The general-purpose analogue computer is not well suited for the solution of algebraic equations. Special-purpose electronic analogue computers, however, have recently been designed and built for this purpose. Such a computer will handle a 14×14 matrix with good accuracy and high speed. We shall briefly describe the solution of simultaneous algebraic equations in this section.

Consider the matrix equation

$$AX = B \tag{9-117}$$

For simplicity, assume that A is square of order 3 and that X and B are column matrices. This equation can be expanded and written as

$$\begin{aligned} a_{11}x_1 + a_{12}x_2 + a_{13}x_3 - b_1 &= 0 \\ a_{21}x_1 + a_{22}x_2 + a_{23}x_3 - b_2 &= 0 \\ a_{31}x_1 + a_{32}x_2 + a_{33}x_3 - b_3 &= 0 \end{aligned} \tag{9-118}$$

Assuming that the equation is scaled such that $a_{ij} \leq 1$, Eq. (9-118) can be solved by the circuit shown in Fig. 9-39. The negative coefficients would require additional amplifiers as sign changers. This circuit is unstable, however, unless certain restrictions are satisfied.

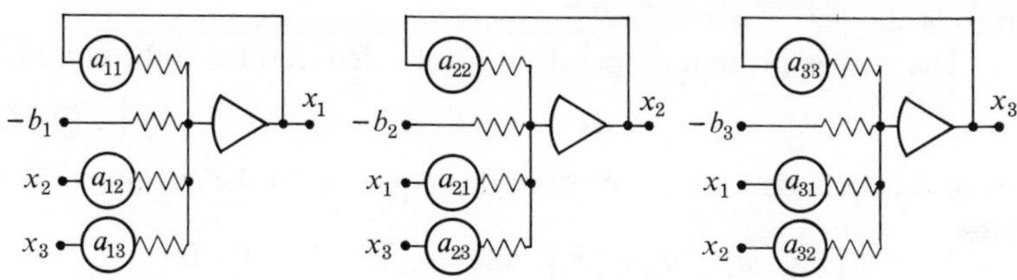

Fig. 9-39. *Computer circuit to solve $AX = B$*

Alternatively, integrators can be used to solve this set of algebraic equations, as shown in Fig. 9-40. The equations being solved by this circuit are

$$\begin{aligned} x_1 &= -\int(a_{11}x_1 + a_{12}x_2 + a_{13}x_3 - b_1)\,dt \\ x_2 &= -\int(a_{21}x_1 + a_{22}x_2 + a_{23}x_3 - b_2)\,dt \\ x_3 &= -\int(a_{31}x_1 + a_{32}x_2 + a_{33}x_3 - b_3)\,dt \end{aligned} \tag{9-119}$$

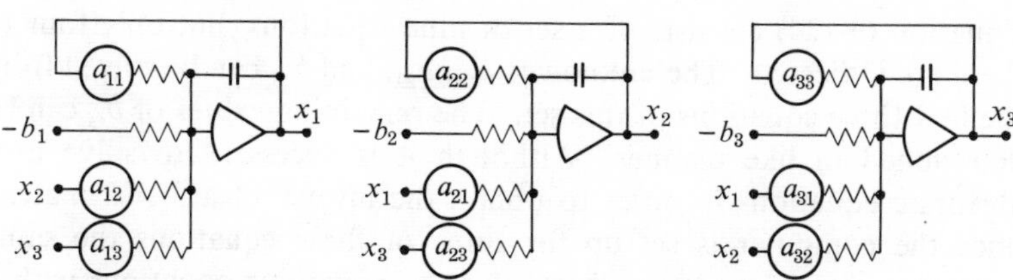

Fig. 9-40. *Computer circuit to solve $AX = B$*

or

$$\begin{aligned} \dot{x}_1 &= -(a_{11}x_1 + a_{12}x_2 + a_{13}x_3 - b_1) \\ \dot{x}_2 &= -(a_{21}x_1 + a_{22}x_2 + a_{23}x_3 - b_2) \\ \dot{x}_3 &= -(a_{31}x_1 + a_{32}x_2 + a_{33}x_3 - b_3) \end{aligned} \tag{9-120}$$

The values of x_i are obtained when their time derivatives become zero, that is, when the outputs of the amplifiers reach their asymptotic values. This method of solving a set of simultaneous algebraic equations is more stable than that using the summers, because the integrators act as low-pass filters which attenuate high-frequency oscillations that might otherwise occur.

It should be noted that Eq. (9-120) can be written as

$$(sI + A)X = B \tag{9-121}$$

where s denotes the time derivative and I is a unit matrix. Equation (9-121) is a set of differential equations. The characteristic equation is formed by equating the determinant of $(sI + A)$ to zero. If all the roots of the characteristic equation have negative real parts, the computer setup is stable.

The matrix inversion operation can be deduced from the equation

$$AA^{-1} = AB = I \tag{9-122}$$

where B is defined as A^{-1}. Writing this equation in the expanded form gives

$$\begin{bmatrix} a_{11} & a_{12} & a_{13} \\ a_{21} & a_{22} & a_{23} \\ a_{31} & a_{32} & a_{33} \end{bmatrix} \begin{bmatrix} b_{11} & b_{12} & b_{13} \\ b_{21} & b_{22} & b_{23} \\ b_{31} & b_{32} & b_{33} \end{bmatrix} = \begin{bmatrix} 1 & 0 & 0 \\ 0 & 1 & 0 \\ 0 & 0 & 1 \end{bmatrix} \tag{9-123}$$

or

$$\begin{aligned} a_{11}b_{11} + a_{12}b_{21} + a_{13}b_{31} &= 1 \\ a_{21}b_{11} + a_{22}b_{21} + a_{23}b_{31} &= 0 \\ a_{31}b_{11} + a_{32}b_{21} + a_{33}b_{31} &= 0 \\ \cdots\cdots\cdots\cdots \\ a_{31}b_{13} + a_{32}b_{23} + a_{33}b_{33} &= 1 \end{aligned} \tag{9-124}$$

Equation (9-124) consists of a set of nine equations, but only four of them are indicated. The unknowns b_{11}, b_{21}, and b_{31} can be solved from the first three equations of this set. The remaining values of b_{ij} can be determined in like manner. Although it is necessary to solve nine algebraic equations in order to obtain the inverse of a 3×3 matrix, once the computer is set up for three of these equations the same computer circuit can be used to solve the remaining equations with a change of input to the amplifiers.

Matrix multiplication can be performed on the computer. Consider the equation

$$\begin{bmatrix} a_{11} & a_{12} & a_{13} \\ a_{21} & a_{22} & a_{23} \end{bmatrix} \begin{bmatrix} b_{11} & b_{12} & b_{13} \\ b_{21} & b_{22} & b_{23} \\ b_{31} & b_{32} & b_{33} \end{bmatrix} = \begin{bmatrix} c_{11} & c_{12} & c_{13} \\ c_{21} & c_{22} & c_{23} \end{bmatrix} \tag{9-125}$$

which can be written as

$$\begin{aligned} a_{11}b_{11} + a_{12}b_{21} + a_{13}b_{31} &= c_{11} \\ a_{11}b_{12} + a_{12}b_{22} + a_{13}b_{32} &= c_{12} \\ a_{11}b_{13} + a_{12}b_{23} + a_{13}b_{33} &= c_{13} \\ \cdots\cdots\cdots\cdots & \\ a_{21}b_{13} + a_{22}b_{23} + a_{23}b_{33} &= c_{33} \end{aligned} \tag{9-126}$$

where c_{ij} are the unknowns. The computer circuit to solve this set of equations is shown in Fig. 9-41. The elements b_{ij} are treated as coefficient potentiometer settings, and the elements a_{ij} are the voltage supplies to the potentiometers. Equation (9-126) consists of a set of six equations. The circuit shows that three of the equations can be solved at a time. The remaining three equations can be solved by changing the values of the supply voltages.

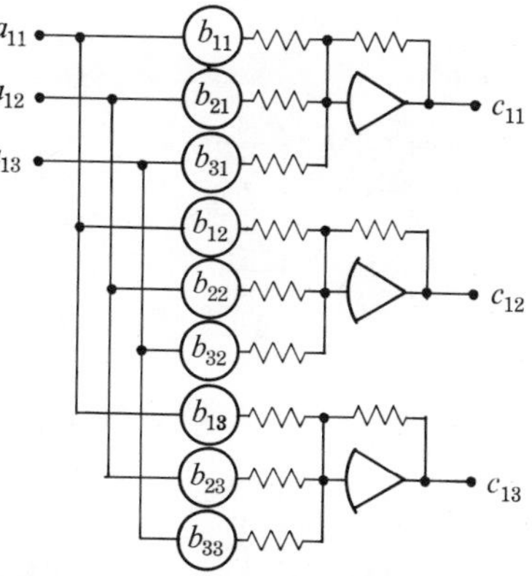

Fig. 9-41. *Computer solution of matrix multiplication*

It was shown in Chap. 7 that, in the study of vibrations, we often have to solve the equation

$$(A - \lambda I)X = 0 \tag{9-127}$$

where λ is a characteristic root of A, I is a unit matrix, and X is a column matrix. For simplicity, assume that A is of order 3. From Eq. (9-127) we have

$$\begin{aligned} (a_{11} - \lambda)x_1 + a_{12}x_2 + a_{13}x_3 &= 0 \\ a_{21}x_1 + (a_{22} - \lambda)x_2 + a_{23}x_3 &= 0 \\ a_{31}x_1 + a_{32}x_2 + (a_{33} - \lambda)x_3 &= 0 \end{aligned} \tag{9-128}$$

This set of equations can be solved by trial and error.

Assume that the errors ε_i are associated with these equations, if the characteristic root λ is not of the proper value.

$$\begin{aligned}(a_{11} - \lambda)x_1 + a_{12}x_2 + a_{13}x_3 &= \varepsilon_1 \\ a_{21}x_1 + (a_{22} - \lambda)x_2 + a_{23}x_3 &= \varepsilon_2 \\ a_{31}x_1 + a_{32}x_2 + (a_{33} - \lambda)x_3 &= \varepsilon_3\end{aligned} \tag{9-129}$$

It can be shown that the computer circuit of Fig. 9-42, with x_1 assumed to be arbitrary, can be used to solve Eq. (9-128) if the values of λ can

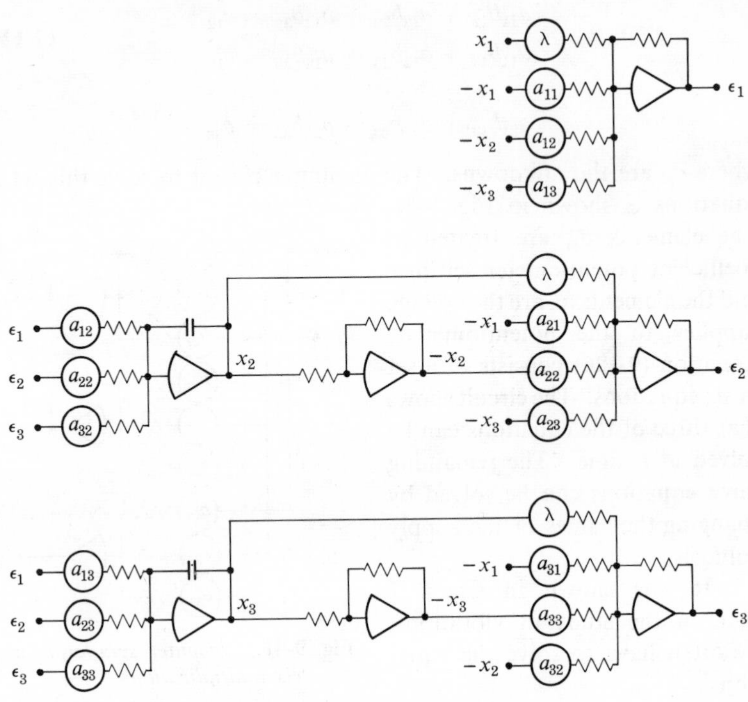

Fig. 9-42. *Computer solution of* $AX = \lambda X$

be obtained from the potentiometer settings. If the λ coefficient potentiometers are ganged together on one shaft, they can be adjusted together until the errors become zero. Thus Eq. (9-128) is solved.

So far, we have assumed that the magnitudes of the matrix elements are less than or equal to unity and that corresponding values can be obtained from the setting of the coefficient potentiometers. If the absolute values of these elements are greater than unity, the equation can be scaled by dividing through by the largest absolute value of the elements. Alternatively, it is expedient to divide instead by the first power of 10 that is greater than the largest coefficient. Having adjusted the coefficients, the equation can then be magnitude scaled so that the output voltages of the amplifiers will be at reasonable levels.

SUGGESTED READING

Jackson, A. S., *Analog Computation* (New York: McGraw-Hill Book Co., Inc., 1960).

Johnson, C. L., *Analog Computer Techniques* (New York: McGraw-Hill Book Co., Inc., 1956).

Korn, G. A., and T. M. Korn, *Electronic Analog Computers* (New York: McGraw-Hill Book Co., Inc., 2nd ed., 1956).

Scott, N. R., *Analog and Digital Computer Technology* (New York: McGraw-Hill Book Co., Inc., 1960), chaps. 1–4.

Soroka, W. W., *Analog Methods in Computation and Simulation* (New York: McGraw-Hill Book Co., Inc., 1954).

Warfield, J. N., *Introduction to Electronic Analog Computers* (New Jersey: Prentice-Hall, Inc., 1959).

Wass, C. A. A., *Introduction to Electronic Analogue Computers* (New York: McGraw-Hill Book Co., Inc., 1956).

APPENDIX A

ANALOGUE COMPUTER: SUGGESTED LABORATORY EXPERIMENTS

Laboratory work is an essential condition for the understanding of the analogue computer and its capabilities. All the experiments suggested can be performed on a relatively simple, inexpensive, commercially available computer. The experiments are selected so as to require no more than 10 amplifiers. The equipment required for solving linear ordinary differential equations with constant coefficients includes the computer, a low-frequency function generator, and a suitable recorder. With the addition of a function multiplier, diodes, and differential relays, the use of the computer can be greatly extended to include the solving of linear ordinary differential equations with variable coefficients and nonlinear ordinary differential equations, and the simulating of problems that are discontinuous in nature.† The x–y-plotter is necessary for recording the phase-plane trajectories of the differential equations.

† The Heath Company has an analogue computer in kit form. The Donner Scientific Co.; Electronic Associates, Inc.; Applied Dynamics, Inc.; and a few others offer small assembled computers at relatively low cost. The Hewlett-Packard low-frequency function generator has a frequency range from 0.008 to 1,200 cps. Sanborn, Brush, and many others offer a variety of recorders. Some companies offer the function multiplier as additional equipment that can be directly plugged into the computer. Some computers have built-in diodes, or the diode may come as plug-in units. It is relatively simple, however, to build a diode and differential relay unit to use with the computer.

Most of the beginning experiments are outlined in fair detail in this appendix. Since the operational procedure may differ with equipment of different make, the procedure section is omitted. Through experience in teaching the laboratory section of a course in vibrations to mechanical engineering students, it is found that the students should be cautioned against (1) omitting the input resistors to the amplifier circuit, (2) connecting the amplifier output directly to ground, and (3) allowing the amplifiers to saturate, that is, to exceed 100 volts.

EXPERIMENT 1. LINEAR OPERATIONS

Preliminary Calculations

Referring to Fig. A-1, complete the "calculated" columns in Table A-1.

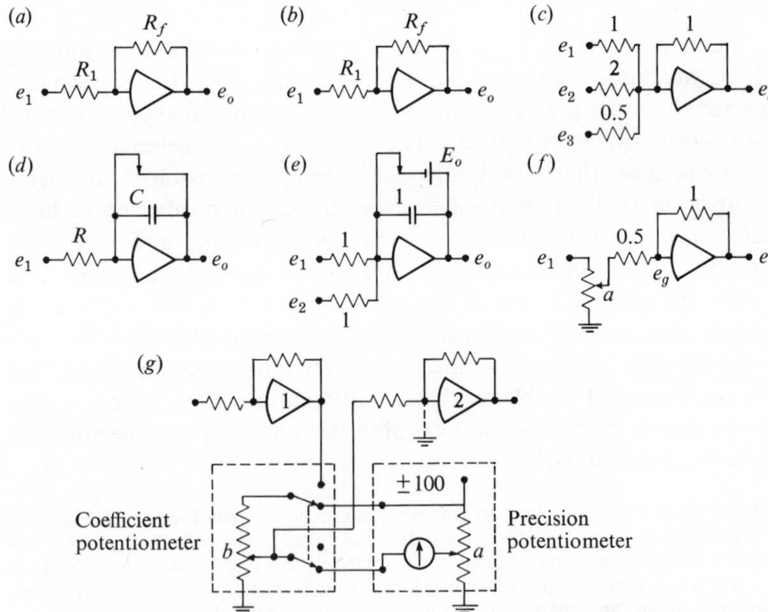

Fig. A-1. *Linear operations: (a) sign inversion; (b) multiplication by a constant; (c) summation; (d) integration with zero initial condition; (e) integration; (f) and (g) potentiometer setting*

TABLE A-1

Linear Operations

(*a*) Sign Inversion

R_f	R_1	e_1	e_o	
			Cal.	Obs.
1	1	40		
10	10	60		

(*b*) Multiplication by a Constant

R_f	R_1	e_1	e_o	
			Cal.	Obs.
2	1	40		
1	0.5	40		
0.5	1	40		

(*c*) Summation

e_1	e_2	e_3	e_o	
			Cal.	Obs.
40	50	−30		
60	40	−50		

(*d*) Integration: $e_o(0) = 0$

R	C	e_1	Time ~ 100 v	
			Cal.	Obs.
1	1	10		
2	1	10		

(*e*) Integration

e_1	e_2	E_o	Time ~ 100 v	
			Cal.	Obs.
10	0	40		
10	−15	−30		

(*f*) Potentiometer Setting

e_1	a	e_o	
		Cal.	Obs.
60	0.745		
80	0.55		

Experiment

The circuits are shown in Fig. A-1.

1. SIGN INVERSION, MULTIPLICATION BY A CONSTANT, AND SUMMATION: With the appropriate computer circuit and input voltages, observe the amplifier output voltage with a meter. Enter observed data in Table A-1.

2. INTEGRATION: Record the amplifier output voltage. Note the time required for the output voltage to reach 100 volts.

3. POTENTIOMETER SETTING: Figure A-1(f) shows a potentiometer placed at the input to an amplifier circuit. Since e_g is essentially at ground potential, the input resistor constitutes a load for the potentiometer. The loading effect is compensated for by setting the potentiometer with the load applied. A potentiometer used in a computer circuit is usually set with a precision potentiometer and a null detector, as shown in Fig. A-1(g). The procedure is to (1) set the wiper a of the precision potentiometer to the desired value, (2) connect a temporary ground (shown dashed) to the load resistor, (3) adjust the wiper b of the coefficient potentiometer until a null is detected, and (4) remove the temporary ground and disconnect the precision potentiometer.

EXPERIMENT 2. FREE VIBRATION: ONE-DEGREE-OF-FREEDOM SYSTEM (see Secs. 2-5 and 2-10)

The equation of motion for the free vibration of a one-degree-of-freedom system is

$$m\ddot{x} + c\dot{x} + kx = 0$$

Defining $c/m = 2\zeta\omega_n$ and $k/m = \omega_n^2$, this equation can be expressed as

$$\ddot{x} + 2\zeta\omega_n\dot{x} + \omega_n^2 x = 0$$

If $0 < \zeta < 1$, the solution of this equation is

$$x = e^{-\zeta\omega_n t}(A_1 \cos \sqrt{1 - \zeta^2}\, \omega_n t + A_2 \sin \sqrt{1 - \zeta^2}\, \omega_n t)$$

or

$$x = Ae^{-\zeta\omega_n t} \sin(\sqrt{1-\zeta^2}\,\omega_n t + \psi)$$

Preliminary Calculations

1. Assume that $\omega_n = 1$ and $\zeta = 0.05$. Corresponding to the initial conditions $x(0) = 60$ and $\dot{x}(0) = 0$, calculate the values of A_1, A_2, A, and ψ. Repeat with the initial conditions $x(0) = 60$ and $\dot{x}(0) = 50$.
2. Calculate the logarithmic decrement.
3. Select a reasonable value of t and calculate the corresponding value of $x(t)$.

Experiment

1. Assume that $\omega_n = 1$ and $0 \leq \zeta \leq 1$. Set up the computer circuit for the simulation of the given system. If the circuit shown in Fig. A-2 is used, the value of ζ is given directly by the potentiometer setting.

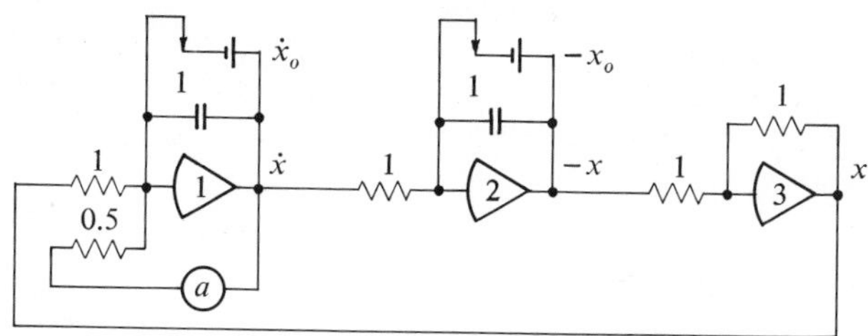

Fig. A-2. *Computer circuit to solve* $\ddot{x} + 2\zeta\dot{x} + x = 0$; $0 \leq \zeta \leq 1$

2. With the initial conditions $x(0) = 60$ and $\dot{x}(0) = 0$, observe $x(t)$ with a meter for a range of values of ζ for $0 \leq \zeta \leq 1.0$. Record $x(t)$ and $\dot{x}(t)$ for $\zeta = 0$ and $\zeta = 0.05$.
3. Repeat Step 2 with the initial conditions $x(0) = 60$ and $\dot{x}(0) = 50$.
4. Change the circuit elements to obtain $\zeta = 2$. With the initial conditions $x(0) = 60$ and $\dot{x}(0) = 50$, record $x(t)$ and $\dot{x}(t)$.

Calculations

1. Determine ζ from the recorded $x(t)$ and $\dot{x}(t)$ curves and compare with the potentiometer setting.

2. With $\zeta = 0.05$, estimate the number of cycles required to reduce the amplitude by a factor of 5. Compare the result with the values shown in Fig. 2-23.
3. Using the same value of t as in Step 3 in the preliminary calculations, observe the value of $x(t)$.

EXPERIMENT 3. STEADY-STATE RESPONSE AND TRANSMISSIBILITY: ONE-DEGREE-OF-FREEDOM SYSTEM (see Secs. 2-5 and 2-12)

The equation of motion of a one-degree-of-freedom system with sinusoidal excitation is

$$m\ddot{x} + c\dot{x} + kx = F \sin \omega t$$

The steady-state response of the system is

$$x = X \sin (\omega t - \phi)$$

where

$$X = \frac{F}{\sqrt{(k - m\omega^2)^2 + (c\omega)^2}} = \frac{F/k}{\sqrt{(1 - r^2)^2 + (2\zeta r)^2}} = \frac{F}{k}\kappa$$

$$\phi = \tan^{-1} \frac{c\omega}{(k - m\omega^2)} = \tan^{-1} \frac{2\zeta r}{1 - r^2}$$

and the frequency ratio r is equal to ω/ω_n.

Experiment

1. Assume that $\omega_n = 1$, $F/m = 20$, $x(0) = \dot{x}(0) = 0$, and $0 \leq \zeta \leq 1$. Set up the circuit for the simulation of the system.
2. With $r = 0.75$, observe $x(t)$ with a meter for a range of values of $0 \leq \zeta \leq 1.0$. Repeat with $r = 1.20$ and $r = 6$.
3. Using $\zeta = 0.2$, record the excitation voltage and $x(t)$ for $r = 0.5$, 1.0, 2.0, and 3.0.
4. The force transmitted to the support of the system is $F_T = kx + c\dot{x}$. Set up the circuit to obtain F_T.
5. Using $\zeta = 0.2$, record the excitation and F_T for $r = 0.5$, 1.0, $\sqrt{2}$, and 3.0.

Calculations

1. Using the recorded data, calculate the magnification factor κ and the phase angle ϕ for each of the frequency ratios indicated. Enter the results in Table A-2(A) and compare the values with those given in Figs. 2-8 and 2-9.
2. Using the recorded data, calculate the transmissibility TR and the phase angle $(\phi - \gamma)$, for each of the frequency ratios indicated. Enter the results in Table A-2(B) and compare the values with those given in Figs. 2-32 and 2-33. (NOTE: The phase angles can be estimated from the relative positions of the curves along the time axis. In some cases it may be difficult to estimate the phase angles accurately.)

TABLE A-2

STEADY-STATE RESPONSE AND TRANSMISSIBILITY: ONE-DEGREE-OF-FREEDOM SYSTEM

r					
X					
κ	a				
	b				
ϕ	a				
	b				

(A)

r					
F_T					
TR	a				
	b				
$\phi - \gamma$	a				
	b				

(B)

a = experimental values.
b = theoretical values indicated in Chapter 2.

EXPERIMENT 4. TIME AND MAGNITUDE SCALING

The equation of motion for the free vibration of a one-degree-of-freedom system is given as

$$\ddot{x} + 1.0\dot{x} + 100x = 0$$

with the initial conditions $x(0) = 60$ and $\dot{x}(0) = 500$.

Preliminary Calculations

1. Time scale the equation so that the system is slowed down by a factor of 10. Magnitude scale the equation and specify the circuit elements for the corresponding computer diagram.
2. Magnitude scale the equation (without time-scale change) and specify the elements for the corresponding computer diagram.

Experiment

1. Set up a computer circuit from Step 1 in the preliminary calculations. Record $x(t)$ and $\dot{x}(t)$.
2. Set up a computer circuit from Step 2 in the preliminary calculations. Record $x(t)$ and $\dot{x}(t)$.
3. Slow down the system from Step 2 by a factor of 5 by time scaling on the computer. Reduce the speed of recording by a factor of 5 and record $x(t)$ and $\dot{x}(t)$.
4. Repeat Step 3, but with a factor of 10.

Calculations

1. Select a reasonable time from the recorded curves of Step 1 in the experiment and calculate $x(t)$ and $\dot{x}(t)$ of the original system.
2. Compare the recorded curves from Steps 2 to 4 in the experiment.
3. Using the same time as Step 1, find $x(t)$ and $\dot{x}(t)$ from the recorded curves from Steps 2 to 4 in the experiment.

EXPERIMENT 5. ONE-DEGREE-OF-FREEDOM SYSTEM: EXCITATION PROPORTIONAL TO ω^2

(see Sec. 2-12)

The equation of motion of a one-degree-of-freedom system with sinusoidal excitation is

$$m\ddot{x} + c\dot{x} + kx = F_{\text{eq}} \sin \omega t$$

where F_{eq} is the magnitude of the equivalent force applied to the mass m. If F_{eq} is proportional to ω^2, this equation can be expressed as

$$\ddot{x} + 2\zeta\omega_n\dot{x} + \omega_n^2 x = b\omega^2 \sin \omega t$$

where b is a constant. Referring to Sec. 2-12, this equation can be used to represent the cases of (1) rotating and reciprocating unbalance [Eq. (2-77)], (2) critical speed of rotating shafts [Eq. (2-80)], and (3) seismic instruments [Eq. (2-93)].

These systems can be simulated by the circuit used in Experiment 3. The quantity $b\omega^2$ is obtained from the appropriate voltage setting of the function generator. Thus the steady-state response curves of Figs. 2-27, 2-38, 2-40, and 2-41 can be observed. Furthermore, with the circuit used in Experiment 3, the force transmitted to the support, as indicated in Fig. 2-34, can be determined.

EXPERIMENT 6. SYSTEMS ATTACHED TO MOVING SUPPORTS: ONE DEGREE OF FREEDOM

(see Case 4, Sec. 2-12)

The equation of motion of a one-degree-of-freedom system attached to a moving support is

$$m\ddot{x}_2 + c\dot{x}_2 + kx_2 = kx_1 + c\dot{x}_1$$

To study the steady-state response of the system to a sinusoidal excitation, we can assume that $\dot{x}_1(t)$ is available from the output of the function generator. Hence the quantity $(kx_1 + c\dot{x}_1)$ can easily be obtained. It was mentioned in Chap. 2 that examples of this type of system are the suspension of vehicles and the mounting of instruments.

EXPERIMENT 7. PRINCIPAL MODES OF VIBRATION

(see Sec. 3-2)

The equations of motion of the vibratory system as shown in Fig. A-3(a) can be expressed as

$$\begin{aligned} m_1\ddot{x}_1 + (k_1 + k_2)x_1 - k_2x_2 &= 0 \\ -k_2x_1 + m_2\ddot{x}_2 + k_2x_2 &= 0 \end{aligned}$$

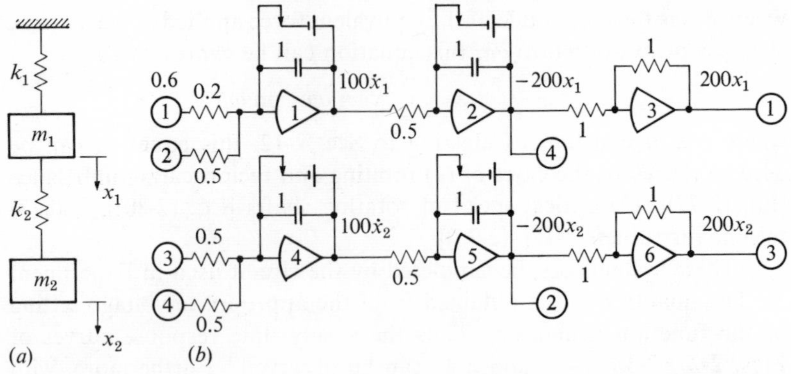

Fig. A-3. *Suggested computer circuit: (a) vibratory system; (b) computer circuit*

The general solutions of these equations are

$$x_1 = A_{11} \cos(\omega_1 t + \psi_1) + A_{12} \cos(\omega_2 t + \psi_2)$$

$$x_2 = \mu_1 A_{11} \cos(\omega_1 t + \psi_1) + \mu_2 A_{12} \cos(\omega_2 t + \psi_2)$$

where $\omega_{1,2}$ are the natural frequencies of the system, $\mu_{1,2}$ are the amplitude ratios of the principal modes of vibration, and A_{11}, A_{12}, ψ_1, and ψ_2 are arbitrary.

Preliminary Calculations

1. Determine A_{11} and A_{12} for the initial conditions $x_1(0) = x_{10}$, $\dot{x}_1(0) = \dot{x}_{10}$, $x_2(0) = x_{20}$, $\dot{x}_2(0) = \dot{x}_{20}$.
2. Determine the initial conditions for the first and second modes of vibration.
3. Assume that $m_1 = m_2 = 1$, $k_1 = 200$, and $k_2 = 400$, and determine $\omega_{1,2}$ and $\mu_{1,2}$.
4. Calculate A_{11}, A_{12}, ψ_1, and ψ_2 for the initial conditions $x_1(0) = 0.25$, $\dot{x}_1(0) = x_2(0) = \dot{x}_2(0) = 0$. Repeat with the initial conditions $x_1(0) = 0.25$, $\dot{x}_1(0) = x_2(0) = 0$, and $\dot{x}_2(0) = 6.0$.
5. Plot $x_1(t)$ and $x_2(t)$ for one of the cases specified in Step 4 for $0 \leq t \leq 1.6$ sec. (An easy way to do this is to sketch the two harmonic components fairly accurately and add the curves graphically.)
6. Rewrite the equations of motion to slow down the problem by a factor of 10. Magnitude scale the equations according to the initial conditions given in Step 4.

Experiment

1. Set up a computer circuit to solve the equations from Step 6 of the preliminary calculations. Figure A-3(*b*) shows a possible circuit.
2. Record $x_1(\tau)$ and $x_2(\tau)$ for the initial conditions

   (*a*) $x_1(0) = 0.25, \quad \dot{x}_1(0) = 0, \quad x_2(0) = 0, \quad \dot{x}_2(0) = 0$

   (*b*) $x_1(0) = 0.25, \quad \dot{x}_1(0) = 0, \quad x_2(0) = 0, \quad \dot{x}_2(0) = 0.6$

3. Record the principal modes of vibration, using (*a*) initial displacements only, (*b*) initial velocities only, and (*c*) initial displacements and velocities.

Calculations

1. Adjusting for the time scale, compare the recorded $x_1(\tau)$ and $x_2(\tau)$ curves with the plotted $x_1(t)$ and $x_2(t)$ curves.
2. Determine $\omega_{1,2}$ and $\mu_{1,2}$ from the recorded curves of the principal modes and compare with the calculated values.

EXPERIMENT 8. BEATING PHENOMENON (see Secs. 1-4 and 3-2)

The equations of motion of a two-degree-of-freedom system without dynamic coupling can be expressed as

$$m_{11}\ddot{q}_1 + k_{11}q_1 + k_{12}q_2 = 0$$
$$m_{22}\ddot{q}_2 + k_{22}q_2 + k_{21}q_1 = 0$$

where q_1 and q_2 are generalized coordinates. The solutions of the equations are

$$q_1 = A_{11} \sin(\omega_1 t + \psi_1) + A_{12} \sin(\omega_2 t + \psi_2)$$
$$q_2 = \mu_1 A_{11} \sin(\omega_1 t + \psi_1) + \mu_2 A_{12} \sin(\omega_2 t + \psi_2)$$

Four examples of two-degree-of-freedom systems are shown in Fig. A-4(*a*) to (*d*). The coupling in the first three examples is quite evident. In (*d*), a rigid mass is attached to a coil spring. The mass has an up-and-down motion and a twisting motion about the longitudinal

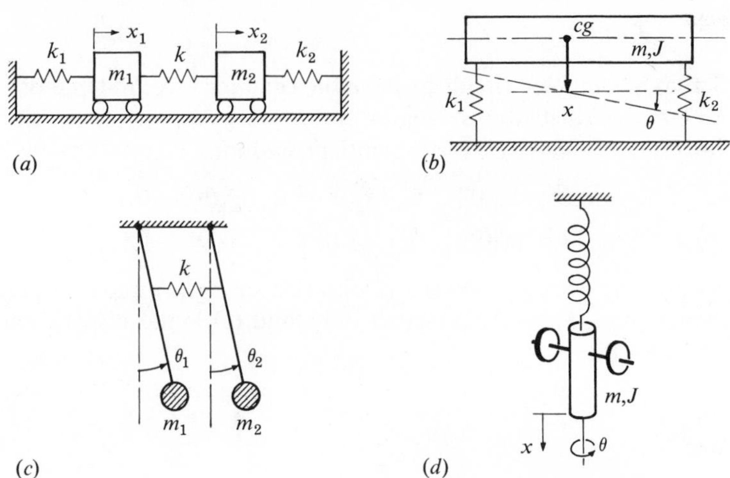

Fig. A-4. *Examples of two-degree-of-freedom systems*

axis of the spring. Coupling exists because, when the spring is elongated, it gives a slight torque, and when twisted, it gives a slight pull. Let the equations of motion of this system be

$$\ddot{x} + 1{,}000x - 100\theta = 0$$
$$\ddot{\theta} + 1{,}000\theta - 100x = 0$$

Preliminary Calculations

1. Explain analytically the conditions under which beating is possible.
2. Calculate the natural frequencies and the amplitude ratios of the principal modes.
3. Rewrite the equations of motion to slow down the problem by a factor of 5. Magnitude scale the equations by assuming $x_{max} = 1$.

Experiment

1. Set up a computer circuit to solve the equations from Step 3 in the preliminary calculations.
2. Record $x(\tau)$ and $\theta(\tau)$. (Recording time should be approximately 30 sec.)
3. Using the appropriate initial conditions, record $x(\tau)$ and $\theta(\tau)$ for the principal modes of vibration.

Calculations

1. Determine the natural frequencies from the recorded data. Compare with the calculated values.
2. Determine the beat period from these natural frequencies. Compare the beat period with the recorded data.

EXPERIMENT 9. SEMIDEFINITE SYSTEMS (see Sec. 3-3)

The equations of motion of a semidefinite system can be expressed as

$$J_1\ddot{\theta}_1 = -k_t(\theta_1 - \theta_2)$$
$$J_2\ddot{\theta}_2 = -k_t(\theta_2 - \theta_1)$$

Set up a computer circuit to solve these equations. With different initial conditions, demonstrate that this system has one of its natural frequencies equal to zero. Figure A-5 shows a possible circuit for this problem.

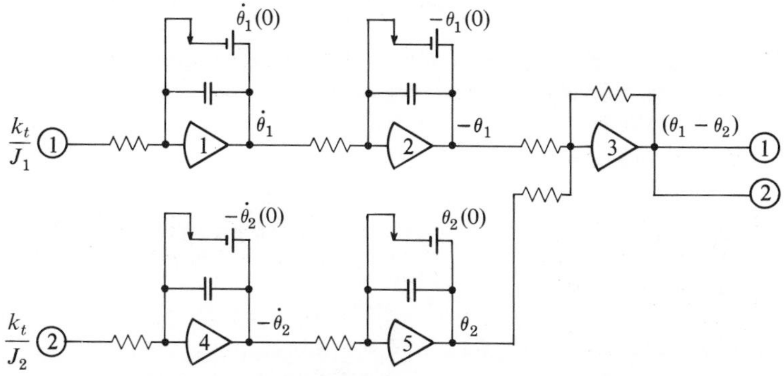

Fig. A-5. *Computer circuit: semidefinite system*

EXPERIMENT 10. DYNAMIC ABSORBER WITH DAMPING

The computer circuit to simulate the dynamic absorber was discussed in Sec. 9-12 (see Example 5). Investigate the effect of the damping coefficient c on the amplitude of oscillation of the mass m_1 over a reasonable frequency range.

EXPERIMENT 11. VEHICLE SUSPENSION

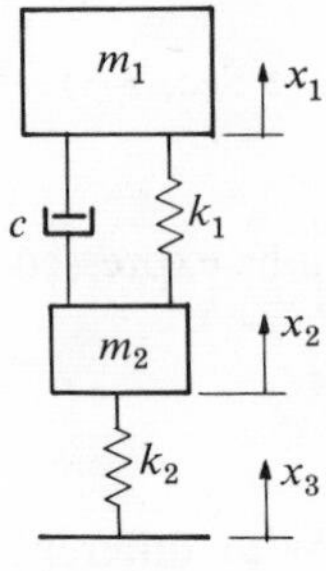

Fig. A-6. *Simplified vehicle-suspension system*

A simplified vehicle-suspension system is shown in Fig. A-6. The mass m_1 represents one fourth of the mass of the vehicle, and m_2 represents half of the wheels and axle combined. The shock absorber and the main spring are indicated as c and k_1, respectively. The equivalent spring constant of the pneumatic tire is assumed to be equal to k_2. The displacement x_3 is the disturbance due to the road roughness. Assume that the equations of motion of the system are

$$2\ddot{x}_1 = -c(\dot{x}_1 - \dot{x}_2) - 100(x_1 - x_2)$$
$$0.2\ddot{x}_2 = -c(\dot{x}_2 - \dot{x}_1) - 100(x_2 - x_1) - 400(x_2 - x_3)$$

If x_3 is a step function of 4 in. in magnitude, determine the value of c such that minimum force is transmitted to m_1.

EXPERIMENT 12. DIODE CHARACTERISTICS AND LIMITING

DIODE CHARACTERISTICS

The characteristics of diodes can be determined from the circuit shown in Fig. A-7(a). If the diode is an ideal switching device, the voltage drop across the resistor R is $e_o = iR = e_i - E_c$, where the bias voltage E_c can be either positive or negative. The voltages e_i and e_o can be plotted conveniently by means of an x–y plotter. Alternatively, it may be expedient to obtain an e_i that is linear with time from an integrator, as shown in Fig. A-7(c). The voltages e_i and e_o can then be recorded by means of a recorder. In this manner we introduce time t as a parameter for recording e_i and e_o. Thus the characteristics of the diode can easily be interpreted from the recorded voltages.

Compare the diode characteristics with that obtained from an idealized diode circuit as shown in Fig. 9-32(a).

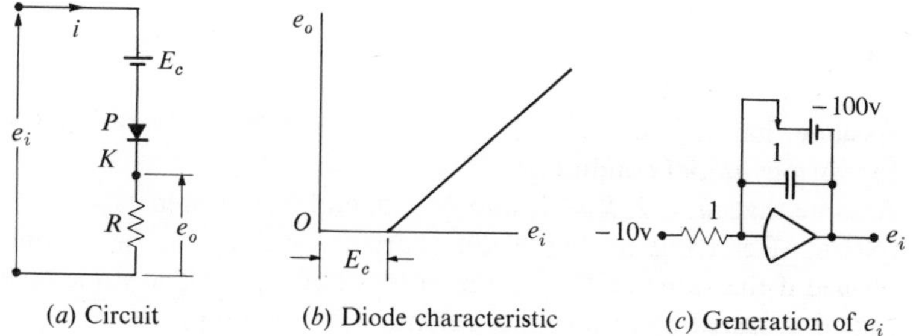

(a) Circuit (b) Diode characteristic (c) Generation of e_i

Fig. A-7. *Characteristics of diodes*

DIODE LIMITING CIRCUITS (see Fig. 9-36)

1. Find the characteristics of the input shunt limiting circuit for various combinations of R_1, R_2, R_3, R_4, E_1, and E_2. Predetermine the characteristic curves through calculations, and verify experimentally. (CAUTION: Do not choose a combination that will cause a short circuit of the voltage supplies.)
2. Predetermine the characteristics of the output series limiting circuit for the various combinations of E_1, E_2, R_2, R_3, and R_4. Note that the values of the resistors can range from zero to infinity. (CAUTION: Do not short-circuit the d-c supplies.)
3. Predetermine the characteristics of the input series limiting circuit for reasonable values of the resistors R_1 to R_4. Repeat, but with R_1 omitted. Verify the characteristics experimentally.

EXPERIMENT 13. COULOMB DAMPING

Coulomb damping in the free vibration of a simple mass-spring system was discussed in Sec. 9-18, and the corresponding circuit diagram was shown in Fig. 9-29(c).

Preliminary Calculations

1. Determine the frequency of oscillation for the free vibration of this system.
2. Show that the amplitude decay per cycle is a constant equal to $4F/k$.

Experiment

1. Assume that $m = 0.5$, $k = 1$, and $F = 5$, and record $x(t)$ and $\dot{x}(t)$ for various initial conditions.
2. Assume that $m = 2$, $k = 1$, and $F = 5$, and repeat Step 1. (NOTE: Referring to Fig. 9-29, sharper limiting can be accomplished if the value of E is on the order of 50 volts. The value of F can be adjusted by means of a potentiometer setting.)

Calculations

1. Determine the frequencies of oscillation from the recorded data and compare with their theoretical values.
2. Determine and compare the amplitude decay per cycle for the two cases.

EXPERIMENT 14. COMBINED DAMPING

Experiment 13 shows that the effect of Coulomb damping in a vibratory system can be studied if the proper function can be generated. Thus, with the use of a multiplier and a differential relay, the effect of velocity-squared damping can be examined. The effect of Coulomb, velocity-squared, and viscous damping can be considered by summing these quantities with an amplifier circuit. In general, any type of damping can be considered, provided the function can be generated.

The effect of damping on forced vibration can be studied if an external excitation is applied to the circuit representing the system under consideration. Hence the discussions in Experiments 13 and 14 are not limited to the study of free vibrations.

EXPERIMENT 15. DEAD-SPACE SIMULATION

A vibratory system with dead space is shown in Fig. 9-30. Select values of m, k_1, and k_2, and set up a computer circuit suitable for measuring $\ddot{x}(t)$, $\dot{x}(t)$, and $x(t)$ of the mass m. Record $\ddot{x}(t)$, $\dot{x}(t)$, and $x(t)$ for various values of the dead space a and b.

EXPERIMENT 16. USE OF DIFFERENTIAL RELAYS

It was mentioned in Sec. 9-19 that differential relays may be used to perform the same function as diodes in a computer circuit. Repeat Experiments 13 and 15 with the use of relays.

EXPERIMENT 17. SIMPLE PENDULUM (see Sec. 2-3)

The equation of motion of a simple pendulum can be expressed as

$$\ddot{\theta} + \frac{g}{L} \sin \theta = 0$$

If $L = 38.6$ in. set up the computer circuit to solve this equation. Find the minimum value of $\dot{\theta}(0)$ such that, once the pendulum is set into motion, it will rotate about its hinge point.

EXPERIMENT 18. NONLINEAR SPRING

The equation of motion of a one-degree-of-freedom system is given as

$$10\ddot{x} + 5\dot{x} + 25x + 0.25x^3 = 0$$

with the initial conditions $x(0) = 10$ and $\dot{x}(0) = 0$. In this equation the spring force is $(25 + 0.25x^2)x$, which is not linear with the spring deformation x. Set up a computer circuit to solve this equation.

EXPERIMENT 19. VAN DER POL'S EQUATION

(see Sec. 9-16)

Solve Van der Pol's equation, Eq. (9-73), on the computer for various values of ε and initial conditions.

EXPERIMENT 20. SYSTEM WITH MECHANICAL STOP

INELASTIC STOP

A mechanical system, as shown in Fig. A-8(*a*), has two limit stops. It may be used to represent a power piston, the motion of which is limited by the cylinder length.

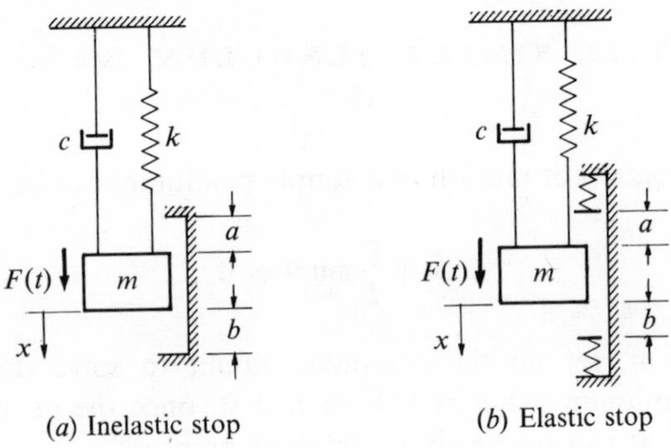

(*a*) Inelastic stop (*b*) Elastic stop

Fig. A-8. *Systems with mechanical stops*

Assume that the excitation force $F(t)$ is sinusoidal. If $m = 1$, $c = 0.4$, and $k = 1$, set up a computer circuit to simulate this system. Record $x(t)$ and $\dot{x}(t)$ for a range of values of gap distances a and b.

ELASTIC STOP

As a variation of this type of physical problem, the mechanical stops may be assumed to be elastic, as shown in Fig. A-8(*b*). Assuming reasonable values for the system parameters, set up a computer circuit to simulate this system. Record $x(t)$ and $\dot{x}(t)$ for a range of values of gap distances a and b.

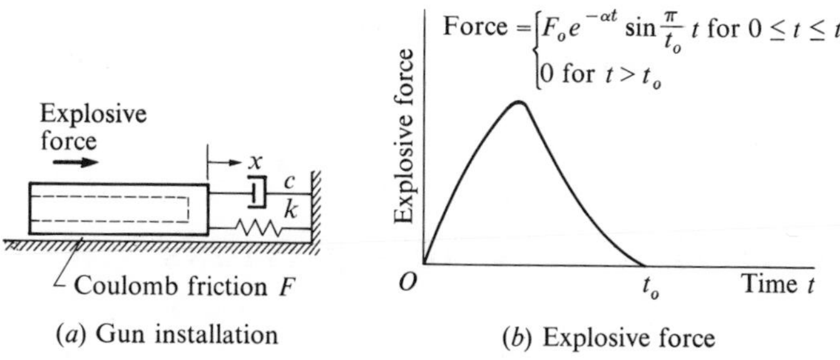

Fig. A-9. *A gun-installation problem*

EXPERIMENT 21. GUN INSTALLATION

A schematic representation of the installation of a large artillery gun is shown in Fig. A-9(*a*). When a shell is fired, the barrel recoils and its motion is braked by the heavy spring. Then the barrel is returned by the spring to its firing position. Assume that (1) the dashpot is constructed so that it offers very little resistance for the recoil stroke of the gun barrel but has a damping coefficient c for the return stroke, (2) Coulomb friction of magnitude F exists between the gun barrel and its guides, and (3) the force exerted on the barrel by the firing explosion is as shown in Fig. A-9(*b*).

Assuming reasonable values for the system parameters, simulate this problem on the computer.

EXPERIMENT 22. SYSTEM WITH BACKLASH

A mechanical system with backlash is shown in Fig. A-10, and a sinusoidal excitation is applied to the mass m_1.

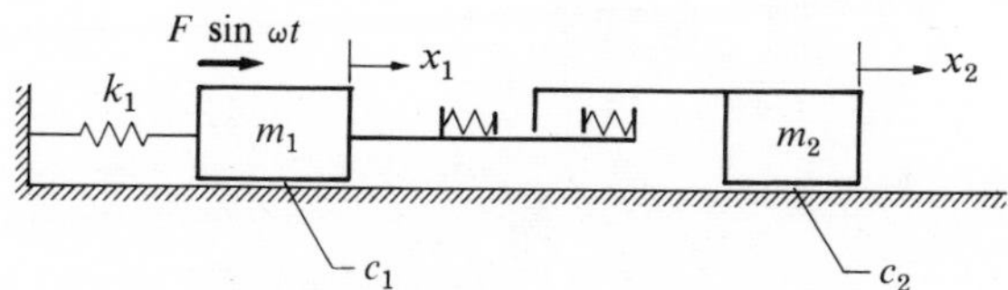

Fig. A-10. *System with backlash*

Write the equations of motion of the system. Assuming reasonable values for the system parameters, simulate this problem on the computer.

EXPERIMENT 23. ALGEBRAIC EQUATIONS

A matrix equation is given as

$$AX = B$$

where

$$A = \begin{bmatrix} 2 & 2 & -1 \\ 2 & 3 & -2 \\ -1 & -2 & 2 \end{bmatrix} \quad B = \begin{bmatrix} 1 \\ -2 \\ 3 \end{bmatrix} \quad \text{and} \quad X = \begin{bmatrix} x_1 \\ x_2 \\ x_3 \end{bmatrix}$$

Use the analogue computer to determine (1) X, (2) A^{-1}, and (3) the characteristic roots of A.

APPENDIX B

LINEAR ORDINARY DIFFERENTIAL EQUATIONS WITH CONSTANT COEFFICIENTS

B-1. INTRODUCTION

A *differential equation* is an equation relating two or more variables in terms of derivatives or differentials. In this appendix we shall limit our discussion to the type of equations used in Chaps. 2 and 3. One-degree-of-freedom systems were discussed in Chap. 2. The typical differential equations used are of the form

$$m\frac{d^2x}{dt^2} + c\frac{dx}{dt} + kx = f(t) \tag{B-1}$$

where x, the unknown, is the *dependent variable* and t the *independent variable*, m, c, and k are constants, and $f(t)$ is a function of t. The variables x and t are related in this equation in terms of derivatives. If x represents displacement and t time, then $m(d^2x/dt^2)$ represents the inertia force, $c(dx/dt)$ the viscous damping force, kx the spring force, and $f(t)$ the excitation force applied to the system.

Multidegree-of-freedom systems were discussed in Chap. 3. Using the notation $\dot{x} = dx/dt$ and $\ddot{x} = d^2x/dt^2$, the typical equations of motion for an undamped two-degree-of-freedom system are of the form

$$\begin{aligned} m_{11}\ddot{x}_1 + k_{11}x_1 + k_{12}x_2 &= f_1(t) \\ m_{22}\ddot{x}_2 + k_{22}x_2 + k_{21}x_1 &= f_2(t) \end{aligned} \tag{B-2}$$

where x_1 and x_2 are the dependent variables and t is the independent variable. Here, we relate two dependent variables and one independent variable in the *simultaneous equations*.

B-2. DEFINITIONS

The types of equations to be considered, as illustrated in Eqs. (B-1) and (B-2), are known as *ordinary differential equations*. There is only one independent variable, and the unknowns, or the dependent variables, are expressed as total derivatives with respect to it. When the unknown is a function of two or more independent variables, partial derivatives are used in the equations, which are called *partial differential equations*.

The *order* of a differential equation is equal to the order of the highest derivative in the equation. Both Eqs. (B-1) and (B-2) are second-order differential equations.

A *linear differential equation* is one in which no terms involving the unknown function or its derivatives appear as products or are raised to a power different from unity. It should be noted that the equation

$$t^2\ddot{x} + t\dot{x} + (t^2 - n^2)x = 0 \tag{B-3}$$

which has the $t^2\ddot{x}$, $t\dot{x}$, and t^2x terms, is linear. The unknown x and its derivatives have exponents of unity, and no products of x and its derivatives exist. The coefficients of x and its derivatives, however, are variables. This type of equation is called a linear ordinary differential equation with *variable coefficients*. No formulas are available for the solution of such equations of order greater than 1. The coefficients of x and its derivatives in Eqs. (B-1) and (B-2) are constants. These are linear ordinary differential equations with *constant coefficients*.

A *nonlinear differential equation* is one which is not linear. Examples of nonlinear equations are

$$m\ddot{x} + c(\dot{x})^2 + kx = f(t) \tag{B-4}$$

$$m\ddot{x} + c\dot{x} + kx^3 = f(t) \tag{B-5}$$

$$\ddot{\theta} + (g/L)\sin\theta = 0 \tag{B-6}$$

$$\ddot{x} + a(x^2 - 1)\dot{x} + x = 0 \tag{B-7}$$

Equation (B-4) is a nonlinear ordinary differential equation with constant coefficients. Comparing with Eq. (B-1), it is noted that the damping force indicated in Eq. (B-4) is proportional to the square of the velocity. Similarly, the spring force in Eq. (B-5) is proportional to x^3. Equation (B-6) describes the motion of a simple pendulum with large amplitudes of oscillation. Equation (B-7) is Van der Pol's equation. No effective general analytical methods have been devised to solve nonlinear differential equations.

In this appendix we shall discuss only linear ordinary differential equations with constant coefficients.

B-3. SOLUTION

The *solution* or integral of a differential equation is a functional relation, not involving derivatives, between the variables implied in the equation. For example, the relation

$$x = C_1 \sin t \tag{B-8}$$

is a solution of the differential equation

$$\ddot{x} + x = 0 \tag{B-9}$$

Differentiating Eq. (B-8) twice with respect to t and substituting in Eq. (B-9) gives

$$-C_1 \sin t + C_1 \sin t = 0 \tag{B-10}$$

and Eq. (B-9) is satisfied. Here, the value of C_1 is arbitrary, and it is called a *constant of integration*. Similarly, it can be shown that $x = C_2 \cos t$ is also a solution of Eq. (B-9). Hence a more general solution of Eq. (B-9) is

$$x = C_1 \sin t + C_2 \cos t \tag{B-11}$$

To generalize this observation, consider the equation

$$\frac{d^n x}{dt^n} + a_1 \frac{d^{n-1}x}{dt^{n-1}} + a_2 \frac{d^{n-2}x}{dt^{n-2}} + \cdots + a_{n-1}\frac{dx}{dt} + a_n x = f(t) \tag{B-12}$$

which is a linear ordinary differential equation with constant coefficients of order n. For brevity, we shall call it an nth-order differential equation. It is stated without proof that the *general solution* of an nth-order differential equation has n arbitrary constants of integration. Solutions obtained from the general solution by giving particular values to these constants are called *particular solutions*.

It is convenient to introduce the *linear differential operator*, denoted by the symbol $L(D)$ or simply L, for this discussion. Let the symbol D be defined in terms of the operations

$$Dx = \frac{d}{dt}x = \frac{dx}{dt}; \quad D^2x = D\,Dx = \frac{d}{dt}\frac{d}{dt}x = \frac{d^2x}{dt^2}; \quad \cdots; \quad D^nx = \frac{d^nx}{dt^n} \tag{B-13}$$

The symbol $D = d/dt$ is called a *linear differential operator* or the *D operator*. It has no physical interpretation unless it is applied to a differentiable function $x(t)$. When it is used with linear ordinary differential equations with constant coefficients, however, it can be manipulated as an algebraic quantity. Thus Eq. (B-12) can be expressed in the form

$$(D^n + a_1D^{n-1} + a_2D^{n-2} + \cdots + a_{n-1}D + a_n)x = f(t) \tag{B-14}$$

This equation can be written more concisely as

$$Lx = f(t) \tag{B-15}$$

where

$$L = (D^n + a_1D^{n-1} + a_2D^{n-2} + \cdots + a_{n-1}D + a_n) \tag{B-16}$$

Since derivatives have the property

$$\frac{d^p}{dt^p}(C_1x_1 + C_2x_2) = C_1\frac{d^p}{dt^p}x_1 + C_2\frac{d^p}{dt^p}x_2$$

the operator L is linear in the respect that

$$L(C_1x_1 + C_2x_2) = C_1Lx_1 + C_2Lx_2 \tag{B-17}$$

Thus the principle of superposition applies to the operator L. An nth-order differential equation, in the form of Eqs. (B-12), (B-14), or (B-15), is called a *nonhomogeneous equation*. If $f(t) = 0$, the corresponding equation

$$Lx = 0 \tag{B-18}$$

is called a *homogeneous equation*.

The general solution of an nth-order differential equation $Lx = f(t)$ consists of (1) a *complementary function*, which is the general solution of the corresponding homogeneous equation $Lx = 0$, and (2) a *particular integral*, which satisfies the nonhomogeneous equation $Lx = f(t)$.

Let x_1 be a solution of $Lx = 0$; that is, $Lx_1 = 0$. Similarly, let $x_2, x_3, \ldots, x_n$ be solutions of $Lx = 0$. From the linear property of the operator L, we have

$$L(C_1x_1 + C_2x_2 + \cdots + C_nx_n) = C_1Lx_1 + C_2Lx_2 + \cdots + C_nLx_n = 0$$

Hence a linear combination of the solutions of $Lx = 0$ is also a solution of $Lx = 0$. It is stated without proof that an nth-order homogeneous equation $Lx = 0$ has n linearly independent solutions;† that is,

$$x_c = C_1x_1 + C_2x_2 + \cdots + C_nx_n \tag{B-19}$$

where x_c is called the complementary function.

Let x_p be the particular integral of $Lx = f(t)$; that is, x_p satisfies the equation $Lx_p = f(t)$. Substituting

$$x = C_1x_1 + \cdots + C_nx_n + x_p = x_c + x_p \tag{B-20}$$

in Eq. (B-15) gives

$$L(x_c + x_p) = 0 + f(t) = f(t)$$

Thus the function $x = (x_c + x_p)$ satisfies Eq. (B-15) and the general solution of a nonhomogeneous equation $Lx = f(t)$ is a linear combination of its complementary function x_c and its particular integral x_p.

B-4. COMPLEMENTARY FUNCTION

The complementary function is the general solution of the homogeneous differential equation

$$\frac{d^nx}{dt^n} + a_1\frac{d^{n-1}x}{dt^{n-1}} + \cdots + a_{n-1}\frac{dx}{dt} + a_nx = 0 \tag{B-21}$$

If s is a constant and $x = e^{st}$ is substituted in this equation, we obtain

$$(s^n + a_1s^{n-1} + \cdots + a_{n-1}s + a_n)e^{st} = 0$$

Since e^{st} cannot be zero for all finite values of t, we have

$$s^n + a_1s^{n-1} + \cdots + a_{n-1}s + a_n = 0 \tag{B-22}$$

Thus, if the value of s is so chosen that it is a root of Eq. (B-22), $x = e^{st}$ is a solution of Eq. (B-21). Equation (B-22) is called the

† The functions $x_1(t), x_2(t), \ldots, x_n(t)$ are *linearly dependent* if there exist n constants, $C_1, C_2, \ldots, C_n$, which are not all zero, such that

$$C_1x_1 + C_2x_2 + \ldots + C_nx_n = 0$$

Linear dependence means that a function of the set can be expressed as a linear combination of the other functions of the set. Often, it is possible to determine by inspection whether the set of functions is linearly dependent. A mathematical test for linear dependence of a set of functions is that their Wronskian vanishes over the interval of the independent variable considered.

auxiliary or the *characteristic equation* of the physical system, for its roots establish the characteristics of the physical system represented by the differential equation. By the fundamental theorem of algebra, Eq. (B-22) has n number of roots. If the n roots are distinct, the general solution of Eq. (B-21) is

$$x = C_1e^{s_1t} + C_2e^{s_2t} + \cdots + C_ne^{s_nt} \tag{B-23}$$

Since the roots in this general discussion can be real, imaginary, or complex, there are, in effect, only two separate cases—roots distinct and roots repeating. We shall confine our discussion to the second-order differential equation, which is the type of equation encountered in Chap. 2. The discussion can easily be generalized. For completeness, the discussion is presented as four separate cases.

Case 1. Roots Real and Distinct

Consider the second-order differential equation

$$m\ddot{x} + c\dot{x} + kx = 0 \tag{B-24}$$

From Eq. (B-22), the corresponding characteristic equation of Eq. (B-24) is

$$ms^2 + cs + k = 0 \tag{B-25}$$

The roots $s_{1,2}$ and the general solution are

$$s_{1,2} = \frac{1}{2m}(-c \pm \sqrt{c^2 - 4mk}) \tag{B-26}$$

$$x = C_1e^{s_1t} + C_2e^{s_2t} \tag{B-27}$$

Let us assume that $c^2 > 4mk$ in order that the roots be real. Since $\sqrt{c^2 - 4mk} < c$, the roots are real and negative. Thus Eq. (B-27) indicates that the function $x(t)$ decreases exponentially with increasing t.

Case 2. Roots Real and Repeating

If the roots are equal, that is, $s_1 = s_2 = s$, the constants C_1 and C_2 in Eq. (B-27) can be combined to give a new arbitrary constant. Thus only one solution is obtained by the prescribed method. From Eq. (B-26), the roots are equal if $c^2 = 4mk$ and $s_{1,2} = -c/2m$.

To obtain the second solution, let us write Eq. (B-24) in the form

$$(D^2 - 2sD + s^2)x = 0 \tag{B-28}$$

where

$$-2s = c/m \quad \text{and} \quad s^2 = \frac{c^2}{4m^2} = \frac{4mk}{4m^2} = k/m$$

From Eq. (B-28), we have

$$(D - s)(D - s)x = 0 \tag{B-29}$$

Defining

$$(D - s)x = y \tag{B-30}$$

Eq. (B-29) can be rewritten as

$$(D - s)y = 0 \tag{B-31}$$

Clearly, $y = e^{st}$ is a solution of this equation. Substituting $y = e^{st}$ in Eq. (B-30) gives

$$(D - s)x = e^{st} \tag{B-32}$$

It can be shown by the method of undetermined coefficients (see Sec. B-5) that the solution of Eq. (B-32) is of the form

$$x = te^{st} \tag{B-33}$$

It can be verified readily, by direct substitution, that Eq. (B-33) is a solution of Eq. (B-28). Thus the general solution is a linear combination of its two solutions; that is,

$$x = (C_1 + C_2 t)e^{st} \tag{B-34}$$

It may be noted that $x(t)$ is finite for finite values of t. Applying L'Hospital's rule, $x(t)$ is zero for t toward infinity.

Case 3. Roots Complex

The roots of the characteristic equation, Eq. (B-25), are complex if $c^2 < 4mk$. Let the conjugate complex roots from Eq. (B-26) be

$$s_{1,2} = -\alpha \pm j\beta \tag{B-35}$$

where $j = \sqrt{-1}$. Since the roots are distinct, the solution of Eq. (B-24) is

$$\begin{aligned} x &= C_1 e^{(-\alpha + j\beta)t} + C_2 e^{(-\alpha - j\beta)t} \\ &= e^{-\alpha t}(C_1 e^{+j\beta t} + C_2 e^{-j\beta t}) \end{aligned}$$

Applying Euler's formula $e^{\pm j\theta} = \cos\theta \pm j\sin\theta$, this equation becomes

$$\begin{aligned} x &= e^{-\alpha t}[(C_1 + C_2)\cos\beta t + j(C_1 - C_2)\sin\beta t] \\ x &= e^{-\alpha t}(A\cos\beta t + B\sin\beta t) \end{aligned} \quad \textbf{(B-36)}$$

Since $x(t)$ is real, the constants A and B in Eq. (B-36) must be real constants, indicating that the constants $C_{1,2}$ associated with $s_{1,2}$ must be complex conjugates.

Generalizing, if a given 6th-order homogeneous differential equation has the distinct complex roots $s_{1,2} = -\alpha_1 \pm j\beta_1$, $s_{3,4} = -\alpha_2 \pm j\beta_2$, and distinct real roots $s_{5,6}$, the general solution is of the form

$$x = e^{-\alpha_1 t}(A_1\cos\beta_1 t + B_1\sin\beta_1 t) + e^{-\alpha_2 t}(A_2\cos\beta_2 t + B_2\sin\beta_2 t) + C_5 e^{s_5 t} + C_6 e^{s_6 t}$$

If the complex roots are repeated, that is, $\alpha_1 = \alpha_2 = \alpha$ and $\beta_1 = \beta_2 = \beta$, and the real roots are distinct, the general solution is of the form

$$x = e^{-\alpha t}[(A_1 + A_2 t)\cos\beta t + (B_1 + B_2 t)\sin\beta t] + C_5 e^{s_5 t} + C_6 e^{s_6 t}$$

The observation can easily be generalized for the nth-order equation.

Case 4. Roots Imaginary

From Eq. (B-26), the roots of the characteristic equation are imaginary if $c = 0$; that is, $s_{1,2} = \pm j\sqrt{k/m} = \pm j\omega_n$. Since the roots are distinct, the general solution of Eq. (B-24) is of the form

$$x = C_1 e^{+j\omega_n t} + C_2 e^{-j\omega_n t} \quad \textbf{(B-37)}$$

Following the method outlined in Case 3, Eq. (B-37) can be expressed as

$$x = A\cos\omega_n t + B\sin\omega_n t \quad \textbf{(B-38)}$$

It is evident that this is a special case of complex roots when the real part of the complex root is zero.

B-5. PARTICULAR INTEGRAL

The particular integral x_p is a solution of the nonhomogeneous differential equation

$$Lx = f(t) \quad \textbf{(B-39)}$$

By definition, x_p does not contain any arbitrary constants. Thus x_p does not include x_c, the general solution of $Lx = 0$, as discussed in the previous section. If x_c is known, an x_p can always be found by the method of variation of parameters. This method, however, is generally very laborious.

In many practical problems the function $f(t)$ in Eq. (B-39) is a sine or cosine, a polynomial, an exponential, or a product or a linear combination of such functions. For these functions a much simpler trial procedure, called the *method of undetermined coefficients*, can be used for finding x_p. This procedure will be discussed in this section. It should be remembered that this method is not generally applicable unless Eq. (B-39) is a differential equation with constant coefficients and $f(t)$ consists of the functions enumerated which possess a finite number of linearly independent derivatives.

When $f(t)$ consists of a number of terms, such as

$$Lx = f_1(t) + f_2(t)$$

the particular integral is $x_p = x_{p1} + x_{p2}$, where $Lx_{p1} = f_1(t)$ and $Lx_{p2} = f_2(t)$. Since the differential operator L is linear, we have

$$Lx_p = L(x_{p1} + x_{p2}) = f_1(t) + f_2(t)$$

The method of undetermined coefficients can be stated in Rules 1 and 2 to follow. It is assumed that x_c, the general solution of $Lx = 0$, has been determined.

Rule 1. The trial function for the evaluation of x_p of Eq. (B-39) is a linear combination with constant undetermined coefficients of $f(t)$ and all of its independent derivatives.

Consider the differential equation

$$\ddot{x} + \omega_n^2 x = F \sin \omega t \tag{B-40}$$

Since the successive derivatives of $\sin \omega t$ consist of $\sin \omega t$ and $\cos \omega t$ only, the trial function is of the form

$$x_p = A \sin \omega t + B \cos \omega t \tag{B-41}$$

The coefficients A and B are as yet to be determined. Substituting Eq. (B-41) in Eq. (B-40) and collecting the sine and cosine terms gives

$$(\omega_n^2 - \omega^2)A \sin \omega t + (\omega_n^2 - \omega^2)B \cos \omega t = F \sin \omega t$$

Thus the values of A and B are

$$A = F/(\omega_n^2 - \omega^2), \quad \text{and} \quad B = 0 \tag{B-42}$$

The method fails if $f(t)$ or one of its independent derivatives is proportional to one of the terms in the complementary function x_c. For example, the complementary function of the equation

$$\ddot{x} + \omega_n^2 x = F \sin \omega_n t \tag{B-43}$$

is $x_c = C_1 \sin \omega_n t + C_2 \cos \omega_n t$. In this case the trial function as expressed in Eq. (B-41) and the results obtained in Eq. (B-42) are not applicable. Let us form a new trial function

$$x_p = At \sin \omega_n t + Bt \cos \omega_n t \tag{B-44}$$

Substituting Eq. (B-44) and Eq. (B-43) and simplifying gives

$$2\omega_n A \cos \omega_n t - 2\omega_n B \sin \omega_n t - \omega_n^2 At \sin \omega_n t - \omega_n^2 Bt \cos \omega_n t \\ + \omega_n^2 At \sin \omega_n t + \omega_n^2 Bt \cos \omega_n t = F \sin \omega_n t$$

Further simplifying and equating the coefficients of the sine and cosine terms yields

$$A = 0, \quad \text{and} \quad B = -F/2\omega_n \tag{B-45}$$

Thus the trial function, as indicated in Eq. (B-44), gives the correct solution. This method is generalized and stated as Rule 2.

Rule 2. When the trial function x_p of Rule 1 has $f(t)$ or its independent derivatives proportional to a term in x_c, a new trial function is substituted. The new trial function is the product of the initial trial function and the lowest integral power of t, such that none of its terms is proportional to terms in x_c.

Example 1. Determine the general solution of the differential equation

$$\ddot{x} - \dot{x} = 3t^2 - 4t + 5 + 2e^t + \sin t \tag{B-46}$$

Solution: Rewriting the differential equation in operator form gives

$$Lx = D(D - 1)x = f_1(t) + f_2(t) + f_3(t) \tag{B-47}$$

where

$$f_1(t) = 3t^2 - 4t + 5, \text{ a polynomial}$$
$$f_2(t) = 2e^t, \text{ an exponential}$$
$$f_3(t) = \sin t, \text{ a sine function}$$

The corresponding homogeneous equation and the characteristic equation are

$$Lx = D(D - 1)x = 0 \tag{B-48}$$
$$s(s - 1) = 0 \tag{B-49}$$

The roots of Eq. (B-49) are $s = 0$ and $s = 1$, and the complementary function is

$$x_c = C_1 + C_2e^t \tag{B-50}$$

By Rule 1, the particular integral due to $f_1(t)$ is a linear combination of $f_1(t)$ and all of its independent derivatives. Thus the trial function is

$$x_{p1} = At^2 + Bt + C$$

Since C_1 is a term in x_c, Eq. (B-50), and it is proportional to C in the above equation, by Rule 2 a new trial function must be substituted. The new trial function is of the form

$$x_{p1} = (At^2 + Bt + C)t \tag{B-51}$$

Substituting Eq. (B-51) in $Lx = f_1(t)$ gives

$$6At + 2B - 3At^2 - 2Bt - C = 3t^2 - 4t + 5$$

Thus the coefficients A, B, and C are determined by the simultaneous algebraic equations

$$-3A = 3; \quad 6A - 2B = -4; \quad 2B - C = 5$$

giving

$$A = -1, \quad B = -1, \quad \text{and} \quad C = -7$$

Thus

$$x_{p1} = -t^3 - t^2 - 7t \tag{B-52}$$

Similarly, by Rule 1, the particular integral due to $f_2(t)$ is Ae^t. Again, this is proportional to C_2e^t in Eq. (B-50). By Rule 2, the new trial function is Ate^t. Substituting this expression in $Lx = f_2(t)$ gives

$$A(te^t + 2e^t - te^t - e^t) = 2e^t$$

Thus the corresponding particular integral is

$$x_{p2} = 2te^t \tag{B-53}$$

By Rule 1, the particular integral due to $f_3(t)$ is of the form

$$x_{p3} = A \sin t + B \cos t$$

Substituting this expression in $Lx = f_3(t)$ gives

$$-A \sin t - B \cos t - A \cos t + B \sin t = \sin t$$

Thus the coefficients are $A = -1/2$ and $B = 1/2$. The particular integral is

$$x_{p3} = -1/2 \sin t + 1/2 \cos t \tag{B-54}$$

From Eqs. (B-50), (B-52), (B-53), and (B-54), the general solution is

$$\begin{aligned} x &= x_c + x_{p1} + x_{p2} + x_{p3} \\ &= C_1 - 7t - t^2 - t^3 + (C_2 + 2t)e^t - (1/2)(\sin t - \cos t) \end{aligned}$$

B-6. SIMULTANEOUS LINEAR DIFFERENTIAL EQUATIONS

Equation (B-2) shows a pair of simultaneous linear differential equations in which two unknowns x_1 and x_2, or the dependent variables, are related to a single independent variable t. Using the linear differential operator L_{11} to denote $(m_{11}D^2 + k_{11})$, etc., these equations can be written in the general form as

$$\begin{aligned} L_{11}x_1 + L_{12}x_2 &= f_1(t) \\ L_{21}x_1 + L_{22}x_2 &= f_2(t) \end{aligned} \tag{B-55}$$

This is a set of simultaneous linear differential equations. It is noted that the D operators are commutative such that $(a_iD^i)(a_jD^j) = (a_jD^j)(a_iD^i) = a_ia_jD^{(i+j)}$, where a_i and a_j are constants and i and j are integers. Thus the L operators with constant coefficients are commutative such that $L_iL_j = L_jL_i$. Hence Eq. (B-55) can be manipulated and solved like an algebraic equation for the unknowns x_1 and x_2. The operators L are treated as coefficients of the unknowns, and $f_1(t)$ and $f_2(t)$ are the given known functions.

A convenient method to solve these equations is by Cramer's rule, which may be stated as follows.† Let $\Delta(L)$ be the determinant of the coefficients of the unknowns. The product $\Delta(L)$ and an unknown equals the determinant $\Delta(L)^*$, which is obtained from $\Delta(L)$ by substituting the known functions in place of the coefficients of the corresponding unknown. Recalling the elementary rules of determinants, from Eq. (B-55) we obtain

$$\begin{aligned} \Delta(L) &= \begin{vmatrix} L_{11} & L_{12} \\ L_{21} & L_{22} \end{vmatrix} = L_{11}L_{22} - L_{12}L_{21} \\ \Delta_1(L)^* &= \begin{vmatrix} f_1 & L_{12} \\ f_2 & L_{22} \end{vmatrix} = L_{22}f_1 - L_{12}f_2 \\ \Delta_2(L)^* &= \begin{vmatrix} L_{11} & f_1 \\ L_{21} & f_2 \end{vmatrix} = L_{11}f_2 - L_{21}f_1 \end{aligned} \tag{B-56}$$

Applying Cramer's rule to Eq. (B-55), we obtain

$$(L_{11}L_{22} - L_{12}L_{21})x_1 = L_{22}f_1 - L_{12}f_2 \tag{B-57}$$

$$(L_{11}L_{22} - L_{12}L_{21})x_2 = L_{11}f_2 - L_{21}f_1 \tag{B-58}$$

† See, for example, L. E. Dickson, *New First Course in the Theory of Equations*, John Wiley & Sons, Inc., New York, 1939, Chap. 9, for the proof of Cramer's rule.

which are linear differential equations each of which involves only one dependent variable. The unknowns x_1 and x_2 can thus be solved by the methods described in the previous sections.

Since Eqs. (B-57) and (B-58) are of the same form as Eq. (B-15), $Lx = f(t)$, the general solution of each of the unknowns must consist of a complementary function and a particular integral. Since Eqs. (B-57) and (B-58) are derived from Eq. (B-55), the solutions of Eq. (B-55) are also solutions of these equations. If $f_1(t)$ and $f_2(t)$ in these equations possess a finite number of linearly independent derivatives, the particular integrals can be obtained from Eqs. (B-57) and (B-58) by the method of undetermined coefficients outlined in Sec. B-5.

It is noted that the homogeneous equations, corresponding to Eqs. (B-57) and (B-58), have the same linear differential operator $\Delta(L)$. Thus these equations possess the same characteristic equation, and their complementary functions are linear combinations of the same functions as indicated in Sec. B-4. If the differential operator $\Delta(L)$ is of order 4, the complementary functions are of the form†

$$x_{c1} = C_{11}e^{s_1t} + C_{12}e^{s_2t} + C_{13}e^{s_3t} + C_{14}e^{s_4t} \qquad \textbf{(B-59)}$$

$$x_{c2} = C_{21}e^{s_1t} + C_{22}e^{s_2t} + C_{23}e^{s_3t} + C_{24}e^{s_4t} \qquad \textbf{(B-60)}$$

where s_i, $i = 1, 2, 3$, and 4, are the roots of the characteristic equation, and C's are arbitrary constants.

The general solutions are $x_1 = x_{c1} + x_{p1}$ and $x_2 = x_{c2} + x_{p2}$, where $x_{p1,2}$ are the particular integrals obtained from Eqs. (B-57) and (B-58). The solutions thus obtained, however, must also satisfy the original equations [Eq. (B-55)] each of which involves both x_1 and x_2. The arbitrary constants are thus related, and the constants in Eqs. (B-59) and (B-60) are not independent of one another.

The arbitrary constants are related through the homogeneous equations corresponding to Eq. (B-55) and Eqs. (B-57) and (B-58), which are

$$L_{11}x_1 + L_{12}x_2 = 0 \qquad \textbf{(B-61)}$$

$$L_{21}x_1 + L_{22}x_2 = 0 \qquad \textbf{(B-62)}$$

and

$$(L_{11}L_{22} - L_{12}L_{21})x_1 = 0 \qquad \textbf{(B-63)}$$

$$(L_{11}L_{22} - L_{12}L_{21})x_2 = 0 \qquad \textbf{(B-64)}$$

† For the type of equations considered [Eq. (B-2)], in which inertia forces are involved, the original equations are of order 2. The differential operator $\Delta(L)$, derived from a pair of such equations, is of order $2 \times 2 = 4$. This order is, in general, the sum of the orders of the equations considered and, in certain cases, may be less than this number.

One method to relate these constants is to solve Eq. (B-63) or (B-64) and then substitute the solution, in the form of Eq. (B-59) or (B-60), in either Eq. (B-61) or (B-62) to determine the second unknown. Alternatively, both Eqs. (B-59) and (B-60) may be substituted in either Eq. (B-61) or (B-62) to determine the relation of these constants.

We use a pair of simultaneous equations to illustrate the method. It is evident that the procedure can be generalized and applied to a set of n simultaneous equations.

Example 2. Find the general solution of the simultaneous equations

$$m_1\ddot{x}_1 + (k_1 + k)x_1 - kx_2 = F\sin\omega t$$
$$-kx_1 + m_2\ddot{x}_2 + (k_2 + k)x_2 = 0$$

Solution: Expressing the equations in operator form, we have

$$\begin{aligned}(m_1D^2 + k_1 + k)x_1 - kx_2 &= F\sin\omega t\\ -kx_1 + (m_2D^2 + k_2 + k)x_2 &= 0\end{aligned} \tag{B-65}$$

The direct application of Eq. (B-56) gives

$$\Delta(L) = m_1m_2D^4 + [(k_1 + k)m_2 + (k_2 + k)m_1]D^2 + k_1k_2 + k(k_1 + k_2)$$
$$\Delta_1(L)^* = (m_2D^2 + k_2 + k)F\sin\omega t - 0 = (k_2 + k - m_2\omega^2)F\sin\omega t$$
$$\Delta_2(L)^* = 0 - (-k)F\sin\omega t = kF\sin\omega t$$

Hence, corresponding to Eqs. (B-57) and (B-58), we obtain

$$\begin{aligned}\Delta(L)x_1 &= (k_2 + k - m_2\omega^2)F\sin\omega t\\ \Delta(L)x_2 &= kF\sin\omega t\end{aligned} \tag{B-66}$$

The particular integrals can be obtained by the method of undetermined coefficients or the impedance method. (See Sec. 2-6, Chap. 2, and Sec. 3-4, Chap. 3.) We shall use the former method for this problem.

From Eq. (B-66), the characteristic equation of the given system is

$$m_1m_2s^4 + [(k_1 + k)m_2 + (k_2 + k)m_1]s^2 + k_1k_2 + k(k_1 + k_2) = 0 \tag{B-67}$$

which is a quadratic algebraic equation in s^2. It can be shown that its roots s^2 are real and negative. (See Sec. 3-2, in Chap. 3.) Defining the roots $s_1^2 = -\omega_1^2$ and $s_2^2 = -\omega_2^2$, the roots of the quartic equation are $\pm j\omega_1$ and $\pm j\omega_2$.

Assuming that ω in $F\sin\omega t$ is not a root of the characteristic equation, from Eq. (B-66) the particular integral x_{p1} is of the form

$$x_{p1} = A_1\sin\omega t + B_1\cos\omega t$$

Substituting this expression in Eq. (B-66) and simplifying gives

$$\Delta(\omega)(A_1 \sin \omega t + B_1 \cos \omega t) = (k_2 + k - m_2\omega^2)F \sin \omega t$$

where

$$\Delta(\omega) = m_1 m_2 \omega^4 - [(k_1 + k)m_2 + (k_2 + k)m_1]\omega^2 + k_1 k_2 + k(k_1 + k_2)$$

Hence

$$A_1 = F(k_2 + k - m_2\omega^2)/\Delta(\omega), \quad B_1 = 0 \tag{B-68}$$

Similarly, the particular integral x_{p2} is of the form

$$x_{p2} = A_2 \sin \omega t + B_2 \cos \omega t$$

It can be shown that

$$A_2 = Fk/\Delta(\omega), \quad B_2 = 0 \tag{B-69}$$

It can be verified, by direct substitution, that these values of x_{p1} and x_{p2} satisfy the original equations, Eq. (B-65).

Analogous to Eqs. (B-61) to (B-64), the homogeneous equations corresponding to Eqs. (B-65) and (B-66) are

$$(m_1 D^2 + k_1 + k)x_1 - kx_2 = 0 \tag{B-70}$$

$$-kx_1 + (m_2 D^2 + k_2 + k)x_2 = 0 \tag{B-71}$$

and

$$\Delta(L)x_1 = 0 \tag{B-72}$$

$$\Delta(L)x_2 = 0$$

Since the roots of the characteristic equation are $s = \pm j\omega_1$ and $\pm j\omega_2$, the complementary function x_{c1} is

$$x_{c1} = C_{11}e^{j\omega_1 t} + C_{12}e^{-j\omega_1 t} + C_{13}e^{j\omega_2 t} + C_{14}e^{-j\omega_2 t} \tag{B-73}$$

Substituting this equation in Eq. (B-70) yields

$$\begin{aligned} x_{c2} = {} & (C_{11}/k)(k_1 + k - m_1\omega_1^2)e^{j\omega_1 t} + (C_{12}/k)(k_1 + k - m_1\omega_1^2)e^{-j\omega_1 t} \\ & + (C_{13}/k)(k_1 + k - m_1\omega_2^2)e^{j\omega_2 t} + (C_{14}/k)(k_1 + k - m_1\omega_2^2)e^{-j\omega_2 t} \end{aligned} \tag{B-74}$$

We have obtained x_{c2} from Eq. (B-70). It can be shown, by direct substitution, that x_{c1} and x_{c2} from Eqs. (B-73) and (B-74) also satisfy Eq. (B-71).

Since $x(t)$ is a real function, it is more convenient to express

Eqs. (B-73) and (B-74) in terms of circular functions. Following the procedure outlined in Case 3, Sec. B-4, we obtain

$$\begin{aligned} x_{c1} &= A_{11} \cos \omega_1 t + B_{11} \sin \omega_1 t \\ &\qquad + A_{12} \cos \omega_2 t + B_{12} \sin \omega_2 t \\ x_{c2} &= \mu_1 A_{11} \cos \omega_1 t + \mu_1 B_{11} \sin \omega_1 t \\ &\qquad + \mu_2 A_{12} \cos \omega_2 t + \mu_2 B_{12} \sin \omega_2 t \end{aligned} \tag{B-75}$$

where

$$\begin{aligned} A_{11} &= C_{11} + C_{12} & A_{12} &= C_{13} + C_{14} \\ B_{11} &= j(C_{11} - C_{12}) & B_{12} &= j(C_{13} - C_{14}) \\ \mu_1 &= (k_1 + k - m_1\omega_1^2)/k & \mu_2 &= (k_1 + k - m_1\omega_2^2)/k \end{aligned}$$

Alternatively, if x_{c2} is first obtained from Eq. (B-72) and it is substituted in Eq. (B-71) to obtain x_{c1}, the resultant equations are identical to Eq. (B-75) except that

$$\mu_1 = k/(k_2 + k - m_2\omega_1^2) \quad \mu_2 = k/(k_2 + k - m_2\omega_2^2)$$

The values of $\mu_{1,2}$ obtained by the two methods must be identical. This can be verified by substituting $\pm j\omega_{1,2}$ in the characteristic equation, Eq. (B-67), and expressing the equation in the form

$$\frac{k_1 + k - m_1\omega_{1,2}^2}{k} = \frac{k}{k_2 + k - m_2\omega_{1,2}^2} \tag{B-76}$$

Thus either procedure will give the same solutions.

SUGGESTED READING

Agnew, R. P., *Differential Equations* (New York: McGraw-Hill Book Co., Inc., 1942), chaps. 5, 8, 9, and 16.

Hildebrand, F. B., *Advanced Calculus for Engineers* (New Jersey: Prentice-Hall, Inc., 1949), chap. 1.

Ince, E. L., *Ordinary Differential Equations* (New York: Dover Publications, Inc., 1956), chaps. 5 and 6.

Wylie, C. R., Jr., *Advanced Engineering Mathematics* (New York: McGraw-Hill Book Co., Inc., 1960), chaps. 3 and 4.

APPENDIX C

DETERMINANTS AND MATRICES

The object of this appendix is to introduce and to review some mathematical techniques that are useful for the discussion of Chap. 7. The treatment is not intended to be comprehensive or rigorous, and the theorems are stated without proofs in order to limit the appendix to a reasonable length. The statements made, however, are illustrated with examples. The appendix is divided into two parts: the first part is devoted to the discussion of determinants and the second part to matrices.

DETERMINANTS

Introduction

The concept of determinants is familiar to us from elementary algebra. For example, in solving a system of two linear algebraic equations

$$\begin{aligned} a_{11}x_1 + a_{12}x_2 &= b_1 \\ a_{21}x_1 + a_{22}x_2 &= b_2 \end{aligned} \tag{C-1}$$

the unknowns x_1 and x_2 can be determined by the relations

$$\begin{aligned} (a_{11}a_{22} - a_{12}a_{21})x_1 &= a_{22}b_1 - a_{12}b_2 \\ (a_{11}a_{22} - a_{12}a_{21})x_2 &= a_{11}b_2 - a_{21}b_1 \end{aligned} \tag{C-2}$$

if $(a_{11}a_{22} - a_{12}a_{21}) \neq 0$. The expression $(a_{11}a_{22} - a_{12}a_{21})$ is called the *determinant* of the coefficients of Eq. (C-1). It is represented by the symbol

$$\begin{vmatrix} a_{11} & a_{12} \\ a_{21} & a_{22} \end{vmatrix} = |a_{ij}| \tag{C-3}$$

where the quantities a_{ij} are called the *elements* of the determinant. The horizontal lines of elements are called the *rows*, and the vertical ones are the *columns*. The first subscript of an element identifies the row, and the second subscript identifies the column.

Definitions

In the introductory example the 2×2 array of quantities in Eq. (C-3) consists of 2^2 elements, and it is called a second-order determinant. Its expansion consists of 2! terms; namely, $a_{11}a_{22}$ and $-a_{12}a_{21}$. Similarly, a third-order determinant consists of 3^2 elements, and its expansion has 3! terms.

$$\begin{vmatrix} a_{11} & a_{12} & a_{13} \\ a_{21} & a_{22} & a_{23} \\ a_{31} & a_{32} & a_{33} \end{vmatrix} = +a_{11}a_{22}a_{33} + a_{12}a_{23}a_{31} + a_{13}a_{21}a_{32} - a_{11}a_{23}a_{32} - a_{12}a_{21}a_{33} - a_{13}a_{22}a_{31}$$

$$= \sum \pm a_{1i_1}a_{2i_2}a_{3i_3} \tag{C-4}$$

In each of the six terms in Eq. (C-4), the first indices of the factors are 1, 2, 3 and the second indices are a permutation of the integers 1, 2, and 3. All the 3! possible permutations are present. The $\pm$ sign of the terms in Eq. (C-4) depends on the number of *inversions* in the permutation of 1, 2, 3 in the second indices. The sign is + if the number of inversions is even and − if the number of inversions is odd. In the given example, we have

Permutation	Number of Inversions
123	0
132	1
213	1
312	2
231	2
321	3

For example, to obtain the permutation 132 from 123, it is necessary to interchange 2 and 3, and the process involves one inversion.

An nth *order* determinant $|A|$ has n^2 elements, and its expansion consists of $n!$ terms.

$$|A| = |a_{ij}| = \begin{vmatrix} a_{11} & a_{12} & \cdots & a_{1n} \\ a_{21} & a_{22} & \cdots & a_{2n} \\ \cdot & \cdot & \cdot & \cdot \\ a_{n1} & a_{n2} & \cdots & a_{nn} \end{vmatrix} = \sum \pm a_{1i_1} a_{2i_2} \cdots a_{ni_n} \tag{C-5}$$

In each of the terms of this expansion, the first indices of the factors are $(1, 2, \ldots, n)$ and the second indices are a permutation of the integers $(1, 2, \ldots, n)$. The summation is to extend over all the $n!$ possible permutations $(i_1, i_2, \ldots, i_n)$ of $(1, 2, \ldots, n)$. The sign is $+$ if the number of inversions of the second indices is even, and $-$ if odd.

The determinant $|A|$ can also be expanded in the form

$$|A| = \sum \pm a_{j_1 1} a_{j_2 2} \cdots a_{j_n n} \tag{C-6}$$

In each of the terms of this expansion, the second indices of the factors are $(1, 2, \ldots, n)$ and the first indices are a permutation of the integers $(1, 2, \ldots, n)$. The summation is to extend over all the $n!$ possible permutations $(j_1, j_2, \ldots, j_n)$ of $(1, 2, \ldots, n)$. The sign is $+$ if the number of inversions of the first indices is even, and $-$ if odd. This form of $|A|$ is called the *expansion by columns*, and that indicated in Eq. (C-5) is called the *expansion by rows*.

Let A be a $n \times n$ square array of quantities as indicated in Eq. (C-7).

$$A = [a_{ij}] = \begin{bmatrix} a_{11} & a_{12} & \cdots & a_{1n} \\ a_{21} & a_{22} & \cdots & a_{2n} \\ \cdot & \cdot & \cdot & \cdot \\ a_{n1} & a_{n2} & \cdots & a_{nn} \end{bmatrix}, \quad A^T = \begin{bmatrix} a_{11} & a_{21} & \cdots & a_{n1} \\ a_{12} & a_{22} & \cdots & a_{n2} \\ \cdot & \cdot & \cdot & \cdot \\ a_{1n} & a_{2n} & \cdots & a_{nn} \end{bmatrix} \tag{C-7}$$

The *transpose* of A, denoted by the symbol A^T, is an array whose columns are the rows of A and whose rows are the columns of A in the proper sequence.

If the elements of the ith row and jth column are deleted from an nth order determinant, Eq. (C-5), the resultant determinant is of $(n - 1)$th order. This $(n - 1)$th order determinant is called the *minor* of the element a_{ij}. The *cofactor*, denoted by the symbol A_{ij}, of the element a_{ij} is the minor of a_{ij} multiplied by $(-1)^{i+j}$.

Properties of Determinants

The following properties of determinants can be verified using the definitions in Eqs. (C-5) and (C-6).

1. The value of a determinant is unchanged if the elements of corresponding rows and columns are interchanged; that is,

$$|A| = |A^T| \tag{C-8}$$

This property was implied in Eqs. (C-5) and (C-6) which stated that a determinant can be expanded by either rows or columns. Thus any property of a determinant stated for rows is also valid when stated for columns of the determinant.

2. If A contains a row of zeros, then $|A| = 0$.

3. If each element of any row of a determinant $|A|$ is multiplied by k, the resultant determinant is $k|A|$.

4. An interchange of any two rows of $|A|$ induces a change in sign in $|A|$.

5. If two rows of $|A|$ are identical, then $|A| = 0$.

6. If each element of a row of $|A|$ is multiplied by k and added to the corresponding elements of another row, the value of the determinant is unchanged.

7. *Simple Laplace Development.* A determinant may be evaluated by a Laplace development on any row or column, that is

$$|A| = \sum_{i=1}^{n} a_{ij}A_{ij} = \sum_{j=1}^{n} a_{ij}A_{ij} \tag{C-9}$$

The elements a_{ij} in Eq. (C-9) must be taken from a single row or a single column of $|A|$.

8. *Differentiation.* The derivative $d\Delta(x)/dx$ of the determinant $\Delta(x)$ is the sum of all determinants obtained when any one column is differentiated and the remaining columns are taken as they stand. This statement is illustrated with a third-order determinant as follows:

$$\Delta(x) = \begin{vmatrix} a_{11}(x) & a_{12}(x) & a_{13}(x) \\ a_{21}(x) & a_{22}(x) & a_{23}(x) \\ a_{31}(x) & a_{32}(x) & a_{33}(x) \end{vmatrix}$$

$$\frac{d\Delta(x)}{dx} = \begin{vmatrix} a'_{11}(x) & a_{12}(x) & a_{13}(x) \\ a'_{21}(x) & a_{22}(x) & a_{23}(x) \\ a'_{31}(x) & a_{32}(x) & a_{33}(x) \end{vmatrix} + \begin{vmatrix} a_{11}(x) & a'_{12}(x) & a_{13}(x) \\ a_{21}(x) & a'_{22}(x) & a_{23}(x) \\ a_{31}(x) & a'_{32}(x) & a_{33}(x) \end{vmatrix}$$
$$+ \begin{vmatrix} a_{11}(x) & a_{12}(x) & a'_{13}(x) \\ a_{21}(x) & a_{22}(x) & a'_{23}(x) \\ a_{31}(x) & a_{32}(x) & a'_{33}(x) \end{vmatrix} \quad \textbf{(C-10)}$$

The prime (′) denotes the differentiation with respect to x. Since each determinant of $d\Delta(x)/dx$ can be expanded in terms of the minors, it is evident that $d\Delta(x)/dx$ is a linear homogeneous function of these minors.

Example 1. Evaluate the given determinant by (*a*) Eq. (C-9), and (*b*) property 6.

$$|A| = \begin{vmatrix} 2 & 4 & 5 \\ 1 & 3 & 2 \\ 3 & 6 & 0 \end{vmatrix}$$

Solution: (*a*) Using the first row, we have

$$|A| = 2\begin{vmatrix} 3 & 2 \\ 6 & 0 \end{vmatrix} - 4\begin{vmatrix} 1 & 2 \\ 3 & 0 \end{vmatrix} + 5\begin{vmatrix} 1 & 3 \\ 3 & 6 \end{vmatrix}$$
$$= 2(0 - 12) - 4(0 - 6) + 5(6 - 9) = -15$$

Using the first column, we have

$$|A| = 2\begin{vmatrix} 3 & 2 \\ 6 & 0 \end{vmatrix} - 1\begin{vmatrix} 4 & 5 \\ 6 & 0 \end{vmatrix} + 3\begin{vmatrix} 4 & 5 \\ 3 & 2 \end{vmatrix}$$
$$= 2(0 - 12) - 1(0 - 30) + 3(8 - 15) = -15$$

(*b*) Using property 6, we multiply the second row by -2 and add to the first row, and multiply the second row by -3 and add to the third row. The resultant determinant is expanded by the first column. The steps are indicated as follows:

$$|A| = \begin{vmatrix} 2 & 4 & 5 \\ 1 & 3 & 2 \\ 3 & 6 & 0 \end{vmatrix} = \begin{vmatrix} 0 & -2 & 1 \\ 1 & 3 & 2 \\ 0 & -3 & -6 \end{vmatrix}$$
$$= 0\begin{vmatrix} 3 & 2 \\ -3 & -6 \end{vmatrix} - 1\begin{vmatrix} -2 & 1 \\ -3 & -6 \end{vmatrix} + 0\begin{vmatrix} -2 & 1 \\ 3 & 2 \end{vmatrix} = -15$$

MATRICES

Introduction

Consider a set of simultaneous linear algebraic equations

$$\begin{aligned} a_{11}x_1 + a_{12}x_2 + a_{13}x_3 &= b_1 \\ a_{21}x_1 + a_{22}x_2 + a_{23}x_3 &= b_2 \\ a_{31}x_1 + a_{32}x_2 + a_{33}x_3 &= b_3 \end{aligned} \tag{C-11}$$

This equation can be written in matrix notation as

$$AX = B \tag{C-12}$$

where A denotes a collection of the coefficients of the unknowns, X a column of the unknowns, and B a column of the known quantities.

$$A = \begin{bmatrix} a_{11} & a_{12} & a_{13} \\ a_{21} & a_{22} & a_{23} \\ a_{31} & a_{32} & a_{33} \end{bmatrix} = [a_{ij}], \quad X = \begin{bmatrix} x_1 \\ x_2 \\ x_3 \end{bmatrix}, \quad B = \begin{bmatrix} b_1 \\ b_2 \\ b_3 \end{bmatrix}$$

A is called the *coefficient matrix* of Eq. (C-11), and X is the *solution vector*.

Assuming that the inverse (reciprocal) of A, denoted by A^{-1}, can be determined, the solution of Eq. (C-12) is

$$X = A^{-1}B \tag{C-13}$$

Although the discussion pertains to algebraic equations, by appropriate transformations, the solution of a set of linear differential equations with constant coefficients can be reduced to the solution of a set of algebraic equations.

Definitions

An *$m \times n$ matrix* is a rectangular array of mn quantities arranged in m rows and n columns. A matrix is usually denoted by the symbol

$$\begin{bmatrix} a_{11} & a_{12} & \cdots & a_{1n} \\ a_{21} & a_{22} & \cdots & a_{2n} \\ \cdot & \cdot & \cdots & \cdot \\ a_{m1} & a_{m2} & \cdots & a_{mn} \end{bmatrix} = [a_{ij}] = [a] = A \tag{C-14}$$

If $m = n$, the array is called a *square matrix* of *order n*. In a square matrix the sloping line of elements extending from a_{11} to a_{nn} is called the *principal diagonal* of the matrix. A square matrix in which all elements not on the principal diagonal are zero is called a *diagonal matrix*. A diagonal matrix in which all elements on the principal diagonal are unity is called a *unit matrix I*. A square matrix in which $a_{ij} = a_{ji}$ is called a *symmetrical matrix*.

A *null matrix* 0 has all its elements equal zero. A 1×1 matrix is called a *scalar*. An $m \times 1$ matrix is called a *column matrix* or *column vector*. A $1 \times n$ matrix is called a *row matrix* or a *row vector*. We shall use the symbol { } to denote a column matrix and [ ] to denote a row matrix.

If the determinant of a square matrix is zero, the matrix is termed *singular*; if the determinant is not zero, the matrix is *nonsingular*. All rectangular matrices are termed singular. The *transpose* A^T of an $m \times n$ matrix A is the $n \times m$ matrix obtained by interchanging the rows and columns of A. The *cofactor* A_{ij} of the element a_{ij} in a square matrix A is the cofactor of a_{ij} in $|A|$. The *adjoint* of a square matrix A is formed by replacing every element of A by its cofactor and transposing. It is denoted by the symbol adj. A.

Matrix Operations

Two matrices $A = [a_{ij}]$ and $B = [b_{ij}]$ are *equal* if and only if $a_{ij} = b_{ij}$ for every i and j.

The *sum*, or difference, of two matrices A and B having the same number of rows and columns is defined by the relation

$$A + B = [a_{ij}] + [b_{ij}] = [a_{ij} + b_{ij}] \qquad \textbf{(C-15)}$$

The operation is commutative, that is, $A + B = B + A$, and associative, that is, $(A + B) + C = A + (B + C)$.

The product of a scalar k and the matrix A is defined by the relation

$$B = kA, \quad \text{or} \quad [b_{ij}] = [ka_{ij}] \qquad \textbf{(C-16)}$$

Each element in B equals k times the corresponding element in A.

The *product* of two matrices $A = [a_{ij}]$ and $B = [b_{ij}]$ is defined by the relations

$$\underset{(m \times p)}{A} \; \underset{(p \times n)}{B} = \underset{(m \times n)}{C} \qquad \textbf{(C-17)}$$

where

$$C = [c_{ij}] = [\sum_k a_{ik} b_{kj}] \qquad \textbf{(C-18)}$$

The element c_{ij} in the ith row and the jth column of C is the sum of the products of the elements in the ith row of A and the corresponding elements in the jth column of B; that is, the product is obtained by row-on-column multiplication. The multiplication is defined only if the number of columns in A is equal to the number of rows in B. Such matrices are called *conformable.* Matrix multiplication is not commutative; that is, in general, $AB \neq BA$. If the matrices are conformable, the multiplication is associative, that is, $(AB)C = A(BC)$, and distributive, that is, $(A + B)C = AC + BC$. If $AB = 0$, a null matrix, it does not imply $A = 0$, or $B = 0$, or $A = B = 0$. For example,

$$A = \begin{bmatrix} 0 & 1 \\ 0 & 1 \end{bmatrix}, \quad B = \begin{bmatrix} 1 & 1 \\ 0 & 0 \end{bmatrix}, \quad AB = \begin{bmatrix} 0 & 0 \\ 0 & 0 \end{bmatrix}, \quad BA = \begin{bmatrix} 0 & 2 \\ 0 & 0 \end{bmatrix}$$

The *transpose* of a product matrix is the product of the transposes of the separate matrices in reversed order. Thus

$$(AB)^T = B^T A^T$$
$$(ABC)^T = C^T B^T A^T$$

If a square matrix A is nonsingular, it possesses an *inverse* (reciprocal) A^{-1} which has the property

$$AA^{-1} = A^{-1}A = I \tag{C-20}$$

where I is a unit matrix. The inverse is defined by the relation

$$A^{-1} = \frac{\text{adj. } A}{|A|} \tag{C-21}$$

The inverse of a product matrix is the product of the inverses of the separate matrices in reversed order. Thus

$$(ABC)^{-1} = C^{-1}B^{-1}A^{-1} \tag{C-22}$$

The *derivative* of a matrix $[a_{ij}(x)]$, when it exists, is defined by the relation

$$\frac{d}{dx}[a_{ij}(x)] = \left[\frac{d}{dx}a_{ij}(x)\right] \tag{C-23}$$

Hence the elements of the derivative matrix are the derivatives of the corresponding elements of the original matrix. The *integral* of a matrix $[a_{ij}(x)]$ is defined as

$$\int [a_{ij}(x)]\, dx = \left[\int a_{ij}(x)\, dx\right] \tag{C-24}$$

Thus the integral exists if the integral of each of the elements of the original matrix exists.

Example 2. Determine the product AB for the given matrices.

$$A = \begin{bmatrix} 1 & 2 & 5 \\ 3 & 1 & 4 \\ 1 & 1 & 2 \end{bmatrix} \qquad B = \begin{bmatrix} 1 & 2 \\ 4 & 1 \\ 6 & 3 \end{bmatrix}$$

Solution: By row-on-column multiplication, we have

$$AB = \begin{bmatrix} 1 & 2 & 5 \\ 3 & 1 & 4 \\ 1 & 1 & 2 \end{bmatrix} \begin{bmatrix} 1 & 2 \\ 4 & 1 \\ 6 & 3 \end{bmatrix} = \begin{bmatrix} 1+8+30 & 2+2+15 \\ 3+4+24 & 6+1+12 \\ 1+4+12 & 2+1+6 \end{bmatrix} = \begin{bmatrix} 39 & 19 \\ 31 & 19 \\ 17 & 9 \end{bmatrix}$$

Example 3. Find the inverse of the given matrix, using Eq. (C-21)

$$A = \begin{bmatrix} 1 & 2 & 5 \\ 3 & 1 & 4 \\ 1 & 1 & 2 \end{bmatrix}$$

Solution: The value of $|A|$ is 4. The array of the cofactors A_{ij} and the adjoint of A are

$$[A_{ij}] = \begin{bmatrix} -2 & -2 & 2 \\ 1 & -3 & 1 \\ 3 & 11 & -5 \end{bmatrix} \qquad \text{adj. } A = \begin{bmatrix} -2 & 1 & 3 \\ -2 & -3 & 11 \\ 2 & 1 & -5 \end{bmatrix}$$

Hence

$$A^{-1} = \frac{\text{adj. } A}{|A|} = \frac{1}{4} \begin{bmatrix} -2 & 1 & 3 \\ -2 & -3 & 11 \\ 2 & 1 & -5 \end{bmatrix} = \begin{bmatrix} -1/2 & 1/4 & 3/4 \\ -1/2 & -3/4 & 11/4 \\ 1/2 & 1/4 & -5/4 \end{bmatrix}$$

Finding the Inverse of a Matrix

We shall describe two additional methods for finding the inverse of a matrix.

USE OF ELEMENTARY OPERATIONS: The inverse of a matrix can be found by using either the elementary row or the elementary column operations. The elementary operations are defined as (1) interchanging two rows (columns) of a matrix, (2) multiplying a row (column) by a nonzero constant, and (3) adding to a row (column) the multiple of another row (column) of the matrix.

Let us illustrate the elementary row operations for finding the inverse of a matrix. First, we write the given array A and a unit matrix I of the same order, and then we perform the same operations on both A and I. A^{-1} is obtained from the transformation of I when A is transformed into a unit matrix. Starting with the arrays:

$$\begin{array}{rrr|rrr} 1 & 2 & 5 & 1 & 0 & 0 \\ 3 & 1 & 4 & 0 & 1 & 0 \\ 1 & 1 & 2 & 0 & 0 & 1 \end{array}$$

Multiply row 1 by -3 and add to row 2; multiply row 1 by -1 and add to row 3:

$$\begin{array}{rrr|rrr} 1 & 2 & 5 & 1 & 0 & 0 \\ 0 & -5 & -11 & -3 & 1 & 0 \\ 0 & -1 & -3 & -1 & 0 & 1 \end{array}$$

Multiply row 3 by 2 and add to row 1; multiply row 3 by -6 and add to row 2:

$$\begin{array}{rrr|rrr} 1 & 0 & -1 & -1 & 0 & 2 \\ 0 & 1 & 7 & 3 & 1 & -6 \\ 0 & -1 & -3 & -1 & 0 & 1 \end{array}$$

Add row 2 to row 3:

$$\begin{array}{rrr|rrr} 1 & 0 & -1 & -1 & 0 & 2 \\ 0 & 1 & 7 & 3 & 1 & -6 \\ 0 & 0 & 4 & 2 & 1 & -5 \end{array}$$

Divide row 3 by 4:

$$\begin{array}{rrr|rrr} 1 & 0 & -1 & -1 & 0 & 2 \\ 0 & 1 & 7 & 3 & 1 & -6 \\ 0 & 0 & 1 & 2/4 & 1/4 & -5/4 \end{array}$$

Add row 3 to row 1; multiply row 3 by -7 and add to row 2:

$$\begin{array}{ccc|ccc} 1 & 0 & 0 & -2/4 & 1/4 & 3/4 \\ 0 & 1 & 0 & -2/4 & -3/4 & 11/4 \\ 0 & 0 & 1 & 2/4 & 1/4 & -5/4 \end{array}$$

The inverse obtained is identical to that shown in Example 3. The use of elementary column operations for finding the inverse of the given matrix is left as an exercise.

USE OF SUBMATRICES: Let the given matrix be partitioned into submatrices as

$$A = \left[\begin{array}{cc|c} 1 & 2 & 5 \\ 3 & 1 & 4 \\ \hline 1 & 1 & 2 \end{array}\right] = \left[\begin{array}{c|c} P & Q \\ \hline R & S \end{array}\right]$$

where P is a 2×2 matrix, Q is 2×1, R is 1×2, and S is 1×1. Assume that A^{-1} is partitioned in like manner so that

$$AA^{-1} = \left[\begin{array}{c|c} P & Q \\ \hline R & S \end{array}\right]\left[\begin{array}{c|c} P_1 & Q_1 \\ \hline R_1 & S_1 \end{array}\right] = \left[\begin{array}{cc|c} 1 & 0 & 0 \\ 0 & 1 & 0 \\ \hline 0 & 0 & 1 \end{array}\right] = I \qquad \textbf{(C-25)}$$

It is evident that the submatrices that enter into the multiplications are conformable. From Eq. (C-25), we obtain

$$PP_1 + QR_1 = I\,(2 \times 2) \qquad \textbf{(C-26)}$$

$$PQ_1 + QS_1 = 0\,(2 \times 1) \qquad \textbf{(C-27)}$$

$$RP_1 + SR_1 = 0\,(1 \times 2) \qquad \textbf{(C-28)}$$

$$RQ_1 + SS_1 = I\,(1 \times 1) \qquad \textbf{(C-29)}$$

Premultiplying Eq. (C-28) by R^{-1} gives

$$P_1 + R^{-1}SR_1 = 0$$

or

$$R^{-1}SR_1 = -P_1$$

Rearranging the last equation yields

$$R_1 = -S^{-1}RP_1 \qquad \textbf{(C-30)}$$

Substituting Eq. (C-30) in Eq. (C-26) and rearranging gives

$$P_1 = [P - QS^{-1}R]^{-1} \qquad \textbf{(C-31)}$$

Premultiplying Eq. (C-27) by P^{-1} and rearranging gives

$$Q_1 = -P^{-1}QS_1 \tag{C-32}$$

Substituting Eq. (C-32) into Eq. (C-29) and rearranging yields

$$S_1 = [S - RP^{-1}Q]^{-1} \tag{C-33}$$

Thus the submatrices P_1, Q_1, R_1, and S_1 can be found by solving Eqs. (C-31) and (C-33) and then Eqs. (C-30) and (C-32).

Applying Eq. (C-31) to the given example, we obtain

$$P_1 = \left[\begin{bmatrix} 1 & 2 \\ 3 & 1 \end{bmatrix} - \begin{bmatrix} 5 \\ 4 \end{bmatrix}[1/2][1 \quad 1]\right]^{-1} = \begin{bmatrix} -3/2 & -1/2 \\ 1 & -1 \end{bmatrix}^{-1} = \begin{bmatrix} -1/2 & 1/4 \\ -1/2 & -3/4 \end{bmatrix}$$

Applying Eq. (C-33) gives

$$S_1 = \left[[2] - [1 \quad 1]\begin{bmatrix} 1 & 2 \\ 3 & 1 \end{bmatrix}^{-1}\begin{bmatrix} 5 \\ 4 \end{bmatrix}\right]^{-1} = [-4/5]^{-1} = [-5/4]$$

Using Eqs. (C-30) and (C-32), R_1 and Q_1 are determined as

$$R_1 = -[1/2][1 \quad 1]\begin{bmatrix} -1/2 & 1/4 \\ -1/2 & -3/4 \end{bmatrix} = [1/2 \quad 1/4]$$

$$Q_1 = -\begin{bmatrix} -1/5 & 2/5 \\ 3/5 & -1/5 \end{bmatrix}\begin{bmatrix} 5 \\ 4 \end{bmatrix}[-5/4] = \begin{bmatrix} 3/4 \\ 11/4 \end{bmatrix}$$

Thus

$$A^{-1} = \left[\begin{array}{c|c} P_1 & Q_1 \\ \hline R_1 & S_1 \end{array}\right] = \left[\begin{array}{cc|c} -1/2 & 1/4 & 3/4 \\ -1/2 & -3/4 & 11/4 \\ \hline 1/2 & 1/4 & -5/4 \end{array}\right]$$

Linear Algebraic Equations

It was shown in Eqs. (C-11) and (C-12) that the solutions of a set of linear algebraic equations can be expressed in matrix notation as

$$AX = B \quad \text{and} \quad X = A^{-1}B$$

where A is the coefficient matrix, X is a column of the unknowns or the solution vector, and B is a column of given quantities. If matrix

A is singular, the concept of the *rank* of a matrix is necessary in order to answer whether the solutions exist.

If matrix A is $m \times n$, square submatrices of orders $1, 2, \ldots, k$ can be formed from the elements of A, where k is equal to the smaller of the numbers m and n. The determinants formed from the elements common to an arbitrary r row and an arbitrary r column of A is called an r-rowed minor of A of order r. If at least one of the r-rowed minors of A is not zero, whereas all the minors of order $> r$ are zero, then matrix A is of *rank* r. If A is a square matrix of order n and rank $(n - 1)$, A is said to be *simply degenerate*, or of degeneracy 1. If A is of rank r and $(n - r)$ is two or greater, A is *multiply degenerate*.

It can be shown that the rank of a matrix is unchanged by the elementary operations discussed in finding the inverse of a matrix. We shall illustrate the use of elementary operations for finding the rank of a matrix.

Example 4. Determine the rank of the matrix

$$A = \begin{bmatrix} 4 & -1 & 1 \\ 2 & -3 & 5 \\ 1 & 1 & -2 \\ 5 & 0 & -1 \end{bmatrix}$$

Solution: A is the given matrix. Matrix A_1 is obtained from A by (1) adding row 3 to row 1, and (2) multiplying row 3 by 3 and adding to row 2. Matrix A_2 is obtained from A_1 by (1) multiplying column 2 by -1 and adding to column 1, and (2) multiplying column 2 by 2 and adding to column 3. Matrix A_3 is obtained from A_2 by multiplying column 3 by 5 and adding to column 1. Matrix A_4 is obtained from A_3 by (1) multiplying row 4 by -1 and adding to row 1, and (2) multiplying row 4 by -1 and adding to row 2. It is evident from A_4 that at least one of the two-rowed minors does not vanish, whereas all the three-rowed minors are identically zero. Hence the given matrix A is of rank 2.

$$A_1 = \begin{bmatrix} 5 & 0 & -1 \\ 5 & 0 & -1 \\ 1 & 1 & -2 \\ 5 & 0 & -1 \end{bmatrix} \qquad A_2 = \begin{bmatrix} 5 & 0 & -1 \\ 5 & 0 & -1 \\ 0 & 1 & 0 \\ 5 & 0 & -1 \end{bmatrix}$$

$$A_3 = \begin{bmatrix} 0 & 0 & -1 \\ 0 & 0 & -1 \\ 0 & 1 & 0 \\ 0 & 0 & -1 \end{bmatrix} \quad A_4 = \begin{bmatrix} 0 & 0 & 0 \\ 0 & 0 & 0 \\ 0 & 1 & 0 \\ 0 & 0 & -1 \end{bmatrix}$$

We shall state four theorems from theory of equations without proof and then illustrate their applications with examples.

Theorem 1. If A is $m \times n$, then a necessary and sufficient condition that the equations $AX = B$ have a solution for X is that the rank of the coefficient matrix A be equal to the rank of the *augmented matrix* $[A,B]$.

Theorem 2. If A is of order n, then a necessary and sufficient condition that the equations $AX = 0$ have a solution other than $X = 0$ is that A be singular.

Theorem 3. If A is $m \times n$ and of rank r, then the equations $AX = 0$ have $(n - r)$ linearly independent solutions, and every solution is a linear combination of them. This set of independent solutions is called a *fundamental* set of the solutions.

Theorem 4. If X_p is some known solution of the consistent system $AX = B$, then every solution of the system is obtained by adding to X_p the general solution of $AX = 0$, that is, a linear combination of the solutions of a fundamental set of solutions of $AX = 0$.

Example 5. Solve the set of equations

$$\begin{aligned} 5x \qquad - z &= 1 \\ 4x - y + z &= 5 \\ 2x - 3y + 5z &= 11 \\ x + y - 2z &= -3 \end{aligned}$$

Solution: The coefficient matrix and the augmented matrix of the equations are

$$A = \begin{bmatrix} 5 & 0 & -1 \\ 4 & -1 & 1 \\ 2 & -3 & 5 \\ 1 & 1 & -2 \end{bmatrix} \qquad [A,B] = \left[\begin{array}{ccc|c} 5 & 0 & -1 & 1 \\ 4 & -1 & 1 & 5 \\ 2 & -3 & 5 & 11 \\ 1 & 1 & -2 & -3 \end{array}\right]$$

The rank of the matrices can be determined by using the elementary row and column operations. It is a saving of effort, however, to use these operations on $[A,B]$ alone, avoiding the addition of multiples of the last column to the other columns. Then the rank of A is unaffected by these operations. It can be shown that $[A,B]$ can be reduced to the form

$$\left[\begin{array}{ccc|c} 0 & 0 & 0 & -1 \\ 0 & 0 & 0 & 0 \\ 0 & 0 & -1 & 0 \\ 0 & 1 & 0 & 0 \end{array}\right]$$

Hence rank $A = 2$ and rank $[A,B] = 3$. From Theorem 1, the unknowns which satisfy all the equations do not exist. The equations are said to be *inconsistent.*

Example 6. Solve the set of equations

$$\begin{aligned} 5x \qquad - z &= 2 \\ 4x - y + z &= 5 \\ 2x - 3y + 5z &= 11 \\ x + y - 2z &= -3 \end{aligned}$$

Solution: Using elementary row and column operations, it can be shown that rank A = rank $[A,B] = 2$ and the system of equations is *consistent.* (See Theorem 1.) We shall first solve for the fundamental set of solutions from the corresponding homogeneous equations and then solve for the nonhomogeneous solutions.

Since A is of rank 2, the corresponding homogeneous equations can be solved from two of the equations. Using the first two equations, we have

$$\begin{aligned} 5x \qquad - z &= 0 \\ 4x - y + z &= 0 \end{aligned}$$

Letting $x = 1$, arbitrary, we obtain

$$\begin{aligned} - z &= -5 \\ -y + z &= -4 \end{aligned}$$

This set of equations is consistent, and the solutions are $z = 5$ and $y = 9$. In the homogeneous system we have $(n - r) = (3 - 2) = 1$.

Hence the only set of fundamental solutions is $\{x,y,z\} = c\{1,9,5\}$, where c is arbitrary. (See Theorem 3.)

To find the solutions of the nonhomogeneous equations, we again use the first two equations of the set and let $x = 0$.

$$-z = 2$$
$$-y + z = 5$$

Hence $z = -2$ and $y = -7$, and the particular solution is $\{0,-7,-2\}$.

From Theorem 4, the general solution of the given system of equations is

$$\{x,y,z\} = c\{1,9,5\} + \{0,-7,-2\}$$

It can be verified by direct substitution that this general solution satisfies all the four given equations.

Modal Column Matrix

We shall consider a particular system of homogeneous equations which is of importance in the study of conservative dynamic systems. Let A be a square matrix of order n and λ a scalar parameter. Let us form a set of homogeneous equations

$$(\lambda I - A)X = 0 \qquad \textbf{(C-34)}$$

where X is a column of unknowns. $(\lambda I - A)$ is a square matrix of order n, and it is called the *characteristic matrix* of A. Equation (C-34) can also be written in the form

$$\lambda X = AX \qquad \textbf{(C-35)}$$

Since Eq. (C-34) is a set of homogeneous linear equations, by Theorem 2 it is necessary that $(\lambda I - A)$ be singular; that is,

$$\Delta(\lambda) = |\lambda I - A| = 0 \qquad \textbf{(C-36)}$$

Equation (C-36) is called the *characteristic equation* of A, the roots of which, $\lambda_1, \lambda_2, \ldots, \lambda_n$, are called the *latent roots* or the *characteristic values* of A. Let us assume that all the roots are distinct.

Each solution vector $X(\lambda_s)$ of Eq. (C-34) is called an *eigen vector* or a *modal column* appropriate to λ_s, $s = 1, 2, \ldots, n$.

$$X(\lambda_s) = \{x_{1s} x_{2s} \ldots x_{ns}\} \qquad \textbf{(C-37)}$$

The modal columns, taken for all the latent roots $\lambda_1, \lambda_2, \ldots, \lambda_n$, form a *modal column matrix* $[x_{ij}]$

$$[x_{ij}] = [X(\lambda_1)X(\lambda_2)\ldots X(\lambda_n)] = \begin{bmatrix} x_{11} \cdots x_{1n} \\ \cdot \quad \cdot \quad \cdot \quad \cdot \quad \cdot \\ x_{n1} \cdots x_{nn} \end{bmatrix} \qquad \text{(C-38)}$$

Appropriate to each λ_s, Eq. (C-35) can be written as

$$\lambda_s X(\lambda_s) = AX(\lambda_s) \qquad \text{(C-39)}$$

Since A is of order n, considering all the latent roots, this relation becomes

$$\begin{bmatrix} x_{11} & \cdots & x_{1n} \\ x_{21} & \cdots & x_{2n} \\ \cdot & \cdot & \cdot \\ x_{n1} & \cdots & x_{nn} \end{bmatrix} \begin{bmatrix} \lambda_1 & 0 & \cdots & 0 \\ 0 & \lambda_2 & \cdots & 0 \\ \cdot & \cdot & \cdot & \cdot \\ 0 & \cdots & 0 & \lambda_n \end{bmatrix}$$

$$= \begin{bmatrix} a_{11} & \cdots & a_{1n} \\ a_{21} & \cdots & a_{2n} \\ \cdot & \cdot & \cdot \\ a_{n1} & \cdots & a_{nn} \end{bmatrix} \begin{bmatrix} x_{11} & \cdots & x_{1n} \\ x_{21} & \cdots & x_{2n} \\ \cdot & \cdot & \cdot \\ x_{n1} & \cdots & x_{nn} \end{bmatrix} \qquad \text{(C-40)}$$

It is observed that Eq. (C-39) is also satisfied if the modal column $X(\lambda_s)$ is multiplied by an arbitrary constant c_s; that is,

$$\lambda_s c_s X(\lambda_s) = Ac_s X(\lambda_s)$$

Hence a modal column is indeterminate to the extent that it can always be multiplied by an arbitrary constant. Let us select a modal column

$$\{\mu_{is}\} = \{\mu_{1s}\mu_{2s}\ldots\mu_{ns}\} = c_s\{x_{1s}x_{2s}\ldots x_{ns}\} \qquad \text{(C-41)}$$

appropriate to λ_s. Forming a modal column matrix $[\mu]$ from the selected modal columns, we have

$$[\mu] = [\mu_{ij}] \equiv \begin{bmatrix} x_{11} & \cdots & x_{1n} \\ x_{21} & \cdots & x_{2n} \\ \cdot & \cdot & \cdot \\ x_{n1} & \cdots & x_{nn} \end{bmatrix} \begin{bmatrix} c_1 & 0 & \cdots & 0 \\ 0 & c_2 & \cdots & 0 \\ \cdot & \cdot & \cdot & \cdot \\ 0 & \cdots & 0 & c_n \end{bmatrix} \qquad \text{(C-42)}$$

Hence a modal column matrix is indeterminate to the extent that it can always be postmultiplied by a nonsingular diagonal matrix of arbitrary constants. It is convenient to denote a modal column matrix by the symbol $[\mu]$ for later considerations.

Using a modal column matrix $[\mu]$, Eq. (C-40) can be expressed as

$$\begin{bmatrix} \mu_{11} & \cdots & \mu_{1n} \\ \mu_{21} & \cdots & \mu_{2n} \\ \cdot & \cdots & \cdot \\ \mu_{n1} & \cdots & \mu_{nn} \end{bmatrix} \begin{bmatrix} \lambda_1 & 0 & \cdots & 0 \\ 0 & \lambda_2 & \cdots & 0 \\ \cdot & \cdot & \cdots & \cdot \\ 0 & \cdots & 0 & \lambda_n \end{bmatrix}$$

$$= \begin{bmatrix} a_{11} & \cdots & a_{1n} \\ a_{21} & \cdots & a_{2n} \\ \cdot & \cdots & \cdot \\ a_{n1} & \cdots & a_{nn} \end{bmatrix} \begin{bmatrix} \mu_{11} & \cdots & \mu_{1n} \\ \mu_{21} & \cdots & \mu_{2n} \\ \cdot & \cdots & \cdot \\ \mu_{n1} & \cdots & \mu_{nn} \end{bmatrix}$$

or

$$[\mu]\Lambda = A[\mu] \tag{C-43}$$

where Λ is a diagonal matrix with λ_s as its diagonal elements. Premultiplying Eq. (C-43) by $[\mu]^{-1}$ gives

$$[\mu]^{-1}A[\mu] = \Lambda \tag{C-44}$$

Postmultiplying Eq. (C-43) by $[\mu]^{-1}$ yields

$$[\mu]\Lambda[\mu]^{-1} = A \tag{C-45}$$

Hence the modal column matrix $[\mu]$ which diagonalizes A can be found by grouping the modal columns of A to form a square matrix. This is possible when all the latent roots of the characteristic equation of A are distinct. When two or more of the latent roots are equal to each other, reduction to the diagonal form is not always possible.

Example 7. Determine the modal column matrix of

$$A = \begin{bmatrix} 2 & -1 & 1 \\ -8 & 3 & 7 \\ -8 & -1 & 11 \end{bmatrix}$$

Solution: The characteristic equation of A is

$$\Delta(\lambda) = \left| \begin{bmatrix} \lambda & 0 & 0 \\ 0 & \lambda & 0 \\ 0 & 0 & \lambda \end{bmatrix} - \begin{bmatrix} 2 & -1 & 1 \\ -8 & 3 & 7 \\ -8 & -1 & 11 \end{bmatrix} \right| = \begin{vmatrix} \lambda - 2 & 1 & -1 \\ 8 & \lambda - 3 & -7 \\ 8 & 1 & \lambda - 11 \end{vmatrix}$$

$$= (\lambda - 2)(\lambda - 4)(\lambda - 10) = 0$$

Hence the latent roots are $\lambda_1 = 2$, $\lambda_2 = 4$, and $\lambda_3 = 10$. From Eq. (C-34) we have

$$\begin{bmatrix} \lambda - 2 & 1 & -1 \\ 8 & \lambda - 3 & -7 \\ 8 & 1 & \lambda - 11 \end{bmatrix} \begin{bmatrix} x_1 \\ x_2 \\ x_3 \end{bmatrix} = \begin{bmatrix} 0 \\ 0 \\ 0 \end{bmatrix}$$

For each latent root λ_s there corresponds a set of homogeneous equations of this form and a set of values of x's.

If $\lambda = \lambda_1 = 2$, we obtain

$$\begin{bmatrix} 0 & 1 & -1 \\ 8 & -1 & -7 \\ 8 & 1 & -9 \end{bmatrix} \begin{bmatrix} x_1 \\ x_2 \\ x_3 \end{bmatrix} = \begin{bmatrix} 0 \\ 0 \\ 0 \end{bmatrix}$$

Since the coefficient matrix of the equations is of rank 2, we use the first two equations to solve for the unknowns and assign $x_3 = 1$ arbitrarily.

$$x_2 = 1$$
$$8x_1 - x_2 = 7$$

The corresponding solutions are $x_1 = 1$ and $x_2 = 1$. The solution vector of the given set of homogeneous equations is

$$\{x_{11} \quad x_{21} \quad x_{31}\} = c_1\{1 \quad 1 \quad 1\}$$

where c_1 is arbitrary. The second subscript indicates that the solution corresponds to $\lambda = \lambda_1$. Since c_1 is arbitrary, the modal column may be selected as $\{1 \quad 1 \quad 1\}$.

Similarly, if $\lambda = \lambda_2 = 4$, the solution vector is $\{x_{12} \quad x_{22} \quad x_{32}\} = c_2\{1 \quad -1 \quad 1\}$. If $\lambda = \lambda_3 = 10$, the solution vector is $\{x_{13} \quad x_{23} \quad x_{33}\} = c_3\{0 \quad 1 \quad 1\}$.

The modal column matrix may be selected as

$$[\mu] = [\mu_{ij}] = \begin{bmatrix} 1 & 1 & 0 \\ 1 & -1 & 1 \\ 1 & 1 & 1 \end{bmatrix}$$

Equations (C-44) and (C-45) can be verified by using the numerical values obtained in this example.

The Lambda Matrix and Its Adjoint Matrix

A square matrix $f(\lambda)$, the elements of which are rational integral functions of a scalar parameter λ, is called a λ matrix. For example, Eq. (C-34) can be written as

$$f(\lambda)X = 0$$

This particular λ matrix is the characteristic matrix of A. The equation

$$\Delta(\lambda) \equiv |f(\lambda)| = 0$$

is called the *determinantal* or the *characteristic equation.*

The adjoint $F(\lambda)$ of $f(\lambda)$ is itself a λ matrix. From the definition of the inverse of a matrix [see Eqs. (C-20) and (C-21)], we obtain

$$F(\lambda)f(\lambda) = f(\lambda)F(\lambda) = I\Delta(\lambda) \tag{C-46}$$

Differentiating Eq. (C-46) with respect to λ gives

$$\frac{d}{d\lambda}[F(\lambda)f(\lambda)] = \frac{d}{d\lambda}[f(\lambda)F(\lambda)] = I\Delta^{(1)}(\lambda) \tag{C-47}$$

or, writing this equation explicitly, we have

$$F^{(1)}(\lambda)f(\lambda) + F(\lambda)f^{(1)}(\lambda) = f^{(1)}(\lambda)F(\lambda) + f(\lambda)F^{(1)}(\lambda) = I\Delta^{(1)}(\lambda) \tag{C-48}$$

The procedure for differentiating a matrix and a determinant was discussed in the previous sections. More generally, differentiating Eq. (C-46) p times gives

$$\frac{d^p}{d\lambda^p}[F(\lambda)f(\lambda)] = \frac{d^p}{d\lambda^p}[f(\lambda)F(\lambda)] = I\Delta^{(p)}(\lambda) \tag{C-49}$$

We shall state the following theorems without proof and then illustrate them with examples:

Theorem 5. If a root λ_s of $\Delta(\lambda) = 0$ is substituted in $f(\lambda)$, the matrix $f(\lambda_s)$ is necessarily singular. $f(\lambda_s)$ is of degeneracy 1 if the roots λ_s are distinct.

Theorem 6. If $f(\lambda_s)$ is of degeneracy 1, its adjoining $F(\lambda_s)$ is of unit rank. $F(\lambda_s)$ can be expressed as the product of a column matrix $\{\mu_{is}\}$ and a row matrix $\lfloor\kappa_{is}\rfloor$ appropriate to λ_s, $i = 1, 2, \ldots, n$; that is

$$F(\lambda_s) = \{\mu_{is}\}\lfloor\kappa_{is}\rfloor \tag{C-50}$$

Theorem 7. If a root λ_s of $\Delta(\lambda) = 0$ is substituted in $f(\lambda)$ and $f(\lambda_s)$ has degeneracy p, at least p of the roots are equal to λ_s.

Theorem 8. If a root λ_s of $\Delta(\lambda) = 0$ is substituted in $f(\lambda)$ and $f(\lambda_s)$ has multiple degeneracy p, the adjoint $F(\lambda)$ and its derivatives up to and including $F^{(p-2)}(\lambda)$ at least are all null for $\lambda = \lambda_s$.

Example 8. *Illustrating Theorems* 5 *to* 8.

Solutions: (*a*) Theorem 5: Distinct roots and $f(\lambda_s)$ simply degenerate. Consider

$$f(\lambda) = \begin{bmatrix} \lambda - 2 & \lambda - 1 & 3 \\ 1 & \lambda - 1 & \lambda - 2 \\ 1 & 0 & \lambda - 5 \end{bmatrix}$$

$$\Delta(\lambda) = (\lambda - 1)(\lambda - 2)(\lambda - 5) = 0$$

The roots are $\lambda_1 = 1$, $\lambda_2 = 2$, and $\lambda_3 = 5$. It can be shown that the matrices

$$f(\lambda_1) = \begin{bmatrix} -1 & 0 & 3 \\ 1 & 0 & -1 \\ 1 & 0 & -4 \end{bmatrix},$$

$$f(\lambda_2) = \begin{bmatrix} 0 & 1 & 3 \\ 1 & 1 & 0 \\ 1 & 0 & -3 \end{bmatrix},$$

$$f(\lambda_3) = \begin{bmatrix} 3 & 4 & 3 \\ 1 & 4 & 3 \\ 1 & 0 & 0 \end{bmatrix}$$

are of rank 2. Hence $f(\lambda_s)$ is simply degenerate.

(*b*) Theorem 6: Simply degenerate and $F(\lambda_s) = \{\mu_{is}\}[\kappa_{is}]$. Consider the λ matrix in part (*a*). The corresponding adjoint matrix is

$$F(\lambda) = \begin{bmatrix} (\lambda-1)(\lambda-5) & -(\lambda-1)(\lambda-5) & (\lambda-1)(\lambda-5) \\ (\lambda-2)-(\lambda-5) & (\lambda-2)(\lambda-5)-3 & 3-(\lambda-2)^2 \\ -(\lambda-1) & (\lambda-1) & (\lambda-1)(\lambda-3) \end{bmatrix}$$

The roots of $\Delta(\lambda) = |f(\lambda)| = 0$ are $\lambda_1 = 1$, $\lambda_2 = 2$, and $\lambda_3 = 5$.

$$F(\lambda_1) = \begin{bmatrix} 0 & 0 & 0 \\ 3 & 1 & 2 \\ 0 & 0 & 0 \end{bmatrix} = \begin{bmatrix} 0 \\ 1 \\ 0 \end{bmatrix} [3 \quad 1 \quad 2]$$

$$F(\lambda_2) = \begin{bmatrix} -3 & 3 & -3 \\ 3 & -3 & 3 \\ -1 & 1 & -1 \end{bmatrix} = \begin{bmatrix} 3 \\ -3 \\ 1 \end{bmatrix} [-1 \quad 1 \quad -1]$$

$$F(\lambda_3) = \begin{bmatrix} 0 & 0 & 0 \\ 3 & -3 & -6 \\ -4 & 4 & 8 \end{bmatrix} = \begin{bmatrix} 0 \\ 3 \\ -4 \end{bmatrix} [1 \quad -1 \quad -2]$$

It is evident that $F(\lambda_s)$ is of unit rank.

(*c*) Theorem 7: Multiple degeneracy and repeated roots. Consider

$$f(\lambda) = \begin{bmatrix} \lambda-1 & \lambda-1 & 0 \\ 0 & \lambda-1 & 0 \\ \lambda-1 & 0 & \lambda-2 \end{bmatrix}$$

$$\Delta(\lambda) = (\lambda-1)^2(\lambda-2) = 0$$

The roots are $\lambda_1 = \lambda_2 = 1$ and $\lambda_3 = 2$.

$$f(\lambda_1) = f(\lambda_2) = \begin{bmatrix} 0 & 0 & 0 \\ 0 & 0 & 0 \\ 0 & 0 & -1 \end{bmatrix}, \quad f(\lambda_3) = \begin{bmatrix} 1 & 1 & 0 \\ 0 & 1 & 0 \\ 1 & 0 & 0 \end{bmatrix}$$

$f(\lambda_1)$ and $f(\lambda_2)$ are of degeneracy 2 for the repeated roots $\lambda_1 = \lambda_2 = 1$.

(*d*) Theorem 8: Multiple degeneracy p and $F^{(p-2)}(\lambda_s) = 0$. Consider

$$f(\lambda) = \begin{bmatrix} \lambda - 1 & 0 & 0 & 0 \\ 0 & \lambda - 1 & 0 & 0 \\ 0 & 0 & \lambda - 1 & 0 \\ 0 & 0 & 0 & \lambda - 2 \end{bmatrix}$$

$$\Delta(\lambda) = (\lambda - 1)^3(\lambda - 2) = 0$$

The roots are $\lambda_1 = \lambda_2 = \lambda_3 = 1$ and $\lambda_4 = 2$. Considering the repeated roots, it is noted that $f(\lambda_s)$ is of degeneracy 3. Hence it is expected that $F^{(3-2)}(\lambda_s) = 0$. The adjoint matrix is

$$F(\lambda) = \begin{bmatrix} (\lambda - 1)^2(\lambda - 2) & 0 & 0 & 0 \\ 0 & (\lambda - 1)^2(\lambda - 2) & 0 & 0 \\ 0 & 0 & (\lambda - 1)^2(\lambda - 2) & 0 \\ 0 & 0 & 0 & (\lambda - 1)^3 \end{bmatrix}$$

It can be verified that for the repeated roots $F(\lambda_s) = F^{(1)}(\lambda_s) = 0$.

Modal Column Matrix and Adjoint Matrix

It was shown that the solutions of Eq. (C-34)

$$(\lambda I - A)X = 0$$

are the modal columns $X(\lambda_s)$ appropriate to λ_s, where λ_s are the roots of $|\lambda I - A| = 0$. Let us rewrite this equation in the form

$$f(\lambda)X = 0 \tag{C-51}$$

Since $X(\lambda_s)$ is a solution of Eq. (C-51), it is obvious that

$$f(\lambda_s)X(\lambda_s) = 0 \tag{C-52}$$

Consider a λ matrix $f(\lambda)$ and assume that λ_s is a distinct root of the equation $\Delta(\lambda) = |f(\lambda)| = 0$. From Eq. (C-46), we have

$$f(\lambda_s)F(\lambda_s) = I\Delta(\lambda_s) = 0 \tag{C-53}$$

From Theorem 5, if λ_s is a simple root, $f(\lambda_s)$ is of degeneracy 1. From Theorem 6, if $f(\lambda_s)$ is simply degenerate, $F(\lambda_s)$ is of unit rank and can be expressed as $F(\lambda_s) = \{\mu_{is}\}[\kappa_{is}]$. Substituting this relation in Eq. (C-53) gives

$$f(\lambda_s)\{\mu_{is}\}[\kappa_{is}] = 0$$

Since neither the column $\{\mu_{is}\}$ nor the row $[\kappa_{is}]$ is null, the last equation indicates that

$$f(\lambda_s)\{\mu_{is}\} = 0 \tag{C-54}$$

Comparing Eqs. (C-52) and (C-54), it is deduced that the modal column $X(\lambda_s)$ is proportional to any nonzero column $\{\mu_{is}\}$ of the adjoint matrix $F(\lambda_s)$; that is,

$$X(\lambda_s) = (\text{constant})[\text{any nonzero column of } F(\lambda_s)] \tag{C-55}$$

Since a modal column is indeterminate to the extent that it can always be multiplied by an arbitrary constant, the constant in Eq. (C-55) is arbitrary, and any multiple of a nonzero column of $F(\lambda_s)$ may be selected as a modal column. Thus a modal column matrix $[\mu]$ can be constructed from the adjoint of the characteristic matrix of A.

When λ_s is a once-repeated root of $\Delta(\lambda) = 0$ and $f(\lambda_s)$ is of degeneracy 2, the adjoint $F(\lambda_s)$ is null. (See Theorem 8.) The modal columns can be obtained from any two independent nonzero columns of $F^{(1)}(\lambda_s)$. It can be shown that in this case $F^{(1)}(\lambda_s)$ is of rank 2. From Eq. (C-48) we have

$$f(\lambda_s)F^{(1)}(\lambda_s) = I\Delta^{(1)}(\lambda_s) = 0 \tag{C-56}$$

$\Delta^{(1)}(\lambda_s)$ necessarily equals zero because $(\lambda - \lambda_s)^2$ is a factor of $\Delta(\lambda)$. Hence $F^{(1)}(\lambda_s)$ satisfies Eq. (C-52).

Example 9. Using the matrix A as given in Example 7, find the modal column matrix from the adjoint matrix $F(\lambda)$.

Solution: The characteristic matrix and its adjoint are

$$f(\lambda) = (\lambda I - A) = \begin{bmatrix} \lambda - 2 & 1 & -1 \\ 8 & \lambda - 3 & -7 \\ 8 & 1 & \lambda - 11 \end{bmatrix}$$

$$F(\lambda) = \begin{bmatrix} (\lambda - 4)(\lambda - 10) & -(\lambda - 10) & (\lambda - 10) \\ -8(\lambda - 4) & (\lambda - 3)(\lambda - 10) & -(7\lambda - 22) \\ -8(\lambda - 4) & -(\lambda - 10) & (\lambda^2 - 5\lambda - 2) \end{bmatrix}$$

From Example 7, the latent roots are $\lambda_1 = 2$, $\lambda_2 = 4$, and $\lambda_3 = 10$. Hence

$$F(\lambda_1) = \begin{bmatrix} 16 & 8 & -8 \\ 16 & 8 & -8 \\ 16 & 8 & -8 \end{bmatrix} = \begin{bmatrix} 1 \\ 1 \\ 1 \end{bmatrix} [16 \quad 8 \quad -8]$$

$$F(\lambda_2) = \begin{bmatrix} 0 & 6 & -6 \\ 0 & -6 & 6 \\ 0 & 6 & -6 \end{bmatrix} = \begin{bmatrix} 1 \\ -1 \\ 1 \end{bmatrix} [0 \quad 6 \quad -6]$$

$$F(\lambda_3) = \begin{bmatrix} 0 & 0 & 0 \\ -48 & 0 & 48 \\ -48 & 0 & 48 \end{bmatrix} = \begin{bmatrix} 0 \\ 1 \\ 1 \end{bmatrix} [-48 \quad 0 \quad 48]$$

The modal column matrix $[\mu]$ may be selected as

$$[\mu] = \begin{bmatrix} 1 & 1 & 0 \\ 1 & -1 & 1 \\ 1 & 1 & 1 \end{bmatrix}$$

Example 10. Determine the modal columns of the characteristic matrix of A

$$f(\lambda) = (\lambda I - A) = \begin{bmatrix} \lambda - 6 & 5 & 0 & 0 \\ 5 & \lambda - 6 & 0 & 0 \\ 0 & 0 & \lambda - 1 & 0 \\ 0 & 0 & 0 & \lambda - 11 \end{bmatrix}$$

Solution: It can be shown that the characteristic equation is

$$\Delta(\lambda) = (\lambda - 1)^2(\lambda - 11)^2 = 0$$

Hence the latent roots are $\lambda_1 = \lambda_2 = 1$ and $\lambda_3 = \lambda_4 = 11$.

$$f(\lambda_1) = f(\lambda_2) = \begin{bmatrix} -5 & 5 & 0 & 0 \\ 5 & -5 & 0 & 0 \\ 0 & 0 & 0 & 0 \\ 0 & 0 & 0 & -10 \end{bmatrix},$$

$$f(\lambda_3) = f(\lambda_4) = \begin{bmatrix} 5 & 5 & 0 & 0 \\ 5 & 5 & 0 & 0 \\ 0 & 0 & 10 & 0 \\ 0 & 0 & 0 & 0 \end{bmatrix}$$

Since λ_1 is a once-repeated root and $f(\lambda_1)$ is of degeneracy 2, the adjoint $F(\lambda_1)$ is null. Similarly, $F(\lambda_3)$ is null.

$$F(\lambda) = (\lambda - 1)(\lambda - 11) \begin{bmatrix} \lambda - 6 & -5 & 0 & 0 \\ -5 & \lambda - 6 & 0 & 0 \\ 0 & 0 & \lambda - 11 & 0 \\ 0 & 0 & 0 & \lambda - 1 \end{bmatrix}$$

It can be verified that

$$F^{(1)}(\lambda_1) = \begin{bmatrix} 50 & 50 & 0 & 0 \\ 50 & 50 & 0 & 0 \\ 0 & 0 & 100 & 0 \\ 0 & 0 & 0 & 0 \end{bmatrix} \qquad F^{(1)}(\lambda_3) = \begin{bmatrix} 50 & -50 & 0 & 0 \\ -50 & 50 & 0 & 0 \\ 0 & 0 & 0 & 0 \\ 0 & 0 & 0 & 100 \end{bmatrix}$$

Thus $F^{(1)}(\lambda_1)$ and $F^{(1)}(\lambda_3)$ are of rank 2, and the modal column matrix may be selected as

$$[\mu] = \begin{bmatrix} 1 & 0 & 1 & 0 \\ 1 & 0 & -1 & 0 \\ 0 & 1 & 0 & 0 \\ 0 & 0 & 0 & 1 \end{bmatrix} \quad \text{and} \quad [\mu]^{-1} = \tfrac{1}{2} \begin{bmatrix} 1 & -1 & 0 & 0 \\ 0 & 0 & -2 & 0 \\ 1 & -1 & 0 & 0 \\ 0 & 0 & 0 & 2 \end{bmatrix}$$

Substituting $[\mu]$ and its inverse in Eq. (C-44) gives

$$[\mu]^{-1}A[\mu] = \Lambda = \begin{bmatrix} 1 & 0 & 0 & 0 \\ 0 & 1 & 0 & 0 \\ 0 & 0 & 11 & 0 \\ 0 & 0 & 0 & 11 \end{bmatrix}$$

SUGGESTED READING

Dickson, L. E., *New First Course in the Theory of Equations* (New York: John Wiley & Sons, Inc., 1939).

Faddeeva, V. N., translated by C. D. Benster, *Computational Methods of Linear Algebra* (New York: Dover Publications, Inc., 1959).

Frazer, R. A., W. J. Duncan, and A. R. Collar, *Elementary Matrices* (London: Cambridge University Press, 1957), chaps. 1 to 4.

Pipes, L. A., *Applied Mathematics for Engineers and Physicists* (New York: McGraw-Hill Book Co., Inc., 2nd ed., 1958), chap. 4.

Wylie, C. R., Jr., *Advanced Engineering Mathematics* (New York: McGraw-Hill Book Co., Inc., 2nd ed., 1960), chaps. 1 and 4.

APPENDIX D

LAPLACE TRANSFORM AND TABLE OF LAPLACE TRANSFORM PAIRS

In this appendix we shall (1) derive the Laplace transform from the generalization of the Fourier transform, (2) indicate that the transformations are special cases of the linear integral transformation, and (3) summarize the Laplace transform pairs in Chap. 8 and provide additional transform pairs for references.

FROM FOURIER TRANSFORM TO LAPLACE TRANSFORM

In Chap. 2 it was shown that a periodic function $f(t)$ can be expressed as a trigonometric series. [See Eq. (2-99), Chap. 2.]

$$f(t) = \frac{a_o}{2} + \sum_{n=1}^{\infty} (a_n \cos n\omega t + b_n \sin n\omega t) \tag{D-1}$$

where

$$a_o = \frac{2}{\tau} \int_o^{\tau} f(t)\, dt$$

$$a_n = \frac{2}{\tau} \int_o^{\tau} f(t) \cos n\omega t\, dt \tag{D-2}$$

$$b_n = \frac{2}{\tau} \int_o^{\tau} f(t) \sin n\omega t\, dt$$

where τ is the period of $f(t)$. Now let us write Eq. (D-1) in its exponential form by expressing the cosine and sine functions as

$$\cos n\omega t = \frac{1}{2}(e^{jn\omega t} + e^{-jn\omega t})$$
$$\sin n\omega t = \frac{1}{2j}(e^{jn\omega t} - e^{-jn\omega t}) \tag{D-3}$$

Substituting Eq. (D-3) in Eq. (D-1) and simplifying, we have

$$f(t) = \alpha_o + \sum_{n=1}^{\infty} (\alpha_n e^{jn\omega t} + \alpha_{-n} e^{-jn\omega t}) \tag{D-4}$$

where

$$\alpha_o = \frac{a_o}{2}, \quad \alpha_n = \frac{1}{2}(a_n - jb_n), \quad \alpha_{-n} = \frac{1}{2}(a_n + jb_n)$$

Noting that the second term in the summation in Eq. (D-4) can be written as

$$\sum_{n=1}^{\infty} \alpha_{-n} e^{-jn\omega t} = \sum_{n=-1}^{-\infty} \alpha_n e^{jn\omega t}$$

Eq. (D-4) can be simplified by changing its limits of summation.

$$f(t) = \sum_{n=-\infty}^{\infty} \alpha_n e^{jn\omega t} \tag{D-5}$$

where

$$\alpha_n = \frac{1}{\tau}\int_o^{\tau} f(t)\, e^{-jn\omega t}\, dt$$

Eq. (D-5) is called the exponential form of the Fourier series. Since $f(t)$ is periodic, α_n can be written as

$$\alpha_n = \frac{1}{\tau}\int_{-\tau/2}^{\tau/2} f(t)\, e^{-jn\omega t}\, dt \tag{D-6}$$

If $f(t)$ is aperiodic, the period τ becomes infinite. Since $\omega = 2\pi/\tau$, as ω approaches zero, n becomes meaningless. Let the period τ be very large and the angular frequency of a typical component in Eq. (D-5) be $k\omega = k2\pi/\tau = \omega_k$. The frequency of the next component is $(k + 1)2\pi/\tau$. Thus

$$\Delta\omega = (k + 1)2\pi/\tau - k2\pi/\tau = 2\pi/\tau$$

Using these relations and combining Eqs. (D-5) and (D-6) gives

$$f(t) = \sum_{k=-\infty}^{\infty} \left(\frac{1}{2\pi} \int_{-\tau/2}^{\tau/2} f(t)\, e^{-j\omega_k t}\, dt\, \Delta\omega \right) e^{j\omega_k t}$$

$$= \frac{1}{2\pi} \sum_{k=-\infty}^{\infty} \left(\int_{-\tau/2}^{\tau/2} f(t)\, e^{-j\omega_k t}\, dt \right) e^{j\omega_k t}\, \Delta\omega$$

As the period τ approaches infinity, $\Delta\omega$ approaches zero, and the discrete variable ω_k becomes a continuous variable ω. Hence the last equation can be expressed as

$$f(t) = \frac{1}{2\pi} \int_{-\infty}^{\infty} g(\omega)\, e^{jt\omega}\, d\omega = \mathscr{F}^{-1}\, g(\omega) \tag{D-7}$$

$$g(\omega) = \int_{-\infty}^{\infty} f(t)\, e^{-j\omega t}\, dt = \mathscr{F}\, f(t) \tag{D-8}$$

Equations (D-7) and (D-8) are called a Fourier transform pair; $g(\omega)$ is the Fourier transform of $f(t)$, and $f(t)$ is the inverse Fourier transform of $g(\omega)$. The transform pair is represented by the symbols as shown in these equations.

Although the Fourier transform is extensively used in communication theory and other problems, the integral, Eq. (D-8), does not exist for certain useful functions. For example, the Fourier transform of the unit step function $u(t)$, defined as $u(t) = 0$ for $t < 0$ and $u(t) = 1$ for $t > 0$, does not exist.

$$g(\omega) = \mathscr{F}\, u(t) = \int_{0}^{\infty} e^{-j\omega t}\, dt = \int_{0}^{\infty} (\cos \omega t - j \sin \omega t)\, dt$$

Neither cosine nor sine can be integrated from zero to infinity.

In a physical problem there must be some instant which can be designated as $t = 0$. Thus we define the unilateral Fourier transform

$$g(\omega) = \int_{0}^{\infty} f(t)\, e^{-j\omega t}\, dt \tag{D-9}$$

If $f(t)$ is discontinuous at $t = 0$, the value of $f(t)$ to be used at $t = 0$ is $f(0+)$.

Let us introduce a *convergence factor* $e^{-\sigma t}$ in the integrand of Eq. (D-9), where σ is a real number. Writing the unilateral Fourier transform $g(\omega)$ as $F(\sigma,\omega)$, we have

$$F(\sigma,\omega) = \int_{0}^{\infty} [f(t)\, e^{-\sigma t}] e^{-j\omega t}\, dt = \int_{0}^{\infty} f(t)\, e^{-(\sigma + j\omega)t}\, dt \tag{D-10}$$

The value of σ should be sufficiently large to make the integral in

Eq. (D-10) converge. More specifically, if $f(t)$ is a single-valued function, and if there is a real number σ such that

$$\lim_{T\to\infty} \int_o^T |f(t)|e^{-\sigma t}\,dt < \infty$$

then Eq. (D-9) can be used to find the transform of $f(t)\,e^{-\sigma t}$ for $t \geq 0$. For convenience, we substitute s for $(\sigma + j\omega)$ and rewrite Eq. (D-10) as

$$F(s) = \int_o^\infty f(t)\,e^{-st}\,dt = \mathscr{L} f(t) \qquad \textbf{(D-11)}$$

$F(s)$ is called the Laplace transform of $f(t)$. When the integral exists, the function $f(t)$ is Laplace transformable. Owing to the convergence factor $e^{-\sigma t}$, the Laplace transform of the unit step function $u(t)$ is $1/s$. It was shown previously that the Fourier transform of $u(t)$ does not exist.

The inverse transform can be obtained from Eq. (D-7).

$$e^{-\sigma t} f(t) = \frac{1}{2\pi} \int_{-\infty}^{\infty} F(\sigma,\omega)\,e^{j\omega t}\,d\omega$$

Multiplying both sides of this equation by $e^{\sigma t}$ gives

$$f(t) = \frac{1}{2\pi} \int_{-\infty}^{\infty} F(\sigma,\omega)\,e^{(\sigma+j\omega)t}\,d\omega$$

By changing the variable of integration from ω to $s = (\sigma + j\omega)$, we have

$$f(t) = \frac{1}{2\pi j} \int_{\sigma-j\infty}^{\sigma+j\infty} F(s)\,e^{ts}\,ds = \mathscr{L}^{-1} F(s) \qquad \textbf{(D-12)}$$

The function $f(t)$ is called the inverse transform of $F(s)$. Equations (D-11) and (D-12) are called a Laplace transform pair.

The foregoing derivation of the Laplace transform pair is more heuristic than rigorous. It may serve to give the reader a better feel for the subject than merely defining the transformation integrals. For a more detailed treatment, the reader should refer to standard texts on the subject.

LINEAR INTEGRAL TRANSFORMATION

A general linear integral transformation of a function $f(x)$ with respect to the kernel $K(p,x)$ is defined by the integral equation

$$F(p) = \int_a^b f(x)K(p,x)\,dx \qquad \textbf{(D-13)}$$

$K(p,x)$ is a known function of p and x. The function $f(x)$ is used instead of $f(t)$ to emphasize that the variable does not have to represent time. The limits a and b can be finite or infinite. In the Laplace transform, a is zero and b is infinite, and the kernel is e^{-px}.

Five different kernels have been used to the solution of boundary-value problems. We shall list the corresponding transforms as follows:

Laplace transform:

$$F(p) = \int_0^\infty f(x)\, e^{-px}\, dx$$

Fourier sine and cosine transform:

$$F(p) = \int_0^\infty f(x) \begin{matrix}\sin\\ \cos\end{matrix}\, px\, dx$$

Complex Fourier transform:

$$F(p) = \int_{-\infty}^{\infty} f(x)\, e^{jpx}\, dx$$

Hankel transform:

$$F(p) = \int_0^\infty f(x)\, xJ_n(px)\, dx$$

where $J_n(px)$ is the Bessel function of the first kind of order n.

Mellin transform:

$$F(p) = \int_0^\infty f(x)\, x^{p-1}\, dx$$

TABLES OF LAPLACE TRANSFORM PAIRS

Table D-1 is a tabulation of operation-transform pairs and Table D-2 of function-transform pairs. All the transform pairs can be derived with the technique presented in Chap. 8. The reader should refer to standard texts on the subject for a more comprehensive listing of the transform pairs.†

† See, for example, F. E. Nixon, *Handbook of Laplace Transformation—Tables and Examples*, Prentice-Hall, Inc., Englewood Cliffs, N.J., 1960.

TABLE D-1

OPERATION-TRANSFORM PAIRS

NUMBER	OPERATION	$f(t)$	$F(s) = \mathscr{L} f(t)$
1	Transform integral	$f(t)$	$\int_0^\infty f(t)\, e^{-st}\, dt$
2	Linearity		
	Constant multiplication	$af(t)$	$aF(s)$
	Addition, subtraction	$f_1(t) \pm f_2(t)$	$F_1(s) \pm F_2(s)$
3	Real differentiation		
	First derivative	$\frac{d}{dt} f(t) = f'(t)$	$sF(s) - f(0+)$
	Second derivative	$\frac{d^2}{dt^2} f(t) = f''(t)$	$s^2F(s) - sf(0+) - f'(0+)$
	nth derivative	$\frac{d^n}{dt^n} f(t)$	$s^nF(s) - s^{n-1}f(0+) - s^{n-2}f'(0+) - \cdots - f^{(n-1)}f(0+)$
4	Real integration		
	Indefinite integral	$\int f(t)\, dt$	$\frac{1}{s}[F(s) + f^{(-1)}(0+)]$
	Definite integral	$\int_0^t f(t)\, dt$	$\frac{1}{s} F(s)$
5	Real translation	$f(t - a)u(t - a)$	$e^{-as} F(s)$
6	Complex translation	$e^{-at} f(t)$	$F(s + a)$
7	Periodic function	$f(t) = f(t + \tau)$	$\frac{1}{1 - e^{-\tau s}} \int_0^\tau f(t)\, e^{-st}\, dt$
8	Convolution	$\int_0^t f_1(t - \tau) f_2(\tau)\, d\tau = f_1(t) * f_2(t)$	$F_1(s)\, F_2(s)$
9	Initial value	$\lim_{t \to 0+} f(t)$	$\lim_{s \to \infty} s\, F(s)$
10	Final value	$\lim_{t \to \infty} f(t)$	$\lim_{s \to 0} s\, F(s)$
11	Complex differentiation	$t f(t)$	$-\frac{d}{ds} F(s)$
12	Complex integration	$\frac{1}{t} f(t)$	$\int_s^\infty F(s)\, ds$
13	Scale change	$f(at)$	$\frac{1}{a} F\left(\frac{s}{a}\right)$

TABLE D-2

Function-Transform Pairs

Number	$F(s)$	$f(t),\ t \geq 0$
1	1	$\delta(t)$
2	$\dfrac{1}{s}$	$u(t)$
3	$\dfrac{1}{s^2}$	t
4	$\dfrac{1}{s^n},\ n = 1, 2, 3, \ldots$	$\dfrac{t^{n-1}}{(n-1)!}$
5	$\dfrac{1}{s+a}$	e^{-at}
6	$\dfrac{1}{(s+a)^2}$	te^{-at}
7	$\dfrac{s}{(s+a)^2}$	$(1-at)e^{-at}$
8	$\dfrac{1}{(s+a)^n},\ n = 1, 2, 3, \ldots$	$\dfrac{1}{(n-1)!}\, t^{n-1}e^{-at}$
9	$\dfrac{1}{(s+a)(s+b)}$	$\dfrac{1}{(b-a)}\,(e^{-at} - e^{-bt})$
10	$\dfrac{s}{(s+a)(s+b)}$	$\dfrac{1}{(a-b)}\,(ae^{-at} - be^{-bt})$
11	$\dfrac{1}{(s+a)(s+b)^2}$	$\dfrac{1}{(a-b)^2}\,e^{-at} + \dfrac{(a-b)t - 1}{(a-b)^2}\,e^{-bt}$
12	$\dfrac{s}{(s+a)(s+b)^2}$	$-\dfrac{a}{(a-b)^2}\,e^{-at} - \dfrac{b(a-b)t - a}{(a-b)^2}\,e^{-bt}$
13	$\dfrac{s^2}{(s+a)(s+b)^2}$	$\dfrac{a^2}{(a-b)^2}\,e^{-at} + \dfrac{b^2(a-b)t + b^2 - 2ab}{(a-b)^2}\,e^{-bt}$
14	$\dfrac{1}{s^2+a^2}$	$\dfrac{1}{a}\sin at$
15	$\dfrac{s}{s^2+a^2}$	$\cos at$
16	$\dfrac{1}{s^2-a^2}$	$\dfrac{1}{a}\sinh at$
17	$\dfrac{s}{s^2-a^2}$	$\cosh at$

TABLE D-2 (cont.)

NUMBER	$F(s)$	$f(t),\ t \geq 0$
18	$\dfrac{1}{s(s^2+a^2)}$	$\dfrac{1}{a^2}(1-\cos at)$
19	$\dfrac{1}{s^2(s^2+a^2)}$	$\dfrac{1}{a^3}(at-\sin at)$
20	$\dfrac{1}{(s+a)(s^2+b^2)}$	$\dfrac{1}{a^2+b^2}\left[e^{-at}+\dfrac{\sqrt{a^2+b^2}}{b}\sin(bt-\phi)\right]$ $\phi=\tan^{-1}(b/a)$
21	$\dfrac{s}{(s+a)(s^2+b^2)}$	$-\dfrac{1}{a^2+b^2}[ae^{-at}-\sqrt{a^2+b^2}\sin(bt+\phi)]$ $\phi=\tan^{-1}(a/b)$
22	$\dfrac{s^2}{(s+a)(s^2+b^2)}$	$\dfrac{1}{a^2+b^2}[a^2e^{-at}-b\sqrt{a^2+b^2}\sin(bt-\phi)]$ $\phi=\tan^{-1}(b/a)$
23	$\dfrac{1}{(s+a)^2+b^2}$	$\dfrac{1}{b}e^{-at}\sin bt$
24	$\dfrac{s+a}{(s+a)^2+b^2}$	$e^{-at}\cos bt$
25	$\dfrac{1}{s[(s+a)^2+b^2]}$	$\dfrac{1}{a^2+b^2}\left[1-\dfrac{1}{b}\sqrt{a^2+b^2}\,e^{-at}\sin(bt+\phi)\right]$ $\phi=\tan^{-1}(b/a)$
26	$\dfrac{1}{s^2[(s+a)^2+b^2]}$	$\dfrac{1}{a^2+b^2}\left[t-\dfrac{2a}{a^2+b^2}+\dfrac{1}{b}e^{-at}\sin(bt+\phi)\right]$ $\phi=\tan^{-1}\dfrac{2ab}{a^2-b^2}$
27	$\dfrac{1}{(s^2+c^2)[(s+a)^2+b^2]}$	$\dfrac{1}{[(d^2-c^2)^2+4a^2c^2]^{1/2}}\left[\dfrac{1}{c}\sin(ct-\phi_1)\right.$ $\left.+\dfrac{1}{b}e^{-at}\sin(bt-\phi_2)\right]$ $d^2=a^2+b^2,\ \phi_1=\tan^{-1}\dfrac{2ac}{a^2+b^2-c^2}$ $\phi_2=\tan^{-1}\dfrac{-2ab}{a^2-b^2+c^2}$

TABLE D-2 (cont.)

NUMBER	$F(s)$	$f(t),\ t \geq 0$
28	$\dfrac{s}{(s^2+c^2)[(s+a)^2+b^2]}$	$\dfrac{1}{[(d^2-c^2)^2+4a^2c^2]^{1/2}}\left[\sin(ct+\phi_1) - \dfrac{\sqrt{a^2+b^2}}{b}e^{-at}\sin(bt+\phi_2)\right]$ $d^2=a^2+b^2,\ \phi_1=\tan^{-1}\dfrac{d^2-c^2}{2ac}$ $\phi_2=\tan^{-1}\dfrac{(d^2-c^2)b}{(d^2+c^2)a}$
29	$\dfrac{1}{(s^2+a^2)^2}$	$\dfrac{1}{2a^3}(\sin at - at\cos at)$
30	$\dfrac{s}{(s^2+a^2)^2}$	$\dfrac{t}{2a}\sin at$
31	$\dfrac{s^2}{(s^2+a^2)^2}$	$\dfrac{1}{2a}(\sin at + at\cos at)$
32	$\dfrac{s^2-a^2}{(s^2+a^2)^2}$	$t\cos at$
33	e^{-as}	$\delta(t-a)$
34	$\dfrac{e^{-as}}{s}$	$u(t-a)$

PROBLEMS

1-1. A harmonic motion is expressed as $x = 0.5 \sin (15\pi t - \pi/3)$, where x is measured in inches, t in seconds, and the phase angle in radians. Determine: (*a*) the frequency and the period of the motion; (*b*) the maximum displacement, velocity, and acceleration; (*c*) the displacement, velocity, and acceleration at $t = 0$. (*d*) Repeat part (*c*) for $t = 0.2$ sec.

1-2. A vibration-measuring instrument indicates that a body is vibrating sinusoidally at a frequency of 30 cycles per second with a maximum acceleration of 3,570 in.-sec^{-2}. What is the amplitude of vibration?

1-3. A harmonic motion is described by the equation $x = X \cos (100t + \psi)$. The initial conditions are $x(0) = 0.1$ in. and $\dot{x}(0) = 50$ in.-sec^{-1}. Find the constants X and ψ. Express the given equation in the form

$$x = A \cos \omega t + B \sin \omega t$$

and determine the constants A, B, and ω.

1-4. Use the algebraic method to find the sum of the harmonic motions $x_1 = 2 \sin (\omega t + \pi/3)$ and $x_2 = 3 \sin (\omega t + 2\pi/3)$. Check the addition graphically.

1-5. Use the algebraic method to find the sum of the harmonic motions $x_1 = 2 \sin (\omega t + \pi/3)$ and $x_2 = -3 \cos (\omega t + 2\pi/3)$. Express the sum in the form $x = X \sin (\omega t + \alpha)$. Check the addition graphically.

1-6. The motion of a particle is described by the equation

$$x = 4 \sin (\omega t + \pi/6).$$

If the motion has two harmonic components, one of which is

$$x_1 = 2 \sin (\omega t - \pi/3),$$

determine the other harmonic component.

1-7. A periodic motion is described by the equation

$$x = 5 \sin 2\pi t + 3 \sin 4\pi t.$$

In a plot of x versus t, sketch the motion for $0 \leq t \leq 2$ sec.

1-8. Determine the sum of the harmonic motions $x_1 = X \cos \omega t$ and $x_2 = (X + \delta) \cos (\omega + \varepsilon)t$, where $\delta \ll X$ and $\varepsilon \ll \omega$. If beating should occur, determine the amplitude and the beat frequency.

1-9. In a sketch of x versus t, plot the motions described by the equations $x_1 = 5 \sin 10\pi t$ and $x_2 = 5 \sin (10\pi t + \pi/4)$ for $0 \leq t \leq 0.4$ sec.

1-10. Sketch the motion described by the equation

$$x = 5e^{-t} \sin (10\pi t + \pi/4)$$

for $0 \leq t \leq 1.2$ sec.

1-11. Sketch the motion described by the equation

$$x = 5e^{-t} \sin (10\pi t + \pi/4) + 7 \sin 2\pi t$$

for $0 \leq t \leq 2$ sec.

1-12. Express the following complex numbers in the exponential form $Ae^{j\theta}$.

(*a*) $1 + j\sqrt{3}$

(*b*) -2

(*e*) $3/(\sqrt{3} - j)$

(*d*) $5j$

(*e*) $3/(\sqrt{3} - j)^2$

(*f*) $(\sqrt{3} + j)(3 + 4j)$

(*g*) $(\sqrt{3} - j)/(3 - 4j)$

(*h*) $[(2j)^2 + 3j + 8]$

1-13. The motion of a particle vibrating in a plane has two perpendicular harmonic components: $x_1 = 2 \sin (\omega t + \pi/6)$ and $x_2 = 2.9 \sin \omega t$. Determine the motion of the particle graphically.

1-14. The motion of a particle vibrating in a plane has two perpendicular harmonic components: $x_1 = 2 \sin (2\omega t + \pi/6)$ and $x_2 = 2.9 \sin \omega t$. Determine the motion of the particle graphically.

2-1. Use the energy method to determine the equations of motion and the natural frequencies of the systems shown in the following figures:

(*a*) Fig. 2-1(*b*), p. 25. Assume that the mass of the torsional bar k_t is negligible.
(*b*) Fig. 2-1(*f*), p. 25. Consider the mass of the uniform rod L.
(*c*) Fig. P2-1(*a*). Assume that there is no slippage between the roller and the surface.
(*d*) Fig. P2-1(*b*). Assume that there is no slippage between the roller and the surface.
(*e*) Fig. P2-1(*c*). Assume that there is no slippage between the cord and the pulley.
(*f*) Fig. P2-1(*d*). Assume that $m_1 > m_2$.
(*g*) Fig. P2-1(*e*). The U-tube is of uniform cross section.
(*h*) Fig. P2-1(*f*). The cross-sectional areas are as indicated.

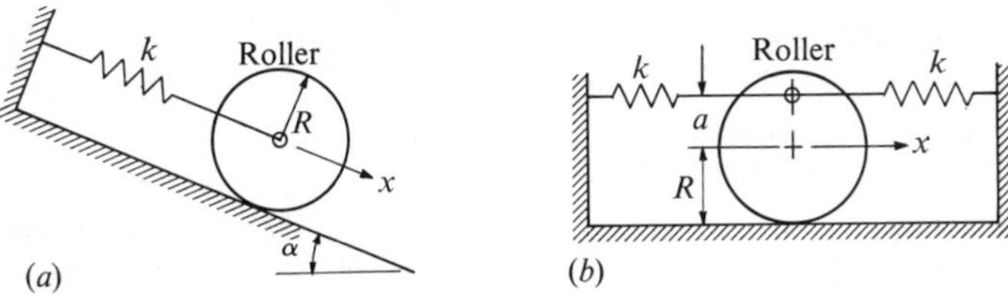

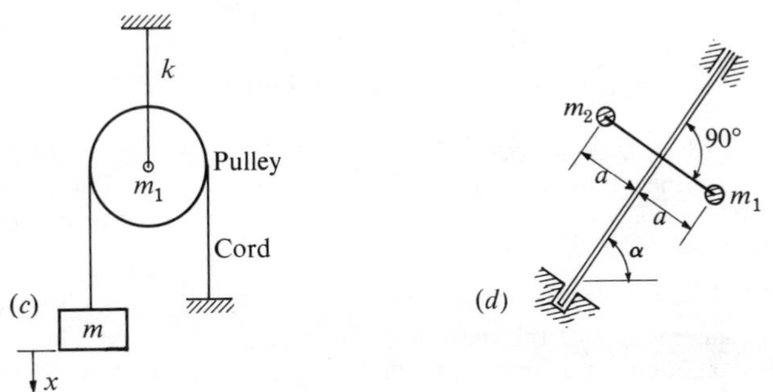

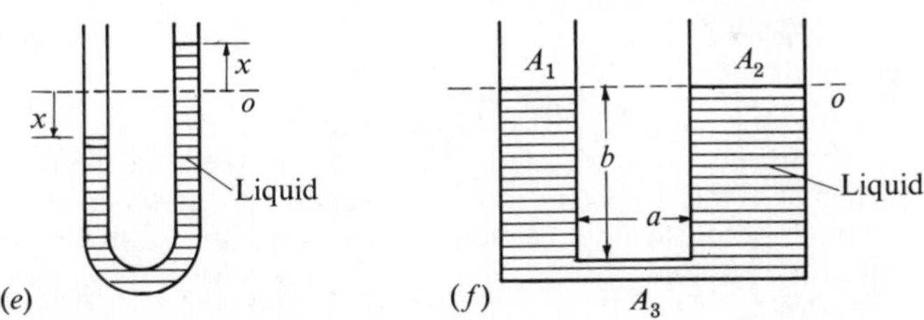

Fig. P2-1. *Vibratory systems*

2-2. A uniform cantilever beam of ρ mass per unit length is shown in Fig. P2-2. Assume that the beam deflection during vibration is the same as its deflection for a concentrated load at its free end; that is

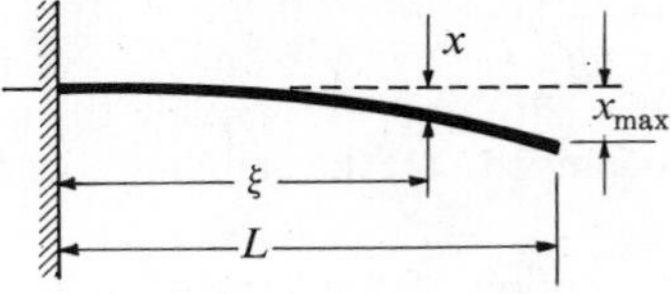

Fig. P2-2. *Cantilever*

$x = \frac{1}{2}x_{max}[3(\xi/L)^2 - (\xi/L)^3]$. Determine the natural frequency of the beam.

2-3. If a mass m is attached to the free end of the beam of Prob. 2-2, determine the natural frequency of this system.

2-4. A simply supported and uniformly loaded beam with a mass m attached at its mid-span is as shown in Fig. P2-3. If the mass of the beam and its

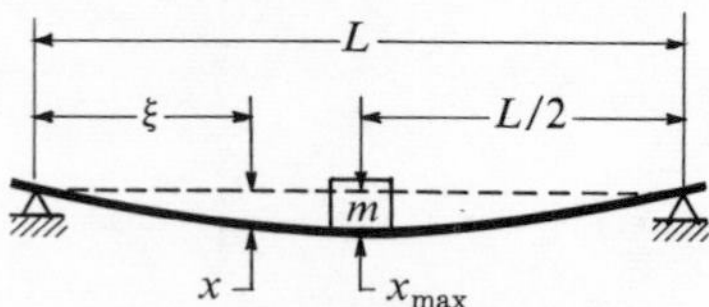

Fig. P2-3. *Simply supported beam*

uniform load is ρ mass per unit length, determine the fundamental frequency of the system. Assume that the deflection during vibration is the same as the static deflection for a concentrated load at mid-span; that is, $x = x_{max}[3(\xi/L) - 4(\xi/L)^3]$, for $0 < \xi < L/2$.

2-5. Repeat Prob. 2-1, using Newton's law of motion.

2-6. The displacement x of the mass m of the system shown in Fig. 2-1(*a*), p. 25, is considered positive in the downward direction. Use Newton's law to derive the equation of motion, considering that: (*a*) the mass m is moving upward but is below its static equilibrium position o; (*b*) m is moving upward but is above o; (*c*) m is moving downward but is above o.

2-7. The equations of motion of vibratory system are given as: (*a*) $m\ddot{x} + c\dot{x} + kx = F \sin(\omega t + \alpha)$; (*b*) $m\ddot{x} + c\dot{x} + kx = F \cos(\omega t + \beta)$. Solve the equations for the steady-state response by the method of undetermined coefficients.

2-8. A machine weighing 50 lb is mounted on springs and dashpots, as shown schematically in Fig. 2-7, p. 38. If the total stiffness of the springs is 100 lb-in.$^{-1}$ and the total damping is 2 lb-sec-in.$^{-1}$, determine the motion x if the initial conditions are: (*a*) $x(0) = 1$ in. and $\dot{x}(0) = 0$; (*b*) $x(0) = 0$ and $\dot{x}(0) = 10$ in.-sec^{-1}; (*c*) $x(0) = 1$ in. and $\dot{x}(0) =$ 10 in.-sec^{-1}.

2-9. Find the steady-state response and the transient motion of the system described in Prob. 2-8 if an excitation force of $20 \cos 35t$ is applied to the mass. Assume that the initial conditions are: (*a*) $x(0) = \dot{x}(0) = 0$; (*b*) $x(0) = 1$ in. and $\dot{x}(0) = 0$; (*c*) $x(0) = 0$ and $\dot{x}(0) = 10$ in.-sec^{-1}; (*d*) $x(0) = 1$ in. and $\dot{x}(0) = 10$ in.-sec^{-1}.

2-10. An excitation of $20 \sin(10t - 30°)$ lb is applied to the mass of a mass-spring system with $m = 38.6$ lb and $k = 40$ lb-in^{-1}, and the initial conditions are $x(0) = \dot{x}(0) = 0$. Determine the motion of the mass if: (*a*) the system is undamped; (*b*) the damping coefficient of the damper is 2 lb-sec-in.$^{-1}$.

2-11. The equation of motion of the system shown in Fig. 2-6, p. 34, is $m\ddot{x} + c\dot{x} + kx = F \sin \omega t$. Representing these forces by rotating vectors (see Fig. 2-15, p. 48), indicate the position of the vectors when: (*a*) the mass m is moving downward but is below its equilibrium position o; (*b*) m is moving upward but is below o; (*c*) m is moving upward but is above o; (*d*) m is moving downward but is above o. Show graphically that the given equation is satisfied for all the positions enumerated.

2-12. The equation of motion of a vibratory system is given as: (*a*) $m\ddot{x} + c\dot{x} + kx = F \sin(\omega t + \alpha)$; (*b*) $m\ddot{x} + c\dot{x} + kx = F \cos(\omega t + \beta)$. Find the steady-state response by the method of mechanical impedance.

2-13. Find the steady-state response of the system described in Prob. 2-8 by the impedance method if an excitation force of $20 \cos 35t$ lb is applied to the mass.

2-14. A mass-spring system with damping has $mg = 3.86$ lb, $c = 0.20$ lb-sec-in.$^{-1}$, and $k = 20$ lb-in.$^{-1}$, and a driving force of $0.75 \cos \omega t$ lb is applied to the mass. (*a*) Use the impedance method to find the steady-state amplitude X and the phase angle ϕ when the excitation frequency is 6, ω_n, and 120 rad-sec^{-1}. (*b*) Use force vector polygons to verify the answer graphically.

2-15. Derive the equations of motion for the systems shown in Fig. P2-4.

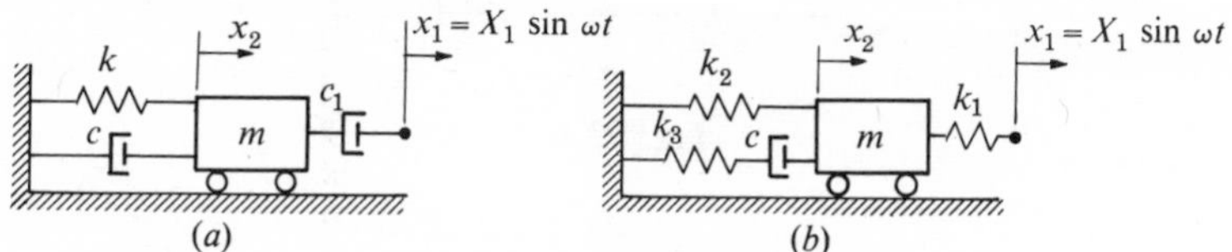

Fig. P2-4. *Vibratory systems*

Solve for the steady-state response by the impedance method.

2-16. Referring to Fig. 2-4, p. 31, if the cylinder weighs 100 lb, $R_1 = 5$ in., and $R = 20$ in., determine the natural frequency of the system.

2-17. For the simple mass-spring system shown in Fig. 2-1(*a*), p. 25, derive an expression relating the static deflection δ_{st} due to the weight of the mass m and the natural frequency f of the system. Plot $\ln f$ versus $\ln \delta_{st}$ for $0.01 < \delta_{st} < 1.0$ in.

2-18. A mass is suspended as shown in Fig. P2-5. If the beam is of negligible

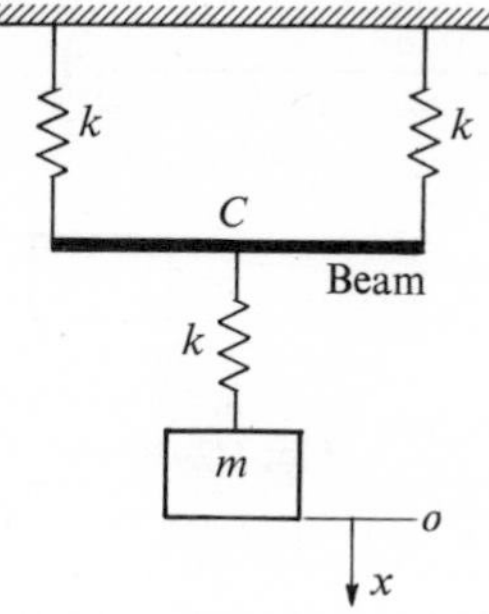

Fig. P2-5

mass, C is its midpoint, and the beam deflection is given by the equation $\delta = PL^3/48EI$, determine the natural frequency of the system.

2-19. Referring to Fig. 2-1(*a*), p. 25, the spring constant is 40 lb-in.$^{-1}$ and the mass weighs 38.6 lb. (*a*) Find the natural frequency of the system. (*b*) If the mass is attached to the midpoint of the spring and the two ends of the spring are securely fixed, determine the natural frequency of this system. (*c*) If the ends of the spring are securely fixed and the mass is attached to some intermediate point of the spring, show that the natural frequency of this system is higher than that of part (*b*).

2-20. A mass particle is supported by two springs as shown in Fig. P2-6. Assuming that the bar is rigid but is of negligible mass, determine the

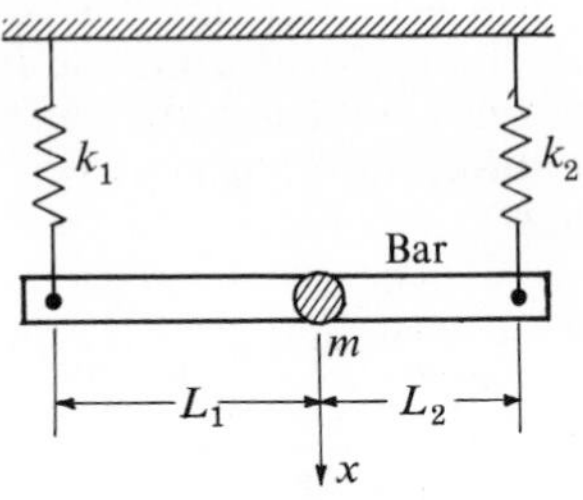

Fig. P2-6

natural frequency of the system if: (*a*) the bar is constrained to remain horizontal while the mass is oscillating vertically; (b) the bar is free to pivot at the hinge points. (*c*) Show that the natural frequency determined in part (*a*) is higher than that of part (*b*).

2-21. Determine the equation of motion of the mass m for the system shown in Fig. P2-7. Assume that the horizontal bar is rigid and is of negligible mass.

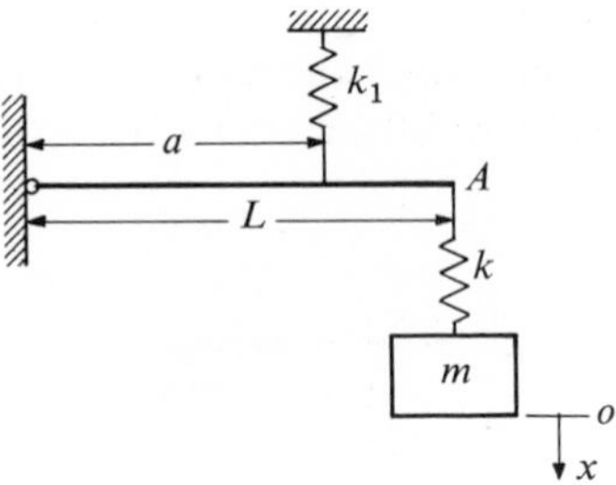

Fig. P2-7

2-22. Fig. P2-8 represents a mechanism in a machine. Determine its equation of motion. Assume that the tension in the spring k_3 is constant.

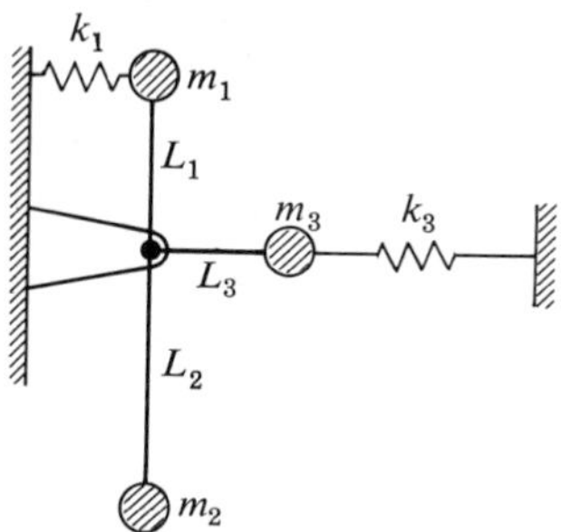

Fig. P2-8

2-23. The mass moment of inertia of a connecting rod of mass m is determined by placing it on a horizontal platform of mass m_1 and timing the periods of oscillation. The platform, as shown in Fig. P2-9, is suspended

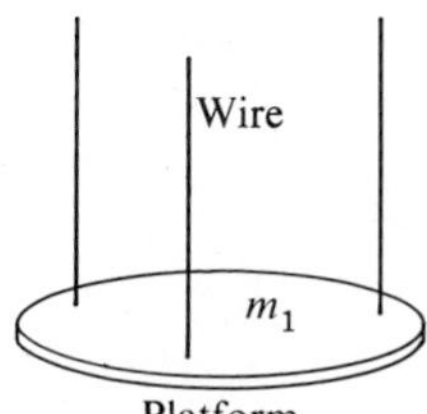

Fig. P2-9

by wires equally spaced and of equal lengths. With the platform empty and an amplitude of 6°, the period is τ_1. With the center of gravity of the connecting rod coinciding with that of the platform and an amplitude of 7°, the period is τ_2. Find the mass moment of inertia of the connecting rod.

2-24. The mass moment of inertia of a large electrical generator of mass m is to be determined by attaching a small mass m_1 at a distance R_1 from its longitudinal axis and timing the period of oscillation. The test setup is as shown in Fig. P2-10. (*a*) Determine the mass moment of inertia of

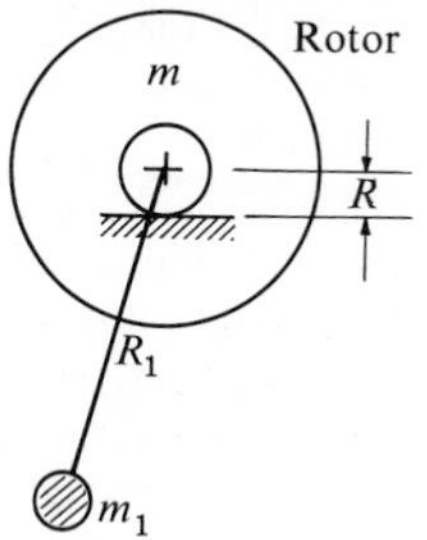

Fig. P2-10

the rotor. (*b*) Show that small variations of R will have the least effect when $R_1 = (m/m_1 + 1)R$.

2-25. Fig. P2-11 is a schematic representation of a machine member. The mass m is constrained to move only in the x-direction. Determine the:

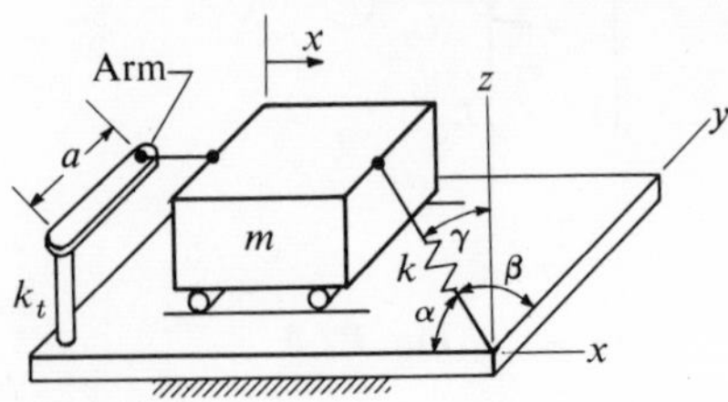

Fig. P2-11

(*a*) equation of motion; (*b*) natural frequency of the system. Assume that the mass of the arm is negligible.

2-26. A machine component is shown schematically in Fig. P2-12 in its

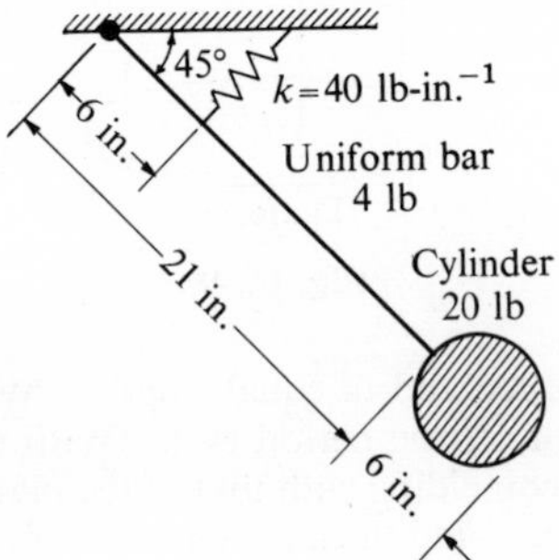

Fig. P2-12

static equilibrium position. Determine the natural frequency of this system by: (*a*) Newton's law of motion; (*b*) the energy method.

2-27. If ζ is appreciably less than unity, the logarithmic decrement δ can be approximated as $\delta = 2\pi\zeta$. (*a*) Plot Eqs. (2-62) and (2-63), p. 59, in a graph of δ versus ζ for $0 < \zeta < 1$. (*b*) From the curves in part (*a*), plot the percent error in ζ versus δ.

2-28. If a vibrating system consists of a weight of 386 lb, a spring of stiffness 400 lb-in.$^{-1}$, and a damper with damping coefficient of 4 lb-sec-in.$^{-1}$, determine the: (*a*) damping factor; (*b*) frequency of oscillation of the system.

2-29. If a vibrating system consists of a weight of 15 lb, a spring of stiffness 30 lb-in.$^{-1}$, and a damper with a damping coefficient of 0.2 lb-sec-in.$^{-1}$, find the: (*a*) damping factor; (*b*) logarithmic decrement; (*c*) ratio of any two consecutive amplitudes.

2-30. From the data given in Prob. 2-8, determine the logarithmic decrement for each of the given sets of initial conditions.

2-31. A rather complicated device can be represented by a single-degree-of-freedom torsional system. It is desired to determine: (*a*) its natural frequency; (*b*) the mass moment of inertia of the rotor; (*c*) the damping required for the system to be critically damped. It is not possible to disassemble the device, however, and the following method is used to obtain the required data: when the rotor is turned 30°, a torque of 3 in.-lb is required to maintain it in this position. When the rotor is held in this position and then released, its swings to $-5.5°$ and then to 1°; the time for each swing is 1 sec. Calculate the required information.

2-32. Derive the equations of motion for the systems shown in Fig. P2-13. Assume the bars to be rigid and weightless.

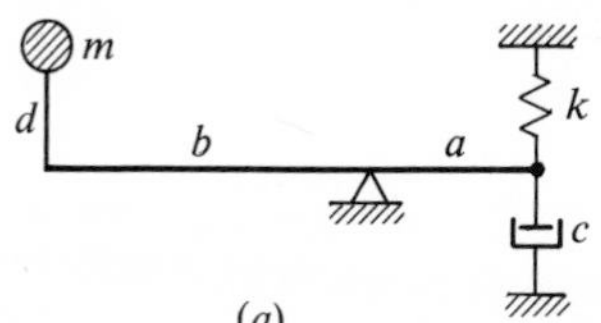

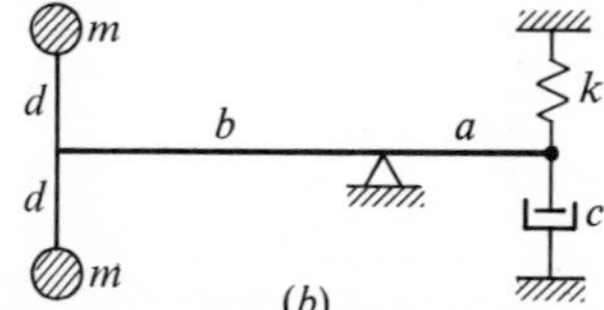

Fig. P2-13

2-33. Derive Eqs. (2-67) and (2-71), p. 62.

2-34. Determine the equation of motion for the system shown in Fig. P2-7 if an excitation force $F \sin \omega t$ is applied to: (*a*) the mass m; (*b*) the free end of the bar A.

2-35. A force $F \sin \omega t$ is applied to the mass of a simple mass-spring system as shown in Fig. 2-1(*a*), p. 25. If $\omega = (1 + \varepsilon)\omega_n$ and the system is initially relaxed, determine the motion of m.

2-36. A sinusoidal displacement $e \sin \omega t$ is applied to the fulcrum of the system as shown in Fig. P2-14. If the horizontal bar is rigid and is of

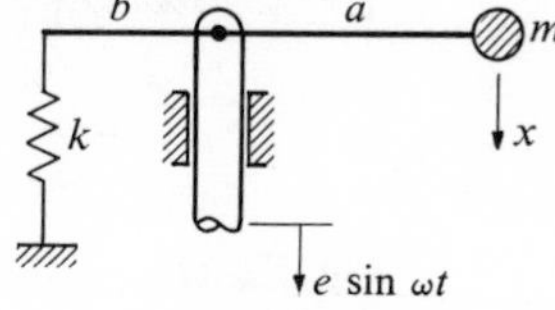

Fig. P2-14

negligible mass, derive the equation of motion for the vertical motion of the mass m.

2-37. The top end of the spring of the system shown in Fig. P2-15 is given a harmonic motion $e \cos \omega t$. Determine the equation of motion of the system.

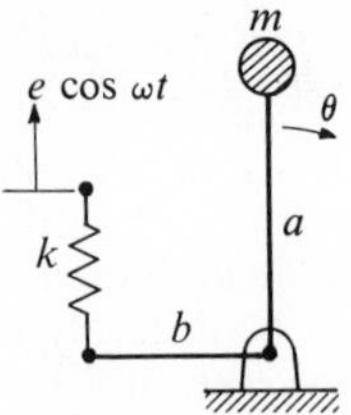

Fig. P2-15

2-38. A mass-spring system is suspended from S, as shown in Fig. P2-16. (*a*) At $t = 0$, S is given a step displacement downward of magnitude y_o. Find the motion of the mass m. (*b*) While the mass is in motion, at

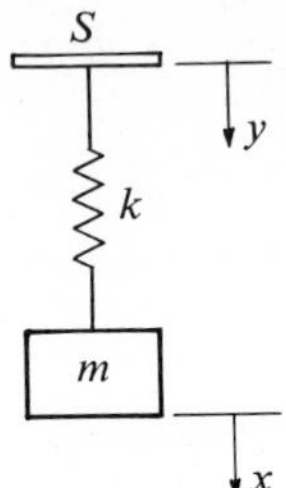

Fig. P2-16

$t = 3\pi/\omega_n$, S is given a step displacement upward of magnitude y_o; that is, S is returned to its original position. Determine the motion of m for $t > 3\pi/\omega_n$.

2-39. A pneumatic actuating mechanism is schematically shown in Fig. P2-17. Assuming that the air pressure is varied sinusoidally, determine

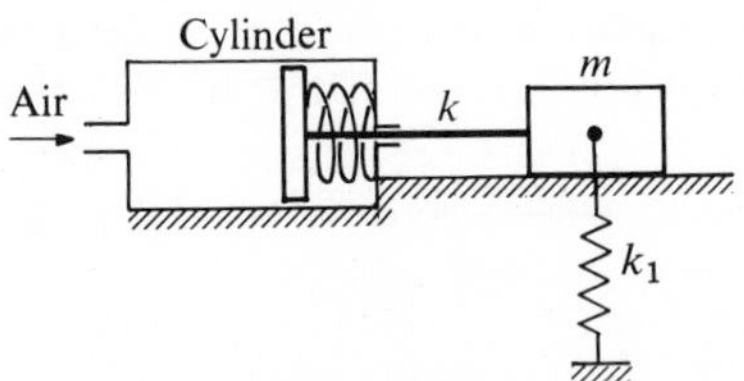

Fig. P2-17

the equation of motion of the system. State the assumptions made in deriving this equation.

2-40. Show analytically that the peak amplitudes of the κ versus r curves in Fig. 2-8, p. 41, occur at $r < 1$. Sketch a locus through the peak amplitudes.

2-41. Show analytically that the peak amplitudes of the curves in Fig. 2-27, p. 68, occur at $r > 1$. Sketch a locus through the peak amplitudes.

2-42. A table for sorting seeds requires a reciprocating motion of 0.02 in. with frequency ranging from 2 to 20 cycles per second. The excitation is provided by an eccentric weight shaker. Together, the table and the shaker weigh approximately 385 lb. (*a*) Propose a scheme for mounting this table. (*b*) Specify the spring constant, the damping coefficient, and the unbalance of the exciter.

2-43. A machine weighing 250 lb has a rotor of 50 lb with 0.20 in. eccentricity. The operating speed is 600 rpm, the machine is mounted on springs with $k = 500$ lb-in.$^{-1}$, damping is negligible, and the unit is constrained to move vertically. (*a*) Determine the dynamic amplitude of the machine. (*b*) Redesign the mounting so that the dynamic amplitude is reduced to one half of the original value, but maintain the same natural frequency.

2-44. A variable-speed counterrotating eccentric weight exciter is used to determine the natural frequency of a machine in its mounting. With the exciter at 1,000 rpm, a stroboscope shows the eccentric weights of the exciter at the top the instant the machine is moving upward through its static equilibrium position. If the total unbalance of the exciter is 10 lb-in., the exciter weighs 50 lb, the machine weighs 1,000 lb, and the displacement amplitude is 0.50 in., determine: (*a*) the natural frequency of the machine and its mounting without the exciter; (*b*) the damping factor of the system.

2-45. A rotating machine for research has an annular clearance of 0.03 in. between the rotor and the stator. The rotor weighs 80 lb with an unbalance of 4 oz-in. It is mounted at mid-span of a round shaft 1 ft in length and supported by two ball bearings. The operating speed ranges from 600 to 6,000 rpm. If the dynamic deflection of the shaft is to be less than 0.003 in., specify the size of the shaft.

2-46. A circular disk weighing 38.6 lb is mounted midway on a 0.75-in.-diameter shaft 2.50 ft in length. The center of gravity of the disk is 0.125 in. from its geometric center. The unit is rotated at 1,000 rpm and the damping factor ζ is estimated to be 0.05. (*a*) Compare the dead-load stress in the shaft with the stress at the operating speed. (*b*) Repeat part (*a*) with a 1-in.-diameter shaft.

2-47. It is proposed to use a three-cylinder two-stroke-cycle diesel engine to drive an electric generator. The operating speed is 600 rpm. The generator consists of a hollow shaft with an 8-in. outside diameter and 4-in. bore, 80 in. long, with a 2.25-ton rotor centrally mounted on the shaft. A preliminary test shows that, when the rotor is suspended horizontally with its axis 3 ft from the point of suspension, the period of lateral oscillation is $2\pi/3$ sec. If you were the consulting engineer, would you approve the proposal?

2-48. Show analytically that the cross-over point of the transmissibility curves in Fig. 2-32, p. 73, occurs at $r = \sqrt{2}$.

2-49. Fig. 2-32 shows that transmissibility $TR = 1$ for frequency ratio $r = \sqrt{2}$. Considering an undamped system and defining reduction in transmissibility as $(1 - TR)$, derive an expression relating the excitation frequency and the static deflection of the system due to its own weight with reduction in TR as parameter. Plot the resultant expression in a ln–ln plot for static deflection ranging from 0.01 to 1.0 in.

2-50. A refrigeration unit weighs 60 lb and operates at 700 rpm. The unit is supported by three equal springs. (*a*) Specify the springs if 10% of the unbalance of the unit is permitted to be transmitted to the foundation. (*b*) Verify the calculation, using the chart obtained in Prob. 2-49.

2-51. A vertical single-stage air compressor weighing 1,000 lb is mounted on springs with $k = 1{,}000$ lb-in.$^{-1}$ and dampers with $\zeta = 0.2$. The rotating parts are well balanced, and the equivalent reciprocating parts weigh 40 lb. The stroke is 8 in. Determine the dynamic amplitude of the vertical motion, the phase angle with respect to the excitation force, the transmissibility, and the force transmitted to the foundation if: (*a*) the compressor is operated at 200 rpm; (*b*) at 600 rpm.

2-52. A rotating machine weighing 100 lb is mounted as shown in Fig. P2-18. The machine has an unbalance of 5 lb-in. and operates at

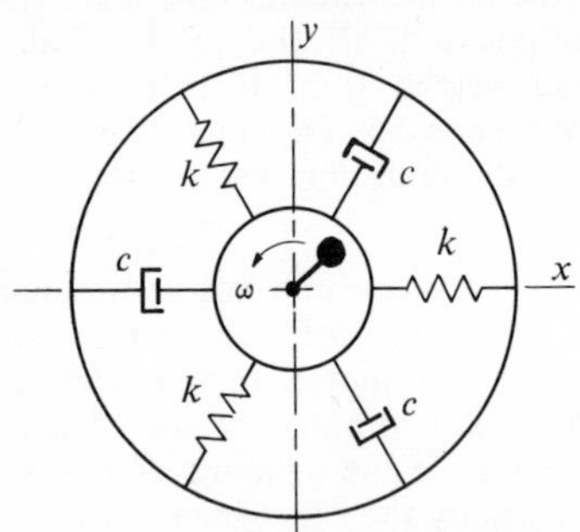

Fig. P2-18

800 rpm. If the dynamic amplitude of the machine is to be not more than 0.25 in. and it is desired to have low transmissibility, specify the springs and the dampers for the mounting.

2-53. An instrument in an aircraft weighs 50 lb. It is to be isolated from engine vibrations ranging from 1,800 to 2,400 cycles per minute. Assuming that damping is negligible, specify the springs of the mounting for 80% isolation.

2-54. A delicate instrument is mounted on a table weighing 500 lb. The table is isolated from the floor by springs and dampers. The effective stiffness of the springs is 120 lb-in.$^{-1}$ and the damping coefficient of the dampers is 20 lb-sec-in.$^{-1}$ If the floor vibrates vertically ± 0.1 in. with a frequency of 10 cycles per second, find the motion of the table.

2-55. A trailer weighs 1 ton fully loaded and 0.2 ton empty. The stiffness of the springs is 600 lb-in.$^{-1}$ and the damping is negligible. If the road surface varies sinusoidally 16 ft-cycle^{-1} with an amplitude of 4 in., find the motion of the trailer when it is traveling at: (*a*) 40 mph; (*b*) 60 mph.

2-56. Repeat Prob. 2-55 if damping is not negligible and $\zeta = 0.75$ when the trailer is fully loaded.

2-57. A vibration pickup (torsiograph) is used to measure the speed fluctuation of a rotating shaft at an average speed of 600 rpm. The pickup consists of a small hollow cylinder weighing 1 lb with a 1.5-in. radius of gyration, mounted coaxially with the shaft and connected to it by a spiral spring. Assume that viscous friction exists between the cylinder and the shaft. The frequency of the fluctuation varies from four to eight times the operating speed of the shaft. Specify the spring constant and the damping coefficient if the pickup is to measure displacement.

2-58. Repeat Prob. 2-57 if the pickup is to measure acceleration.

2-59. Fig. P2-19 shows a vibrometer for measuring the rectilinear motion $x(t)$. The pivot constrains the pendulum to oscillate in the plane of the

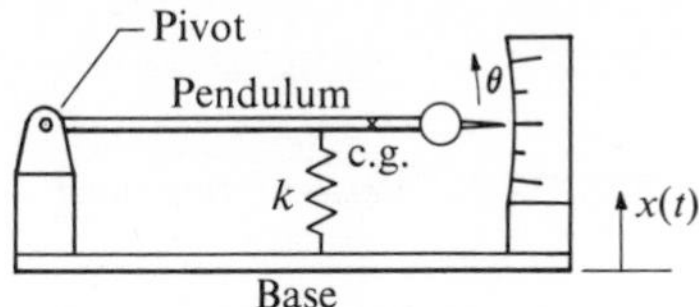

Fig. P2-19

paper, and viscous damping exists at the pivot. Derive the equation of motion of the system.

2-60. A vibrometer, as shown schematically in Fig. 2-37, p. 78, is used to measure the vibrations at the base of a variable-speed machine. The vibrations consist of two harmonic components, which are due to the primary and secondary inertia forces. It is desired to keep the amplitude distortion at not more than 4%. The operating speed of the machine ranges from 500 to 1,500 rpm. Determine the natural frequency of the vibrometer if: (*a*) damping is negligible; (*b*) the damping factor $\zeta = 0.60$.

2-61. An accelerometer with $\zeta = 0.60$ is used to measure the vibrations described in Prob. 2-60. The amplitude distortion is to be kept at not more than 4%. (*a*) From Fig. 2-40, p. 81, determine the natural frequency of the accelerometer. (*b*) If the machine is operated at 1,000 rpm, find the amplitude distortion of the harmonics. (*c*) From Fig. 2-41 determine the phase distortion of the harmonics and calculate the phase shift in units of time.

2-62. Find the Fourier series expansion of the functions shown in Fig. P2-20.

2-63. If the dynamic system shown in Fig. 2-44, p. 87, is actuated by a cam, with a profile as shown in Fig. P2-20(*c*), determine the steady-state response of the system. Assume that the mass weighs 386 lb, $k_1 = k =$ 40 lb-in.$^{-1}$, $c =$ 10 lb-sec-in.$^{-1}$, total cam lift $=$ 2 in., and cam speed $=$ 60 rpm.

2-64. Derive the transmissibility equations for the systems shown in Fig. P2-4. (See Prob. 2-15.)

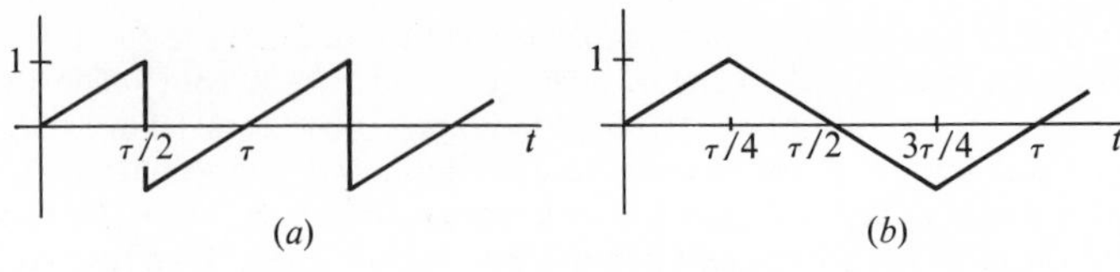

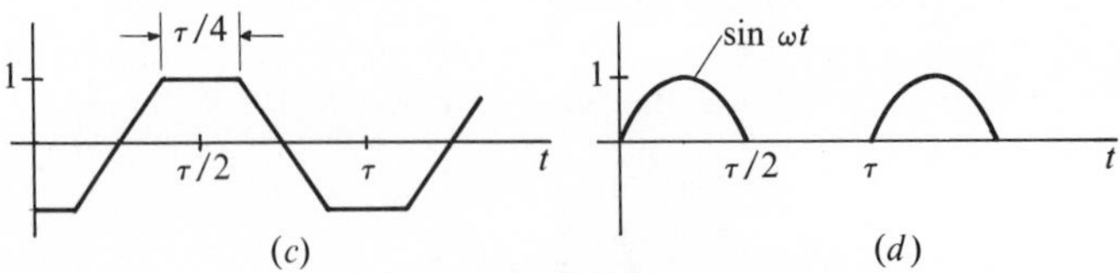

Fig. P2-20. *Periodic functions*

3-1. Find the natural frequencies of the system shown in Fig. 3-1, p. 96, for the given data:

	m_1(lb)	m_2(lb)	k_1(lb-in.$^{-1}$)	k_2(lb-in.$^{-1}$)	k(lb-in.$^{-1}$)
(*a*)	3.86	3.86	4	4	6
(*b*)	3.86	7.72	4	14	6
(*c*)	3.86	3.86	16	16	24

3-2. Find the initial conditions necessary to obtain the principal modes for a two-degree-of-freedom system.

3-3. From the data given in Prob. 3-1, find the motions $x_1(t)$ and $x_2(t)$ for the initial conditions $x_1(0) = 1$ and $\dot{x}_1(0) = x_2(0) = \dot{x}_2(0) = 0$. Plot $x_1(t)$ and $x_2(t)$ for $0 < t < 1$ sec.

3-4. Repeat Prob. 3-3 for the initial conditions $x_1(0) = \dot{x}_1(0) = \dot{x}_2(0)$ and $x_2(0) = 1$.

3-5. A uniform bar weighing 386 lb, supported by two springs, is shown in Fig. 3-6, p. 105. If $L = L_1 + L_2 = 60$ in., $k_1 = 100$ lb-in.$^{-1}$, and $k_2 = 110$ lb-in.$^{-1}$, and the initial conditions are $x(0) = 1$ in. and $\dot{x}(0) = \theta(0) = \dot{\theta}(0) = 0$, find the motions $x(t)$ and $\theta(t)$.

3-6. Determine the motions $\theta_1(t)$, $\theta_2(t)$, and $\theta_3(t)$ for the vibratory system shown in Fig. 3-8, p. 108, if $J_1 = 2J_2$, $J_2 = 2J_3$, $k_{t1} = 2k_{t2}$, and $k_{t2} = 2k_{t3}$.

3-7. The systems illustrated in Fig. P3-1 are shown in their static equilibrium positions. Specify the coordinates, write the equations of motion, and find the frequency equation for each of the systems:

(*a*) A double pendulum.
(*b*) The arm is horizontal in its static equilibrium position.

(*c*) The pendulums are identical and springs are unstressed.
(*d*) A double compound pendulum.
(*e*) The system is a schematic representation of an overhead crane.
(*f*) The system is constrained to move in the plane of the paper.
(*g*) A uniform bar is supported by a horizontal shaft. The bar is horizontal in its static equilibrium position. It can move vertically as well as rotate in a vertical plane.

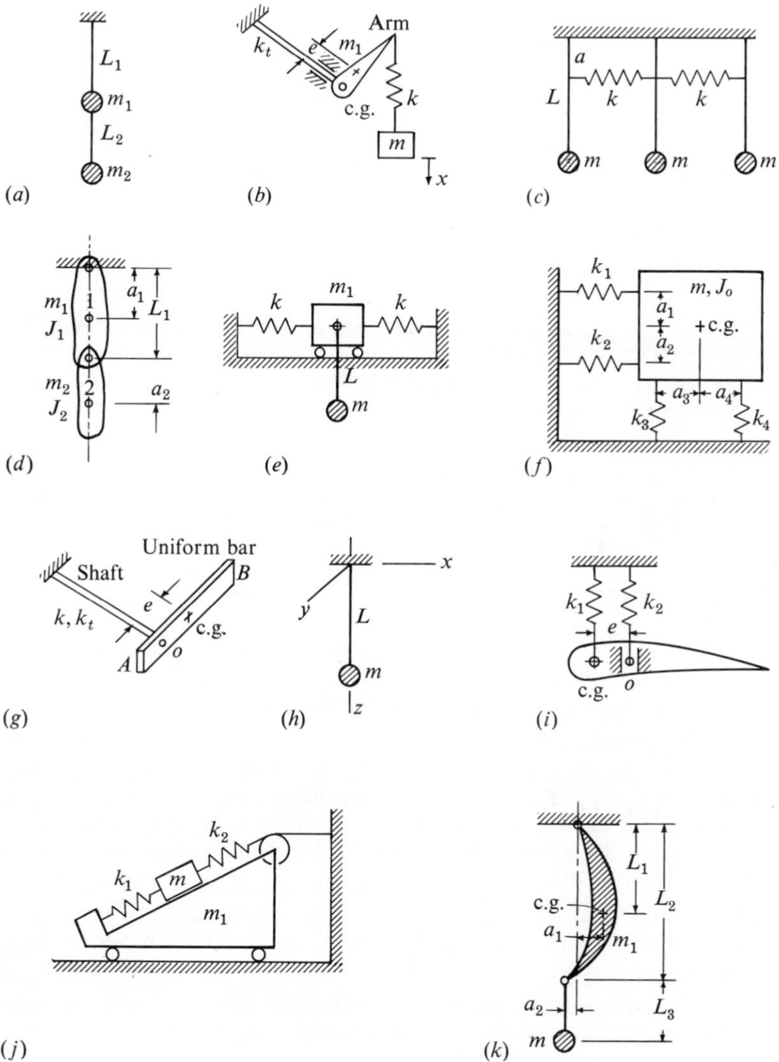

Fig. P3-1. *Vibratory systems*

(*h*) A spherical pendulum.
(*i*) The system represents an airfoil being tested in a wind tunnel.
(*j*) Assume that there is no friction between m and m_1.
(*k*) The pendulums are constrained to move in the plane of the paper.

3-8. A semidefinite system is shown in Fig. 3-11, p. 111. Determine the motions $\theta_1(t)$ and $\theta_2(t)$ for the initial conditions: (*a*) $\theta_1(0) = \theta_{10}$, $\theta_2(0) = \theta_{20}$, and $\dot{\theta}_1(0) = \dot{\theta}_2(0) = 0$; (*b*) $\theta_1(0) = \dot{\theta}_1(0) = \theta_2(0) = 0$ and $\dot{\theta}_2(0) = \dot{\theta}_{20}$.

3-9. For the semidefinite system shown in Fig. 3-12, p. 112, if $J_1 = 10$ in.-lb-sec^2, $J_2 = J_3 = 2J_1$, $k_{t1} = 2 \times 10^5$ in.-lb-rad^{-1}, and $k_{t2} = 2k_{t1}$, find the natural frequencies of the system and draw the principal mode curves.

3-10. (*a*) From the data given in Prob. 3-9, if the initial conditions are $\theta_1(0) = 0.1$ rad, $\theta_2(0) = \theta_3(0) = 0$, and the system is released with zero initial velocity, find the motions $\theta_1(t)$, $\theta_2(t)$, and $\theta_3(t)$. (*b*) Repeat the problem if the initial conditions are $\dot{\theta}_1(0) = 10$ rad-sec^{-1}, $\dot{\theta}_2(0) = \dot{\theta}_3(0) = 0$, and the system has zero initial displacement.

3-11. A geared system is shown in Fig. 3-13(*a*), p. 114. Derive the equivalent systems as shown in Fig. 3-13(*b*) and (*c*) from energy considerations.

3-12. For the geared system shown in Fig. 3-13(*a*), assuming that the inertial effect of the pinion and gear is negligible and that $J_1 = 1.0$ in.-lb-sec^2, $J_2 = 4J_1$, $k_{t1} = 2 \times 10^5$ in.-lb-rad^{-1}, $k_{t2} = 4k_{t1}$, and pinion-to-gear-speed ratio = 3:1, find the natural frequency of the system: (*a*) referring to shaft 1; (*b*) referring to shaft 2.

3-13. From the data given in Prob. 3-12, if a variable-speed prime mover having four impulses per revolution is attached to shaft 1, determine the speed of the prime mover at which the system is in resonance. Find the critical speeds of shafts 1 and 2.

3-14. Repeat Prob. 3-13, if the prime mover is attached to shaft 2.

3-15. For the geared system shown in Fig. 3-13(*a*), p. 114, assuming that $J_1 = 1.0$ in.-lb-sec^2, $J_2 = 4J_1$, $J_3 = 0.1$ in.-lb-sec^2, $J_4 = 20\,J_3$, $k_{t1} = 2 \times 10^5$ in.-lb-rad^{-1}, $k_{t2} = 4k_{t1}$, and the pinion-to-gear-speed ratio = 3:1, find the natural frequency of the system: (*a*) referring to shaft 1; (*b*) referring to shaft 2.

3-16. Referring to Fig. 3-17, p. 122, assuming that $m_1 = m_2 = m_3$, $L_1 = L_2 = L_3 = L_4$, and an excitation force $F \sin \omega t$ is applied to m_2, derive: (*a*) the frequency equation; (*b*) expressions for the steady-state response of the system. Plot the steady-state amplitude versus frequency curves similar to those shown in Fig. 3-19, p. 125.

3-17. If a horizontal excitation force $F \sin \omega t$ is applied to the mass m_1 of the system shown in Fig. P3-1(*e*), find the condition for which m_1 is stationary.

3-18. Fig. 3-21, p. 131, shows a dynamic absorber with damping. Assuming that, for optimum design, the amplitudes of X_1 are equal at the intersections of curves 1 and 2, show that the relation $k_2/k_1 = m_1 m_2/(m_1 + m_2)^2$ must be satisfied for optimum design.

3-19. An air compressor weighing 600 lb is mounted as shown in Fig. 3-15, p. 119. If the normal operating speed is 1,750 rpm and the resonant frequencies should be at least ±20% from the operating speed: (*a*) specify k_1, k_2, and m_2. (*b*) What is the amplitude of oscillation of m_2?

3-20. A torque $T \sin \omega t$ is applied to J_1 of the torsional system shown in Fig. P3-2(*a*). If $J_1 = 5$ in.-lb-sec^2, $k_{t1} = 5 \times 10^6$ in.-lb-rad^{-1}, $T =$

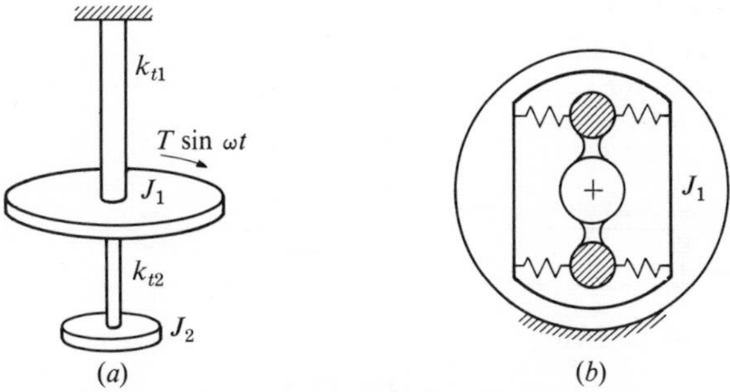

Fig. P3-2. *Torsional dynamic absorber*

2,000 in.-lb, and $\omega = 10^3$ rad-sec^{-1}, specify the size of the absorber J_2 and k_{t2} such that the resonant frequencies are 20% from the excitation frequency.

3-21. Repeat Prob. 3-20 if the configuration shown in Fig. P3-2(*b*) is used for the damper.

3-22. A company crates it products for shipping as shown in Fig. P3-3(*a*). The skid is to be securely mounted on a truck. Experience indicates that

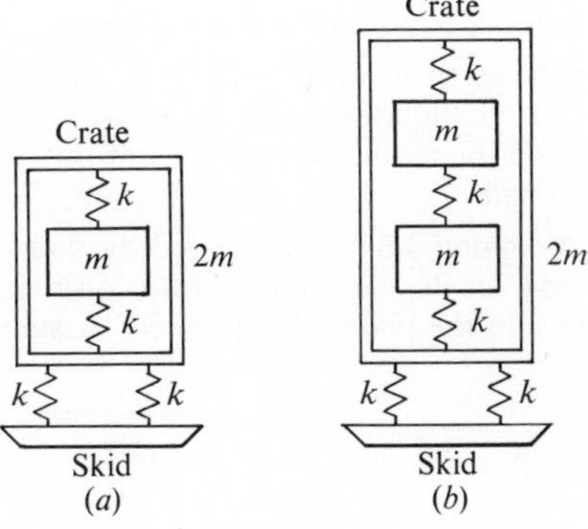

Fig. P3-3

this method of crating is satisfactory. To cut the shipping cost, it is proposed to place two items in a crate as shown in Fig. P3-3(*b*). Would you approve this proposed method of crating?

3-23. For the systems shown in Figs. 3-1, 3-6, and 3-8 (pp. 96, 105, and 108): (*a*) Find the influence coefficients. (*b*) Write the dynamic equations. (*c*) Show that the corresponding frequency equations can be reduced to that as shown in Eq. (3-6), p. 97; Example 4, p. 104; and Example 6, p. 108; respectively.

3-24. Find the influence coefficients and write the dynamic equations for the systems shown in Figs. P3-1(*a*) to P3-1(*c*).

3-25. Repeat Prob. 3-16 by the method of influence coefficients.

3-26. Find the influence coefficients necessary for writing the dynamic equations for each of the systems shown in Fig. P3-4. Assume that the beams are of negligible mass.

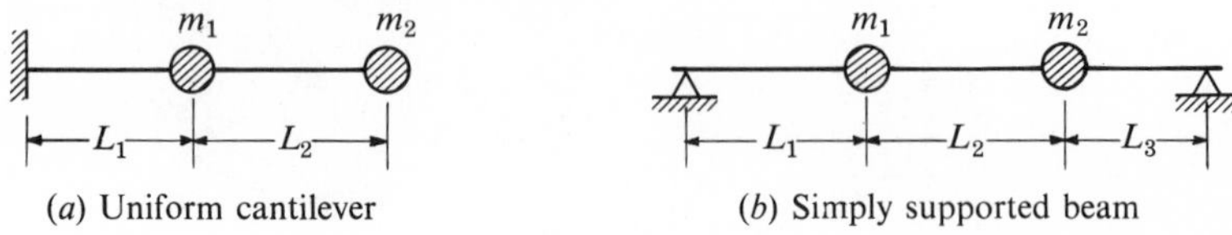

(*a*) Uniform cantilever (*b*) Simply supported beam

Fig. P3-4

3-27. A uniform bar is supported by a horizontal shaft as shown in Fig. P3-1(*g*). The bar is horizontal in its static equilibrium position. The bar can move vertically and rotate in a vertical plane. Write the equations of motion and the frequency equation of the system if the vertical displacement is measured from: (*a*) the center of gravity c.g.; (*b*) the point *o*; (*c*) the point *A*; (*d*) the point *B*. (*e*) Show that the same frequency equation is obtained for each of the cases enumerated.

3-28. For the double pendulum shown in Fig. P3-1(*a*), if $m_1 = m_2$, $L_1 = L_2$, and x_1 and x_2 are the horizontal displacements of m_1 and m_2, respectively, assuming small oscillations, use the coordinates x_1 and x_2 to set up the equations of motion, and find the natural frequencies of the system.

3-29. From the data given in Prob. 3-28, if θ_1 and θ_2 are the angular displacements of the pendulums, write the equations of motion in terms of θ_1 and θ_2 and find the natural frequencies of the system. Determine the principal coordinates.

3-30. A mass *m*, weighing 3 lb, is suspended by three springs, as shown in Fig. 3-28, p. 144, in its static equilibrium position. The mass is constrained to move in the plane of the paper. Assume that

Item	*1*	*2*	*3*
α, deg	45	120	−90
β, deg	−45	30	180
k, lb-in.$^{-1}$	2	5	7

where the angles α and β are measured counterclockwise with respect to the *x*- and *y*-axes, respectively. If $x(0) = 1$ in. and $\dot{x}(0) = y(0) = \dot{y}(0) = 0$, find the resultant motion of *m*. Find the principal coordinates of the system.

4-1. Using the Lagrangian method, determine the equations of motion for small oscillations for the systems shown in the following figures:

(*a*) Fig. P3-1(*b*).
(*b*) Fig. P3-1(*c*).
(*c*) Fig. P3-1(*d*).
(*d*) Fig. P3-1(*e*).
(*e*) Fig. P3-1(*j*).
(*f*) Fig. P3-1(*k*).
(For parts (*a*) to (*f*), see Prob. 3-7.)
(*g*) Fig. P3-2(*a*). (See Prob. 3-20.)
(*h*) Fig. 3-27(*a*) to (*c*), p. 139.

4-2. The system shown in Fig. P4-1 consists of a mass m, two rollers J_1 and J_2, and a number of springs. As shown, the system is in its static equilibrium position, and the springs k_1 and k_2 are under compression.

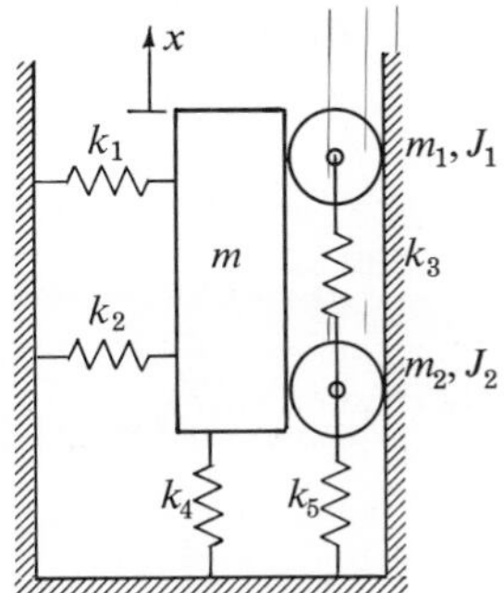

Fig. P4-1

Assuming that there is no slippage between the rollers, the mass m, and the fixed surface, find the natural frequency of the system.

4-3. A uniform bar of mass m and length L is hinged at its top o, as shown in Fig. P4-2. The system is rotated with constant angular velocity Ω,

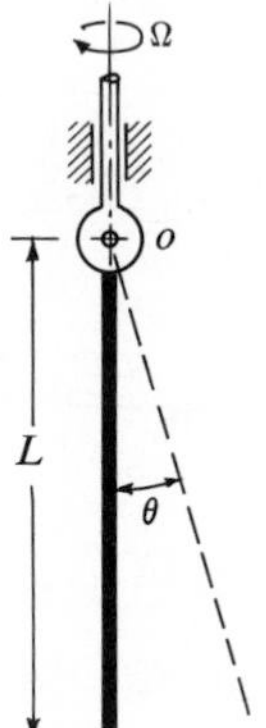

Fig. P4-2

and the bar assumes a dynamic equilibrium position θ as indicated. Find the equation of motion of the system when the bar is disturbed from equilibrium.

4-4. A pendulum is suspended from a carriage of mass m_1, as shown in Fig. P4-3. Determine the equation of motion of the system. Assume that the mass of the pendulum rod L is negligible.

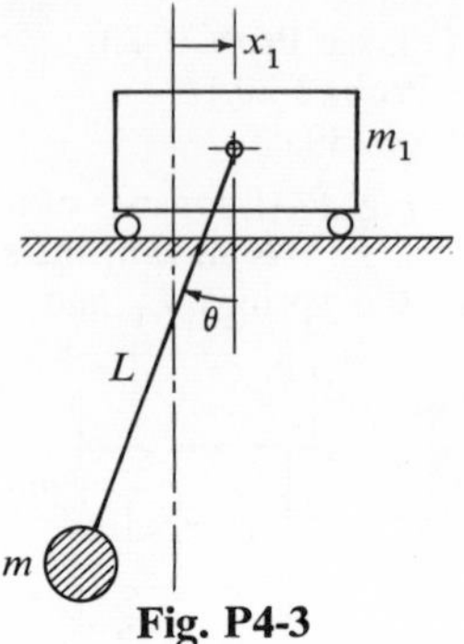

Fig. P4-3

4-5. A semidefinite system is shown in Fig. P4-4. Determine its equation of motion.

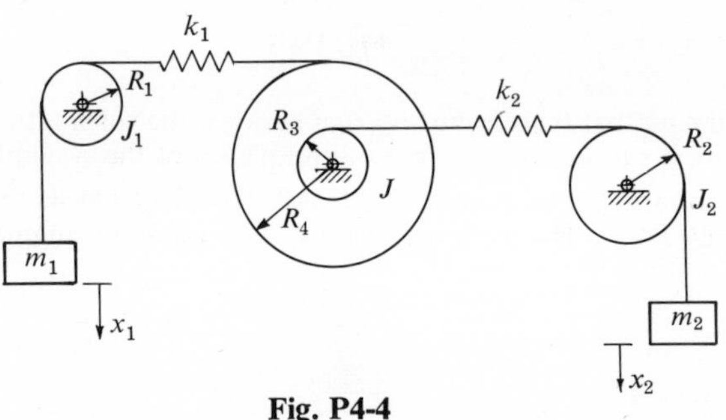

Fig. P4-4

4-6. A horizontal rotating disk of mass m_1 and mass moment of inertia J carries a mass-spring pendulum with mass m and spring k, as shown in Fig. P4-5. The mass is constrained to move radially. Assuming that the free length of the spring is a, there is no friction between m and its guides, and a torque $T_o \sin \omega t$ is applied to the disk, determine the equations of motion of the system.

4-7. For the system described in Prob. 4-6, if the disk rotates with constant angular velocity and the mass m is disturbed from its dynamic equilibrium position, find the frequency of oscillation of m.

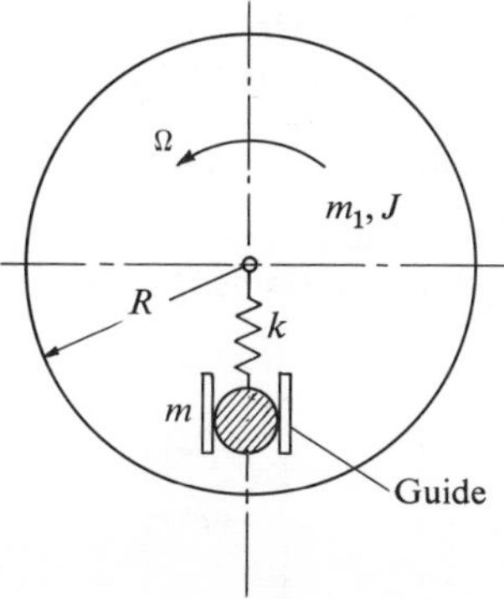

Fig. P4-5

4-8. The disk described in Prob. 4-6 is placed vertically on its edge on a horizontal surface. Thus the equilibrium position of the system is when the pendulum is vertical with m below the center of the disk. Assuming that the disk is constrained to roll in a vertical plane, find the equations of motion of the system.

4-9. The disk described in Prob. 4-6 is placed vertically on its edge on a curved surface of radius $4R$. Assume that the disk is constrained to roll on the curved surface in a vertical plane and that the equilibrium position of the system is when the pendulum is vertical with m below the center of the disk. Derive the equations of motion of the system.

4-10. A flyball governor is shown in Fig. P4-6. Assume that friction in the

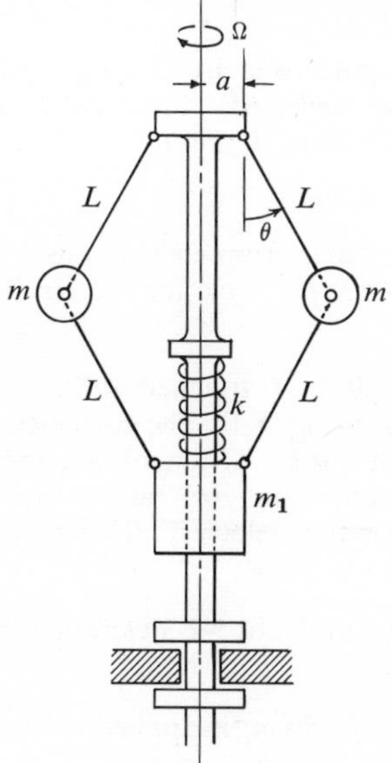

Fig. P4-6. *Flyball governor*

system is negligible and that the spring k is unstressed when the arms L are vertical. If the angular velocity Ω is constant and the corresponding dynamic position of the arms is θ_o, derive the equations for small oscillations about the dynamic equilibrium position.

4-11. Each of two uniform rods of mass m is hinged to a disk J, as shown in Fig. P4-7. The rods are constrained to move in the θ_1 and θ_2 directions,

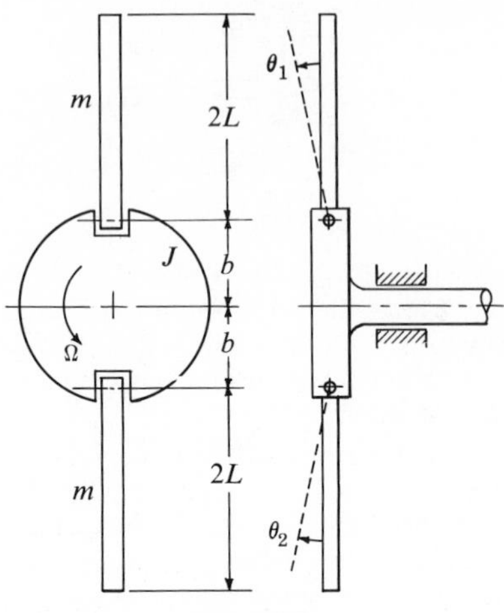

Fig. P4-7

as indicated. The angular velocity Ω is high, so the gravitational effect on the rods may be neglected. Find the natural frequencies of the system for small oscillations θ_1 and θ_2.

5-1. In Fig. 5-1, p. 176, show, from energy considerations, that the rate of amplitude decay is $4F/k$ per cycle for the free vibration with Coulomb damping.

5-2. A body weighing 20 lb is attached to a spring with a stiffness of 40 lb-in.$^{-1}$, as shown in Fig. 5-1. The coefficient of friction between the body and the support is 0.15. If the initial conditions are $x(0) = 1.0$ in. and $\dot{x}(0) = 0$, determine the: (*a*) rate of decay of the displacement amplitude; (*b*) maximum velocity; (*c*) rate of decay of the velocity amplitude.

5-3. From the data given in Prob. 5-2, determine the position at which the body stops.

5-4. If a harmonic force $F_o \sin \omega t$ is applied to the mass of the system, as shown in Fig. 5-1, use Eq. (5-3), p. 175, to show that the energy dissipation per cycle is $4FX$.

5-5. Consider a one-degree-of-freedom system with Coulomb friction in which a harmonic force $F_o \sin \omega t$ is applied to the mass. Assuming that $4F/\pi F_o < 1$, show that: (*a*) the transmissibility TR is infinite at resonance; (*b*) $TR < 1$ for the frequency ratio $r > \sqrt{2}$; (*c*) $TR > 1$ for $r < \sqrt{2}$.

5-6. Derive Eq. (5-17), p. 178.

5-7. A harmonic excitation $F_o \sin \omega t$ is applied to the mass of a one-degree-of-freedom system with velocity-squared damping. Determine the resonance amplitude from energy considerations.

5-8. It is proposed to use velocity-squared damping in a seismic instrument. (See Fig. 2-37, p. 78.) Discuss the merits of this proposal.

5-9. A machine weighing 772 lb with an eccentric weight of 38.6 lb at 4 in. radius is mounted on springs and a damper. The damper consists of a cylinder and a piston 6 in.^{2} in diameter, with a nozzle for the passage of the damping fluid. The density of the damping fluid is $\rho = 60$ lb-ft^{-3}. The natural frequency of the system is 5 cycles per second. Assuming that the equivalent viscous-damping factor $\zeta_{eq} = 0.2$ at resonance, determine the: (*a*) resonance amplitude; (*b*) diameter of the nozzle if the pressure drop across the nozzle is $p = (\rho/2)(\text{velocity})^2$, where (velocity) is that at the throat of the nozzle.

5-10. From the data given in Prob. 5-9, determine the steady-state amplitude at the frequency ratio $r = 2$. Compare the amplitude with that for viscous damping.

5-11. Determine the magnitudes and angular orientations of the required balance weights in planes 1 and 2 as shown in Fig. 5-4, p. 181. Assume that $R_1 = R_2 = 1$ in.

	Unbalance (lb-in.)			Clockwise Orientation (deg.)			Dimension (in.)					
	A	B	C	A	B	C	a_1	a_2	b_1	b_2	c_1	c_2
(*a*)	3	4	0	0	120		13	6	8	11		
(*b*)	4	7	2	0	135	195	10	4	7	7	5	9

5-12. A flexible shaft with an unbalance A of mass m_1 at a radius R_1 from the undeflected axis of the shaft is shown in Fig. 5-6(*a*), p. 182. Two equal masses at radius R_2 are placed diametrically opposite to A to balance the system. Assuming that the masses are equally spaced, find the weight of each of the balancing masses for a given operating frequency ω.

5-13. A four-bar linkage is shown in Fig. P5-1. Find the weights W_3 and W_4 that are required to eliminate the vertical and horizontal shaking

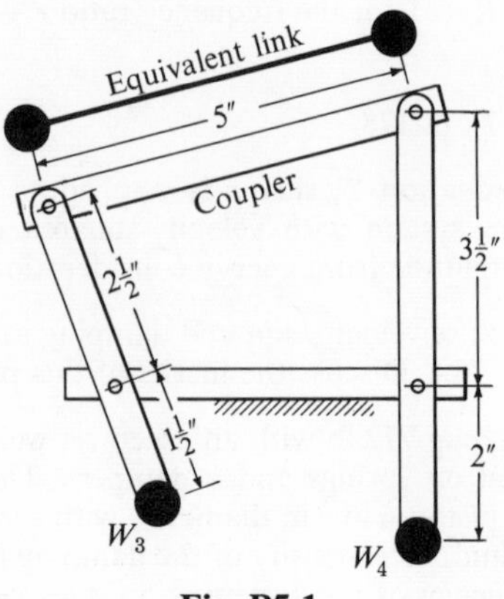

Fig. P5-1

forces. The linkage is made from 3/8- by 3/4-in. steel bars. Assume that the weight of steel is 0.28 lb-in^{-3}. Work from the center lines and neglect the small masses that are added or subtracted at the pin joints.

5-14. A single-cylinder engine is shown schematically in Fig. P5-2. Assume

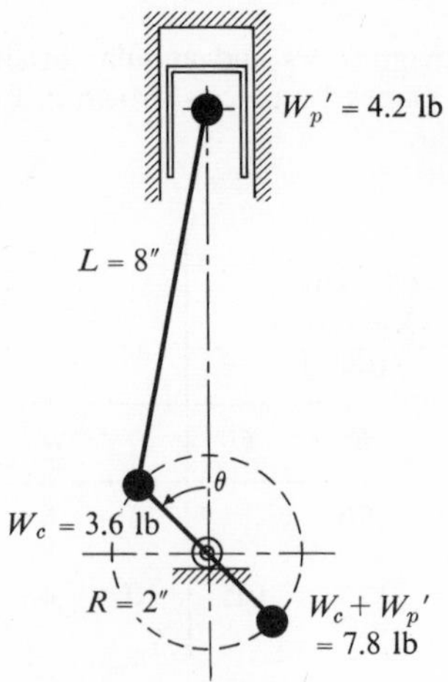

Fig. P5-2

that the counterweight $(W_c + W'_p)$ is placed at the crank radius. (*a*) Show that the vertical shaking force is minimized by this arrangement and that the maximum horizontal shaking force occurs at $\theta = 90$ and

270°. (*b*) For an engine speed of 1,800 rpm, find the vertical and horizontal shaking forces for $\theta = 0$, 90, and 180°. (*c*) Assume that the vertical and horizontal shaking forces are made equal in magnitude by changing W'_p at the counterbalance. Find the new value for W'_p. (*d*) Using W'_p determined in part (*c*), find the vertical and horizontal shaking forces for $\theta = 0$, 90, and 180°.

5-15. Show that the natural frequency obtained from Eq. (5-44), p. 191, is higher than that of the corresponding system in Fig. 5-11, p. 189, with a concentrated mass at the end of the shaft.

5-16. Derive: (*a*) Eq. (5-47), p. 192; (*b*) Eq. (5-48).

5-17. A 10-in.-diameter and 1-in.-thick solid steel rotating disk is attached to a 1-in. solid steel shaft as shown in Fig. 5-11, p. 189. Assume that $L = 8$ in. and $bL = 16$ in. (*a*) Using the curves in Fig. 5-13, p. 193, estimate the natural frequency of the system. (*b*) If the weight of the disk is unchanged but the natural frequency is to be raised to 2,800 rpm, use the curves in Fig. 5-13 to estimate the dimensions of the disk. (*c*) If the mass moment of inertia of the disk is unchanged but the natural frequency is to be raised to 2,800 rpm, use Eq. (5-44), p. 191, to find the dimensions of the disk. Assume that steel weighs 0.28 lb.-in.$^{-3}$.

5-18. In Fig. 5-11, if $b = 0$, the system becomes a fixed-end cantilever with a rotating disk. Assume that $L = 5$ in., shaft diameter = 3/4 in., disk diameter = 8 in., disk thickness = 2 in., and steel weighs 0.28 lb-in.$^{-3}$. (*a*) Estimate the natural frequency of the system. (*b*) If the weight of the disk is unchanged but the natural frequency is to be raised to 5,000 rpm, estimate the dimensions of the disk. (*c*) If the mass moment of inertia of the disk is unchanged but the natural frequency is to be raised to 5,000 rpm, find the dimensions of the disk.

5-19. A 14-in-diameter, 1.5-in.-thick solid steel disk is attached to a 1-in.-diameter steel shaft as shown in Fig. 5-12, p. 192. Assume that $L = 12$ in., $aL = 3$ in., and steel weighs 0.28 lb-in.$^{-3}$. Find the natural frequency of the system: (*a*) using the curves in Fig. 5-13; (*b*) using Eq. (5-44).

5-20. Two identical bifilar-type centrifugal pendulum absorbers are attached to a disk rotating at 900 rpm. (See Figs. 5-14 and 5-15, pp. 194 and 196.) If the magnitude of the disturbing torque $T_o = 10{,}000$ lb-in., $n = 3$, $d_2 = 1$ in., amplitude of oscillation $\Theta = 10°$, and $R + L = 8$ in., determine: (*a*) L; (*b*) the weight of each pendulum; (*c*) the minimum clearance between the pendulum and the disk.

5-21. A centrifugal pendulum absorber is shown in Fig. 5-15, p. 196. Space and balancing conditions require that the total pendulum weight be 24 lb and that $R + L = 5.5$ in. Other data are: rpm = 1,500, $n = 4$, $T_o = 2{,}400$ lb-in., and $d_2 = 1.25$ in. (*a*) Determine R, L, d_1, and Θ. (*b*) Assuming that the speed is reduced to 1,000 rpm, determine Θ.

5-22. Derive Eq. (5-56), p. 199, by expanding the determinant shown in Eq. (5-55).

5-23. Find the natural frequencies of the branched-geared system shown in Fig. 5-18(*b*), p. 201. The parameters in the lb-in.-sec units are

	J_1	J_2	J_3	J_g	k_{t1}	k_{t2}	k_{t3}
(*a*)	25	20	10	40	1×10^6	2×10^6	2×10^6
(*b*)	25	20	10	40	2.5×10^6	2×10^6	1×10^6

5-24. A branched-geared system is shown in Fig. P5-3. The given parameters

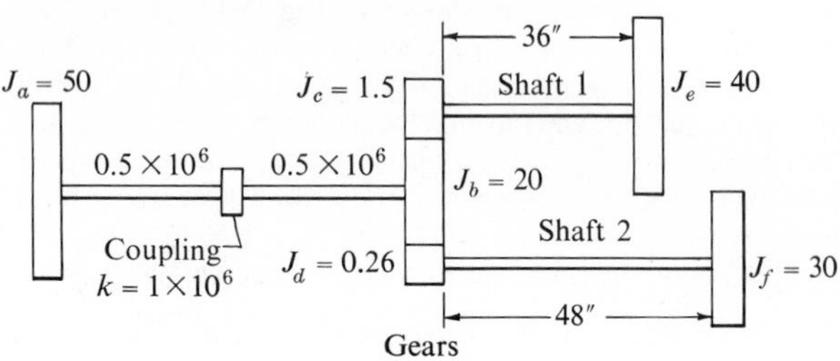

Fig. P5-3

are in the lb-in.-sec units. Assume that the inertial effect of the shafts and the coupling is negligible. The speed ratio of the gears $J_b:J_c = 1:2$ and $J_b:J_d = 1:3$. (*a*) Specify the diameters of shafts 1 and 2 for nodal drive. (*b*) Find the natural frequencies of the system.

5-25. A steel disk weighing 386 lb is mounted on a 4-in. outside-diameter and 3-in. inside-diameter steel shaft, as shown in Fig. P5-4. (*a*) Neglect-

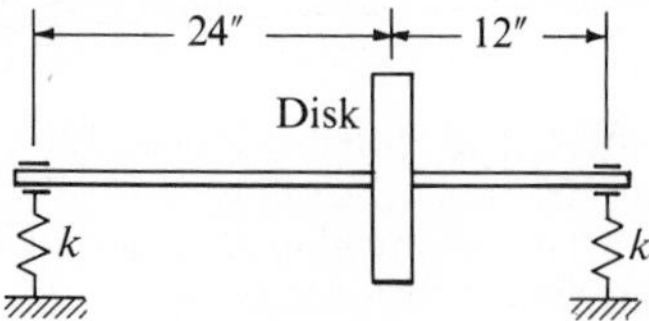

Fig. P5-4

ing the flexibility of the bearing supports, find the critical speed of the assembly. (*b*) If the bearings are flexible with a spring constant $k = 4 \times 10^5$ lb-in.$^{-1}$ in any direction normal to the shaft axis, find the change in critical speed. Assume that the mass of the shaft and the gyroscopic effect of the disk are negligible.

5-26. A steel disk weighing 300 lb is attached to a 3-in.-diameter steel shaft, as shown in Fig. P5-4. The bearing supports are flexible with a spring constant $k = 3 \times 10^5$ lb-in.$^{-1}$ in the horizontal direction normal to the shaft axis but are essentially rigid in the vertical direction. Find the critical speed of the shaft.

5-27. A shaft carrying two rotating disks is shown in Fig. P5-5. Assume that the bearing supports are rigid and that the inertial effect of the

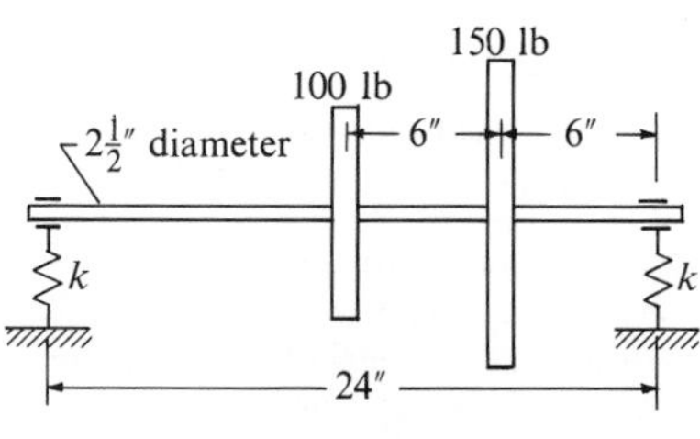

Fig. P5-5

shaft and the gyroscopic effect of the disks are negligible. Find the critical speeds of the assembly.

5-28. A shaft of bending stiffness EI, carrying three rotating disks, is shown in Fig. P5-6. Assume that the mass of the shaft and the gyroscopic

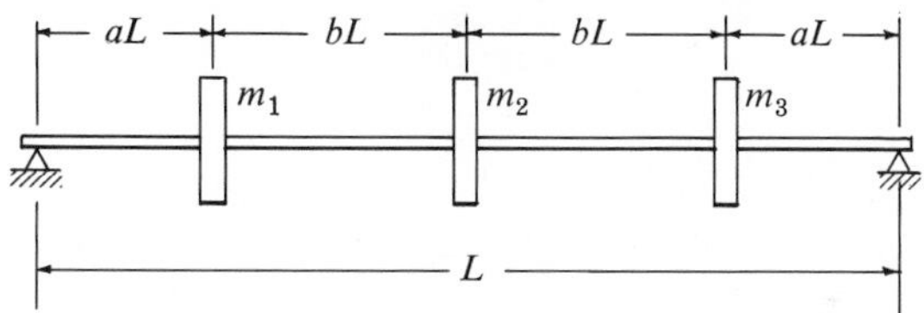

Fig. P5-6

effect of the disks are negligible. If $m_1 = m_3 = 2m_2$ and $a = b$, find the critical speeds of the assembly.

5-29. A shaft of bending stiffness EI, carrying three rotating disks, is shown in Fig. P5-6. Assume that the mass of the shaft and the gyroscopic effect of the disks are negligible. If $m_1 = m_2 = m_3$ and $b = 2a$, find the critical speeds of the assembly.

5-30. A continuous shaft of negligible mass, carrying two disks, is shown in Fig. P5-7. Determine its critical speeds.

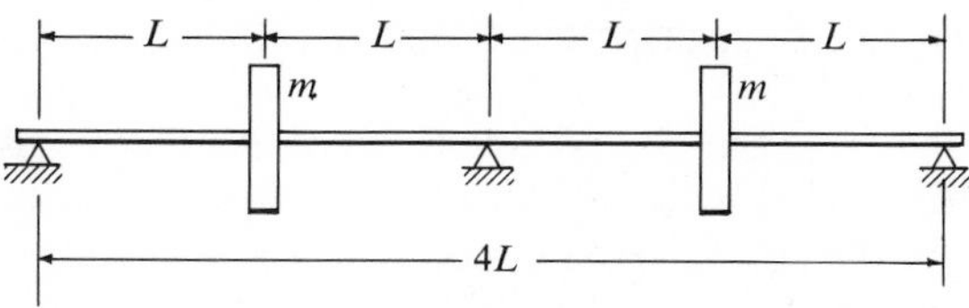

Fig. P5-7

5-31. The first critical speed of a shaft of negligible mass, carrying several disks, can be estimated from the Rayleigh equation

$$\omega = \sqrt{g(\sum W_i\delta_i / \sum W_i\delta_i^2)}$$

where W_i is the weight of the disk and δ_i is the static deflection of the shaft at location i along the shaft. Derive the Rayleigh equation for a system consisting of three disks.

5-32. Use the Rayleigh equation to estimate the first critical speed of the systems described in Probs. 5-27, 5-28, 5-29, and 5-30.

5-33. In the system described in Prob. 5-27, assume that the spring supports are flexible with a spring constant $k = 4 \times 10^5$ lb-in.$^{-1}$ in the vertical direction, but that these are essentially rigid in the horizontal direction normal to the shaft axis. Estimate the first critical speed of the assembly.

5-34. A turbogenerator set, as schematically shown in Fig. P5-8, is to be operated at 3,600 rpm. Assume that the system is symmetrical and that

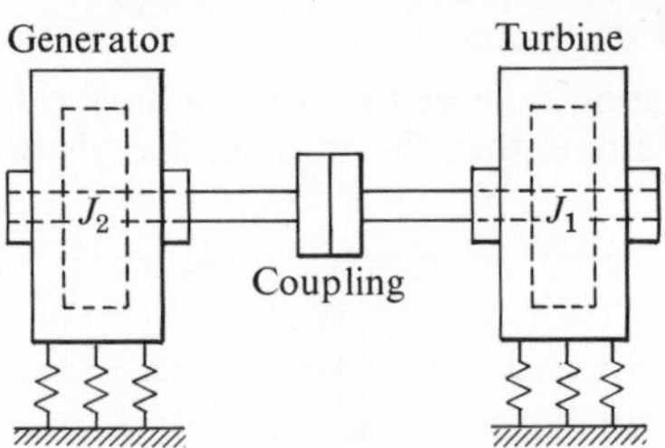

Fig. P5-8

friction is negligible. The supports are flexible in the vertical direction with $k = 2 \times 10^5$ lb-in.$^{-1}$ at each end, but they are essentially rigid in the horizontal direction. (*a*) Estimate the first critical speed of the assembly. (*b*) Find the natural frequencies for torsional vibration. The other data are:

Rotor: 100 lb each, $J_1 = J_2 = 6$ in.-lb-sec^2
Housing: 400 lb each
Coupling: 50 lb, $J = 1$ in.-lb-sec^2
Shaft: Length, 36 in.; deflection, 0.001 in. under 50 lb at midpoint

6-1. A 100-lb disk is attached 8 in. from one end of a 1-in.-diameter, 24-in.-long, steel shaft. The bearings at each end of the shaft are considered as simple supports. Determine the critical speed of the system. Neglect the weight of the shaft.

6-2. Determine the critical speed of the simply supported, hollow (1-in. outside-diameter, 0.75-in. inside-diameter), steel shaft shown in Fig. P6-1. Neglect the weight of the shaft.

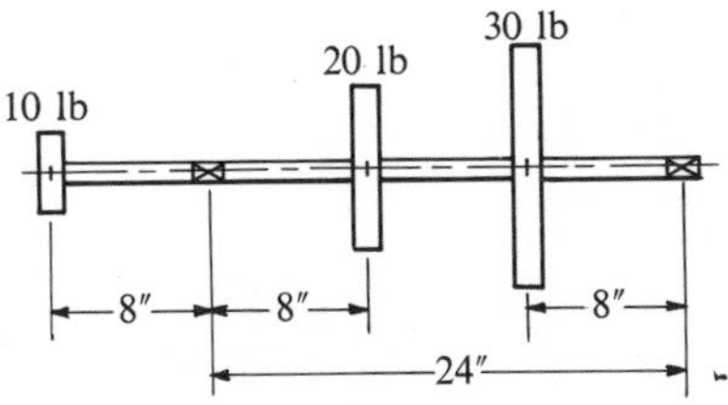

Fig. P6-1

6-3. Determine the fundamental frequency of lateral vibration for a simply supported beam of length L, uniform stiffness EI, subjected to its own uniformly distributed weight w lb per unit length. (*a*) Assume that the dynamic deflection curve is given by the following sine curve:

$$y = y_m \sin \pi x/\mathrm{L}$$

where y_m is the deflection at the center of the span. (*b*) Assume that the dynamic deflection curve is given by the static deflection curve

$$y = 3.2y_m[(x/L) - 2(x/L)^3 + (x/L)^4]$$

where $y_m = 5wL^4/384EI$. (*c*) Compare the results from parts (*a*) and (*b*), and comment on the relative accuracy of the two frequencies.

6-4. Subdivide the beam in Prob. 6-3 into three sections of equal length. Consider the weight of each section to act as a concentrated load at the midpoint of the section, and calculate the corresponding static deflection curve. Using Eq. (6-4), p. 210, calculate the fundamental frequency and compare it with the results of Prob. 6-3.

6-5. Determine the fundamental frequency of lateral vibration for a fixed-ended beam of length L, uniform stiffness EI, subjected to its own uniformly distributed weight w lb per unit length. (*a*) Assume that the deflection curve is given by the equation

$$y = y_m \sin^2 \pi x/L$$

where y_m is the deflection at the midpoint of the span. (*b*) Assume that the dynamic deflection curve is given by the static deflection curve

$$y = 16y_m[(x/L) - (x/L)^2]^2$$

where $y_m = wL^4/384EI$. (*c*) Compare the results of parts (*a*) and (*b*) and comment on the relative accuracy of the two assumed dynamic deflection curves.

6-6. Using graphical integration and Eq. (6-4), p. 210, determine the critical speed of the steel shaft and disk system of Fig. P6-2. Neglect the weight of the shaft and consider the bearings as simple supports.

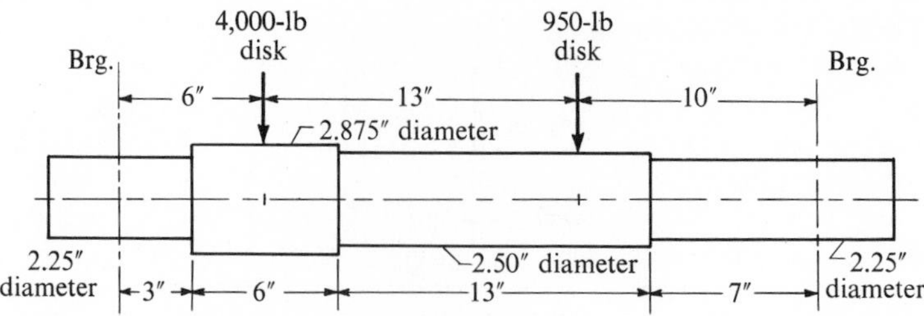

Fig. P6-2

6-7. Using Holzer's method, determine the fundamental and second mode frequency of torsional vibration for the three-disk system of Fig. 6-5, p. 219. Locate the nodes. Assume that $J_1 = J_2 = J_3 = 10$ in.-lb-sec^2 and $k_{t1} = k_{t2} = 1{,}000$ in.-lb-rad^{-1}.

6-8. Apply Holzer's tabular method to determine the fundamental frequency of the multimass rectilinear system shown in Fig. P6-3. Using the

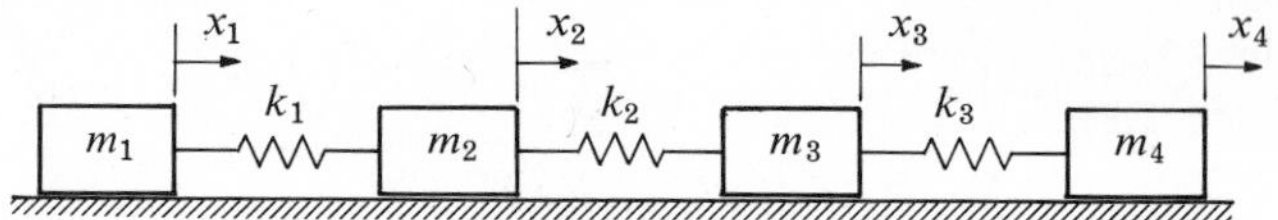

Fig. P6-3

in.-lb-sec units, the system parameters are given as $m_1 = 2$, $m_2 = 4$, $m_3 = 6$, $m_4 = 8$, and $k_1 = k_2 = k_3 = 2{,}000$.

6-9. Using Holzer's tabular method, determine the fundamental frequency of the multidisk system of Fig. P6-4. Using the in.-lb-sec units, the

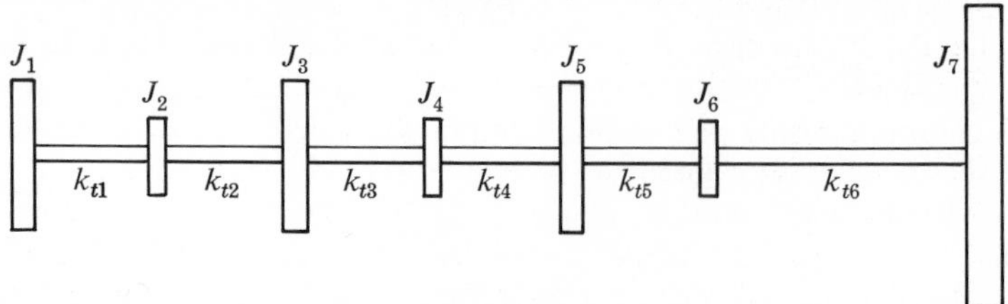

Fig. P6-4

system parameters are given as $J_1 = 3{,}000$, $J_2 = 500$, $J_3 = 3{,}000$, $J_4 = 200$, $J_5 = 4{,}000$, $J_6 = 500$, $J_7 = 40{,}000$, $k_{t1} = 600 \times 10^6$, $k_{t2} = 600 \times 10^6$, $k_{t3} = 600 \times 10^6$, $k_{t4} = 400 \times 10^6$, $k_{t5} = 500 \times 10^6$, and $k_{t6} = 10 \times 10^6$.

6-10. It is stated that the graph of the "external torque" versus ω^2, as as shown in Fig. 6-6, p. 221, is essentially a plot of the characteristic function of the system versus ω^2. Verify the statement for a three-disk–two-shaft system by: (*a*) expanding the determinant representing the characteristic function; (*b*) obtaining an expression for "external torque" versus ω^2; (*c*) comparing the two resultant expressions.

6-11. Using Holzer's method, determine the fundamental and second mode frequencies of the torsional system shown in Fig. 6-8, p. 225. Using the in.-lb-sec units, the system parameters are given as $J_1 = 4$, $J_2 = 3$, $J_3 = 2$, $k_{t1} = 2 \times 10^6$, $k_{t2} = 1 \times 10^6$, and $k_{t3} = 1.5 \times 10^6$.

6-12. Use Holzer's tabular method to determine the fundamental, second mode, and third mode frequencies for the torsional system shown in Fig. P6-5. The system parameters are given as $J_1 = J_2 = J_3 = 2$ in.-lb-

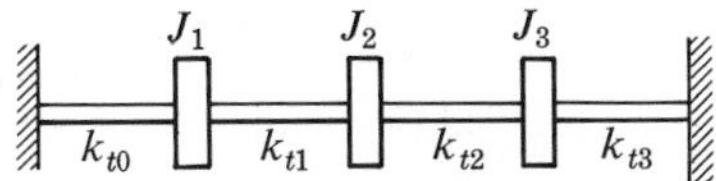

Fig. P6-5

sec^2 and $k_{t0} = k_{t1} = k_{t2} = k_{t3} = 4 \times 10^4$ in.-lb-rad^{-1}. (HINT: Assume a trial frequency. J at the fixed end is infinite, however, and the corresponding θ is zero. If $\theta_1 = 1$ radian, the torque in shaft k_{t0} is $-k_{t0}$.)

6-13. If a harmonically varying torque $T_o \sin \omega t$ is applied to the disk J_2 of the system shown in Fig. 6-5, p. 219, show that for steady-state harmonic motion of the disks: (*a*) the sum of the inertia torques $\sum J\omega^2\Theta$ plus the external torque T_o equals zero; (*b*) the amplitude Θ_3 of disk J_3 can be determined from the expression

$$\Theta_3 = \Theta_2 - (1/k_{t2})\left(\sum_{i=1}^{2} J_i\omega^2\Theta_i + T_o\right)$$

6-14. Using the results of Prob. 6-13, determine the steady-state amplitude of each disk of the system shown in Fig. 6-5, p. 219, when an external torque 1,000 sin 5t in.-lb is applied to the disk J_2. Assume that $J_1 = J_2 = J_3 = 10$ in.-lb-sec^2 and $k_{t1} = k_{t2} = 1{,}000$ in.-lb-rad^{-1}. (HINT: Let the displacement of J_1 be Θ_1. Complete the Holzer table in terms of Θ_1. Equate $\sum J\omega^2\Theta + T_o = 0$.)

6-15. From the data given in Example 4 and Fig. 6-7(*a*), p. 221, determine the steady-state amplitude of each disk when an external torque 100,000 sin 200t in.-lb is applied to disk J_2.

6-16. Use the Holzer method to find the lowest natural frequency of the six-disk branched system shown in Fig. P6-6. The system parameters in

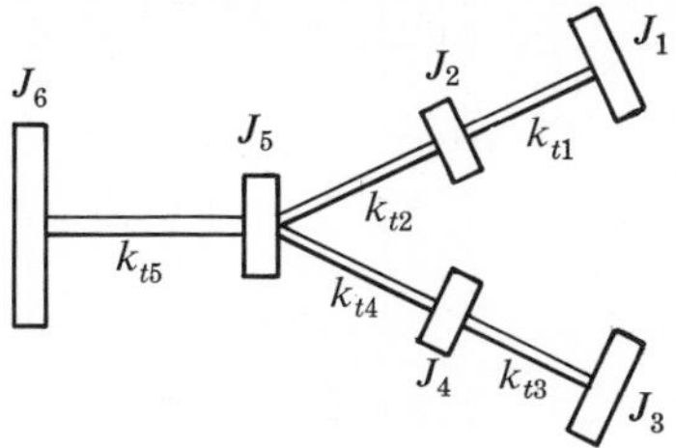

Fig. P6-6

the in.-lb-sec units are $J_1 = 5, J_2=10, J_3 = 5, J_4 = 10, J_5 = 10, J_6 = 20$, $k_{t1} = 100$, $k_{t2} = 200$, $k_{t3} = 200$, $k_{t4} = 100$, $k_{t5} = 50$. (HINT: Start at J_1 by assuming $\Theta_1 = 1$ radian and proceed to the junction disk J_5. Repeat at J_3 by assuming $\Theta_3 = 1$ radian and again proceed to J_5. The value of Θ_5 calculated from either branch must be the same. Adjust Θ_3 until Θ_5 has a unique value, and then continue along main branch to J_6.)

7-1. Evaluate the following determinants:

$$(a)\ \begin{vmatrix} 1 & -2 & 1 \\ 2 & 0 & 5 \\ 1 & 1 & 3 \end{vmatrix} \qquad (b)\ \begin{vmatrix} 1 & 4 & 2 \\ -2 & 4 & 0 \\ 1 & 12 & 5 \end{vmatrix}$$

$$(c)\ \begin{vmatrix} -4 & 3 & 1 \\ 2 & 3 & 3 \\ 0 & 6 & 5 \end{vmatrix} \qquad (d)\ \begin{vmatrix} 0 & 4 & 5 \\ -6 & 6 & 3 \\ 1 & 2 & 3 \end{vmatrix}$$

$$(e)\ \begin{vmatrix} 1 & -2 & 3 \\ 2 & 0 & -3 \\ 1 & 1 & 1 \end{vmatrix} \qquad (f)\ \begin{vmatrix} 1 & 3 & 1 \\ 4 & -3 & 3 \\ 3 & -2 & 1 \end{vmatrix}$$

7-2. Evaluate the following determinants:

$$(a)\ \begin{vmatrix} 4 & 2 & 7 & 4 \\ 1 & 2 & 3 & 5 \\ -1 & 3 & 4 & 6 \\ 4 & 6 & -5 & 3 \end{vmatrix} \qquad (b)\ \begin{vmatrix} 3 & 10 & 5 & 11 \\ 0 & 11 & 5 & 7 \\ -1 & 6 & 3 & 4 \\ 3 & 9 & 9 & -1 \end{vmatrix}$$

7-3. Find A^{-1} and B^{-1} of the given matrices A and B by: (*a*) elementary row operations; (*b*) elementary column operations: (*c*) submatrices; (*d*) the definition $A^{-1} = \text{adj. } A/|A|$.

$$A = \begin{bmatrix} 2 & 6 & 3 \\ 1 & 2 & -1 \\ 3 & 4 & 2 \end{bmatrix} \qquad B = \begin{bmatrix} 1 & -2 & 1 \\ 2 & -2 & 7 \\ -1 & 0 & -5 \end{bmatrix}$$

7-4. From the data given in Prob. 7-3, find the product AB and $(AB)^{-1}$. Verify that $(AB)(AB)^{-1} = I$.

7-5. Solve the following algebraic equations:

$$(a)\ \begin{bmatrix} 1 & 3 & 4 \\ 1 & 1 & 3 \\ 2 & 3 & 7 \end{bmatrix} \begin{bmatrix} x \\ y \\ z \end{bmatrix} = \begin{bmatrix} 6 \\ 5 \\ 10 \end{bmatrix} \qquad (b)\ \begin{bmatrix} 1 & 2 & 1 \\ 2 & 5 & 2 \\ 3 & 6 & 4 \end{bmatrix} \begin{bmatrix} x \\ y \\ z \end{bmatrix} = \begin{bmatrix} 1 \\ 4 \\ 5 \end{bmatrix}$$

$$(c)\ \begin{bmatrix} 1 & 3 & 4 \\ 1 & 1 & 2 \\ 3 & 4 & 7 \end{bmatrix} \begin{bmatrix} x \\ y \\ z \end{bmatrix} = \begin{bmatrix} 6 \\ 4 \\ 13 \end{bmatrix} \qquad (d)\ \begin{bmatrix} 1 & 3 & 4 \\ 4 & 6 & 1 \\ 3 & 5 & 2 \end{bmatrix} \begin{bmatrix} x \\ y \\ z \end{bmatrix} = \begin{bmatrix} 4 \\ 7 \\ 6 \end{bmatrix}$$

$$(e)\ \begin{bmatrix} 1 & 3 & 4 \\ 1 & 1 & 2 \\ 3 & 4 & 7 \end{bmatrix} \begin{bmatrix} x \\ y \\ z \end{bmatrix} = \begin{bmatrix} 6 \\ 4 \\ 10 \end{bmatrix} \qquad (f)\ \begin{bmatrix} 1 & -2 & 4 \\ 4 & -8 & 16 \\ 3 & -6 & 12 \end{bmatrix} \begin{bmatrix} x \\ y \\ z \end{bmatrix} = \begin{bmatrix} 3 \\ 12 \\ 9 \end{bmatrix}$$

7-6. Determine the modal column matrices $[\mu]$ of

(*a*) $A = \begin{bmatrix} 8 & 0 & -2 \\ 1 & 3 & -1 \\ 5 & -1 & 1 \end{bmatrix}$ (*b*) $A = \begin{bmatrix} 5 & -1 & -1 \\ 1 & 1 & 1 \\ 0 & -2 & 6 \end{bmatrix}$ (*c*) $A = \begin{bmatrix} 2 & 0 & 0 \\ -4 & 4 & 0 \\ -2 & 1 & 2 \end{bmatrix}$

7-7. Find $[\mu]^{-1}$ for each of the matrices given in Prob. 7-6. Verify the relations: (*a*) $[\mu]^{-1}A[\mu] = \Lambda$; (*b*) $[\mu]\Lambda[\mu]^{-1} = A$.

7-8. The equation of motion in matrix notation is given as

$$M\{\ddot{q}\} + K\{q\} = 0.$$

Derive this equation for a three-degree-of-freedom system by substituting $T = \frac{1}{2}[\dot{q}]M\{\dot{q}\}$ and $U = \frac{1}{2}[q]K\{q\}$ in Lagrange's equation $(d/dt)(\partial T/\partial \dot{q}_j) + \partial U/\partial q_j = 0$.

7-9. In the vibratory system shown in Fig. 7-3, p. 243, if

	m_1	m_2	k_1	k	k_2
(*a*)	4	1	20	4	2
(*b*)	4	2	4	4	2
(*c*)	3	1	6	3	2

determine the motions x_1 and x_2 and the initial conditions necessary for the principle modes.

7-10. In the vibratory system shown in Fig. 3-27(*a*), p. 139, if $m = 1$, $J_{c.g.} = 1{,}000$, $k_1 = 60$, $k_2 = 40$, $L_1 = 60$, and $L_2 = 40$: (*a*) Write the equations of motion in the form $M\{\ddot{q}\} + K\{q\} = 0$. (*b*) Find the dynamic matrix $H = M^{-1}K$. (*c*) Determine the principal modes.

7-11. Repeat Prob. 7-10 for the system shown in Fig. 3-27(*b*).

7-12. Repeat Prob. 7-10 for the system shown in Fig. 3-27(*c*).

7-13. Determine the natural frequencies and the modal column matrix of the system shown in Fig. P7-1.

	m_1	m_2	m_3	k_1	k_2	k_3	k_4	k_5	k_6
(*a*)	1	1	1	2	4	6	4	8	2
(*b*)	2	6	4	2	2	2	4	6	4

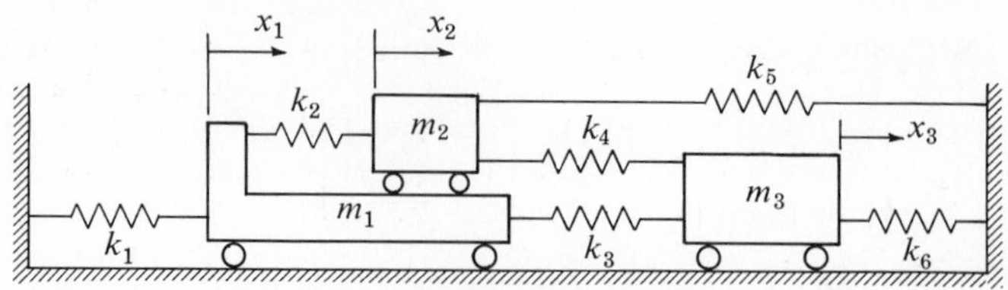

Fig. P7-1

7-14. Determine the natural frequencies and the motions θ_1, θ_2, and θ_3 of the system shown in Fig. P6-5. $J_1 = 6, J_2 = 6, J_3 = 9, k_{t0} = 6, k_{t1} = 18$, $k_{t2} = 18$, and $k_{t3} = 18$.

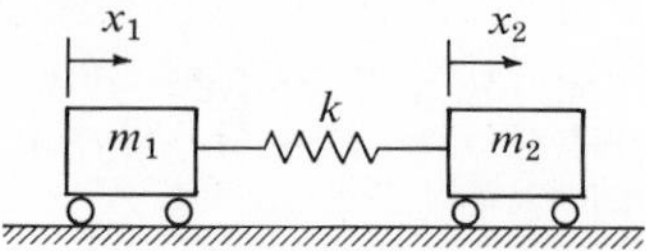

Fig. P7-2

7-15. A semidefinite system is shown in Fig. P7-2. If the initial conditions are $x_1(0) = x_{10}$, $\dot{x}_1(0) = \dot{x}_{10}$, $x_2(0) = x_{20}$, and $\dot{x}_2(0) = \dot{x}_{20}$, find the motions x_1 and x_2.

7-16. In the semidefinite system shown in Fig. 7-4, p. 246, if $J_1 = J_3 = 8$, $J_2 = 2$, and $k_{t1} = k_{t2} = 8$, find the motions q_1, q_2, and q_3 for the initial conditions

	$q_1(0)$	$q_2(0)$	$q_3(0)$	$\dot{q}_1(0)$	$\dot{q}_2(0)$	$\dot{q}_3(0)$
(*a*)	2	1	0	0	0	0
(*b*)	9	0	9	0	0	0
(*c*)	0	8	−2	0	0	0
(*d*)	−4	−8	6	0	0	0
(*e*)	1	0	0	0	0	0
(*f*)	1	0	0	1	0	0

7-17. Find the principal modes of the semidefinite system shown in Fig. 7-4 if

	J_1	J_2	J_3	k_{t1}	k_{t2}
(*a*)	2	5	8	54	36
(*b*)	4	4	16	24	32

7-18. A mass weighing 3.86 lb and suspended by three springs is shown in Fig. 3-28, p. 144, in its static equilibrium position. The mass is constrained to move in the plane of the paper. If $k_1 = k_2 = k_3 = 4$ lb-in.$^{-1}$, $\alpha_1 = 45°$, and the system is symmetrical about the axis of spring k_1, determine the principal modes of vibration.

7-19. Show that $[\delta_{ij}] = [k_{ij}]^{-1}$ for the systems shown in: (*a*) Fig. 3-1, p. 96; (*b*) Fig. 3-6, p. 105; (*c*) Fig. 3-8, p. 108.

7-20. Show that $[\delta_{ij}] = [k_{ij}]^{-1}$ for the system shown in Fig. P3-4(*a*). Assume that $L_1 = L_2$.

7-21. A double pendulum is as shown in Fig. P3-1(*a*). If $L_1 = L_2$ and $m_1 = m_2$, use the angular displacements to describe the configuration of the system, and find the natural frequencies and principal modes by the matrix-iteration method.

7-22. Use the matrix-iteration method to find the natural frequencies and the principal modes of the system shown in Fig. 7-3, p. 243. Assume that $m_1 = 4$, $m_2 = 1$, $k_1 = 20$, $k_2 = 4$, and $k_3 = 2$. [See Prob. 7-9(*a*).]

7-23. Use the matrix-iteration method to find the natural frequencies and the principal modes of the systems shown in Fig. 3-27(*a*), (*b*), and (*c*), p. 139. Assume that $m = 1$, $J_{c.g.} = 1{,}000$, $k_1 = 60$, $k_2 = 40$, $L_1 = 60$, and $L_2 = 40$. (See Probs. 7-10 to 7-12.)

7-24. Repeat Prob. 7-14 by the matrix-iteration method.

7-25. If a force of 8 lb is suddenly applied to the mass m_1 of the systems described in Prob. 7-9(*a*), find the particular integrals x_1 and x_2 by the methods discussed in this chapter.

8-1. Find the $\mathscr{L}$-transforms of the following functions by direct integration.

(*a*) t^2
(*b*) $\cos at$
(*c*) $\sinh at$
(*d*) $e^{-bt} \sin at$
(*e*) $e^{-bt} \cos at$
(*f*) $(1 - at)e^{-at}$

8-2. In the system shown in Fig. P7-1, if

	m_1	m_2	m_3	k_1	k_2	k_3	k_4	k_5	k_6
(*a*)	1	1	1	2	4	6	4	8	2
(*b*)	2	6	4	2	2	2	4	6	4

and m_1 is displaced by an amount x_{10} and the system is released with zero initial velocity, find $X_1(s)$, $X_2(s)$, and $X_3(s)$.

8-3. If an excitation torque $T \sin \omega t$ is applied to J_2 of the system shown in Fig. P6-5 and the initial conditions are zero, find $\Theta_1(s)$, $\Theta_2(s)$, and $\Theta_3(s)$. Assume that $J_1 = 6$, $J_2 = 6$, $J_3 = 9$, $k_{t0} = 6$, $k_{t1} = 18$, $k_{t2} = 18$, and $k_{t3} = 18$.

8-4. Find the inverse transform of the following functions:

(a) $\dfrac{s+1}{s^2+s-12}$ (b) $\dfrac{1}{s(s^2+5s+6)}$

(c) $\dfrac{10s+24}{s^2+4s+8}$ (d) $\dfrac{17s^2+14s+85}{s(s^2+2s+17)}$

(e) $\dfrac{8}{s(s+2)^2}$ (f) $\dfrac{3s^3+10s^2+26s+30}{(s^2+4)(s^2+2s+10)}$

8-5. In the system shown in Fig. 7-3, p. 243, if $m_1 = 4$, $m_2 = 1$, $k_1 = 20$, $k = 4$, and $k_2 = 2$, and the initial conditions are

	$x_1(0)$	$x_2(0)$	$\dot{x}_1(0)$	$\dot{x}_2(0)$
(a)	1	2	0	0
(b)	0	0	1	−2

find the motions x_1 and x_2

8-6. If an excitation force 2 sin 5t is applied to the mass m_1 of the system described in Prob. 8-5 and the system is initially at rest, find x_1 and x_2.

8-7. Repeat Prob. 8-6 if the excitation force is 2 sin 2t.

8-8. An excitation 20 sin (10t − 30°) is applied to a mass weighing 38.6 lb. The mass is supported by a spring with $k = 40$ lb-in.$^{-1}$ in parallel with a damper with $c = 2$ lb-sec-in.$^{-1}$. Find the motion of the mass. Assume that the mass is initially at rest.

8-9. An excitation pulse of duration a is applied to a mass-spring system shown in Fig. P8-1. The wave form of the pulse is shown in Fig. P8-1(b).

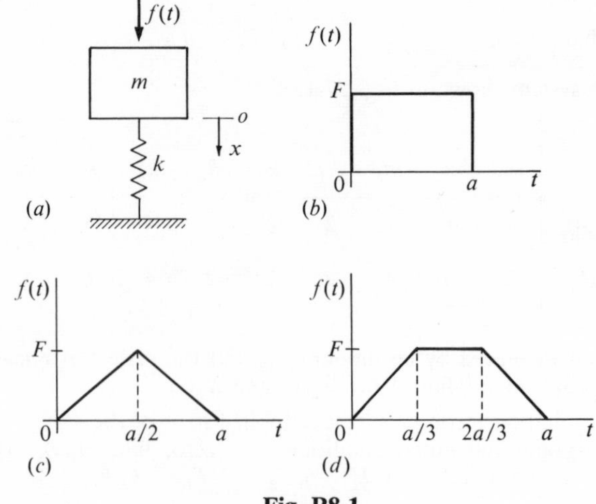

Fig. P8-1

Find the response x for $t > a$ if the pulse duration is equal to (*a*) $\pi/2\omega$, and (*b*) $2\pi/\omega$, where $\omega = \sqrt{k/m}$ is the natural frequency of the system.

8-10. Repeat Prob. 8-9 for the pulse as shown in Fig. P8-1(*c*).

8-11. Repeat Prob. 8-9 for the pulse as shown in Fig. P8-1(*d*).

8-12. Fig. P8-2 shows a mass-spring system, one end of which is actuated

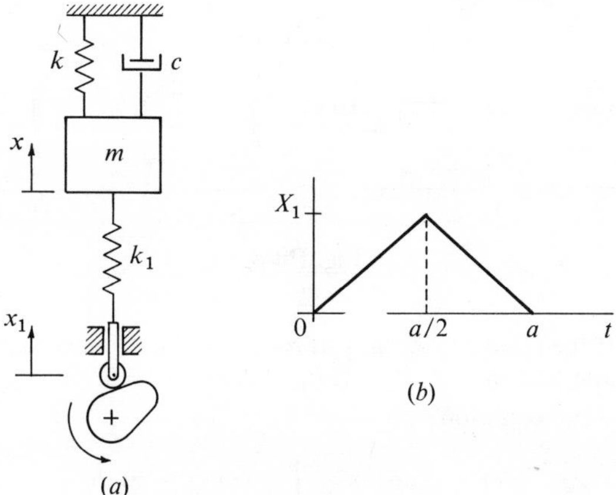

Fig. P8-2

by a cam. The idealized cam motion is shown in Fig. P8-2(*b*). Find the response x for one rotation of the cam.

8-13. A body m, mounted as shown in Fig. P8-3, is dropped on a hard floor.

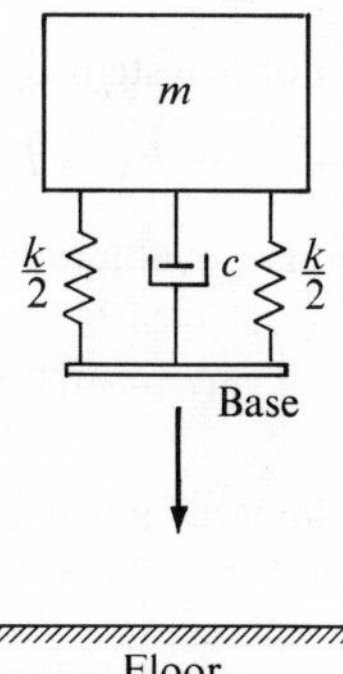

Fig. P8-3

When the base first contacts the floor, the spring is unstressed and the body has dropped through a height 5 ft. Find the acceleration of m. If $m = 38.6$ lb, $k = 10$ lb-in.$^{-1}$, and $c = 0.5$ lb-sec-in.$^{-1}$, determine the maximum acceleration of m.

8-14. Fig. P8-2 shows a mass-spring system, one end of which is actuated by a cam. The idealized cam motion is shown in Fig. P8-2(*b*). Neglecting the effect of the damper c, find the displacement of the mass if at $t = 0$ the cam is given a constant velocity ω_i, where $\omega_i = \omega/4$ and $\omega = \sqrt{(k_1 + k)/m}$. Will the system be in resonance?

8-15. A mass-spring system, as shown in Fig. P8-4, is excited by a series of

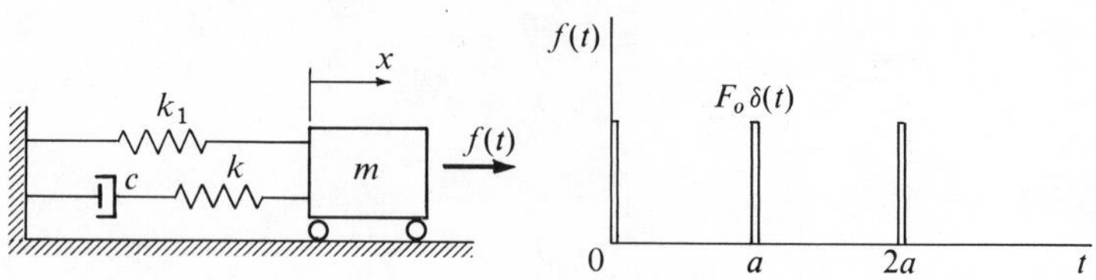

Fig. P8-4

blows. If the system is initially at rest, find the displacement of the mass m. Assume that $m = 1$, $k = 1$, $k_1 = 2$, and $c = 0.75$.

8-16. Derive the equation

$$h(t) = f(0)c(t) + \int_0^t f'(\tau)c(t - \tau)\, d\tau$$

where $c(t)$ is the step response of a system and $h(t)$ is the response to an arbitrary excitation $f(t)$.

8-17. Show that $h(t)$ given in Prob. 8-16 can be expressed in the form

$$h(t) = \frac{d}{dt}\int_0^t f(\tau)c(t - \tau)\, d\tau$$

8-18. The equation of motion of a system is given as

$$m\ddot{x} + c\dot{x} + kx = F_o u(t)$$

If $g(t)$ is the impulse response, evaluate $x(t)$ by the formula

$$x(t) = \int_0^t g(t - \tau)F_o u(\tau)\, d\tau$$

8-19. Use the superposition integral given in Prob. 8-16 to evaluate $x(t)$ when the equation of motion of the system is $m\ddot{x} + c\dot{x} + kx = Ft$.

8-20. Use the formula $x(t) = \int_0^t g(\tau)f(t - \tau)\, d\tau$ to evaluate $x(t)$ when the equation of motion of the system is $m\ddot{x} + c\dot{x} + kx = F\cos\omega t$.

8-21. Use the convolution integral $\int_0^t f_1(\tau)f_2(t - \tau)\, d\tau$ to evaluate

$$f_1(t)*f_2(t) = t*e^{-at}$$

graphically.

8-22. Find the initial displacement $x(0+)$ and the initial velocity $\dot{x}(0+)$ of $x(t)$ specified by

$$\mathscr{L}x(t) = X(s) = \frac{(s+3)(3s+4)}{(s+1)(s+2)^2}$$

8-23. Determine the final value $x(\infty)$ of $x(t)$ as specified in Prob. 8-22.

8-24. The switch S in the network shown in Fig. P8-5 is closed at $t = 0$.

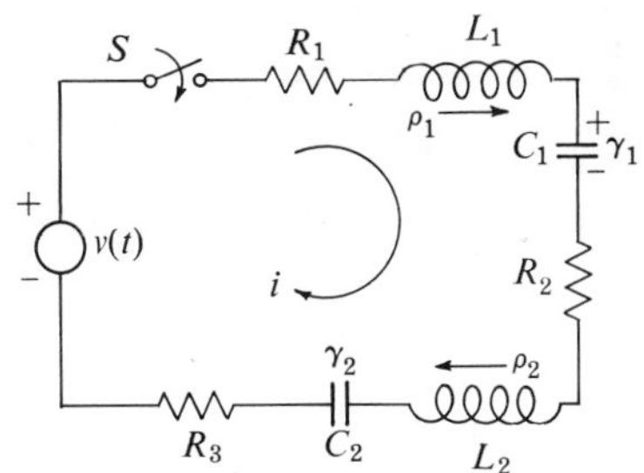

Fig. P8-5

The initial conditions are as indicated. Set up the loop equation and the subsidiary equation of the network.

8-25. The switch S in the network shown in Fig. P8-6 is closed at $t = 0$.

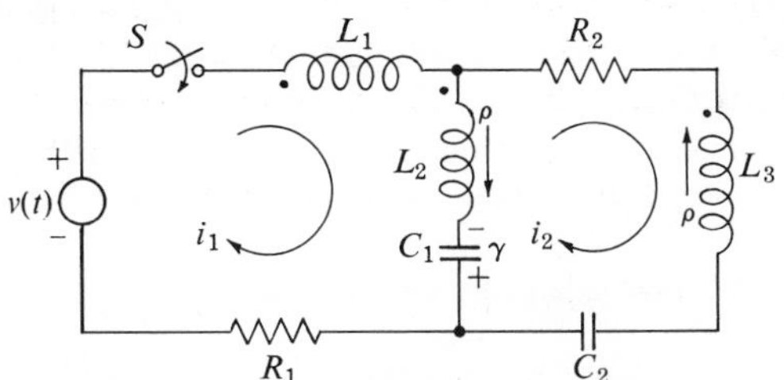

Fig. P8-6

The initial conditions are as indicated. Assume that mutual inductance exists between L_1, L_2, and L_3. Write the subsidiary equations for i_1 and i_2.

8-26. The switch S in the network shown in Fig. P8-7 is opened at $t = 0$. (*a*) If the network is initially relaxed, find $V(s)$. (*b*) If $i(t)$ is a constant current source, find the rate of change of v at $t = 0+$.

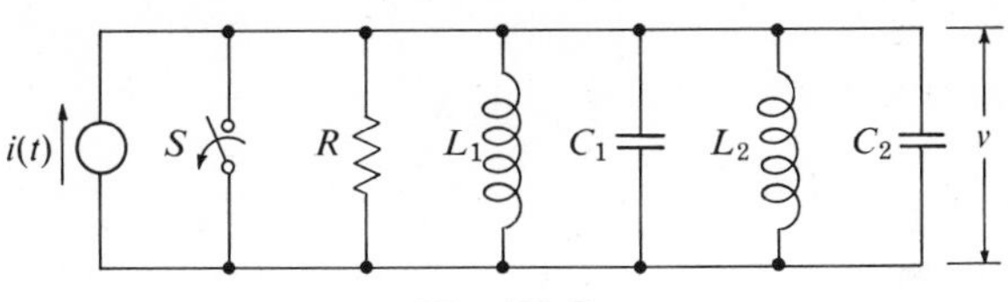

Fig. P8-7

8-27. Neglecting the initial conditions, derive the dual networks of the circuits shown in Figs. P8-5 and P8-8.

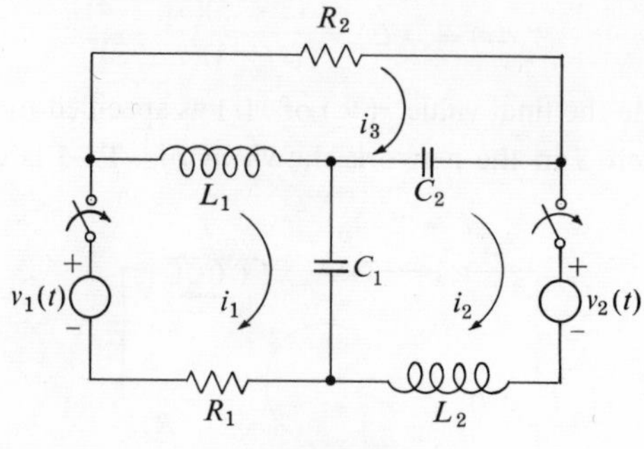

Fig. P8-8

8-28. Sketch the mechanical network of the system shown in Fig. 8-24, p. 325.

8-29. Sketch the mechanical network of the system shown in Fig. P8-2. Draw the analogous electrical networks, using: (*a*) force-current analogy; (*b*) force-voltage analogy.

8-30. Draw the analogous electrical network of the system shown in Fig. P7-1, using: (*a*) force-current analogy; (*b*) force-voltage analogy.

8-31. Using the torque-current analogy, obtain the electrical-analogue network of the system shown in Fig. 8-25, p. 326.

8-32. A geared system is shown in Fig. 8-25. (*a*) Reduce the system to an equivalent system referring to shaft k_{t1}. Obtain the corresponding electrical-analogue networks, using: (*b*) torque-voltage analogy; (*c*) torque-current analogy.

8-33. The masses, springs, and damper, as shown in Fig. P8-9, are connected

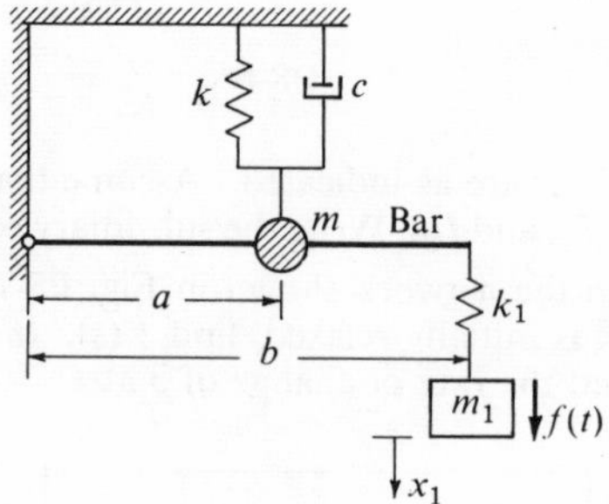

Fig. P8-9

by a rigid bar of negligible mass. Obtain the corresponding electrical-analogue network, using: (*a*) force-voltage analogy; (*b*) force-current analogy.

8-34. Derive Eq. (8-89), p. 329, for the electromechanical system from energy considerations.

8-35. Draw the analogous circuits for the electromechanical system discussed in Example 28, p. 330, assuming that: (*a*) the electrical and the mechanical elements are coupled through an ideal transformer; (*b*) all elements are referred to the primary; (*c*) all elements are referred to the secondary.

8-36. The system discussed in Example 28 is used as vibration pickup. The magnet constitutes the base of the pickup, and it is given a prescribed vertical motion $x_1(t)$. The mass m is constrained to move vertically. Draw an analogous circuit of this system.

9-1. Determine the transfer function e_o/e_i of the circuit shown in Fig. 9-1, p. 340, if: (*a*) $Z_i = R_1$ and Z_f consists of R and C in series; (*b*) $Z_i = R_1$ and Z_f consists of R and C in parallel; (*c*) Z_i has R_1 and C_1 in series and Z_f consists of R and C in series; (*d*) Z_i has R_1 and C_1 in parallel and Z_f consists of R and C in parallel.

9-2. Set up the unscaled computer circuit to solve the following differential equations:

(*a*) $10\dot{x} + x = 0,\ x(0) = 25$

(*b*) $10\dot{x} + 2.75x = 0,\ x(0) = -50$

(*c*) $5\dot{x} + x = f(t),\ x(0) = 0$

(*d*) $\ddot{x} + 4x = 0,\ x(0) = 80,\ \dot{x}(0) = 0$

(*e*) $m\ddot{x} = -c(\dot{x} - \dot{y}) - k(x - y),\ x(0) = \dot{x}(0) = 0,$
and $y(t)$ is a prescribed motion. Assume that $\dot{y}(t)$ is available.

9-3. A vibratory system is described by the equation

$$\ddot{x} + 6\dot{x} + 400x = 0$$

with the initial conditions $x(0) = 0.10$ and $\dot{x}(0) = 1.0$. (*a*) Rewrite the equation to slow down the problem by a factor of 10. (*b*) Set up a computer circuit to solve the equation.

9-4. A vibratory system is described by the equation $\ddot{x} + 4\dot{x} + 100x = 0$ with the initial conditions $x(0) = 0.5$ and $\dot{x}(0) = -1$. (*a*) Set up a computer circuit to solve the equation. (*b*) Slow down the problem by a factor of 10 by changing the time scale of the computer circuit.

9-5. Set up a computer circuit to solve the given equation and speed up the problem by a factor of 5.

$$\ddot{x} + 0.6\dot{x} + x = 2 \sin 1.2t, \quad x(0) = \dot{x}(0) = 0$$

9-6. Set up an unscaled computer circuit to solve the simultaneous equations

$$\dot{y} + x = z(t)$$
$$\dot{y} + y + \dot{x} - x = 0$$

where $z(t)$ is a prescribed function of time.

9-7. Set up a computer circuit to solve the simultaneous equations

$$10\ddot{x} + 500x - 5{,}920\theta = 0$$
$$-5{,}920x + 3.25 \times 10^4\ddot{\theta} + 2.67 \times 10^6\theta = 0$$

with the initial conditions $x(0) = 1$ and $\dot{x}(0) = \theta(0) = \dot{\theta}(0) = 0$.

9-8. Fig. P9-1 is a schematic representation of a trailer traveling at constant

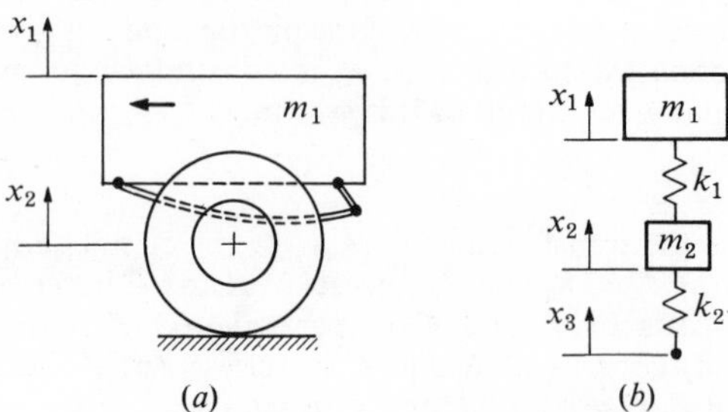

Fig. P9-1. *A trailer and its equivalent system*

speed. m_1 is the mass of the trailer body, m_2 is the mass of the axle assembly, k_1 is the stiffness of the trailer spring, k_2 is the stiffness of the tires, and x_3 is due to the variation of the road surface. The equations of motion of the equivalent system shown in Fig. P9-1(*b*) are

$$4\ddot{x}_1 = -400(x_1 - x_2)$$
$$0.3\ddot{x}_2 = -400(x_2 - x_1) - 800(x_2 - x_3)$$
$$x_3 = 3u(t)$$

If the initial conditions are $x_1(0) = \dot{x}_1(0) = x_2(0) = \dot{x}_2(0) = 0$, set up a computer circuit to solve the given equations.

9-9. Solve the Rayleigh equations on the computer.

$$\ddot{x} + \varepsilon(\dot{x}^2/3 - 1)\dot{x} + x = 0$$

Assume that $0 < \varepsilon < 1$ and $x_m = 3$.

9-10. In the system described in Prob. 9-8, assume that a double-acting shock absorber c is placed between m_1 and m_2 in parallel with spring k_1. The damping force is $c(\dot{x}_1 - \dot{x}_2)$ if $(\dot{x}_1 - \dot{x}_2) > 0$ and $2c(\dot{x}_1 - \dot{x}_2)/3$ if $(\dot{x}_1 - \dot{x}_2) < 0$. It is desired to investigate the effect of the damper for c varying between 0 and 20. Set up a computer circuit to simulate the system.

9-11. An excitation force, the wave form of which is as shown in Fig. P9-2(*a*), is applied to the mass of the system shown in Fig. P9-2(*b*). (*a*) With the use of diodes in a computer circuit, generate the excitation function and specify the computer elements necessary to obtain the period τ. (*b*) Set up a computer circuit to simulate the problem.

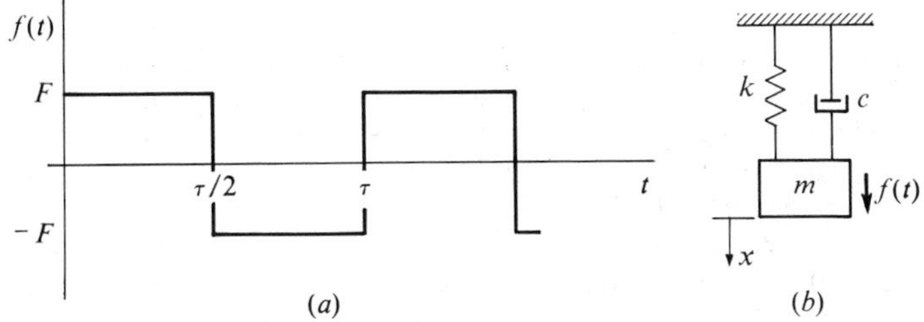

Fig. P9-2

9-12. Generate the function as shown in Fig. P9-2(*a*) with the use of a differential relay in a computer circuit.

9-13. A mechanism is actuated by a cam, as shown in Fig. P9-3(*a*). The idealized cam motion is illustrated in Fig. P9-3(*b*), and the total cam

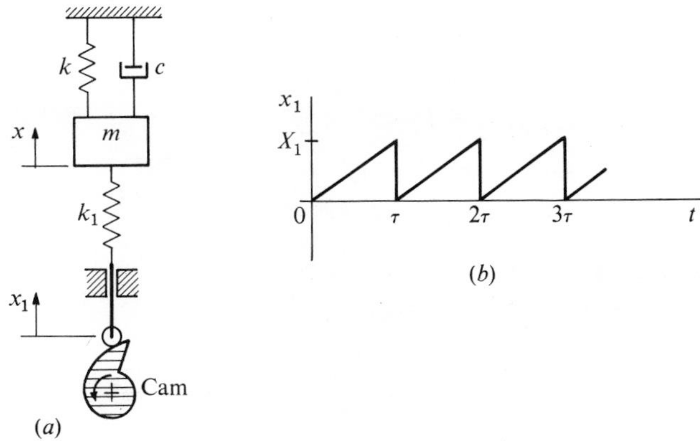

Fig. P9-3

lift is X_1. Devise a computer circuit to simulate this system, and specify the computer elements necessary to obtain the period τ.

9-14. Figure P9-4 represents a trailer in which m_1 is the mass of the trailer

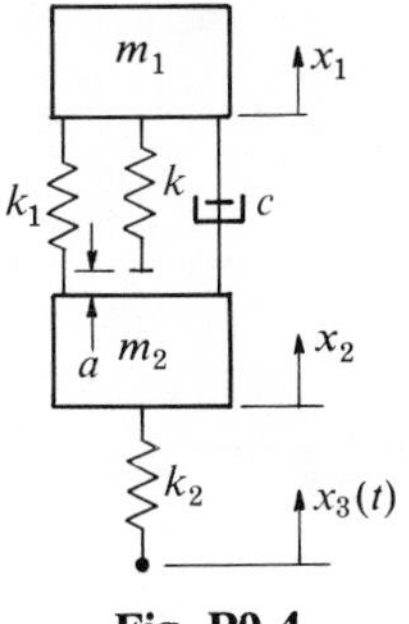

Fig. P9-4

body, m_2 is the mass of the wheel and axle assembly, k_2 is the spring constant of the tires, and $x_3(t)$ is the excitation due to the road condition. At equilibrium, the clearance between the auxiliary spring k and m_2 is a. The damping force due to the shock absorber is $c(\dot{x}_1 - \dot{x}_2)$ when $(\dot{x}_1 - \dot{x}_2) > 0$ and $c^*(\dot{x}_1 - \dot{x}_2)$ when $(\dot{x}_1 - \dot{x}_2) < 0$. Devise a computer circuit to simulate the problem.

9-15. Devise a computer circuit to simulate the system described in Problem 8-13. (*a*) Neglect the effect of the damper c. (*b*) Consider the effect of the damper c.

INDEX